VOLUME ONE HUNDRED AND TWENTY TWO

CURRENT TOPICS IN DEVELOPMENTAL BIOLOGY

T-box Genes in Development

CURRENT TOPICS IN DEVELOPMENTAL BIOLOGY

"A meeting-ground for critical review and discussion of developmental processes"
A.A. Moscona and Alberto Monroy (Volume 1, 1966)

VOLUME ONE HUNDRED AND TWENTY TWO

CURRENT TOPICS IN DEVELOPMENTAL BIOLOGY

T-box Genes in Development

Edited by

MANFRED FRASCH
Friedrich-Alexander University of Erlangen-Nürnberg, Erlangen, Germany

ELSEVIER

ACADEMIC PRESS
An imprint of Elsevier
elsevier.com

Academic Press is an imprint of Elsevier
50 Hampshire Street, 5th Floor, Cambridge, MA 02139, United States
525 B Street, Suite 1800, San Diego, CA 92101-4495, United States
The Boulevard, Langford Lane, Kidlington, Oxford OX5 1GB, United Kingdom
125 London Wall, London, EC2Y 5AS, United Kingdom

First edition 2017

ISBN: 978-0-12-801380-9
ISSN: 0070-2153

For information on all Academic Press publications
visit our website at https://www.elsevier.com/

Publisher: Zoe Kruze
Acquisition Editor: Zoe Kruze
Editorial Project Manager: Shellie Bryant
Production Project Manager: Vignesh Tamil
Cover Designer: Greg Harris

Typeset by SPi Global, India

CONTENTS

Contributors *xi*
Preface *xiii*

1. Evolution and Classification of the T-Box Transcription Factor Family **1**

A. Sebé-Pedrós and I. Ruiz-Trillo

1. An Updated Phylogenetic Classification of the T-Box Gene Family 2
2. Premetazoan T-Box Genes 3
3. Metazoan T-Box Classes and Function 6
4. Functional Conservation of Premetazoan and Early-Metazoan Brachyury Homologs 10
5. Ancestral Conserved Role of Brachyury in Morphogenetic Movements 15
6. The Evolution of T-Box Regulation 17

Acknowledgments 20
References 20

2. The Remarkably Diverse Family of T-Box Factors in *Caenorhabditis elegans* **27**

P.G. Okkema

1. Introduction 28
2. T-Box Genes in *C. elegans* and Related Nematodes 29
3. DNA-Binding Specificity of *C. elegans* T-Box Factors 34
4. Genetic Analyses of T-Box Factor Function in *C. elegans* 37
5. Posttranslational Mechanisms Affecting T-Box Factor Activity 47
6. Concluding Remarks 48

Acknowledgments 49
References 49

3. T-Box Genes and Developmental Gene Regulatory Networks in Ascidians **55**

A. Di Gregorio

1. Introduction 57
2. Brachyury and the Temporal Control of Notochord Specification, Development, and Differentiation 59
3. Tbx2/3: An Essential Mediator of Brachyury Function In the Notochord 66

4. The Lineage-Specific Duplication of *Tbx6* Genes in Ascidians 68
5. Tbx6a 69
6. Tbx6b, Tbx6c, and the Interpretation of the Maternally Inherited Muscle Determinant 70
7. Tbx6b and Tbx6c in Heart Development 73
8. Tbx1/10 and the Evolutionary Origins of the Pharyngeal Muscles 74
9. Tbx15/18/22 (VegTR) and the Establishment of Muscle Cell Identity 75
10. Maternal *T* and Tail Development 76
11. Tbx20: In or Out of the Ascidian Heart? 77
12. Transcriptional Regulation of T-Box Genes Expression 78
13. Concluding Remarks 83
Acknowledgments 84
References 84

4. Eomesodermin—At Dawn of Cell Fate Decisions During Early Embryogenesis 93

S. Probst and S.J. Arnold

1. Introduction 94
2. Early *Eomes* Functions in the Trophectoderm 95
3. *Eomes* in the Visceral Endoderm Promotes Formation of the Embryonic AP Axis 98
4. *Eomes* Functions During the Formation of Germ Layers 99
5. *Eomes* Transcriptional Activities During Germ Layer Formation 106
6. Control of *Eomes* Expression During Gastrulation Onset 109
Acknowledgments 110
References 111

5. Cooperation Between T-Box Factors Regulates the Continuous Segregation of Germ Layers During Vertebrate Embryogenesis 117

G.E. Gentsch, R.S. Monteiro, and J.C. Smith

1. Controversy of Germ Layer Segregation: Holmdahl's Blastema vs Vogt's Predetermination Model 118
2. Multiple Functions of T-Box Factors in Development and Disease 120
3. Early Embryonic Roles of T-Box Factors in Segregating and Differentiating Germ Layers in Vertebrates 126
4. Implications for Reverse Engineering ES Cells 141
5. Summary 143
Acknowledgments 143
References 143

6. T-Box Genes in *Drosophila* Mesoderm Development 161
I. Reim, M. Frasch, and C. Schaub

1. Introduction 162
2. T-Box Genes in Visceral and Gonadal Mesoderm Development 163
3. T-Box Genes in *Drosophila* Cardiogenesis 169
4. The Roles of the *Drosophila* Orthologs of *Tbx1* (*org-1*) and *Tbx20* (*mid* and *H15*) During Embryonic Somatic Muscle Assignments 177
5. The Role of *Drosophila Tbx1* (*org-1*) in Lineage Plasticity, Reprogramming, and Commitment During Adult Myogenesis 182
6. Concluding Remarks and Outlook 185

Acknowledgments 187
References 187

7. *TBX5*: A Key Regulator of Heart Development 195
J.D. Steimle and I.P. Moskowitz

1. *TBX5*, a Member of the T-Box Family of Transcription Factors 196
2. *TBX5* Expression 198
3. *TBX5* Haploinsufficiency: Holt–Oram Syndrome 203
4. Animal Model of Holt–Oram Syndrome 205
5. *TBX5* in Cardiac Morphologic Development 205
6. *TBX5* in Cardiac Conduction System Development 207
7. Homozygous *Tbx5* Mutations Reveal Novel Roles of TBX5 207
8. The TBX5 Gene Regulatory Network 209
9. Concluding Remarks 214

References 214

8. Tbx1: Transcriptional and Developmental Functions 223
A. Baldini, F.G. Fulcoli, and E. Illingworth

1. Tbx1: Structure and Function(s) 224
2. Developmental Functions 228
3. Concluding Remarks and Future Perspectives 237

Acknowledgment 238
References 238

9. T-Box Genes in the Kidney and Urinary Tract 245
A. Kispert

1. Introduction 246
2. T-Box Genes, an Evolutionary Conserved Family of Developmental Regulators 249

3. *Tbx18* in Ureter and Kidney Development 253
4. *Tbx1* in Kidney Development 262
5. *Tbx20* in Bladder/Cloacal Development 264
6. *Tbx2* and *Tbx3*, a Closely Related Pair of T-Box Transcription Factor Genes in Kidney and Bladder Development 266
7. Conclusion and Outlook 270
Acknowledgment 270
References 271

10. Control of Neuronal Development by T-Box Genes in the Brain 279
A.B. Mihalas and R.F. Hevner

1. Discovery of T-Box Gene Expression in the Developing Brain and Neural Retina 280
2. Expression of Tbr1 Subfamily T-Box Genes in Brain and Neural Retina 283
3. Functions of Tbr1 Subfamily T-Box Genes in Brain and Neural Retina 290
4. Regulation of T-Box Genes 296
5. Downstream Target Genes Regulated by T-Box Genes in Cerebral Cortex 303
6. Human Disorders Caused by T-Box Gene Mutations in Brain 305
7. Summary 306
Acknowledgment 306
References 306

11. T-Box Genes in *Drosophila* Limb Development 313
G.O. Pflugfelder, F. Eichinger, and J. Shen

1. Orthology Between *Drosophila* and Mammalian Tbx Genes 314
2. Embryonic Origin of the *Drosophila* Limb Imaginal Discs: The Role of Doc 316
3. Pattern Formation in the Larval Leg Disc 319
4. Tbx Genes in *Drosophila* Leg Development 320
5. Pattern Formation and Proliferation Control in the Larval Wing Disc 323
6. Tbx Genes in *Drosophila* Wing Development 325
7. Tbx Gene Function in Haltere Development 340
8. Conclusion 342
Acknowledgments 342
References 342

12. The Roles of T-Box Genes in Vertebrate Limb Development 355
C.J. Sheeba and M.P.O. Logan

1. Introduction 356
2. The *Tbx2/3/4/5* Subfamily Members Are Significant Players in Limb Development 357

3. The Paralogous T-Box Genes, *Tbx2* and *Tbx3* Have Distinct Functions During Limb Development 361
4. *Tbx4* and *Tbx5*, The Markers of the Hindlimb and Forelimb 367
5. Tbx1 Family Members Have Functions in Limb Chondrogenic Precursors 372
6. *Brachyury* (*T*) Has a Role in AER Maturation 373
7. *Eomes* (*Tbr2*), a Marker for Digit 4 374
8. Future Perspectives 374
References 375

13. T-Box Genes in Human Development and Disease 383

T.K. Ghosh, J.D. Brook, and A. Wilsdon

1. Introduction 384
2. *T* (Brachyury)/Chordoma 384
3. *TBX1*/22q11.2 Deletion Syndrome (DiGeorge Syndrome/Velocardiofacial Syndrome) 390
4. *TBX3*/Ulnar–Mammary Syndrome 392
5. *TBX4*/Small Patella Syndrome 394
6. *TBX5*/Holt–Oram Syndrome/CHD 396
7. *TBX6*/Spondylocostal Dysostosis 398
8. *TBX15*/Cousin Syndrome 399
9. *TBX19*/Isolated ACTH Deficiency 400
10. *TBX20*/Congenital Heart Disease 401
11. *TBX22*/Cleft Palate With or Without Ankyloglossia, X-Linked (CPX) 403
12. Concluding Remarks 405
Acknowledgment 405
References 405

Index 417

CONTRIBUTORS

S.J. Arnold
Institute of Experimental and Clinical Pharmacology and Toxicology, Faculty of Medicine, University of Freiburg; BIOSS Centre for Biological Signalling Studies, Albert-Ludwigs-University of Freiburg, Freiburg, Germany

A. Baldini
University of Naples, Federico II; Institute of Genetics and Biophysics of the CNR, Naples, Italy

J.D. Brook
School of Life Sciences, Queens Medical Centre, University of Nottingham, Nottingham, United Kingdom

A. Di Gregorio
New York University College of Dentistry, New York, NY, United States

F. Eichinger
Institute of Genetics, Johannes Gutenberg University, Mainz, Germany

M. Frasch
Friedrich-Alexander University of Erlangen-Nürnberg, Erlangen, Germany

F.G. Fulcoli
Institute of Genetics and Biophysics of the CNR, Naples, Italy

G.E. Gentsch
Developmental Biology Laboratory, The Francis Crick Institute, Mill Hill Laboratory, London, United Kingdom

T.K. Ghosh
School of Life Sciences, Queens Medical Centre, University of Nottingham, Nottingham, United Kingdom

R.F. Hevner
Center for Integrative Brain Research, Seattle Children's Research Institute; University of Washington School of Medicine, Seattle, WA, United States

E. Illingworth
University of Salerno, Fisciano, Italy

A. Kispert
Institut für Molekularbiologie, Medizinische Hochschule Hannover, Hannover, Germany

M.P.O. Logan
King's College London, London, United Kingdom

A.B. Mihalas
Center for Integrative Brain Research, Seattle Children's Research Institute, Seattle, WA, United States

R.S. Monteiro
Developmental Biology Laboratory, The Francis Crick Institute, Mill Hill Laboratory, London, United Kingdom

I.P. Moskowitz
University of Chicago, Chicago, IL, United States

P.G. Okkema
University of Illinois at Chicago, Molecular, Cell & Developmental Biology Group (MC567), Chicago, IL, United States

G.O. Pflugfelder
Institute of Genetics, Johannes Gutenberg University, Mainz, Germany

S. Probst
Institute of Experimental and Clinical Pharmacology and Toxicology, Faculty of Medicine, University of Freiburg; BIOSS Centre for Biological Signalling Studies, Albert-Ludwigs-University of Freiburg, Freiburg, Germany

I. Reim
Friedrich-Alexander University of Erlangen-Nürnberg, Erlangen, Germany

I. Ruiz-Trillo
Institut de Biologia Evolutiva (CSIC-Universitat Pompeu Fabra); Universitat de Barcelona; Institució Catalana de Recerca i Estudis Avançats, Barcelona, Spain

C. Schaub
Friedrich-Alexander University of Erlangen-Nürnberg, Erlangen, Germany

A. Sebé-Pedrós
Weizmann Institute of Science, Rehovot, Israel

C.J. Sheeba
King's College London, London, United Kingdom

J. Shen
China Agricultural University, Beijing, China

J.C. Smith
Developmental Biology Laboratory, The Francis Crick Institute, Mill Hill Laboratory, London, United Kingdom

J.D. Steimle
University of Chicago, Chicago, IL, United States

A. Wilsdon
School of Life Sciences, Queens Medical Centre, University of Nottingham, Nottingham, United Kingdom

PREFACE

The intricate processes of embryo and tissue development rely on precisely controlled programs of gene expression, which act primarily at the level of transcriptional regulation. During the past ~30 years, it has become apparent that many of the key transcription factors that participate in these developmental programs are members of distinct multiprotein families, which contain DNA-binding domains with closely related primary sequences, tertiary structures, and DNA recognition sites. After the first genes encoding such factors were cloned, the facile isolation of additional family members and their subsequent analysis opened the door to gain rapidly unprecedented insights into a wide variety of developmental processes in which these genes participate.

This volume features a major exemplar of these multigene families, namely, the family of T-box genes. These genes encode proteins with a DNA-binding domain of ~180 amino acids in length, termed T-box domain or T-domain, and are known to play preeminent roles as developmental regulators throughout the animal kingdom.

The founding member of the T-box gene family is indeed one of the first developmental control genes that were identified via forward genetic screens. In 1927, Nadine Dobrovolskaïa-Zavadskaïa at the Pasteur Laboratory in Paris described an X-ray-induced mouse mutation with a *short tail* phenotype, which she named *T* (for Tail; the eponym of the multigene family); she also showed that this phenotype is caused by a heterozygous mutation in a gene that she designated as *Brachyury* (Greek: *brakhús*, short; *ourá*, tail) (Dobrovolskaïa-Zavadskaïa, 1927). Subsequent studies revealed that in the homozygous condition the embryonic lethal phenotype of *T* is accompanied by the absence of the notochord in the presumptive tail region (Gluecksohn-Schoenheimer, 1938). The final breakthrough was the cloning of the gene in the lab of Bernhard Herrmann (Wilkinson, Bhatt, & Herrmann, 1990), followed by the demonstration that it is evolutionarily conserved (Smith, Price, Green, Weigel, & Herrmann, 1991) and is a member of a multigene family that encodes DNA-binding factors (Bollag et al., 1994; Kispert & Herrmann, 1993; Müller & Herrmann, 1997; Pflugfelder, Roth, & Poeck, 1992).

In the intervening years, complete sets of T-box genes were identified in all major animal model organisms as well as in humans. The advances of genomics provided access to the T-box gene sets in many additional species. We have learned that these genes participate in a broad spectrum of key developmental events in these organisms, and functional analysis of T-box genes has often provided pivotal insights into regulatory mechanisms governing these events. In light of these functions and also their frequent dosage sensitivity, already noted by Dobrovolskaïa-Zavadskaïa, it is not surprising that T-box genes also play important roles in human disease. This volume presents an update on the current knowledge on the functions of T-box genes in embryonic and tissue development in a broad range of species and on their contributions to human disease.

The first three chapters review the known roles of the complete sets of T-box genes in several different organisms, whereas Chapters 4–12 focus on the functions of specific subsets of T-box genes in the development of particular tissues, organs, and body parts. Lastly, Chapter 13 presents an overview of human T-box genes that have been linked to congenital diseases.

In Chapter 1, Sebé-Pedrós and Ruiz-Trillo provide an update on the evolution and phylogenetic classification of the T-box gene family, highlighting the gain and loss of specific family members during evolution. Notably, *Brachyury* is the most ancient T-box gene. A major focus of the authors is on T-box genes in nonmetazoan and early metazoan taxa, including zoosporic fungal groups, specific protists that are closely related to animals, as well as sponges, Ctenophorans, and Cnidarians. In Chapter 2, Okkema summarizes our knowledge on the 22 members of the T-box gene family in the nematode *C. elegans*. The emphasis is on the nine genetically characterized representatives, which exert diverse functions during early blastomere fate specification, cell migration, apoptosis, sex determination, and the development of muscles, hypodermal tissues, and neurons. In Chapter 3, Di Gregorio summarizes the developmental functions of T-box genes in ascidians, nonvertebrate chordates that have emerged as powerful developmental model systems in recent years. Genetic circuits in which T-box genes act during the development of the notochord, muscles, heart, and neurons as well as the molecular mechanisms by which some of these T-box genes are controlled by upstream regulators are described.

Many T-box genes have particularly prominent roles during the development of specific body parts, tissues, and organs. The analysis of their expression and function has led to an enhanced understanding of the developmental controls during the formation of these structures and cell types. In Chapter 4, this point is illustrated by Probst and Arnold, who describe the roles of *Eomesodermin* (aka *Eomes/Tbr2*) during very early stages of vertebrate embryogenesis until gastrulation. With a main focus on mouse, the authors discuss the consecutive roles of *Eomes* in the development of extraembryonic tissues (trophectoderm and visceral endoderm) and its key regulatory inputs toward the formation of anterior mesoderm and definitive endoderm. In Chapter 5, Gentsch, Monteiro, and Smith further illuminate the complex regulatory networks by which specific T-box genes control the segregation of germ layers during vertebrate embryogenesis, as exemplified in the frog *Xenopus* and in mouse. The mechanisms discussed provide an up-to-date explanation of the early findings on *Brachyury* functions mentioned above and depict the mechanisms by which Brachyury acts in cooperation with T-box domain factors of the Eomes, Tbx6, and VegT/ Tbx16 subfamilies during gastrulation and embryonic axis elongation.

The following two chapters focus on the roles of T-box genes in tissue formation and organogenesis during subsequent stages of mesoderm development. In Chapter 6, Reim, Frasch, and Schaub describe the critical functions of *Drosophila* T-box genes corresponding to the *T*, *Tbx1*, *Tbx6*, and *Tbx20* subfamilies during mesoderm patterning, specification and differentiation of particular visceral, body wall, and heart-associated muscles, as well as cardiogenesis. We illustrate that, particularly with respect to heart development, a very detailed knowledge of the regulatory circuits has been established in which T-box genes and their upstream and downstream regulators are integral players. In Chapter 7, Steimle and Moskowitz continue this theme and focus on the central role of *Tbx5* during consecutive stages of vertebrate heart development, especially in mouse although fish, frog, and chick and humans are also discussed. During early development of the heart tube, the critical *Drosophila* factors belong to the *Tbx6* subfamily, whereas it is *Tbx5* that plays a key role in vertebrates; however, it is clear that many of the mechanisms have been conserved, including the cooperation of these cardiogenic T-box domain factors with cardiogenic factors of the Nkx and GATA transcription factor families. In Chapter 8, Baldini, Fulcoli, and Illingworth review the knowledge about the molecular action and developmental functions of *Tbx1*. A focus is on the major roles of this gene in controlling normal formation of derivatives of the cardiopharyngeal mesoderm, which include heart tissues derived from the second heart field, great arteries, and various head and neck muscles.

In Chapter 9, Kispert discusses the roles of T-box genes during the development of the excretory system in mouse, zebrafish, and frog. He describes that *Tbx18* is a crucial regulator of mammalian ureter development by regulating the differentiation and maintenance of the ureteric mesenchyme. The functions of *Tbx20* in vertebrate heart development and in the development of the cloaca in zebrafish are also described, as are those of the closely related *Tbx2* and *Tbx3* genes in kidney and bladder development. In the final chapter on the roles of T-box genes in the development of specific tissues and cell types, Chapter 10, Mihalas and Hevner provide an overview of the functions of T-box genes during the development of specific neurons in the brain. The authors focus on the roles of T-box genes of the *Eomes* subfamily, namely, *Tbr1*, *Tbr2*, and *T-bet/Tbx21*, in neuronal development, which includes the migration and differentiation of glutamergic neuronal lineages in the cerebral cortex, olfactory bulb, hippocampal dentate gyrus, cerebellum, and adult subventricular zone.

The next two chapters deal with the roles of T-box genes in limb development. Although the limbs of arthropods and vertebrates are not considered to be homologous structures, they do share some similarities in their developmental programs, including the involvement of related T-box genes. In Chapter 11, Pflugfelder, Eichinger, and Shen summarize the knowledge on the roles of *Drosophila* T-box genes in the development of wings and legs. The authors describe the well-characterized essential role of the *Tbx2* ortholog *omb* in wing development, the roles of *Tbx6*-related *Dorsocross* genes in epithelial remodeling within the developing wing, and the function of *Tbx20* family members in the dorsoventral patterning of the legs. As described by Sheeba and Logan in Chapter 12, T-box genes of the *Tbx2* subfamily also play critical roles in the development of vertebrate limbs, both for the outgrowth of limb buds and their proper patterning. Several members of the *Tbx1* subfamily, as well as *Brachyury* and *Eomes/Tbr2*, also contribute to distinct aspects of limb development. The authors present an overview of the complex signaling and transcriptional networks in which these T-box genes are known to act during vertebrate limb development.

In several of the aforementioned chapters, the authors already implicate specific T-box genes, which function as developmental regulators, in human disease. Chapter 13 by Ghosh, Brook, and Wilsdon is dedicated to human T-box genes and their roles in a broad spectrum of human congenital diseases. The authors provide an overview of the developmental disorders and illustrate the parallels between the disease pathologies and the developmental defects seen in knockdown or knockout situations of the respective orthologs in mouse and nonmammalian models.

The collection of reviews in this volume aims to give the readers a glimpse of the evolutionary relationships, including both similarities and modifications, of T-box gene functions in different species, to acquire up-to-date information on the involvement of T-box genes in their favorite developmental processes and organisms, and to illustrate the vast progress that has been made by studying the genes from this family in developmental biology and human genetics. Because T-box genes participate in so many different aspects of development, it was not possible to cover adequately all of them in a single volume. Particular aspects that are underrepresented or missing, such as the roles of T-box genes in stem cell development, cardiac patterning by additional family members, or the development of the innate and adaptive immune system, have been covered by excellent recent reviews (Lazarevic, Glimcher, & Lord, 2013; Papaioannou, 2014). The next decade, until the centennial of the first publication on T-box genes, will undoubtedly unveil many more exciting discoveries. Additional developmental events will be identified in which T-box genes play crucial roles. Newly developed approaches will allow defining upstream regulators and functional downstream targets much more fully, and additional important protein partners and protein modifications will be identified. All these data can then be used to generate more complete and reliable models of the T-box gene-based regulatory networks. Finally, it is anticipated that all this information will ultimately be beneficial for designing remedies that can alleviate human diseases caused by mutations in various T-box genes.

Manfred Frasch

REFERENCES

Bollag, R. J., Siegfried, Z., Cebra-Thomas, J. A., Garvey, N., Davison, E. M., & Silver, L. M. (1994). An ancient family of embryonically expressed mouse genes sharing a conserved protein motif with the *T* locus. *Nature Genetics*, 7, 383–389.

Dobrovolskaïa-Zavadskaïa, N. (1927). Sur la mortification spontanée de la queue chez la Souris nouveau-née et sur l'existence d'un caractère (facteur) héréditaire, non viable. *Comptes Rendus des Seances de la Societe de Biologie et de Ses Filiales*, *97*, 114–116.

Gluecksohn-Schoenheimer, S. (1938). The development of two tailless mutants in the house mouse. *Genetics*, *23*, 573–584.

Kispert, A., & Herrmann, B. G. (1993). The *Brachyury* gene encodes a novel DNA binding protein. *The EMBO Journal*, *12*, 3211–3220.

Lazarevic, V., Glimcher, L. H., & Lord, G. M. (2013). T-bet: A bridge between innate and adaptive immunity. *Nature Reviews. Immunology*, *13*, 777–789.

Müller, C. W., & Herrmann, B. G. (1997). Crystallographic structure of the T domain-DNA complex of the Brachyury transcription factor. *Nature*, *389*, 884–888.

Papaioannou, V. E. (2014). The T-box gene family: Emerging roles in development, stem cells and cancer. *Development*, *141*, 3819–3833.

Pflugfelder, G. O., Roth, H., & Poeck, B. (1992). A homology domain shared between *Drosophila optomotor-blind* and mouse *Brachyury* is involved in DNA binding. *Biochemical and Biophysical Research Communications*, *186*, 918–925.

Smith, J. C., Price, B. M., Green, J. B., Weigel, D., & Herrmann, B. G. (1991). Expression of a *Xenopus* homolog of *Brachyury* (*T*) is an immediate-early response to mesoderm induction. *Cell*, *67*, 79–87.

Wilkinson, D. G., Bhatt, S., & Herrmann, B. G. (1990). Expression pattern of the mouse *T* gene and its role in mesoderm formation. *Nature*, *343*, 657–659.

CHAPTER ONE

Evolution and Classification of the T-Box Transcription Factor Family

A. Sebé-Pedrós*[,1], I. Ruiz-Trillo[†,‡,§]
*Weizmann Institute of Science, Rehovot, Israel
[†]Institut de Biologia Evolutiva (CSIC-Universitat Pompeu Fabra), Barcelona, Spain
[‡]Universitat de Barcelona, Barcelona, Spain
[§]Institució Catalana de Recerca i Estudis Avançats, Barcelona, Spain
[1]Corresponding author: e-mail address: arnau.sebe-pedros@weizmann.ac.il

Contents

1. An Updated Phylogenetic Classification of the T-Box Gene Family 2
2. Premetazoan T-Box Genes 3
3. Metazoan T-Box Classes and Function 6
   3.1 Brachyury 6
   3.2 Eomes/Tbrain 6
   3.3 Tbx7 7
   3.4 Tbx8 7
   3.5 Tbx2/3 8
   3.6 Tbx4/5 8
   3.7 Tbx1/15/20 9
   3.8 Tbx6 9
4. Functional Conservation of Premetazoan and Early-Metazoan Brachyury Homologs 10
5. Ancestral Conserved Role of Brachyury in Morphogenetic Movements 15
6. The Evolution of T-Box Regulation 17
Acknowledgments 20
References 20

Abstract

T-box proteins are key developmental transcription factors in Metazoa. Until recently they were thought to be animal specific and many T-box classes were considered bilaterian specific. Recent genome data from both early-branching animals and their closest unicellular relatives have radically changed this scenario. Thus, we now know that T-box genes originated in premetazoans, being present in the genomes of some extant early-branching fungi and unicellular holozoans. Here, we update the evolutionary classification of T-box families and review the evolution of T-box function in early-branching animals (sponges, ctenophores, placozoans, and cnidarians) and nonmodel bilaterians. We show that concomitant with the origin of Metazoa, the T-box family radiated into the major known T-box classes. On the other hand, while functional studies are

Current Topics in Developmental Biology, Volume 122
ISSN 0070-2153
http://dx.doi.org/10.1016/bs.ctdb.2016.06.004

still missing for many T-box classes, the emerging picture is that T-box genes have key roles in multiple aspects of development and in adult terminal cell-type differentiation in different animal lineages. A paradigmatic example is that of Brachyury, the founding member of the T-box family, for which several studies indicate a widely conserved role in regulating cell motility in different animal lineages and probably even before the advent of animal multicellularity. Overall, we here review the evolutionary history of T-box genes from holozoans to animals and discuss both their functional diversity and conservation.

1. AN UPDATED PHYLOGENETIC CLASSIFICATION OF THE T-BOX GENE FAMILY

The T-box genes are essential developmental transcription factors (TFs) in Metazoa (Papaioannou, 2001, 2014; Showell, Binder, & Conlon, 2004; Smith, 1999; Wilson & Conlon, 2002). This family is characterized by an evolutionary conserved DNA-binding domain of 180–200 amino acids, known as the T-box domain. The discovery of the first T-box genes (T/Brachyury) in mouse (Herrmann, Labeit, Poustka, King, & Lehrach, 1990), shown to be involved in mesoderm formation, was soon followed by the cloning of T/Brachyury in *Xenopus* (Smith, Price, Green, Weigel, & Herrmann, 1991) and zebrafish (Schulte-Merker, Ho, Herrmann, & Nusslein-Volhard, 1992) and the identification and characterization of members of other T-box classes (Bollag et al., 1994). A T/Brachyury ortholog was soon identified in nonvertebrate species, in particular in insects (Kispert, Herrmann, Leptin, & Reuter, 1994), where it was shown to have conserved expression in developing hindgut. It was also soon found that T/Brachyury encoded a DNA-binding protein involved in transcriptional regulation (Stott, Kispert, & Herrmann, 1993). The first T-box gene described outside Bilateria came in 1999, in which a Brachyury ortholog was identified in the cnidarian *Hydra magnipapillata* (Technau & Bode, 1999). In the early 2000s, several T-box genes were identified in diverse taxa among the earliest-branching metazoan lineages: Brachyury and Tbx2/3 in sponges (Adell, Grebenjuk, Wiens, & Müller, 2003; Adell & Müller, 2005; Larroux et al., 2006, 2008; Manuel, Le Parco, & Borchiellini, 2004), placozoans (Martinelli & Spring, 2003), and ctenophores (Martinelli & Spring, 2005; Yamada, Pang, Martindale, & Tochinai, 2007). Finally, in 2011 T-box genes were identified for the first time outside of animals, in both the filasterean *Capsaspora owczarzaki* (a close unicellular relative of animals) and in the chytrid fungi *Spizellomyces punctatus* (an early-branching fungal species) (Sebé-Pedrós, de Mendoza, Lang, Degnan, & Ruiz-Trillo, 2011).

Here, we expand our previous survey and evolutionary classification of T-box TF (Sebé-Pedrós et al., 2013). To this end, we analyzed recently sequenced genomes/transcriptomes of key nonmetazoan and early-metazoan taxa (see Supplementary Table 1 (http://dx.doi.org/10.1016/bs.ctdb.2016.06.004)), including seven new ctenophore transcriptomes (Moroz et al., 2014), seven sponge transcriptomes (Riesgo, Farrar, Windsor, Giribet, & Leys, 2014), and five new early-branching fungal genomes (Chang et al., 2015), as well as an increased taxon sampling of bilaterian animals, like the hemichordate *Saccoglossus kowalevski* (Simakov et al., 2015), the myriapod *Strigamia maritima* (Chipman et al., 2014), or early-branching vertebrates like the coelacanth *Latimeria chalumnae* (Amemiya et al., 2013), the elephant shark *Callorhinchus milii* (Venkatesh et al., 2014), and the lamprey *Petromyzon marinus* (Smith et al., 2013). In total, we surveyed 153 complete genomes/transcriptomes representing all major eukaryotic clades, as well as a dense representation of the major metazoan lineages, with particular attention to early-branching animals. The results are summarized in Fig. 1, detailed in Supplementary Table 1, and discussed in the following sections.

2. PREMETAZOAN T-BOX GENES

Our new analysis shows that T-box genes are found in different zoosporic fungal groups, including the Cryptomycota *Rozella allomycis* (representing the earliest-branching fungal lineage (Chang et al., 2015)), the Neocallimastigomycota *Piromyces* sp., the chytrids *S. punctatus* and *Gonapodya prolifera*, and the Zygomycota *Rhizophagus irregularis* and *Mortierella verticillata*. We did not find any T-box genes outside the Opisthokonta, that is, the clade that comprises animals, fungi, and their unicellular relatives (Cavalier-Smith, 2003; Ruiz-Trillo, Roger, Burger, Gray, & Lang, 2008; Torruella et al., 2012). Thus, the T-box family evolved at the root of the Opisthokonta.

Within fungi, T-box genes were secondarily lost in Dikarya (that includes the two major groups of fungi, Ascomycota and Basidiomycota, and most known fungal species, like yeasts and most mushroom-forming fungi) and also in several early-branching fungal lineages, the exact number depending on the still poorly resolved deep nodes of the fungal tree (e.g., three extra losses according to the tree topology of Chang et al., 2015).

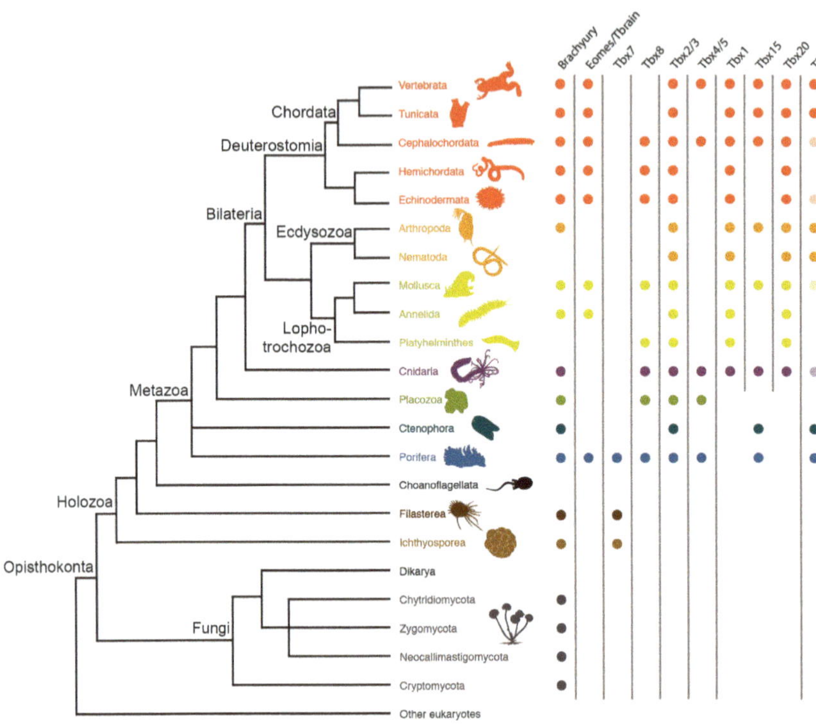

Fig. 1 Phylogenetic distribution of T-box classes. For details, see Supplementary Table 1.

Although highly divergent (Sebé-Pedrós et al., 2013), these fungal T-box genes cluster with the Brachyury family. This indicates that Brachyury is the founding member of the T-box class, being already present before the divergence of animals and fungi.

T-box genes are also present in the genomes of ichthyosporeans and filastereans, two groups of protists closely related to animals. Ichthyosporea have coenocytic development (Mendoza, Taylor, & Ajello, 2002; Suga & Ruiz-Trillo, 2013) and represent the earliest-branching clade inside Holozoa (animals plus three unicellular animal relative lineages) (Torruella et al., 2012). T-box genes are found in the six ichthyosporean taxa for which genomic or transcriptomic data are available, with up to seven copies per genomes (like in *Sphaeroforma arctica*). These ichthyosporean T-box sequences are extremely divergent and lack most of the known functional T-box domain amino acids, but some clearly group inside the Brachyury family (Fig. 1), while others belong to the recently defined Tbx7 family (Sebé-Pedrós et al., 2013). This means that at the root Holozoa (before the divergence of animals and their immediate unicellular relatives) already two T-box classes existed: Brachyury and Tbx7. Similarly, we identified T-box genes from Brachyury and Tbx7 families in the genomes of the two known filasterean species: *C. owczarzaki* and *Ministeria vibrans*. *M. vibrans* is a marine free-living filopodiated amoeba with a flagellar stalk, while *C. owczarzaki* is a fresh-water filopodiated amoeba with an aggregative multicellular stage (Sebé-Pedrós et al., 2013). *C. owczarzaki* has the most conserved nonmetazoan Brachyury ortholog. It has most of the T-box key DNA-binding and dimerization amino acids, as well as conserved exclusive amino acid motifs of the Brachyury class (Sebé-Pedrós et al., 2013). Moreover, it also seems to be functionally conserved with its animal counterparts, as was shown in functional assays (see later). Finally, T-box genes were secondarily lost in the choanoflagellate lineage (Sebé-Pedrós et al., 2011, 2013).

An interesting finding among premetazoan T-box genes is a T-box with two T-domains present in *C. owczarzaki*. T-box TFs are in general composed of a single central T-domain, with the only known exception of the MGA family, a Tbx6 paralog in vertebrates (see later) with both a T-domain and a basic helix–loop–helix (bHLH) zipper domain (Hurlin, Steingrìmsson, Copeland, Jenkins, & Eisenman, 1999). Interestingly, *Capsaspora* has a large protein (1260 amino acids) with two full T-domains (one central and one C-terminal), which was verified by RT-PCR and RACE-PCR (Sebé-Pedrós et al., 2011). When analyzed separately, these two T-domains cluster inside the Tbx7 family, and one of them groups with a partial sequence of *M. vibrans* (the other filasterean species). This suggests that this unique "double T-box" configuration emerged at the root of the Filasterea. Although this "double T-box" configuration is not found in other species, the presence of two DNA-binding domains is not uncommon in other eukaryotic TF families. It has been hypothesized that multiple DNA-binding domains can increase the length and diversity of DNA motifs recognizable by the limited number of DNA-binding domain families (Charoensawan, Wilson, & Teichmann, 2010; Itzkovitz, Tlusty, & Alon, 2006). Whether this or other explanations account for the presence of this T-box gene in *C. owczarzaki* remains to be elucidated.

In summary, our expanded genomic survey confirms that T-box is an ancient TF that most likely originated early within the Opisthokonts and is present in diverse unicellular opisthokont lineages. T-box genes were subsequently secondarily lost in most fungal lineages and in choanoflagellates. Brachyury is the most ancient T-box class, while the Tbx7 class originated in the common ancestor of Holozoa. The remaining classes, which are discussed later, appear to be animal specific.

3. METAZOAN T-BOX CLASSES AND FUNCTION

Our updated dataset (Fig. 1, Supplementary Table 1) shows that two major T-box classes emerged before Metazoa (Brachyury and Tbx7), others at the root of Metazoa (Eomes, Tbx1/15/20, Tbx2/3, Tbx4/5, Tbx6, and Tbx8), while some others originated at the Cnidaria + Bilateria clade or within Bilateria. The nonbilaterian sponges, which are an early (if not the earliest) branching animal lineage, have representatives of most T-box genes classes (except those specific to Bilateria or Cnidaria + Bilateria). Depending on the phylogenetic status of sponges this may suggest different scenarios. If sponges are the sister group to the rest of Metazoa, the so-called Porifera-sister hypothesis (Pick et al., 2010; Pisani et al., 2015), then the data will suggest a quick radiation of the T-box family concomitant with the origin of animal multicellularity, followed by secondary loss of some families in ctenophores. Alternatively, if ctenophores are the sister group to the rest of Metazoa (Moroz et al., 2014; Ryan et al., 2013), the so-called Ctenophora sister, then the radiation probably occurred after the divergence of ctenophores from sponges and the rest of animals (even though secondary loss in the lineage leading to extant ctenophores cannot be ruled out). In any case, it seems probable that an expansion of T-box genes occurred somehow early in animal evolution, followed by extensive neo- and subfunctionalization of T-box genes (see later). Below we revise the presence of the different T-box gene families in different animal phyla, plus their inferred function in some nonmodel species.

3.1 Brachyury

The Brachyury T-box class is the founding member of the family. It is present in nonmetazoan taxa and in the genomes of all major metazoan lineages, with the exception of the analyzed taxa representing Nematoda and Platyhelminthes (Fig. 1). In vertebrates, the Brachyury class diversified into the paralog subclasses Tbx19 and Brachyury. The evolution of Brachyury function is discussed later.

3.2 Eomes/Tbrain

The Eomes/Tbrain class is found in sponges, in particular the calcarean sponges *Sycon ciliatum and Leucosolenia complicata*, and in some protostomes and deuterostomes. This class has been lost in Placozoa and Cnidaria, as well as in Ctenophora (under the sponges-first hypothesis), and in some bilaterian clades, such as Platyhelminthes, Arthropoda, and Nematoda. The Eomes class gave rise to the paralog subclasses Tbx21, Eomes, and Tbrain in vertebrates.

In the calcarean sponge *S. ciliatum*, Eomes is strongly expressed in the oocytes, but not during development. In adult sponges, Eomes is expressed around the opening of the oscular sphincter (Leininger et al., 2014). In the annelid *Hydroides elegans*, Eomes is expressed in animal cap blastomeres and, later, it becomes restricted to the apical tuft cells of the early trochophore larvae (Arenas-Mena, 2008). This pattern of expression in the apical region seems conserved in the tornaria larvae of the hemichordate *Ptychodera flava* (Tagawa, Humphreys, & Satoh, 2000). The authors suggested that this expression pattern might represent an evolutionary link between the apical sensory organ of nonchordate larvae and the vertebrate forebrain. In *Ptychodera*, Eomes is also expressed weakly during development around the blastopore. In the embryo of the sea urchin *Strongylocentrotus purpuratus*, Eomes (ske-T) is expressed only in the skeletogenetic mesenchyme (Croce, Lhomond, Lozano, & Gache, 2001). Finally, in the cephalochordate *Branchiostoma floridae*, Eomes is expressed in the axial and paraxial mesendoderm in early larvae, but no anterior neural domain of expression (similar to that in vertebrate forebrain) is detected in amphioxus (Horton & Gibson-Brown, 2002).

3.3 Tbx7

The Tbx7 class is present in the unicellular filastereans and ichthyosporeans (see earlier), as well as in sponges. Among sponges, Tbx7 is present in the genomes of the calcarean sponges *S. ciliatum* and *L. complicata* and the demosponge *Amphimedon queenslandica*. Therefore, this T-box class was secondarily lost in all other metazoans, under the Porifera-sister hypothesis.

In *Sycon*, Tbx7 is expressed in oocytes and during cleavage and early-stage preinversion embryos. In adult sponges, it is found in cells of the mesohyl (the space between the external pinacoderm and the internal choanoderm, filled with extracellular matrix), particularly at the tips of the uppermost radial chambers (Leininger et al., 2014).

3.4 Tbx8

The Tbx8 class is found in diverse sponges (the demosponges *A. queenslandica* and *Cliona celata*, the calcareans *S. ciliatum* and *L. complicata*, and the homoscleromorph *Oscarella carmela*), placozoans, cnidarians, platyhelminthes, mollusks, cephalochordates, and ambulacrarians (Echinodermata and Hemichordata). This class was secondarily lost in ctenophores (under the Porifera-sister hypothesis), annelids, tunicates, and vertebrates, and all the ecdysozoans examined (Nematoda and Arthropoda). It is one of the only four families present in Platyhelminthes, together with Tbx2/3, Tbx1, and Tbx20 (these three previously identified in the planarian *Schmidtea polychroa* (Martín-Durán & Romero, 2011)).

3.5 Tbx2/3

This is the most widespread T-box family in animals (Fig. 1), present in all the major lineages examined. The class diverged into Tbx2 and Tbx3 at the root of vertebrates. Unlike most T-box TFs, Tbx2/3 class contains the only examples of T-box acting as transcriptional repressors (He, Wen, Campbell, Wu, & Rao, 1999).

Tbx2/3 expression patterns have been studied for several species. In the demosponge *Suberites domuncula*, Tbx2/3 is expressed in isolated cells of the mesohyl of adults, suggesting a possible role in terminal cell-type differentiation (Adell & Müller, 2005). In the placozoan *Trichoplax adhaerens*, Tbx2/3 is expressed in the periphery of attached animals (Martinelli & Spring, 2003). In the ctenophore *Mnemiopsis leyidi*, Tbx2/3 is expressed in the ctene rows and the apical organ region, suggesting a possible common role in sensory organ formation, as Tbx2/3 is also expressed in the eye of *Drosophila* and chordates (Yamada et al., 2007).

Among bilaterians, the planarian *S. polychroa* (Platyhelminthes) has three Tbx2/3 paralogs that show distinct expression patterns in the embryo (Martín-Durán & Romero, 2011): Tbx2/3a is expressed in the gut, Tbx2/3b in parenchymatic cells (both dorsal and ventral), and Tbx2/3c is found in the embryonic brain. Both in the hemichordate *S. kowalevski* and the annelid *H. elegans*, Tbx2/3 is expressed in the dorsal side of the embryo (Arenas-Mena, 2013; Lowe et al., 2006) and, similarly, Tbx2/3 is expressed in the aboral side of different sea urchin species, including *Paracentrotus lividus* (Croce, Lhomond, & Gache, 2003), *S. purpuratus* (Chen, Luo, & Su, 2011), and *Lytechinus variegatus* (Gross, Peterson, Wu, & McClay, 2003). These findings suggest a conserved role of Tbx2/3 in dorsoventral patterning in bilaterians (Arenas-Mena, 2013).

3.6 Tbx4/5

The Tbx 4/5 class, in contrast, is sparsely distributed due to secondary losses in all protostomes analyzed plus in ctenophores and in ambulacrarians (echinoderms + hemichordates). It is present in sponges, placozoans, cnidarians, cephalochordates, and vertebrates (Fig. 1). The Tbx4/5 class diversified into Tbx4 and Tbx5 subclasses in vertebrates. Nothing is known about the expression patterns of Tbx4/5 in nonbilaterians. In amphioxus, it has been shown to be expressed only in the decentralized cardiac domain of the adults (and not during development), suggesting a common role of Tbx4/5 in cardiogenesis in cephalochordates and vertebrates (Pascual-Anaya et al., 2013).

3.7 Tbx1/15/20

The ancestral Tbx1/15/20 class is present in ctenophores and sponges, and it diversified into three classes (Tbx1, Tbx15, and Tbx20) at the root of Cnidaria + Bilateria. Tbx1 and Tbx20 are present in all major cnidarian and bilaterian lineages examined here, while Tbx15 has been lost in platyhelminthes, annelids, nematodes, and ambulacrarians (and also in some specific taxa in other groups, see Supplementary Table 1). At the root of vertebrates, Tbx1 further diversified into Tbx1 and Tbx10, and Tbx20 diversified into Tbx18, Tbx20, and Tbx22. Interestingly, this Tbx20 diversification occurred after the divergence of the lamprey lineage.

In the ctenophore *M. leyidi*, Tbx1/15/20 shows mesendodermal expression and transient expression along the edge of the blastopore in a biradial pattern (Yamada et al., 2007). In the planarian *S. polychroa*, Tbx1 is localized in discrete dorsal cells, while Tbx20 is in the body margin and in the ventral nerve cords (Martín-Durán & Romero, 2011). Finally, in amphioxus Tbx15 is expressed in the mesendoderm during the gastrula stage and, later, in the forming somites, suggesting a conserved role in chordate segmentation (Beaster-Jones, Horton, Gibson-Brown, Holland, & Holland, 2006). In contrast, amphioxus Tbx20 is expressed, like Tbx4/5, in the precursors of the myocardium, suggesting a conserved role in heart development in chordates (Belgacem, Escande, Escriva, & Bertrand, 2011).

3.8 Tbx6

Finally, the Tbx6 class has been traditionally difficult to identify by phylogeny, with many genes presumptively classified as Tbx6 not grouping together in phylogenies (Holstien et al., 2010; Larroux et al., 2008; Sebé-Pedrós et al., 2013). With the addition of new species, we here recover the monophyly of this class (although with weak nodal support) and, for the first time, assign sponge and ctenophore sequences to this group. Thus, we push the origin of this class to the root of Metazoa. This expanded Tbx6 class includes sponge sequences that were traditionally classified within the sponge-specific TbxPor class (Holstien et al., 2010; Sebé-Pedrós et al., 2013). Additionally, the Tbx6 class includes sequences from nematodes and arthropods (like the *Drosophila* Dorsocross (Doc) genes) and from tunicates and vertebrates. Some cnidarian and bilaterian genes were previously considered as Tbx6 in other studies (Belgacem et al., 2011; Paps, Holland, & Shimeld, 2012; Pascual-Anaya et al., 2013), but in our analysis they do not clearly cluster within Tbx6.

The Tbx6 class diversified into three subclasses in vertebrates: Tbx6, MGA, and VegT, the latter being secondarily lost in mammals. The vertebrate MGA subclass contains genes with both T-domain and a basic bHLH zipper domain (Hurlin et al., 1999), being, together with the *Capsaspora* Double-T-box (see earlier), the only T-box genes having additional DNA-binding domains.

In the ctenophore *M. leyidi*, Tbx6 (TbxD) functions in ectodermal development of the tentacles (Yamada et al., 2007). In amphioxus, Tbx6 is expressed in the tail epidermis, in some neurons, and in the unsegmented paraxial mesoderm, suggesting a conserved role in posterior mesoderm specification in chordates (Belgacem et al., 2011).

In summary, the T-box TF family has a highly dynamic evolutionary history, with multiple secondary losses along evolution (with the exception of Tbx2/3, present in all metazoan lineages), some fast-evolving members (for example, in sponges and ichthyosporeans), expansions (such as three paralogous eumetazoan classes related to the ancestral Tbx1/15/20), and major structural rearrangements, such as the double T-domain found in *Capsaspora* or the T-domain/bHLH domain fusion in MGA T-box subclass in vertebrates.

4. FUNCTIONAL CONSERVATION OF PREMETAZOAN AND EARLY-METAZOAN BRACHYURY HOMOLOGS

An intriguing question is whether there is some conserved function of any of the T-box classes between premetazoans and metazoans or between bilaterian and nonbilaterian animals. Heterologous experiments, in which a gene is expressed in a different species, have been used to analyze the evolutionary conservation of T-box genes. For example, Satoh et al. showed that ectopic overexpression of different deuterostome (i.e., tunicate, amphioxus, acorn worm, and sea urchin) Brachyury orthologs in the embryos of the tunicate *Ciona intestinalis* had similar effects in inducing the differentiation of notochord cells (Satoh, Harada, & Satoh, 2000). This indicated high conservation of Brachyury target specificity in between these taxa. In another study, Marcellini et al. used overexpression in *Xenopus* animal caps to demonstrate the specific conserved ability to induce mesoderm of even more evolutionarily distant Brachyury orthologs, including those from the annelid *Platynereis dumerilii* and the cnidarian *H. magnipapillata* (Marcellini, Technau, Smith, & Lemaire, 2003).

An alternative to ectopic overexpression is the use of dominant-negative constructs to analyze the repressive phenotype. For example, *Xenopus* embryos injected with an mRNA encoding a dominant-negative form of Brachyury (XBra_En, consisting of a C-terminal fusion of the Engrailed repressor to Brachyury) showed defective gastrulation and impairment of muscle development (Conlon, Sedgwick, Weston, & Smith, 1996). Using this approach, Yamada et al. studied the functional conservation of the Brachyury ortholog of the ctenophore *M. leyidi* (Yamada, Martindale, Fukui, & Tochinai, 2010). In particular, they injected *Xenopus* embryos with a *M. leyidi.* Bra_En construct and this caused similar defects in gastrulation and mesoderm induction to those observed with the Xbra_En construct and, therefore, suggesting conservation of these distant Brachyury orthologs. Additionally, they analyzed the specific induction of downstream targets of XBra (wnt11, sox17) and also of targets of other T-box genes not activated by Brachyury (chordin, goosecoid) (Conlon, Fairclough, Price, Casey, & Smith, 2001; Xanthos, Kofron, Wylie, & Heasman, 2001). *M. leyidi* Brachyury specifically activated XBra targets, but not the non-XBra targets, revealing high conservation of target specificity between these two distant homologs.

The discovery of T-box genes, and in particular of Brachyury orthologs, outside Metazoa prompted the study of the functional conservation of these nonmetazoan Brachyury genes. To this end, we used coinjection of a dominant-negative XBra_En together with Brachyury mRNA to show, quite surprisingly, that the Brachyury ortholog of the unicellular filasterean *C. owczarzaki* was able to rescue gastrulation and mesoderm induction in *Xenopus* embryos as efficiently as the endogenous Xbra (Fig. 2) (Sebé-Pedrós et al., 2013), thus roughly mimicking the endogenous *Xenopus* Brachyury function (Fig. 2).

Next we evaluated the induction, upon mRNA injection, of downstream targets of XBra (sox17, endodermin, wnt11, and wnt8) and also targets of XVegT (Tbx6) that are known not to be recognized by XBra (such as chordin and pintallavis). Interestingly, the premetazoan *C. owczarzaki* Brachyury ortholog strongly activated all examined T-box targets, not only

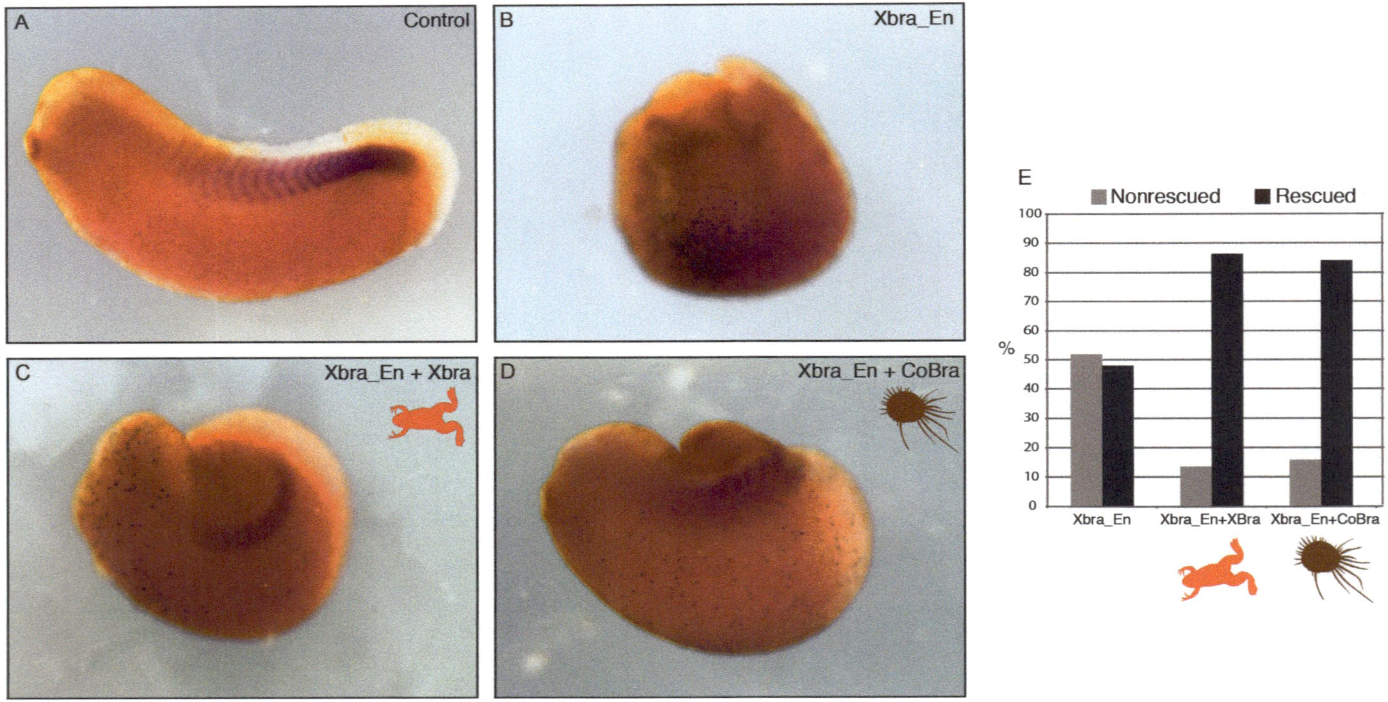

Fig. 2 Rescue assays in *Xenopus laevis* with the *Capsaspora owczarzaki* Brachyury ortholog. (A) Wild-type embryo showing complete trunk formation and full MyoD expression. (B) Embryo injected with Xbra_En (dominant-negative construct), showing no trunk formation and no MyoD expression. (C) Xbra_En-injected embryo rescued by coinjection with endogenous Xbra. (D) Xbra_En-injected embryo rescued by coinjection with *Capsaspora* Brachyury. (E) Barplot summarizing the results of the different control and rescue experiments. *Adapted from Sebé-Pedrós, A., Ariza-Cosano, A., Weirauch, M. T., Leininger, S., Yang, A., Torruella, G., … Ruiz-Trillo, I. (2013). Early evolution of the T-box transcription factor family.* Proceedings of the National Academy of Sciences of the United States of America, 110*(40), 16050–16055. doi:10.1073/pnas.1309748110.*

those XBra-specific. In contrast, Brachyury orthologs of the calcarean sponge *S. ciliatum* and the cnidarian *Nematostella vectensis* fully mimic the specific behavior of endogenous XBra, e.g., not activating chordin (Fig. 3). Protein-binding microarray analysis showed, though, that the DNA-binding motif preference of *C. owczarzaki* Brachyury and those of animals is the same and, in fact, that different T-box families have very similar binding motifs (Fig. 4). This similarity in the binding motifs could explain partial

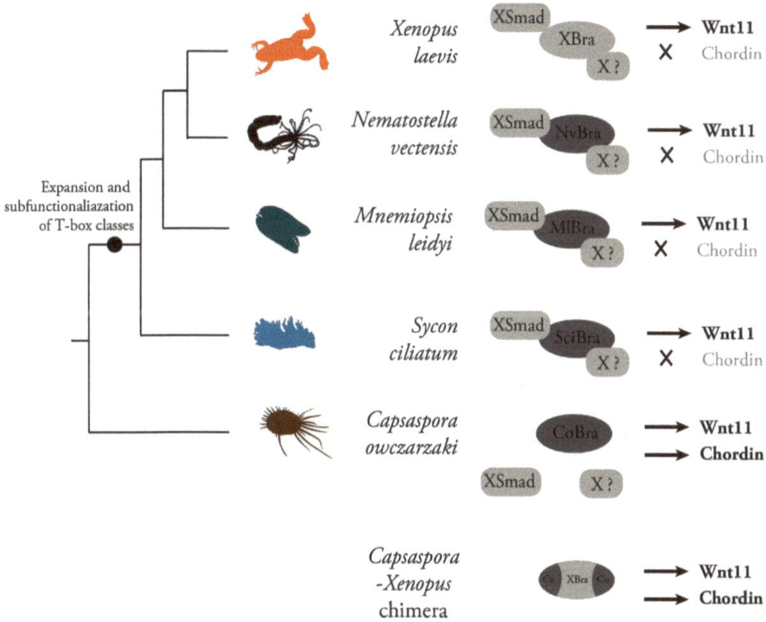

Fig. 3 Functional conservation of Brachyury orthologs in heterologous expression assays. Early-metazoan Brachyury orthologs (from the sponge *Sycon ciliatum* and the cnidarian *Nematostella vectensis*) produce the same molecular phenotype as *Xenopus* Brachyury (activation of Wnt11 and no activation of Chordin). In contrast, nonmetazoan Brachyury orthologs (from the filasteran *Capsaspora owczarzaki*) activate targets of multiple T-box classes, not only of *Xenopus* Brachyury. A chimeric construct with the N- and C-terminal domains (involved in protein–protein interactions) of CoBra and the central T-domain (involved in DNA binding) of XBra shows the same molecular phenotype as wild-type CoBra. These results suggest that the *trans* regulatory interactions between Brachyury and cofactors like Smad (and probably other unknown cofactors, shown as "X?") were established at the onset of Metazoa. *Results from Sebé-Pedrós, A., Ariza-Cosano, A., Weirauch, M. T., Leininger, S., Yang, A., Torruella, G., … Ruiz-Trillo, I. (2013). Early evolution of the T-box transcription factor family.* Proceedings of the National Academy of Sciences of the United States of America, 110*(40), 16050–16055. doi:10.1073/pnas.1309748110.*

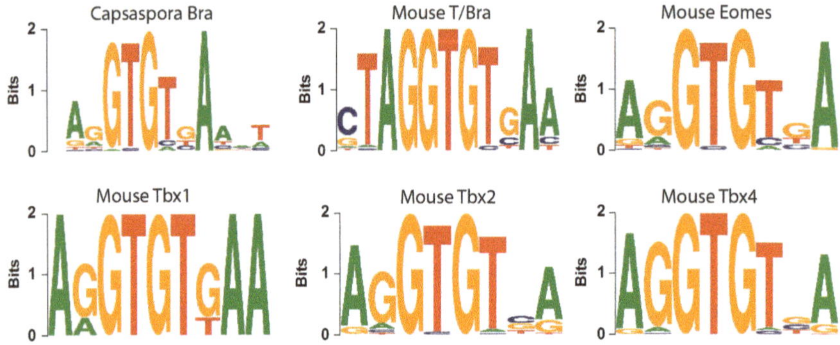

Fig. 4 Highly conserved T-box binding motifs. Protein-binding microarray experiments reveal that the DNA-binding preferences of the mouse and *Capsaspora* Brachyury orthologs are almost identical. Also different mouse T-box classes show similar motifs, indicating conservation across the whole T-box transcription factor family. *Adapted from Sebé-Pedrós, A., Ariza-Cosano, A., Weirauch, M. T., Leininger, S., Yang, A., Torruella, G., … Ruiz-Trillo, I. (2013). Early evolution of the T-box transcription factor family.* Proceedings of the National Academy of Sciences of the United States of America, 110*(40), 16050–16055. doi:10.1073/pnas.1309748110.*

inter-T-box class functional conservation, for example, that observed in rescue experiments (Croce et al., 2003). Croce et al. found that a dominant-negative Tbx2/3 (*coquillette*) in the sea urchin *P. lividus* (Echinodermata) could be rescued not only by coinjecting wild-type Tbx2/3 mRNA but also partially by Brachyury and Eomes mRNA.

Moreover, induction experiments using *C. owczarzaki*-*Xenopus* Brachyury chimeras (i.e., swapping N-terminal, central T- and C-terminal domains) showed that the N- and C-terminal domains, but not the central T-domain, are responsible for the specificity of Bra function (Fig. 3). In line with previous studies (Bielen et al., 2007; Marcellini, 2006; Marcellini et al., 2003), this result indicates that N- and C-terminal regions are essential for Brachyury specificity, by mediating cofactor interactions, e.g., with Smad proteins (Marcellini, 2006).

In summary, the heterologous expression studies of distant Brachyury orthologs suggest that subfunctionalization of Brachyury class was well established at the onset of Metazoa, as evidenced by the ability of sponge and ctenophore Brachyury to mimic endogenous *Xenopus* Brachyury function. The establishment of new cofactor interactions was, probably, an important mechanism in this subfunctionalization process, which occurred concomitantly to the radiation of T-box classes at the root of Metazoa.

5. ANCESTRAL CONSERVED ROLE OF BRACHYURY IN MORPHOGENETIC MOVEMENTS

Heterologous expression experiments tell us about the ability of distant orthologs to mimic endogenous functions. This indicates a certain level of biochemical functional conservation, but tells us nothing about the gene function in its original context. For this, it is necessary to examine expression patterns in developing and adult animals and, when possible, perform perturbation experiments. Since the discovery of Brachyury expression in invaginating hindgut cells in *Drosophila* (Kispert et al., 1994), multiple authors, working on different species, have proposed a conserved role for Brachyury in morphogenetic movements (Gross & McClay, 2001; Scholz & Technau, 2003; Tada & Smith, 2000; Tagawa, Humphreys, & Satoh, 1998; Yamada et al., 2010, 2007). Here, we review the current evidence about Brachyury function in different lineages.

In sponges, Brachyury function has been studied only in two species. Adell et al. used immunostainings to evaluate the expression of Brachyury in different culture stages of the demosponge *S. domuncula* (Adell & Müller, 2005). The highest levels of Brachyury protein were detected in adherent aggregates of cells, after sponge dissociation. Hence, the authors proposed a possible role for *S. domuncula*-Bra in morphogenetic movements, through the regulation of cell motility and adhesion.

In the calcarean sponge *S. ciliatum*, the two Brachyury paralogs are expressed in the oocytes and in the micromere cells of postinversion embryos (Leininger et al., 2014). In adults of *S. ciliatum*, Brachyury is expressed in the choanocytes. The authors propose that this expression pattern (together with other markers) indicates a possible homology of the micromers/choanoderm and eumetazoan endomesoderm (Leininger et al., 2014).

In ctenophores, Brachyury is expressed around the blastopore in *M. leyidi* (Yamada et al., 2007), after the blastopore is formed. It is also found in cells of the invaginating tentacular bulbs and in the floor of the apical organ. In another key study, Yamada et al. injected *M. leyidi* embryos with morpholino to knockdown Brachyury, which resulted in a failure to invaginate the ectodermal cells surrounding the blastopore (Yamada et al., 2010). This effect was rescued by coinjection of the morpholino with endogenous Brachyury mRNA. The results of this study strongly suggest a primitive conserved role of Brachyury in morphogenetic movements, such as those involved in gastrulation.

In placozoans, Brachyury is expressed in discrete groups of cells in the periphery of adult animals, suggesting a role in cell differentiation (Martinelli & Spring, 2003).

In cnidarians, the role of Brachyury has been extensively studied in several species. In the anemone *N. vectensis*, Brachyury is expressed around the blastopore in gastrulating embryos and in the developing mesenteries, although not in the adult muscle cells (Scholz & Technau, 2003). In contrast, in *Podocoryne carnea*, Brachyury is expressed in the adult muscle cells (Spring et al., 2002). During the development of the coral *Acropora digitifera*, Brachyury is expressed around the blastopore, in cells moving inward during gastrulation (Hayward, Grasso, Saint, Miller, & Ball, 2015). Finally, *H. magnipapillata* has two Brachyury paralogs that are differentially expressed (Bielen et al., 2007; Technau & Bode, 1999). HyBra1 is expressed in the endoderm of the hypostome, the tissue surrounding the adult mouth, while HyBra2 is expressed in the ectoderm of the hypostome. *H. magnipapillata* lacks classical gastrulation with a well-defined blastopore (instead, embryos of *H. magnipapillata* undergo multipolar ingression); therefore, circumblastoporal expression of Brachyury cannot be evaluated.

In multiple bilaterian lineages, the circumblastoporal expression of Brachyury is conserved, for example, in annelids (Lartillot, Lespinet, Vervoort, & Adoutte, 2002), echinoderms (Croce, Lhomond, & Gache, 2001; Gross & McClay, 2001; Peterson, Harada, Cameron, & Davidson, 1999; Rast, Cameron, Poustka, & Davidson, 2002), hemichordates (Lowe et al., 2006; Tagawa et al., 1998), priapulids (Martín-Durán, Janssen, Wennberg, Budd, & Hejnol, 2012), and cephalochordates (Onai et al., 2009). In the annelid *H. elegans* Brachyury is expressed in the invaginating blastomers that lead to gastrulation (Arenas-Mena, 2013). Later in development, Brachyury expression is retained in the hindgut and/or the foregut in different lineages (reviewed by Hejnol & Martín-Durán, 2015). For example, Brachyury is expressed both in mouth and anus in echinoderms, hemichordates, annelids, and molluscs, but only in the hindgut in arthropods and priapulids (nematodes have lost Brachyury) (Hejnol & Martín-Durán, 2015). Brachyury is also known to be a key regulator of notochord development in tunicates and cephalochordates (Katikala et al., 2013; Onai et al., 2009). Further support for the role of Brachyury in morphogenetic movements comes from key experiments by Gross et al. in the sea urchin *L. variegatus* (Gross & McClay, 2001). Like in cnidarians, ctenophores, and many bilaterians, LvBra is expressed around the blastopore and, in later stages, in the stomodeum and the anal region of the pluteus larva hindgut. Interestingly, blocking of Brachyury function (by injection of a dominant-negative LvBra_EN construct) completely abolished gastrulation movements, but it did not affect the expression of endodermal and mesodermal marker genes. Similarly, morpholino knock-down of Brachyury also blocks gastrulation in another sea urchin species, *S. purpuratus* (Rast et al., 2002).

What about the function of Brachyury in nonmetazoans? A recent comparative transcriptomic analysis of two life stages of the ichthyosporean *Creolimax fragrantissima* showed that several of the highly divergent Brachyury paralogs in this species were upregulated in the amoeboid dispersal stage (compared with the multinucleated, cell-walled, osmotrophic stage) (De Mendoza, Suga, Permanyer, Irimia, & Ruiz-Trillo, 2015). Additionally, in an analysis of the regulatory genome of the filasterean *C. owczarzaki*, the downstream network of Brachyury was inferred (Fig. 5) (Sebé-Pedrós et al., 2016). Interestingly, multiple gene orthologs are conserved between the *Capsaspora* Brachyury and the mouse Brachyury downstream target networks (Lolas, Valenzuela, Tjian, & Liu, 2014). These conserved orthologs are enriched in functions associated to cell motility, amoeboid movement, and actin cytoskeleton (Fig. 5). This result points to an ancestral role for Brachyury in regulating a core network of genes associated with cell motility, a function that was already present before the advent of animal multicellularity.

Overall, in the past two decades extensive evidence has accumulated in different animal species that support a conserved ancestral involvement of Brachyury in morphogenetic movement, with recent evidence even suggesting a premetazoan role of Brachyury in regulating cell motility. Moreover, we have seen examples of a myriad of additional roles of Brachyury in other developmental and adult contexts.

6. THE EVOLUTION OF T-BOX REGULATION

The evolution of TF function goes beyond the diversification of families/classes and the number of paralog members. Specific TF binding to DNA sites, either in enhancer or promoter elements, depends on multiple layers of regulation, including TF translocation to the nucleus, interaction of other TFs and cofactors, TF-binding affinities to specific sequences, and the chromatin context (nucleosome occupancy and modifications, DNA methylation, and chromatin folding) of these sequences (Spitz & Furlong, 2012). Evolutionary changes affecting TF function can be in *trans* or in *cis*. The ones

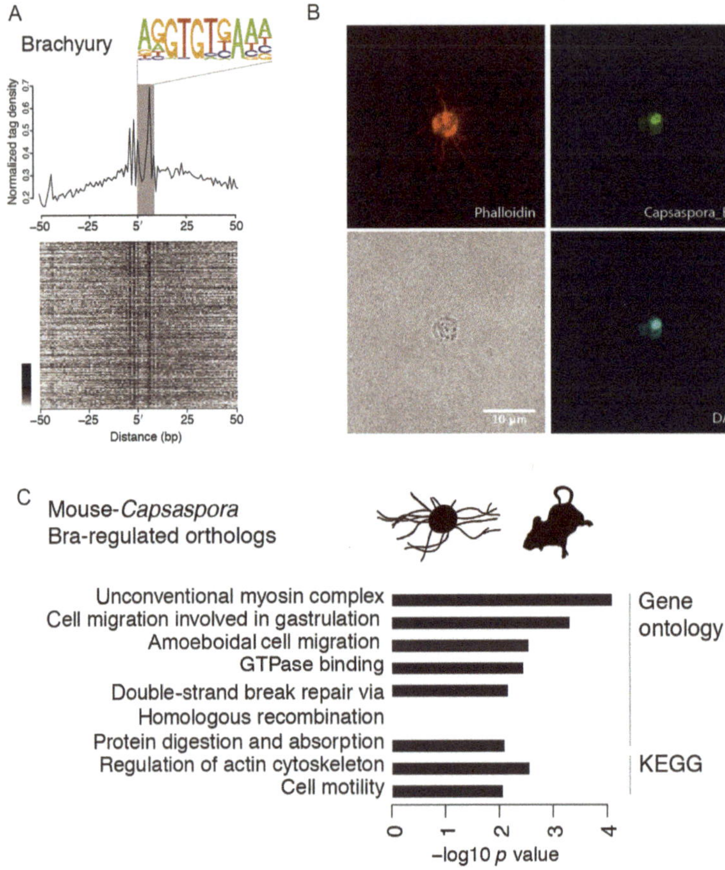

Fig. 5 A unicellular Brachyury regulatory network. (A) Plot of read density centered around *Bra* motifs (see Fig. 4) and heatmap of signal around individual sites in *Capsaspora* ATAC-seq experiments. (B) *Capsaspora* filopodial stage cell stained with phalloidin (*red*, actin cytoskeleton), DAPI (*blue*, nucleus) and *Capsaspora-Brachyury* antibody (*green*). Notice *Bra* localization in the nucleus. (C) Enriched GO terms and KEGG pathways among genes associated with *Bra* regulatory sites with shared orthologs regulated by *Bra* in mouse. *Adapted from Sebé-Pedrós et al. (2016).*

in *trans* affect the primary TF coding sequence, while the ones in *cis* affect regulatory sequences controlling the expression of the TF itself or changes in the TF DNA-binding sites in the genome, which results in new downstream targets controlled by a TF and, therefore, the rewiring of the TF network.

An example of *trans* changes in T-box function is the case of Brachyury cofactor interaction (Marcellini, 2006; Marcellini et al., 2003). It is known that many T-box TFs act together with other TFs and cofactors. For example, in mammals, Tbx5 interacts through its T-domain with the homeobox TF Nkx2.5 and also Gata4 during cardiomyocyte differentiation, Tbx2 with Rb1 protein through its C-terminal region, and Tbx18 interacts with the TF Pax3 in the regulation of AP somite polarity (reviewed by Papaioannou, 2014). In the case of Brachyury, heterologous overexpression experiments (described earlier) suggested that Brachyury target specificity in metazoans arose through changes in cofactor interactions (such as Smad, and probably others), affecting both the N- and C-terminal domains of the protein. These interactions were probably established at the onset of Metazoa. Instead, a premetazoan Brachyury ortholog (that of *C. owczarzaki*) behaves as a "panT-box" gene, strongly activating downstream targets of different animal T-box classes (in this case, Brachyury and Tbx6). This *trans* regulatory change occurred concomitantly with the explosive diversification of T-box classes at root of Metazoa, resulting in the subfunctionalization of rapidly duplicated T-box TFs.

Trans changes can also affect the DNA-binding specificities of a TF, for example, restricting its ability to bind to particular sites (Hudson et al., 2015). An example in the T-box family is found in Eomes/Tbrain of echinoderms. Jarvela et al. reported that the Eomes orthologs of sea urchin (*S. purpuratus*) and sea star (*Patiria miniata*) have differences in their secondary binding motifs, which also differ from the secondary motifs in vertebrate Eomes (Jarvela et al., 2014). These differences likely derive from evolutionary changes in the DNA-contacting amino acids and result in important differences in the role of Eomes in the development of these two echinoderms: in the sea urchin Eomes functions in skeletogenesis (see earlier), while in the sea star Eomes has roles in the endomesoderm and also in the ectoderm (Jarvela et al., 2014).

An example of *cis* changes affecting the domain of expression of T-box TFs is found in Tbx4/5 of chordates (Minguillón, Gibson-Brown, & Logan, 2009). Tbx4/5 is expressed in the cardiac region of the cephalochordate amphioxus, suggesting a conserved role in myogenesis in chordates; while in vertebrates Tbx4 and Tbx5 paralogs have well-studied roles in limb development. Heterologous rescue experiments in mouse by Minguillón et al. showed that amphioxus Tbx4/5 is able to induce limb growth, suggesting that no major changes in Tbx4/5 proteins occurred during the vertebrate transition. Instead, the authors propose that the newly evolved *cis*-regulatory elements (in this case the LPM enhancer) changed the Tbx4/5 expression domain, providing the basis for the acquisition of paired appendages during vertebrate evolution. A similar example of change in enhancer function is the one reported by Infante et al. in snake Tbx4 regulation. In snakes, this gene lost Tbx4 hindlimb expression through changes in the HLEB enhancer that is known to drive hindlimb and genitalia Tbx4 expression in mouse (Infante et al., 2015).

Finally, changes in the *cis*-regulatory target sequences throughout the genome can cause rapid evolution in the downstream network of a TF (Sorrells & Johnson, 2015). In a recent study, Lolas et al. analyzed the Brachyury downstream network in mouse (Lolas et al., 2014), and compared it to those known for zebrafish (Morley et al., 2009) and *Xenopus* (Gentsch et al., 2016). By comparing orthologous targets genes between species, they showed relatively little conservation (~10–15% of the genes) in the Brachyury downstream network of these vertebrate species. This shows the rapid divergence to the Brachyury network, likely mostly through changes in the Bra *cis*-target sequences in the genome.

ACKNOWLEDGMENTS

We thank José M. Martín-Durán for critical discussion about T-box function in different bilaterian lineages and Meritxell Antó for help with the figures. This work was supported by an Institució Catalana de Recerca i Estudis Avançats contract, a European Research Council Consolidator grant (ERC-2012-Co-616960), and grant BFU2014-57779-P from Ministerio de Economía y Competitividad (MINECO) to I.R.-T., funded partly by Fondo Europeo de Desarrollo regional (FEDER). We also acknowledge financial support from Secretaria d'Universitats i Recerca del Departament d'Economia i Coneixement de la Generalitat de Catalunya (Project 2014 SGR 619).

REFERENCES

Adell, T., Grebenjuk, V. A., Wiens, M., & Müller, W. E. G. (2003). Isolation and characterization of two T-box genes from sponges, the phylogenetically oldest metazoan taxon. *Development Genes and Evolution*, *213*(9), 421–434. http://dx.doi.org/10.1007/s00427-003-0345-5.

Adell, T., & Müller, W. E. G. (2005). Expression pattern of the Brachyury and Tbx2 homologues from the sponge Suberites domuncula. *Biology of the Cell/under the Auspices of the European Cell Biology Organization*, *97*(8), 641–650. http://dx.doi.org/10.1042/BC20040135.

Amemiya, C. T., Alföldi, J., Lee, A. P., Fan, S., Philippe, H., Maccallum, I., … Lindblad-Toh, K. (2013). The African coelacanth genome provides insights into tetrapod evolution. *Nature*, *496*(7445), 311–316. http://dx.doi.org/10.1038/nature12027.

Arenas-Mena, C. (2008). The transcription factors HeBlimp and HeT-brain of an indirectly developing polychaete suggest ancestral endodermal, gastrulation, and sensory cell-type specification roles. *Journal of Experimental Zoology. Part B, Molecular and Developmental Evolution*, *310*, 567–576. http://dx.doi.org/10.1002/jez.b.21225.

Arenas-Mena, C. (2013). Brachyury, Tbx2/3 and sall expression during embryogenesis of the indirectly developing polychaete Hydroides elegans. *International Journal of Developmental Biology*, *57*(1), 73–83. http://dx.doi.org/10.1387/ijdb.120056ca.

Beaster-Jones, L., Horton, A. C., Gibson-Brown, J. J., Holland, N. D., & Holland, L. Z. (2006). The amphioxus T-box gene, AmphiTbx15/18/22, illuminates the origins of chordate segmentation. *Evolution & Development*, *8*(2), 119–129. http://dx.doi.org/10.1111/j.1525-142X.2006.00083.x.

Belgacem, M. R., Escande, M. L., Escriva, H., & Bertrand, S. (2011). Amphioxus Tbx6/16 and Tbx20 embryonic expression patterns reveal ancestral functions in chordates. *Gene Expression Patterns*, *11*(3–4), 239–243. http://dx.doi.org/10.1016/j.gep.2010.12.006.

Bielen, H., Oberleitner, S., Marcellini, S., Gee, L., Lemaire, P., Bode, H. R., … Technau, U. (2007). Divergent functions of two ancient Hydra Brachyury paralogues suggest specific roles for their C-terminal domains in tissue fate induction. *Development*, *134*(23), 4187. http://dx.doi.org/10.1242/dev.010173.

Bollag, R. J., Siegfried, Z., Cebra-Thomas, J. A., Garvey, N., Davison, E. M., & Silver, L. M. (1994). An ancient family of embryonically expressed mouse genes sharing a conserved protein motif with the T locus. *Nature Genetics*, 7(3), 383–389.

Cavalier-Smith, T. (2003). Phylogeny of Choanozoa, Apusozoa, and other Protozoa and early eukaryote megaevolution. *Journal of Molecular Evolution*, *56*(5), 540–563. http://dx.doi.org/10.1007/s00239-002-2424-z.

Chang, Y., Wang, S., Sekimoto, S., Aerts, A., Choi, C., Clum, A., … Berbee, M. (2015). Phylogenomic analyses indicate that early fungi evolved digesting cell walls of algal ancestors of land plants. *Genome Biology and Evolution*, 7(6), 1590–1601. http://dx.doi.org/10.1093/gbe/evv090.

Charoensawan, V., Wilson, D., & Teichmann, S. a. (2010). Genomic repertoires of DNA-binding transcription factors across the tree of life. *Nucleic Acids Research*, *38*(21), 7364–7377. http://dx.doi.org/10.1093/nar/gkq617.

Chen, J.-H., Luo, Y.-J., & Su, Y.-H. (2011). The dynamic gene expression patterns of transcription factors constituting the sea urchin aboral ectoderm gene regulatory network. *Developmental Dynamics*, *240*(1), 250–260. http://dx.doi.org/10.1002/dvdy.22514.

Chipman, A. D., Ferrier, D. E. K., Brena, C., Qu, J., Hughes, D. S. T., Schröder, R., … Richards, S. (2014). The first myriapod genome sequence reveals conservative arthropod gene content and genome organisation in the centipede Strigamia maritima. *PLoS Biology*, *12*(11), e1002005. http://dx.doi.org/10.1371/journal.pbio.1002005.

Conlon, F. L., Fairclough, L., Price, B. M., Casey, E. S., & Smith, J. C. (2001). Determinants of T box protein specificity. *Development*, *128*(19), 3749–3758.

Conlon, F. L., Sedgwick, S. G., Weston, K. M., & Smith, J. C. (1996). Inhibition of Xbra transcription activation causes defects in mesodermal patterning and reveals autoregulation of Xbra in dorsal mesoderm. *Development*, *122*(8), 2427–2435.

Croce, J., Lhomond, G., & Gache, C. (2001a). Expression pattern of Brachyury in the embryo of the sea urchin Paracentrotus lividus. *Development Genes and Evolution*, *211*(12), 617–619. http://dx.doi.org/10.1007/s00427-001-0200-5.

Croce, J., Lhomond, G., & Gache, C. (2003). Coquillette, a sea urchin T-box gene of the Tbx2 subfamily, is expressed asymmetrically along the oral-aboral axis of the embryo and is involved in skeletogenesis. *Mechanisms of Development*, *120*(5), 561–572. S0925477303000224 [pii].

Croce, J., Lhomond, G., Lozano, J. C., & Gache, C. (2001b). ske-T, a T-box gene expressed in the skeletogenic mesenchyme lineage of the sea urchin embryo. *Mechanisms of Development*, *107*(1–2), 159–162. S0925477301004701 [pii].

De Mendoza, A., Suga, H., Permanyer, J., Irimia, M., & Ruiz-Trillo, I. (2015). Complex transcriptional regulation and independent evolution of fungal-like traits in a relative of animals. *eLife*, *4*, e08904.

Gentsch, G. E., Owens, N. D. L., Martin, S. R., Piccinelli, P., Faial, T., Trotter, M. W. B., … Smith, J. C. (2016). In vivo T-box transcription factor profiling reveals joint regulation of embryonic neuromesodermal bipotency. *Cell Reports*, *4*(6), 1185–1196. http://dx.doi.org/10.1016/j.celrep.2013.08.012.

Gross, J. M., & McClay, D. R. (2001). The role of Brachyury (T) during gastrulation movements in the sea urchin Lytechinus variegatus. *Developmental Biology*, *239*, 132–147. http://dx.doi.org/10.1006/dbio.2001.0426.

Gross, J. M., Peterson, R. E., Wu, S.-Y., & McClay, D. R. (2003). LvTbx2/3: A T-box family transcription factor involved in formation of the oral/aboral axis of the sea urchin embryo. *Development*, *130*(9), 1989–1999.

Hayward, D. C., Grasso, L. C., Saint, R., Miller, D. J., & Ball, E. E. (2015). The organizer in evolution–gastrulation and organizer gene expression highlight the importance of Brachyury during development of the coral, Acropora millepora. *Developmental Biology*, *399*(2), 337–347. http://dx.doi.org/10.1016/j.ydbio.2015.01.006.

He, M., Wen, L., Campbell, C. E., Wu, J. Y., & Rao, Y. (1999). Transcription repression by Xenopus ET and its human ortholog TBX3, a gene involved in ulnar-mammary syndrome. *Proceedings of the National Academy of Sciences of the United States of America*, *96*(18), 10212–10217. http://dx.doi.org/10.1073/pnas.96.18.10212.

Hejnol, A., & Martín-Durán, J. M. (2015). Getting to the bottom of anal evolution. *Zoologischer Anzeiger—A Journal of Comparative Zoology*, *256*, 61–74. http://dx.doi.org/10.1016/j.jcz.2015.02.006.

Herrmann, B. G., Labeit, S., Poustka, A., King, T. R., & Lehrach, H. (1990). Cloning of the T gene required in mesoderm formation in the mouse. *Nature*, *343*(6259), 617–622.

Holstien, K., Rivera, A., Windsor, P., Ding, S., Leys, S. P., Hill, M., & Hill, A. (2010). Expansion, diversification, and expression of T-box family genes in Porifera. *Development Genes and Evolution*, *220*(9–10), 251–262. http://dx.doi.org/10.1007/s00427-010-0344-2.

Horton, A. C., & Gibson-Brown, J. J. (2002). Evolution of developmental functions by the eomesodermin, T-brain-1, Tbx21 subfamily of T-box genes: Insights from amphioxus. *The Journal of Experimental Zoology*, *294*(2), 112–121. http://dx.doi.org/10.1002/jez.10151.

Hudson, W. H., Kossmann, B. R., Mitchelle, I., De Vera, S., Chuo, S., Weikum, E. R., … Ortlund, E. A. (2015). Distal substitutions drive divergent DNA specificity among paralogous transcription factors through subdivision of conformational space. *Proceedings of the National Academy of Sciences of the United States of America*, *113*(2), 1–6. http://dx.doi.org/10.1073/pnas.1518960113.

Hurlin, P. J., Steingrìmsson, E., Copeland, N. G., Jenkins, N. A., & Eisenman, R. N. (1999). Mga, a dual-specificity transcription factor that interacts with Max and contains a T-domain DNA-binding motif. *The EMBO Journal*, *18*(24), 7019–7028. http://dx.doi.org/10.1093/emboj/18.24.7019.

Infante, C. R., Mihala, A. G., Park, S., Wang, J. S., Johnson, K. K., Lauderdale, J. D., & Menke, D. B. (2015). Shared enhancer activity in the limbs and phallus and functional divergence of a limb-genital cis-regulatory element in snakes. *Developmental Cell*, *35*(1), 107–119. http://dx.doi.org/10.1016/j.devcel.2015.09.003.

Itzkovitz, S., Tlusty, T., & Alon, U. (2006). Coding limits on the number of transcription factors. *BMC Genomics*, 7, 239. http://dx.doi.org/10.1186/1471-2164-7-239.

Jarvela, A. M., Brubaker, L., Vedenko, A., Gupta, A., Armitage, B. A., Bulyk, M. L., & Hinman, V. F. (2014). Modular evolution of DNA-binding preference of a Tbrain transcription factor provides a mechanism for modifying gene regulatory networks. *Molecular Biology and Evolution*, *31*(10), 2672–2688. http://dx.doi.org/10.1093/molbev/msu213.

Katikala, L., Aihara, H., Passamaneck, Y. J., Gazdoiu, S., José-Edwards, D. S., Kugler, J. E., … Di Gregorio, A. (2013). Functional Brachyury binding sites establish a temporal read-out of gene expression in the Ciona notochord. *PLoS Biology, 11*(10), e1001697. http://dx.doi.org/10.1371/journal.pbio.1001697.

Kispert, A., Herrmann, B. G., Leptin, M., & Reuter, R. (1994). Homologs of the mouse Brachyury gene are involved in the specification of posterior terminal structures in Drosophila, Tribolium, and Locusta. *Genes and Development, 8*, 2137–2150. http://dx.doi.org/10.1101/gad.8.18.2137.

Larroux, C., Fahey, B., Liubicich, D., Hinman, V. F., Gauthier, M., Gongora, M., … Degnan, B. M. (2006). Developmental expression of transcription factor genes in a demosponge: Insights into the origin of metazoan multicellularity. *Evolution & Development, 8*(2), 150–173. http://dx.doi.org/10.1111/j.1525-142X.2006.00086.x.

Larroux, C., Luke, G. N., Koopman, P., Rokhsar, D. S., Shimeld, S. M., Degnan, B. M., & Shimeld, M. (2008). Genesis and expansion of metazoan transcription factor gene classes. *Molecular Biology and Evolution, 25*(5), 980–996. http://dx.doi.org/10.1093/molbev/msn047.

Lartillot, N., Lespinet, O., Vervoort, M., & Adoutte, A. (2002). Expression pattern of Brachyury in the mollusc Patella vulgata suggests a conserved role in the establishment of the AP axis in Bilateria. *Development, 129*(6), 1411–1421.

Leininger, S., Adamski, M., Bergum, B., Guder, C., Liu, J., Laplante, M., … Adamska, M. (2014). Developmental gene expression provides clues to relationships between sponge and eumetazoan body plans. *Nature Communications, 5*, 3905. http://dx.doi.org/10.1038/ncomms4905.

Lolas, M., Valenzuela, P. D. T., Tjian, R., & Liu, Z. (2014). Charting Brachyury-mediated developmental pathways during early mouse embryogenesis. *Proceedings of the National Academy of Sciences of the United States of America, 111*, 4478–4483. http://dx.doi.org/10.1073/pnas.1402612111.

Lowe, C. J., Terasaki, M., Wu, M., Freeman, R. M., Runft, L., Kwan, K., … Gerhart, J. (2006). Dorsoventral patterning in hemichordates: Insights into early chordate evolution. *PLoS Biology, 4*(9), e291. http://dx.doi.org/10.1371/journal.pbio.0040291.

Manuel, M., Le Parco, Y., & Borchiellini, C. (2004). Comparative analysis of Brachyury T-domains, with the characterization of two new sponge sequences, from a hexactinellid and a calcisponge. *Gene, 340*(2), 291–301. http://dx.doi.org/10.1016/j.gene.2004.07.021.

Marcellini, S. (2006). When Brachyury meets Smad1: The evolution of bilateral symmetry during gastrulation. *Bioessays, 28*(4), 413–420. http://dx.doi.org/10.1002/bies.20387.

Marcellini, S., Technau, U., Smith, J., & Lemaire, P. (2003). Evolution of Brachyury proteins: Identification of a novel regulatory domain conserved within Bilateria. *Developmental Biology, 260*(2), 352–361.

Martín-Durán, J. M., Janssen, R., Wennberg, S., Budd, G. E., & Hejnol, A. (2012). Deuterostomic development in the protostome Priapulus caudatus. *Current Biology, 22*(22), 2161–2166. http://dx.doi.org/10.1016/j.cub.2012.09.037.

Martín-Durán, J. M., & Romero, R. (2011). Evolutionary implications of morphogenesis and molecular patterning of the blind gut in the planarian Schmidtea polychroa. *Developmental Biology, 352*(1), 164–176. http://dx.doi.org/10.1016/j.ydbio.2011.01.032.

Martinelli, C., & Spring, J. (2003). Distinct expression patterns of the two T-box homologues Brachyury and Tbx2/3 in the placozoan Trichoplax adhaerens. *Development Genes and Evolution, 213*(10), 492–499. http://dx.doi.org/10.1007/s00427-003-0353-5.

Martinelli, C., & Spring, J. (2005). T-box and homeobox genes from the ctenophore Pleurobrachia pileus: Comparison of Brachyury, Tbx2/3 and Tlx in basal metazoans and bilaterians. *FEBS Letters, 579*(22), 5024–5028. http://dx.doi.org/10.1016/j.febslet.2005.08.008.

Mendoza, L., Taylor, J. W., & Ajello, L. (2002). The class mesomycetozoea: A heterogeneous group of microorganisms at the animal-fungal boundary. *Annual Review of Microbiology*, *56*, 315–344. http://dx.doi.org/10.1146/annurev.micro.56.012302.160950.

Minguillón, C., Gibson-Brown, J. J., & Logan, M. P. (2009). Tbx4/5 gene duplication and the origin of vertebrate paired appendages. *Proceedings of the National Academy of Sciences of the United States of America*, *106*(51), 21726–21730. http://dx.doi.org/10.1073/pnas.0910153106.

Morley, R. H., Lachani, K., Keefe, D., Gilchrist, M. J., Flicek, P., Smith, J. C., & Wardle, F. C. (2009). A gene regulatory network directed by zebrafish No tail accounts for its roles in mesoderm formation. *Proceedings of the National Academy of Sciences of the United States of America*, *106*(10), 3829–3834. http://dx.doi.org/10.1073/pnas.0808382106.

Moroz, L. L., Kocot, K. M., Citarella, M. R., Dosung, S., Norekian, T. P., Povolotskaya, I. S., … Kohn, A. B. (2014). The ctenophore genome and the evolutionary origins of neural systems. *Nature*, *510*(7503), 109–114. http://dx.doi.org/10.1038/nature13400.

Onai, T., Lin, H. C., Schubert, M., Koop, D., Osborne, P. W., Alvarez, S., … Holland, L. Z. (2009). Retinoic acid and Wnt/beta-catenin have complementary roles in anterior/posterior patterning embryos of the basal chordate amphioxus. *Developmental Biology*, *332*(2), 223–233. http://dx.doi.org/10.1016/j.ydbio.2009.05.571. S0012-1606(09)00901-4 [pii]\r.

Papaioannou, V. E. (2001). T-box genes in development: From hydra to humans. *International Review of Cytology*, *207*, 1–70.

Papaioannou, V. E. (2014). The T-box gene family: Emerging roles in development, stem cells and cancer. *Development*, *141*, 3819–3833. http://dx.doi.org/10.1242/dev.104471.

Paps, J., Holland, P. W. H., & Shimeld, S. M. (2012). A genome-wide view of transcription factor gene diversity in chordate evolution: Less gene loss in amphioxus? *Briefings in Functional Genomics*, *11*(2), 177–186. http://dx.doi.org/10.1093/bfgp/els012.

Pascual-Anaya, J., Albuixech-Crespo, B., Somorjai, I. M. L., Carmona, R., Oisi, Y., Alvarez, S., … Garcia-Fernàndez, J. (2013). The evolutionary origins of chordate hematopoiesis and vertebrate endothelia. *Developmental Biology*, *375*(2), 182–192. http://dx.doi.org/10.1016/j.ydbio.2012.11.015.

Peterson, K. J., Harada, Y., Cameron, R. A., & Davidson, E. H. (1999). Expression pattern of Brachyury and Not in the sea urchin: Comparative implications for the origins of mesoderm in the basal deuterostomes. *Developmental Biology*, *207*(2), 419–431. http://dx.doi.org/10.1006/dbio.1998.9177.

Pick, K. S., Philippe, H., Schreiber, F., Erpenbeck, D., Jackson, D. J., Wrede, P., … Wörheide, G. (2010). Improved phylogenomic taxon sampling noticeably affects nonbilaterian relationships. *Molecular Biology and Evolution*, *27*(9), 1983–1987. http://dx.doi.org/10.1093/molbev/msq089.

Pisani, D., Pett, W., Dohrmann, M., Feuda, R., Rota-Stabelli, O., Philippe, H., … Wörheide, G. (2015). Genomic data do not support comb jellies as the sister group to all other animals. *Proceedings of the National Academy of Sciences of the United States of America*, *112*(50), 15402–15407. http://dx.doi.org/10.1073/pnas.1518127112.

Rast, J. P., Cameron, R. A., Poustka, A. J., & Davidson, E. H. (2002). Brachyury target genes in the early sea urchin embryo isolated by differential macroarray screening. *Developmental Biology*, *246*(1), 191–208. http://dx.doi.org/10.1006/dbio.2002.0654.

Riesgo, A., Farrar, N., Windsor, P. J., Giribet, G., & Leys, S. P. (2014). The analysis of eight transcriptomes from all poriferan classes reveals surprising genetic complexity in sponges. *Molecular Biology and Evolution*, *31*(5), 1102–1120. http://dx.doi.org/10.1093/molbev/msu057.

Ruiz-Trillo, I., Roger, A. J., Burger, G., Gray, M. W., & Lang, B. F. (2008). A phylogenomic investigation into the origin of metazoa. *Molecular Biology and Evolution*, *25*(4), 664–672. http://dx.doi.org/10.1093/molbev/msn006.

Ryan, J. F., Pang, K., Schnitzler, C. E., Nguyen, A.-D., Moreland, R. T., Simmons, D. K., ... Baxevanis, A. D. (2013). The genome of the ctenophore Mnemiopsis leidyi and its implications for cell type evolution. *Science, 342*(6164), 1242592. http://dx.doi.org/10.1126/science.1242592.

Satoh, G., Harada, Y., & Satoh, N. (2000). The expression of nonchordate deuterostome Brachyury genes in the ascidian Ciona embryo can promote the differentiation of extra notochord cells. *Mechanisms of Development, 96*(2), 155–163.

Scholz, C. B., & Technau, U. (2003). The ancestral role of Brachyury: Expression of NemBra1 in the basal cnidarian Nematostella vectensis (Anthozoa). *Development Genes and Evolution, 212*(12), 563–570. http://dx.doi.org/10.1007/s00427-002-0272-x.

Schulte-Merker, S., Ho, R. K., Herrmann, B. G., & Nusslein-Volhard, C. (1992). The protein product of the zebrafish homologue of the mouse T gene is expressed in nuclei of the germ ring and the notochord of the early embryo. *Development (Cambridge, England), 116*(4), 1021–1032.

Sebé-Pedrós, A., Ariza-Cosano, A., Weirauch, M. T., Leininger, S., Yang, A., Torruella, G., ... Ruiz-Trillo, I. (2013). Early evolution of the T-box transcription factor family. *Proceedings of the National Academy of Sciences of the United States of America, 110*(40), 16050–16055. http://dx.doi.org/10.1073/pnas.1309748110.

Sebé-Pedrós, A., Ballaré, C., Parra-Acero, H., Chiva, C., Tena, J. J., Sabidó, E., ... Ruiz-Trillo, I. (2016). The dynamic regulatory genome of *Capsaspora* and the origin of animal multicellularity. *Cell, 165*(5), 1224–1237. http://dx.doi.org/10.1016/j.cell.2016.03.034.

Sebé-Pedrós, A., de Mendoza, A., Lang, B. F., Degnan, B. M., & Ruiz-Trillo, I. (2011). Unexpected repertoire of metazoan transcription factors in the unicellular holozoan Capsaspora owczarzaki. *Molecular Biology and Evolution, 28*(3), 1241–1254. http://dx.doi.org/10.1093/molbev/msq309.

Showell, C., Binder, O., & Conlon, F. L. (2004). T-box genes in early embryogenesis. *Developmental Dynamics, 229*(1), 201–218. http://dx.doi.org/10.1002/dvdy.10480.

Simakov, O., Kawashima, T., Marlétaz, F., Jenkins, J., Koyanagi, R., Mitros, T., ... Gerhart, J. (2015). Hemichordate genomes and deuterostome origins. *Nature, 527*(7579), 459–465. http://dx.doi.org/10.1038/nature16150.

Smith, J. (1999). T-box genes: What they do and how they do it. *Trends in Genetics, 15*(4), 154–158. http://dx.doi.org/10.1016/S0168-9525(99)01693-5.

Smith, J. J., Kuraku, S., Holt, C., Sauka-Spengler, T., Jiang, N., Campbell, M. S., ... Li, W. (2013). Sequencing of the sea lamprey (Petromyzon marinus) genome provides insights into vertebrate evolution. *Nature Genetics, 45*(4), 415–421. http://dx.doi.org/10.1038/ng.2568. 421e1–2.

Smith, J. C., Price, B. M., Green, J. B., Weigel, D., & Herrmann, B. G. (1991). Expression of a Xenopus homolog of Brachyury (T) is an immediate-early response to mesoderm induction. *Cell, 67*(1), 79–87.

Sorrells, T. R., & Johnson, A. D. (2015). Making sense of transcription networks. *Cell, 161*, 714–723.

Spitz, F., & Furlong, E. E. M. (2012). Transcription factors: From enhancer binding to developmental control. *Nature Reviews. Genetics, 13*(9), 613–626. http://dx.doi.org/10.1038/nrg3207.

Spring, J., Yanze, N., Jösch, C., Middel, A. M., Winninger, B., & Schmid, V. (2002). Conservation of Brachyury, Mef2, and Snail in the myogenic lineage of jellyfish: A connection to the mesoderm of bilateria. *Developmental Biology, 244*(2), 372–384. http://dx.doi.org/10.1006/dbio.2002.0616.

Stott, D., Kispert, A., & Herrmann, B. G. (1993). Rescue of the tail defect of Brachyury mice. *Genes & Development, 7*(2), 197–203.

Suga, H., & Ruiz-Trillo, I. (2013). Development of ichthyosporeans sheds light on the origin of metazoan multicellularity. *Developmental Biology*, *377*, 284–292. http://dx.doi.org/10.1016/j.ydbio.2013.01.009.

Tada, M., & Smith, J. C. (2000). Xwnt11 is a target of Xenopus Brachyury: Regulation of gastrulation movements via Dishevelled, but not through the canonical Wnt pathway. *Development*, *127*(10), 2227–2238.

Tagawa, K., Humphreys, T., & Satoh, N. (1998). Novel pattern of Brachyury gene expression in hemichordate embryos. *Mechanisms of Development*, *75*(1–2), 139–143. http://dx.doi.org/10.1016/S0925-4773(98)00078-1.

Tagawa, K., Humphreys, T., & Satoh, N. (2000). T-brain expression in the apical organ of hemichordate tornaria larvae suggests its evolutionary link to the vertebrate forebrain. *Journal of Experimental Zoology*, *288*(1), 23–31. http://dx.doi.org/10.1002/(SICI)1097-010X(20000415)288:1<23::AID-JEZ3>3.0.CO;2-H.

Technau, U., & Bode, H. (1999). HyBra1, a Brachyury homologue, acts during head formation in Hydra. *Development*, *126*, 999–1010.

Torruella, G., Derelle, R., Paps, J., Lang, B. F., Roger, A. J., Shalchian-Tabrizi, K., & Ruiz-Trillo, I. (2012). Phylogenetic relationships within the opisthokonta based on phylogenomic analyses of conserved single-copy protein domains. *Molecular Biology and Evolution*, *29*(2), 531–544.

Venkatesh, B., Lee, A. P., Ravi, V., Maurya, A. K., Lian, M. M., Swann, J. B., … Warren, W. C. (2014). Elephant shark genome provides unique insights into gnathostome evolution. *Nature*, *505*(7482), 174–179. http://dx.doi.org/10.1038/nature12826.

Wilson, V., & Conlon, F. L. (2002). The T-box family. *Genome Biology*, *3*(6). reviews 3008.1–3008.7.

Xanthos, J. B., Kofron, M., Wylie, C., & Heasman, J. (2001). Maternal VegT is the initiator of a molecular network specifying endoderm in Xenopus laevis. *Development*, *128*(2), 167–180.

Yamada, A., Martindale, M. Q., Fukui, A., & Tochinai, S. (2010). Highly conserved functions of the Brachyury gene on morphogenetic movements: Insight from the early-diverging phylum Ctenophora. *Developmental Biology*, *339*(1), 212–222. http://dx.doi.org/10.1016/j.ydbio.2009.12.019.

Yamada, A., Pang, K., Martindale, M. Q., & Tochinai, S. (2007). Surprisingly complex T-box gene complement in diploblastic metazoans. *Evolution and Development*, *9*(3), 220–230. http://dx.doi.org/10.1111/j.1525-142X.2007.00154.x.

CHAPTER TWO

The Remarkably Diverse Family of T-Box Factors in *Caenorhabditis elegans*

P.G. Okkema[1]

University of Illinois at Chicago, Molecular, Cell & Developmental Biology Group (MC567), Chicago, IL, United States

[1]Corresponding author: e-mail address: okkema@uic.edu

Contents

1. Introduction 28
2. T-Box Genes in *C. elegans* and Related Nematodes 29
3. DNA-Binding Specificity of *C. elegans* T-Box Factors 34
4. Genetic Analyses of T-Box Factor Function in *C. elegans* 37
   4.1 A Network of T-Box Factors Function in *C. elegans* Pharyngeal Development 37
   4.2 *tbx-2* Has Additional Functions in Neuronal Development and Sensory Adaptation 40
   4.3 The Tbx-1 Subfamily Members MAB-9 and MLS-1 Function Postembryonically to Specify Ectodermal and Muscle Cell Fates 41
   4.4 The TBX-8 and TBX-9 Paralogs Play Redundant Roles in Hypodermal and Body Wall Muscle Morphogenesis 43
   4.5 The Divergent T-Box Factor SEA-1 Functions as an Autosomal Counting Element to Determine Sex in *C. elegans* 45
   4.6 The Functions of Other *C. elegans* T-Box Genes Are Only Beginning to Be Characterized 46
5. Posttranslational Mechanisms Affecting T-Box Factor Activity 47
6. Concluding Remarks 48
Acknowledgments 49
References 49

Abstract

The nematode *Caenorhabditis elegans* is a simple metazoan animal that is widely used as a model to understand the genetic control of development. The completely sequenced *C. elegans* genome contains 22 T-box genes, and they encode factors that show remarkable diversity in sequence, DNA-binding specificity, and function. Only three of the *C. elegans* T-box factors can be grouped into the conserved subfamilies found in other organisms, while the remaining factors are significantly diverged and unlike those in most other animals. While some of the *C. elegans* factors can bind canonical T-box binding elements, others bind and regulate target gene expression through distinct

Current Topics in Developmental Biology, Volume 122
ISSN 0070-2153
http://dx.doi.org/10.1016/bs.ctdb.2016.08.005

sequences. The nine genetically characterized T-box factors have varied functions in development and morphogenesis of muscle, hypodermal tissues, and neurons, as well as in early blastomere fate specification, cell migration, apoptosis, and sex determination, but the functions of most of the *C. elegans* T-box factors have not yet been extensively characterized. Like T-box factors in other animals, interaction with a Groucho-family corepressor and posttranslational SUMOylation have been shown to affect *C. elegans* T-box factor activity, and it is likely that additional mechanisms affecting T-box factor activity will be discovered using the effective genetic approaches in this organism.

1. INTRODUCTION

T-box factors are an ancient and highly conserved family of transcription factors found in all animal taxa (reviewed in Papaioannou, 2014). As described throughout this volume, they play crucial roles in development, where they function in processes such as germ-layer specification, mesodermal development, and organogenesis. This family of factors is distinguished by a conserved 180–200 amino acid DNA-binding domain termed the T-box, and these factors are often grouped into specific subfamilies based on sequence conservation in this DNA-binding domain. T-box factors are encoded by a moderately sized, multigene family in all animals with a completely sequenced genome, including 17 T-box genes in mammalian genomes.

This review focuses on T-box factors in the nematode *Caenorhabditis elegans*. *C. elegans* was originally selected by Sydney Brenner over 50 years ago with the express purpose of developing a simple model system to study the molecular mechanisms controlling animal development and function of the nervous system (Brenner, 2003). Its stereotyped development, anatomical simplicity, ease of genetic manipulation, and transparent body make it ideal for genetic and microscopic studies, while the compact genome makes it excellent for molecular analyses (reviewed in Corsi, Wightman, & Chalfie, 2015). Despite its anatomical simplicity, *C. elegans* has distinct cell and tissue types that are found in other more complex organisms, including muscles, neurons, gut, and a reproductive system. It is now widely used to study a diverse range of biological process, including developmental control, cell division, aging, neural function, behavior, apoptosis, and evolution (see http://www.wormbook.org for reviews of these and other topics).

Nematodes are a phylum of animals characterized by a similar body organization with an elongated, thin, and unsegmented body. There are approximately 30,000 known nematode species that include both free-living and parasitic forms, but the number of extant nematode species is estimated to be one million or more (Kiontke & Fitch, 2013). Because of their similar morphologies, nematodes were initially thought to compose a basally diverged phylum. However, molecular analyses have consistently shown nematodes to be protostomes and members of the Ecdysozoan clade, which includes Drosophila and other arthropods (Blaxter, 1998). Nematodes are more distantly related to the deuterostomes, which include humans and other vertebrates, but 7663 of the 20,205 *C. elegans* protein-coding genes (~40%, WS250; http://www.wormbase.org) have clear orthologs in the human genome (Shaye & Greenwald, 2011).

2. T-BOX GENES IN *C. elegans* AND RELATED NEMATODES

The completely sequenced *C. elegans* genome contains 22 T-box genes (Table 1) (Reinke, Krause, & Okkema, 2013). The first of these genes were identified in the mid-1990s using low-stringency hybridizations and degenerate PCRs (Agulnik et al., 1996; Agulnik, Ruvinsky, & Silver, 1997), and the remainder was progressively identified in genetic studies and through analysis of the genome sequence (Consortium, 1998; Kostas & Fire, 2002; Woollard & Hodgkin, 2000). Two of these genes, *tbx-30* and *tbx-42*, are identical and encode indistinguishable T-box proteins, but the remaining genes are unique. The majority of *C. elegans* T-box genes produce a single transcript and do not encode alternative protein isoforms. The exception is the *tbx-7* gene, which produces a low abundance transcript containing an alternative first exon that results in an additional 36-amino acid peptide at the TBX-7 N-terminus (Hillier et al., 2009) (http://www.wormbase.org; WS250).

The *C. elegans* T-box genes are distributed on four of the six *C. elegans* chromosomes, and there is a notable enrichment of these genes on chromosome III, which contains half of the T-box genes in the genome (Fig. 1). Fourteen of the *C. elegans* T-box genes are located in closely spaced pairs, and four of these pairs contain highly related paralogs, including the identical *tbx-30/tbx-42* genes mentioned above and the *tbx-8/tbx-9*, *tbx-39/tbx-40*, and *tbx-31/tbx-32* paralog pairs (Figs. 1 and 2). These pairs were likely formed by relatively recent gene duplications, and, as discussed later, the *tbx-8/tbx-9* pair has partially redundant functions in development.

The T-box factors in *C. elegans* have evolved very differently than those in other organisms. Only three of the *C. elegans* T-box proteins fall into the

Table 1 *C. elegans* T-Box Genes and Loss-of-Function Phenotypes

Gene Public Name	WormBase Gene ID	Reported Mutant or RNAi Phenotypes	References
mab-9	WBGene00003106	Male tail morphology variant, backward locomotion variant	Chisholm and Hodgkin (1989), Hodgkin (1983), and Pocock et al. (2008)
mls-1	WBGene00003376	Uterine muscle mis-specification	Kostas and Fire (2002)
sea-1	WBGene00004750	Sex determination and dosage compensation defects	Powell, Jow, and Meyer (2005)
tbx-11	WBGene00006547	Gastrulation variant	Sawyer et al. (2011)
tbx-2	WBGene00006543	Anterior pharynx variant, neuronal development variant	Miyahara, Suzuki, Ishihara, Tsuchiya, and Katsura (2004), Roy Chowdhuri, Crum, Woollard, Aslam, and Okkema (2006), Singhvi, Frank, and Garriga (2008), and Smith and Mango (2007)
tbx-30	WBGene00006549	Transgene expression increased	Pocock, Ahringer, Mitsch, Maxwell, and Woollard (2004)
tbx-31	WBGene00006550	Embryonic lethal[a]	
tbx-32	WBGene00006551	None reported[a]	
tbx-33	WBGene00006552	Embryonic lethal, apoptosis variant	Green et al. (2011), Kamath et al. (2003), Simmer et al. (2003), and Sonnichsen et al. (2005)
tbx-34	WBGene00006553	None reported[a]	
tbx-35	WBGene00006554	Embryonic arrest, posterior pharynx variant	Broitman-Maduro, Lin, Hung, and Maduro (2006) and Broitman-Maduro et al. (2009)

tbx-36	WBGene00006555	Transgene expression increased	Ackerman and Gems (2012)
tbx-37	WBGene00006556	Embryonic lethal, anterior pharynx variant	Good et al. (2004)
tbx-38	WBGene00006557	Embryonic lethal, anterior pharynx variant	Good et al. (2004)
tbx-39	WBGene00006558	None reported[a]	
tbx-40	WBGene00006559	None reported[a]	
tbx-41	WBGene00006560	None reported[a]	
tbx-42	WBGene00022000	None reported[a]	
tbx-43	WBGene00044798	No data	
tbx-7	WBGene00006544	None reported[a]	
tbx-8	WBGene00006545	Body wall muscle morphology variant, epithelial variant	Andachi (2004) and Pocock et al. (2004)
tbx-9	WBGene00006546	Body wall muscle morphology variant, epithelial variant	Andachi (2004) and Pocock et al. (2004)

[a]Data reported in WormBase (WS252) (http://www.wormbase.org).

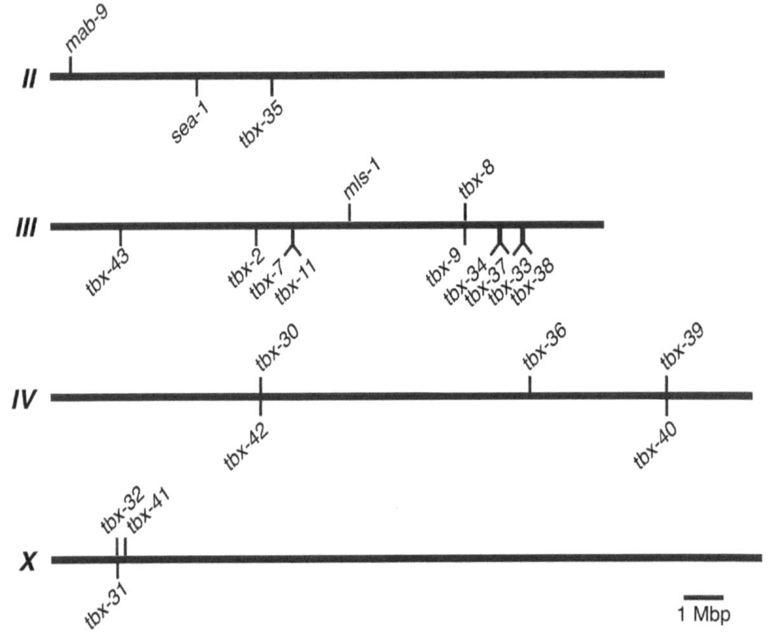

Fig. 1 Genomic locations of *C. elegans* T-box genes. Physical map of the *C. elegans* genome indicating the locations of T-box genes. Genes drawn above the line are transcribed from *left* to *right*, whereas genes below the lines are transcribed from *right* to *left*. *C. elegans* has six chromosomes (autosomes I, II, III, IV, and V, and the sex chromosome X), but only chromosomes containing T-box genes are shown. Scale bar indicates 10^6 bp.

conserved T-box subfamilies (Fig. 2) (Papaioannou, 2014; Pocock et al., 2004). Of these, *C. elegans* TBX-2 is the sole member of the Tbx-2 subfamily, while MAB-9 and MLS-1 are members of the Tbx-1 subfamily. Another of the *C. elegans* factors TBX-7 is related to a divergent factor ascidian T2, but this factor is not conserved in other animals (Papaioannou, 2001; Pocock et al., 2004). Notably, *C. elegans* lacks members of the Tbx-6, Tbr, and T/Brachyury subfamilies, which are all present in other bilaterian genomes (Papaioannou, 2014). Likewise, *C. elegans* lacks the Tbx-7/8 subfamily recently identified in sponges and other nonmetazoan animals (Sebe-Pedros et al., 2013). These observations suggest *C. elegans* members of these subfamilies have been lost during nematode evolution or that they have diverged significantly due to a rapid evolutionary rate in the *C. elegans* genome, ancient divergence of nematodes, or both (Kiontke & Fitch, 2005).

The remaining *C. elegans* T-box genes are not members of any of the conserved subfamilies, and most of these fall into a separate clade of divergent T-box factors (Fig. 2). Much of the sequence divergence in these

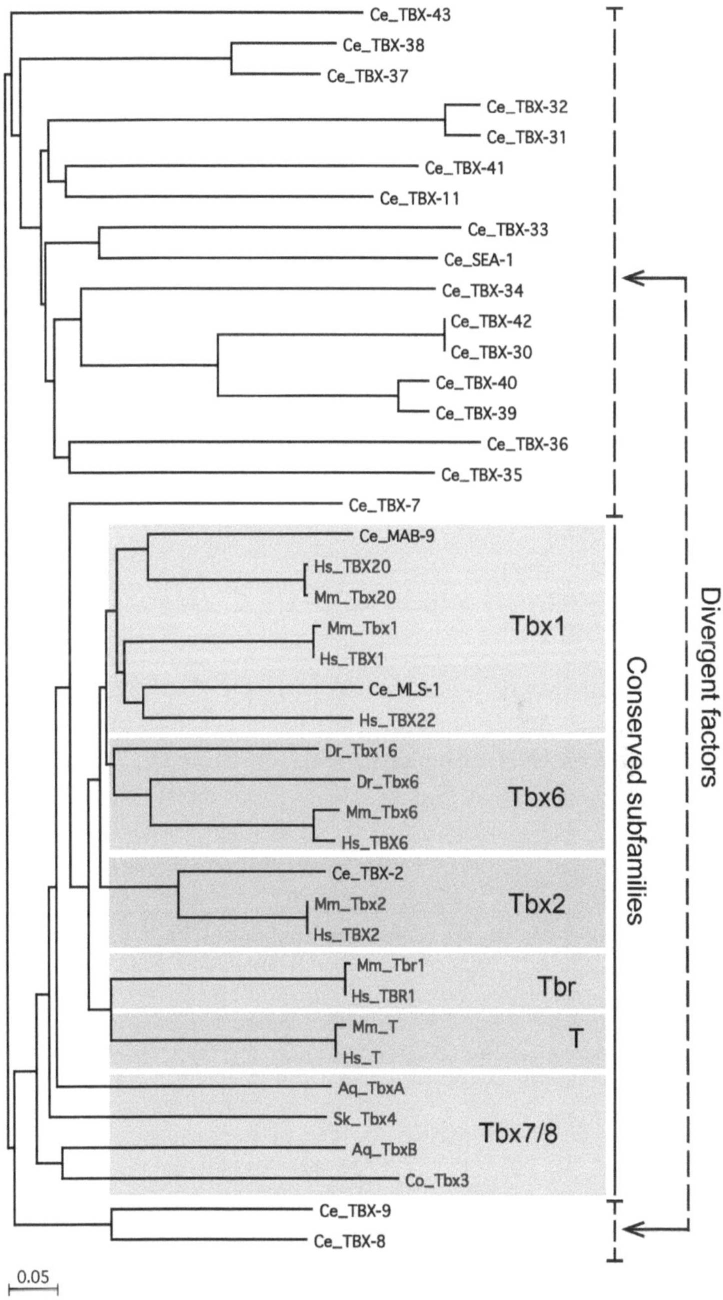

Fig. 2 Phylogenetic tree of the T-box factor family in *C. elegans*. Phylogeny of *C. elegans* (Ce) and selected human (Hs), mouse (Mm), zebrafish (Dr), the sponge *A. queenslandica* (Aq), the hemichordate *S. kowalevskii* (Sk), and the single-celled filasterean *C. owczarzaki*

(Continued)

factors has occurred near the C-terminus of the T-box, in regions that make extensive contacts with bound DNA in structurally characterized T-box factors (Fig. 3) (Coll et al., 2002; El Omari et al., 2012; Muller & Herrmann, 1997; Stirnimann, Ptchelkine, Grimm, & Muller, 2010). Orthologs of many of these divergent *C. elegans* factors are found in the partially or completely sequenced genomes of other Caenorhabditis species. Interestingly, these other Caenorhabditis species have even larger T-box factor families than *C. elegans*, indicating T-box genes have rapidly expanded and/or contracted during the evolution of this genus (Table 2). These species share very similar overall morphologies but show levels of molecular divergence comparable to that seen among mammalian species (Kiontke & Fitch, 2005). It will be exciting to explore the evolutionary plasticity of T-box genes in these nematodes and to understand the role these genes play in generating specific morphological differences among these organisms.

3. DNA-BINDING SPECIFICITY OF *C. elegans* T-BOX FACTORS

Most-characterized T-box factors bind DNA sequences related to AGGTGTGA, and this is often referred to as a T-box binding element (TBE) (Castellanos, Xie, Zheng, Cvekl, & Morrow, 2014; Conlon, Fairclough, Price, Casey, & Smith, 2001; Jolma et al., 2013; Kispert & Herrmann, 1993; Papaioannou, 2014). PCR-based selections for T-box binding sites generally identify TBE dimers with variable orientation and

Fig. 2—Cont'd (Co) T-box factors. Conserved subfamilies (*gray boxes*) are defined based on previous analyses (Papaioannou, 2014; Sebe-Pedros et al., 2013). Divergent factors found in *C. elegans* are indicated by *dashed brackets*. Tree is derived from a T-coffee alignment of the T-box DNA-binding domains from the following proteins (accession) (Notredame, Higgins, & Heringa, 2000): Aq_TbxA (ACA04753), Aq_TbxB (ACA04754), Ce_MAB-9 (NP_493750), Ce_MLS-1 (NP_498640), Ce_SEA-1 (NP_494611), Ce_TBX-11 (NP_498317), Ce_TBX-2 (NP_498088), Ce_TBX-30 (NP_500749), Ce_TBX-31 (NP_508308), Ce_TBX-32 (NP_508304), Ce_TBX-33 (NP_499506), Ce_TBX-34 (NP_499441), Ce_TBX-35 (NP_495059), Ce_TBX-36 (NP_502266), Ce_TBX-37 (NP_499444), Ce_TBX-38 (NP_499526), Ce_TBX-39 (NP_502851), Ce_TBX-40 (NP_502852), Ce_TBX-41 (NP_508343), Ce_TBX-42 (NP_500749), Ce_TBX-43 (NP_001040883), Ce_TBX-7 (CTQ86631), Ce_TBX-8 (NP_499287), Ce_TBX-9 (NP_499286), Co_Tbx3 (ADX60052), Dr_Tbx16 (AAI65213), Dr_Tbx6 (Q8JIS6), Hs_T (O15178), Hs_TBR1 (Q96RJ2), Hs_TBX1 (Q96RJ2), Hs_TBX2 (Q13207), Hs_TBX20 (Q9UMR3), Hs_TBX22 (CAI43070), Hs_TBX6 (Q8TAS4), Mm_T (NP_033335), Mm_Tbr1 (Q64336), Mm_Tbx1 (P70323), Mm_Tbx2 (Q60707), Mm_Tbx20 (Q9ES03), Mm_Tbx6 (P70327), and Sk_Tbx4 (XP_002733999).

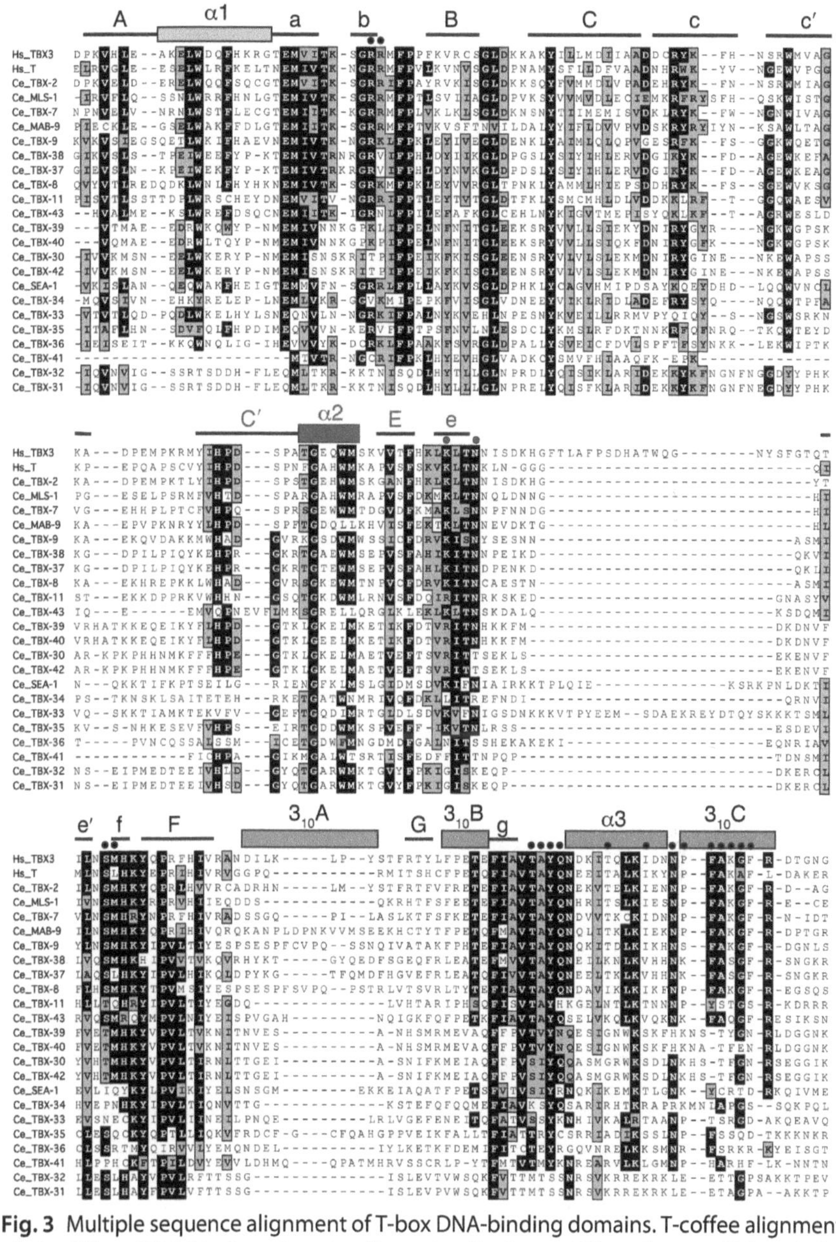

Fig. 3 Multiple sequence alignment of T-box DNA-binding domains. T-coffee alignment of the T-box DNA-binding domains from the *C. elegans* T-box factors (Ce) and the human TBX-3 and T/Brachyury (Hs), with secondary structure elements for human TBX-3 indicated above the sequences (Coll, Seidman, & Muller, 2002; Notredame et al., 2000). Strands in β-sheets are indicated as *lines*, and α-helices are indicated as *boxes*. *Dots* above sequences indicate residues that contact DNA in human TBX-3. The *C. elegans* factors are ordered by decreasing amino acid similarity to TBX-3. Note that many *C. elegans* factors are diverged at the T-box C-terminus in the loop following the g strand and in the α3 and 3_{10}C helices, which make extensive contacts with DNA in TBX-3. Identical or similar amino acids are highlighted in *black* or *gray*, respectively. Protein accession numbers are the same as described in Fig. 2.

Table 2 Caenorhabditis T-Box Genes

Species	Genome Size (Mbp)[a]	# T-Box Genes Identified by PFam (PF00907)	# T-Box Genes Identified by Interpro (IPR001699)
C. elegans	100.3	21	22
C. remanei	145.4	70	73
C. briggsae	108.4	40	43
C. japonica	166.3	44	46
C. brenneri	190.4	56	57

[a]GenBank assembly accessions: *C. elegans* (GCA_000002985.3), *C. remanei* (GCA_000149515.1), *C. japonica* (GCA_000147155.1), *C. briggsae* (GCA_000004555.3), *C. brenneri* (GCA_000143925.2).

spacing. Structural studies of T-box proteins bound to TBE dimers have identified a so-called dimerization loop within the T-box DNA-binding domain (Muller & Herrmann, 1997), but the surface area of this dimerization interface is often small and may not provide a biologically relevant protein interface (Coll et al., 2002; El Omari et al., 2012; Stirnimann et al., 2010). Moreover, functional binding sites in promoters are typically monomeric TBEs, and it therefore remains controversial whether these factors bind dimerized TBEs in vivo (Papaioannou, 2014).

Binding sites have been identified for several *C. elegans* T-box factors through both in vivo promoter analyses and in vitro assays, and these analyses reveal a surprising diversity in T-box DNA-binding specificity. The conserved Tbx-2 subfamily member TBX-2 binds and represses its own promoter in vivo through two separate monomeric sites similar to the canonical TBE (Milton & Okkema, 2015). Likewise, the divergent T-box factors TBX-38, TBX-43, TBX-35, and SEA-1 can bind monomeric sites similar to the canonical TBE in vitro (Broitman-Maduro et al., 2009; Farboud, Nix, Jow, Gladden, & Meyer, 2013; Narasimhan et al., 2015). However, SEA-1 binds and activates expression of its target *xol-1* through five sites that have little or no similarity to the TBE sequence, making it unclear whether SEA-1 binds TBEs in vivo (Farboud et al., 2013). Finally the divergent factors TBX-33, TBX-39, and TBX-40 bind novel sequences unrelated to the canonical TBE in vitro (Narasimhan et al., 2015). In the conserved mammalian T-box factors that have been structurally characterized, specific contacts with DNA are largely made with amino acids at the C-terminus of the T-box, and in particular human TBX-3 residues in the g-α3 region and in the 3_{10}C helix make base-specific contacts in the minor groove (Coll et al., 2002). These regions are not conserved in SEA-1, TBX-33, TBX-39, and TBX-40 (Fig. 3), consistent with the distinct DNA-binding specificities among these divergent T-box factors.

4. GENETIC ANALYSES OF T-BOX FACTOR FUNCTION IN *C. elegans*

4.1 A Network of T-Box Factors Function in *C. elegans* Pharyngeal Development

The *C. elegans* pharynx is a neuromuscular feeding organ located at the anterior end of the digestive tract. It is a complex organ consisting of 80 cells, including muscle cells, several types of neurons, and a variety of structural and secretory cells (Albertson & Thomson, 1976). The entire pharynx is produced during embryogenesis, and there are no additional cells added after hatching (Sulston & Horvitz, 1977; Sulston, Schierenberg, White, & Thomson, 1983). The pharynx shares a number of functional and developmental similarities with the heart in other species. Most notable is the requirement for the *Nkx-2.5* ortholog *ceh-22* in pharyngeal muscle development (Haun, Alexander, Stainier, & Okkema, 1998; Okkema, Ha, Haun, Chen, & Fire, 1997). The recent recognition that the pharyngeal mesoderm in chordates contributes to the heart also suggests that a conserved evolutionary relationship exists for pharyngeal and cardiac muscle development among various animal types (reviewed in Diogo et al., 2015). However, the evolutionary relationship between pharyngeal muscle in *C. elegans* and cardiac muscle in other species remains controversial, and the *C. elegans* pharynx is often referred to as foregut, which is traditionally described as an endodermal derivative, or as ectoderm (Mango, 2007).

The pharynx is produced from descendants of two blastomeres present in the early embryo called ABa and MS (Fig. 4A) (Sulston et al., 1983). Both of these blastomeres are polyclonal precursors that contribute to a variety of cell types both inside and outside the pharynx. MS is largely a mesodermal precursor, and, while ABa is typically considered an ectodermal precursor, mesodermal fates are induced in specific ABa descendants (Hashimshony, Feder, Levin, Hall, & Yanai, 2015; Priess, Schnabel, & Schnabel, 1987; Priess & Thomson, 1987). ABa descendants contribute primarily the anterior half of the pharynx, while MS descendants contribute to the posterior pharynx (Fig. 4A). Notably, both ABa and MS produce a mix of pharyngeal cell types (e.g., pharyngeal muscles or neurons), and there are no clonal precursors of specific pharyngeal cell types.

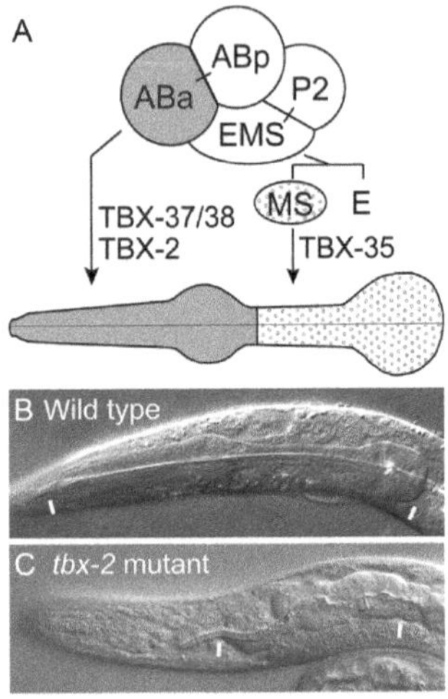

Fig. 4 T-box factors involved in *C. elegans* pharyngeal development. (A) Schematic diagram of the 4-cell *C. elegans* embryo indicating blastomere names and regions that these blastomeres contribute to in the mature pharynx (Sulston et al., 1983). ABa gives rise to cells that predominantly contribute to the anterior pharynx (*gray shading*); EMS divides to produce MS, which contributes to the posterior pharynx (*dots*), and E, which produces the gut. T-box factors required for ABa- and MS-derived pharynx are indicated and are discussed in the text. (B and C) DIC micrographs of L1 animals showing the pharynx in wild type (B) and a *tbx-2(ok529)* null mutant lacking ABa-derived pharyngeal muscles (C). The extent of pharyngeal tissue is indicated (*white bars*).

Early specification mechanisms in *C. elegans* generally establish blastomere identity, which defines the stereotyped pattern of cell divisions within the descendants of these blastomeres, while later mechanisms specify organ and cell-type identity within these descendent cells. A variety of elegant genetic and molecular studies have shown that ABa and MS fates are specified early in development by completely different mechanisms (reviewed in Maduro, 2006; Mango, 2007). ABa fate is specified conditionally through signaling mediated by the Notch-family receptor GLP-1 (Priess et al., 1987). In contrast, MS fate is specified using a combination of Wnt signals and a network of autonomously functioning transcription factors expressed from the maternal and zygotic genomes (Bowerman, Draper, Mello, & Priess, 1993; Lin, Thompson, & Priess, 1995; Maduro, Lin, & Rothman, 2002; Maduro, Meneghini, Bowerman, Broitman-Maduro, & Rothman, 2001). Mutations affecting these early specification mechanisms typically lead to defects that have recognizable effects on either anterior or posterior pharynx development. Subsequently, the later mechanisms that specify pharyngeal precursors and specific pharyngeal cell types converge and function in the descendants of both ABa and MS, and these later events depend on factors such as the pharynx specification factor PHA-4 and the pharyngeal muscle specification factor CEH-22 (Horner et al., 1998; Kalb et al., 1998; Mango, Lambie, & Kimble, 1994; Okkema & Fire, 1994; Okkema et al., 1997).

T-box factors play essential roles at multiple points in the development of both ABa- and MS-derived pharynx (Fig. 4A). ABa blastomere specification in the early embryo depends on the divergent T-box factors TBX-37 and TBX-38 (Fig. 2). These are closely related and genetically redundant factors that function in parallel to GLP-1/Notch signaling to specify the fate of all ABa descendants, and *tbx-37* and *tbx-38* double mutants lack anterior pharynx and other cell types derived from ABa (Good et al., 2004). Because of the widespread defects in ABa descendants, these double mutant embryos exhibit severe morphological defects and embryonic lethality. *tbx-37* and *tbx-38* are expressed very transiently in all eight cells descended from ABa at the 24-cell stage of embryogenesis. Expression of these genes is repressed in descendants of the ABp blastomere by members of the REF-1 family of bHLH transcriptional repressors to prevent inappropriate blastomere specification and pharyngeal development in ABp descendants, and this repression is the key event distinguishing ABa from ABp (Good et al., 2004; Mello, Draper, & Priess, 1994; Neves & Priess, 2005).

In contrast to the early function of TBX-37 and TBX-38, the conserved factor TBX-2 functions later in embryogenesis to specifically promote pharyngeal muscle fate among the descendants of ABa (Roy Chowdhuri et al., 2006; Smith & Mango, 2007) (Fig. 4A). TBX-2 is a member of the Tbx-2 subfamily of T-box factors, and it is most closely related to the mammalian TBX-2 and TBX-3 transcriptional repressors (Fig. 2) (Pocock et al., 2004). *C. elegans tbx-2* is expressed in a dynamic pattern through embryonic to adult stages (Miyahara et al., 2004; Roy Chowdhuri et al., 2006; Singhvi et al., 2008; Smith & Mango, 2007), and, while the *tbx-2* expression pattern has not yet been completely described, it is expressed strongly in the embryonic pharynx based on in situ hybridizations and using *tbx-2::gfp* reporter genes (Kohara, 2001; Roy Chowdhuri et al., 2006; Smith & Mango, 2007). *tbx-2* null mutants lack the pharyngeal muscles derived from ABa, but they retain the muscles derived from MS (Fig. 4B and C). The ultimate fate of these missing muscle cells is unknown, but they do not undergo apoptosis and have been suggested to arrest development or undergo dedifferentiation (Smith & Mango, 2007). Homozygous *tbx-2* mutant embryos elongate normally and hatch, but because they do not contain a complete pharynx, they are unable to feed and arrest as first stage L1 larvae. This phenotype is specific for the pharyngeal muscles, as other ABa-derived pharyngeal cell types develop normally (Roy Chowdhuri et al., 2006; Smith & Mango, 2007). Thus TBX-2 functions as a pharyngeal muscle specification factor that has a more restricted function in embryogenesis than TBX-37 and TBX-38, which specify identity of all ABa descendants.

Finally, MS blastomere fate and the production of posterior pharynx derived from MS depend on the divergent factor TBX-35 (Fig. 4A) (Broitman-Maduro et al., 2006). TBX-35 is expressed in MS and its descendants, where its expression is activated by the redundant GATA-type factors MED-1 and MED-2. TBX-35 directly activates expression of the NK-2-like homeobox gene *ceh-51*, and TBX-35 and CEH-51 function together in a feed-forward network to promote pharyngeal and mesodermal fates in cells derived from MS (Broitman-Maduro et al., 2009). *tbx-35* and *ceh-51* single mutants exhibit incompletely penetrant defects, but *tbx-35; ceh-51* double mutants completely lack MS-derived posterior pharynx and other MS-derived muscles and mesodermal tissues. In these double mutants, the fate of MS is transformed to that of the related embryonic blastomere called C, which does not contribute to the pharynx. *tbx-35; ceh-51* double mutants exhibit elongation defects and do not hatch, and they arrest as morphologically abnormal embryos containing anterior pharynx derived from ABa but not posterior pharynx normally derived from MS.

4.2 *tbx-2* Has Additional Functions in Neuronal Development and Sensory Adaptation

In addition to its role in pharyngeal muscle development discussed earlier, *C. elegans tbx-2* has additional roles in neuronal development and adaptation to sensory stimulation. In the embryo, *tbx-2* functions to promote neuronal migration and to inhibit apoptosis in two bilaterally symmetric pairs of hermaphrodite-specific neurons (HSNs) and phasmid B neurons (PHBs) (Singhvi et al., 2008). Each HSN and PHB are sister cells produced by the division of a HSN/PHB precursor in the tail of the developing embryo (Sulston et al., 1983). In hermaphrodite embryos, the HSNs migrate to the mid body, where they will eventually differentiate as serotonergic motor neurons that innervate the vulval muscles for egg laying, while in males the HSNs undergo apoptotic cell death. The PHBs remain in the tail and differentiate in both sexes as ciliated chemosensory neurons. TBX-2 has two distinct functions in HSN and PHB development (Singhvi et al., 2008).

First, it promotes appropriate migration of the HSNs, and *tbx-2* mutants have misplaced HSNs. Second, TBX-2 inhibits apoptosis of the PHBs, and mutants lack PHB cells because they inappropriately die. Both of these phenotypes are partially penetrant in *tbx-2* null mutants and can be enhanced by mutations in other genes, including those affecting the winged-helix factor HAM-1 and the Hox factor EGL-5, suggesting that TBX-2 functions in parallel with other regulatory mechanisms promoting neuronal development.

In adults, TBX-2 is necessary for adaptation to olfactory cues sensed by the pair of AWC chemosensory neurons (Miyahara et al., 2004). The AWCs are ciliated amphid neurons located in the head, and they mediate chemotaxis toward volatile odorants, such as benzaldehyde and isoamyl alcohol (reviewed in Bargmann, 2006). In wild-type animals, prolonged exposure to these odorants leads to adaptation and decreased chemotactic response (Colbert & Bargmann, 1995). Viable hypomorphic *tbx-2* mutants respond normally to these odorants, but they do not adapt, and they remain attracted to these odorants even after prolonged exposure (Miyahara et al., 2004). These mutations do not affect AWC differentiation or morphology, suggesting they specifically affect gene expression necessary for adaptation in fully differentiated AWC cells. The two hypomorphic mutant alleles characterized in these studies are both missense mutations that affect a lysine residue (FH$\mathbf{K_{164}}$LKL) located between the E and e strands of the T-box (Fig. 3). This region is conserved among vertebrate T-boxes, and by analogy to the solved structure of the human TBX-3 T-box (Coll et al., 2002), K_{164} is solvent exposed and located near DNA-contacting residues. These mutations could affect either DNA binding or interaction with co-factors that are specifically required for adaptation in the AWCs. In this latter case, it will be interesting to identify proteins interacting with this region of TBX-2.

4.3 The Tbx-1 Subfamily Members MAB-9 and MLS-1 Function Postembryonically to Specify Ectodermal and Muscle Cell Fates

The *C. elegans* genome encodes two members of the Tbx-1 subfamily of T-box factors, MAB-9 and MLS-1 (Fig. 2), and both of these play important roles in specifying cell fates during larval development. MAB-9 is most closely related to vertebrate TBX-20, which is known to have crucial roles in heart development but has also been implicated in nervous system development (Song et al., 2006; Stennard et al., 2005; Woollard & Hodgkin, 2000). MLS-1 is most closely related to TBX-22, which is mutated in human X-linked cleft palate with ankyloglossia (Braybrook et al., 2001; Kostas & Fire, 2002).

C. elegans MAB-9 plays important roles specifying the fate of a subset of epithelial cells in the hindgut and in controlling axonal guidance in the nervous system. *mab-9* mutants were initially isolated based on abnormalities in the male tail, which is a highly complex structure containing sensory and structural components necessary for mating with hermaphrodites (Hodgkin, 1983; Lints & Hall, 2009a). Much of the male tail is produced from four epithelial cells located in the hindgut of the newly hatched L1 larva called B, Y, F, and U (Fig. 5) (Sulston, Albertson, & Thomson, 1980; Sulston & Horvitz, 1977). These cells are present in both hermaphrodites and males, but they are sex-specific blast cells, which in wild-type males undergo a stereotyped pattern of cell divisions to produce sensory and structural components of the tail. In *mab-9* mutant males, the B and F blast cells are improperly specified and are transformed to fates resembling those of Y and U, respectively (Chisholm & Hodgkin, 1989). This transformation is most striking in the B blast cell lineage, where the timing and orientation of cell divisions, and the ultimate differentiation pattern of cells closely resemble those of Y. The transformation of B and F results in severe defects in the male tail, and these animals are completely unable to mate (Chisholm & Hodgkin, 1989; Hodgkin, 1983). In hermaphrodites, B and F do not divide but rather contribute to the dorsal posterior wall of the rectum. However, the hermaphrodite B cell also appears morphologically more similar to the Y cell in *mab-9* mutants, and this likely underlies a severe constipation phenotype in *mab-9* mutant hermaphrodites (Chisholm & Hodgkin, 1989; Woollard & Hodgkin, 2000). *mab-9* mutants also exhibit defects in the posterior cuticle which is produced by the hypodermis, and this contributes to male tail defects and decreased susceptibility to infection by bacterial pathogens (Appleford, Gravato-Nobre, Braun, & Woollard, 2008).

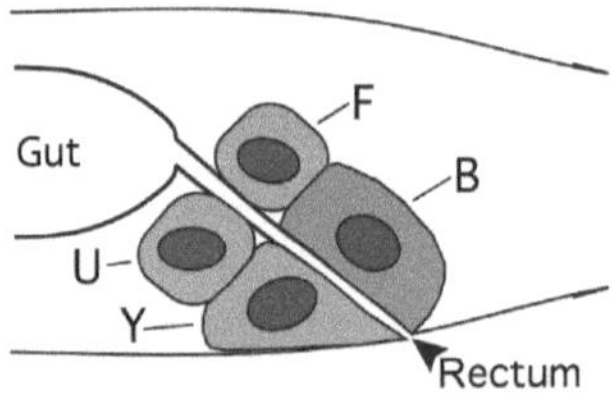

Fig. 5 Epithelial cells in the L1 hindgut. Schematic diagram of the L1 *C. elegans* hindgut indicating B, Y, F, and U cells discussed in the text. *Adapted from Lints, R., Altun, Z. F., & Hall, D. H. (2009). Alimentary system, rectum and anus.* WormAtlas.

Thus, MAB-9 is essential to specify the fates of specific cells in the hindgut of both sexes, and it is not a sex-specific differentiation gene. Finally, *mab-9* mutants also exhibit uncoordinated movement when backing up, which results from defects in axonal guidance for a subset of motor neurons located in the ventral nerve cord (Pocock et al., 2008). Both the defects in the male tail and in axonal guidance are rescued by expression of a chimeric MAB-9 protein containing the human TBX-20 T-box DNA-binding domain, suggesting these factors have similar DNA-binding preferences. *mab-9::gfp* reporter genes are expressed in the B and F cells and their descendants in both males and hermaphrodites and in ventral cord motor neurons produced during embryogenesis, as well as in the posterior hypodermis underlying the cuticle (Pocock et al., 2008; Woollard & Hodgkin, 2000). *mab-9* expression is repressed in additional posterior cells by the *even-skipped*-like homeodomain factor VAB-7, and this repression is suggested to be crucial in preventing deleterious effects resulting from widespread *mab-9* expression (Pocock et al., 2004).

A second Tbx-1 subfamily factor MLS-1 specifies the fate of several non-striated uterine muscles in the sex myoblast (SM) lineage in *C. elegans* hermaphrodites (Kostas & Fire, 2002). In wild-type animals, the two SMs divide during the L3 and L4 larval stages to produce eight uterine muscles and eight vulval muscles (Sulston & Horvitz, 1977). The uterine muscles encircle the uterus and contract to move fertilized eggs toward the vulva, while the vulval muscles connect to the lips of the vulva to open the vulva during egg laying (Lints & Hall, 2009b). In *mls-1* null mutants, the two SMs are produced and divide normally, but all eight uterine muscles are transformed to a vulval muscle fate, resulting in the loss of all uterine muscles and formation of excess vulval muscles (Kostas & Fire, 2002). *mls-1* reporter genes are expressed in uterine muscle precursors and somewhat later in a subset of the vulval muscles and some of the enteric muscles, and this expression is directly activated by the *C. elegans* Twist ortholog HLH-8. Ectopic *mls-1* expression outside of these cells transforms multiple mesodermal cell types to a uterine muscle fate. Thus, MLS-1 functions as a uterine muscle fate determinant.

4.4 The TBX-8 and TBX-9 Paralogs Play Redundant Roles in Hypodermal and Body Wall Muscle Morphogenesis

The divergent T-box factors TBX-8 and TBX-9 are closely related to each other (Fig. 2) and have largely redundant functions in patterning and morphogenesis of hypodermal and body wall muscle cells during embryonic development (Andachi, 2004; Pocock et al., 2004). While *tbx-8* and *tbx-9* single mutants have relatively mild morphological defects, animals with both of these genes knocked down exhibit highly penetrant late embryo and early larval lethality. In *tbx-9* and *tbx-8* double mutants, defects are apparent in the dorsal and lateral hypodermal cells, which contribute to the outer epidermal surface of the worm and in the body wall muscles necessary for movement. The dorsal hypodermal cells fail to intercalate while the lateral hypodermal cells do not adhere properly to form continuous rows of seam cells (Altun & Hall, 2009a; Andachi, 2004; Pocock et al., 2004). Similarly, body wall muscles exhibit migratory defects resulting in gaps in the body wall musculature (Altun & Hall, 2009b; Andachi, 2004; Pocock et al., 2004). Both the hypodermal and body wall muscle cells affected in these double mutants are born and differentiate, indicating that TBX-9 and TBX-8 have roles in morphogenesis rather than cell fate specification.

tbx-9::gfp and *tbx-8::gfp* reporters are expressed in largely overlapping patterns in the dorsal and lateral hypodermis and in body wall muscle cells prior to migration (Araya et al., 2014; Pocock et al., 2004), and expression in both of these tissues suggests that TBX-9 and TBX-8 function autonomously in these tissues. *tbx-8* and *tbx-9* are proposed to function in a gene regulatory network with the *even-skipped* related gene *vab-7* and the T-box genes *tbx-30* and *mab-9* to properly pattern posterior regions in the embryo (Fig. 6) (Pocock et al., 2004). *vab-7* is required for normal patterning of posterior hypodermis and body wall muscle (Ahringer, 1996), and as discussed earlier, it inhibits *mab-9* expression outside of the hindgut in these posterior regions. *vab-7* expression is in turn activated by TBX-8 and TBX-9 in posterior regions of the embryo but repressed anteriorly by the divergent factor TBX-30. Consensus TBEs present in the *vab-7* promoter are necessary for anterior repression, but TBX-30 is highly diverged in the T-box residues

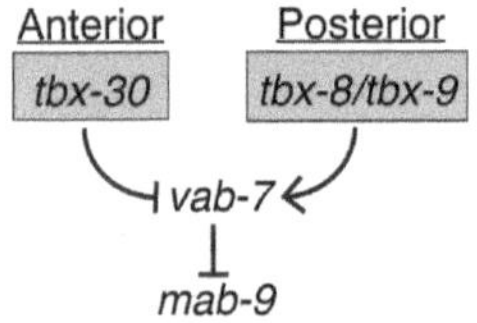

Fig. 6 Gene regulatory network including T-box genes and *vab-7*. Pathway of genetic interactions between T-box genes and the *even-skipped*-related gene *vab-7*. *Lines* with *arrowheads* or *bars* indicate activation or repression, respectively. *Adapted from Pocock, R., Ahringer, J., Mitsch, M., Maxwell, S., & Woollard, A. (2004). A regulatory network of T-box genes and the* even-skipped *homologue* vab-7 *controls patterning and morphogenesis in* C. elegans. Development, 131, *2373–2385.*

that contact DNA (Fig. 3), and it may not bind these consensus TBEs. Thus, while it is not clear whether these regulatory interactions are direct, this network shares organizational similarity with the networks of interacting T-box factors that function in vertebrate heart patterning (reviewed in Greulich, Rudat, & Kispert, 2011).

4.5 The Divergent T-Box Factor SEA-1 Functions as an Autosomal Counting Element to Determine Sex in *C. elegans*

C. elegans occurs in two sexes—hermaphrodites and males—and sex is determined by the ratio of X chromosomes to sets of autosomes (X/A ratio) (Madl & Herman, 1979). Both hermaphrodites and males contain two sets of autosomes, but hermaphrodites have two X chromosomes (XX:AA; X/A = 1), while males have only a single X chromosome (X:AA; X/A = 0.5). This ratio is assessed in early embryogenesis through discrete X-chromosome signaling elements (XSEs) and autosomal signaling elements (ASEs) that function antagonistically to regulate expression of the master sex determination switch gene *xol-1* (reviewed in Meyer, 2010). XSEs and ASEs are protein-coding genes that encode *trans*-acting factors that inhibit or activate *xol-1* expression, respectively. In XX hermaphrodites, the higher dose of XSEs relative to ASEs leads to *xol-1* repression, while in males, the lower dose of XSEs promotes *xol-1* expression. *xol-1* subsequently regulates cascades of downstream genes to specify sexual fate and to regulate dosage compensation mechanisms that allow equal expression of X-linked genes in the two sexes. XSEs were identified first and include the transcription factors SEX-1 and CEH-39 that target the *xol-1* promoter, as well as the RNA-binding protein FOX-1 that affects *xol-1* mRNA splicing (Carmi, Kopczynski, & Meyer, 1998; Gladden & Meyer, 2007; Hodgkin, Zellan, & Albertson, 1994; Nicoll, Akerib, & Meyer, 1997; Skipper, Milne, & Hodgkin, 1999).

In comparison, the ASEs were more difficult to identify, but the divergent T-box factor SEA-1 proved to be the first ASE discovered in *C. elegans* (Fig. 2) (Powell et al., 2005). A *sea-1* loss-of-function mutant was identified in a genetic screen for suppressors of the XX-specific lethal phenotype of animals with a reduced dose of XSEs, and elegant genetic analyses demonstrated that *sea-1* exhibits all the characteristics predicted of an ASE. SEA-1 is a dose-dependent activator of *xol-1* expression, and it functions cumulatively with a second identified ASE, SEA-2, which encodes a C_2H_2-zinc finger factor (Farboud et al., 2013; Powell et al., 2005). SEA-1 binds sites in the *xol-1* promoter region, and as discussed earlier, these sites do not resemble canonical TBEs, indicating that SEA-1 exhibits a divergent DNA-binding specificity. SEA-1-binding sites do not overlap those identified for the XSEs SEX-1 and CEH-39, suggesting that binding site competition does not underlie the antagonistic activities of these XSEs and ASEs, and these factors are hypothesized to recruit chromatin modifying factors with reciprocal activities on *xol-1* promoter activity (Farboud et al., 2013). Understanding how SEA-1 functions in a dose-dependent manner with other ASEs and XSEs will likely serve as a useful paradigm for understanding the dose dependency and haploinsufficient phenotypes observed for many mammalian T-box genes (reviewed in Papaioannou, 2014).

4.6 The Functions of Other *C. elegans* T-Box Genes Are Only Beginning to Be Characterized

The developmental functions of other *C. elegans* T-box genes have not yet been extensively characterized. Genome-wide transcriptome analyses have shown that most of these genes are expressed in temporally dynamic patterns during embryogenesis (Gerstein et al., 2010; Hashimshony et al., 2015; Levin, Hashimshony, Wagner, & Yanai, 2012). The exceptions to this generalization are *tbx-41* and *tbx-34*, which are expressed at only very low levels in embryos. *tbx-41* expression remains low throughout the life cycle, but *tbx-34* is strongly expressed in late larvae and adults, and it is particularly enriched in L4 males, indicating it may play a role in sperm production or the proliferation of the germ line, which occurs in these stages.

Large-scale RNAi screens have not revealed obvious phenotypes for T-box genes outside of those discussed in the previous sections (Table 1). However, more recent specialized screens have begun to reveal functions for some of these T-box genes. For example, an RNAi screen in a sensitized genetic background showed that *tbx-11* knockdown results in partially penetrant gastrulation defects and embryonic lethality (Sawyer et al., 2011). Likewise, *tbx-36(RNAi)* results in increased expression of a ferritin gene *gfp* reporter, suggesting this T-box factor may function to regulate iron homeostasis (Ackerman & Gems, 2012). Finally, *tbx-33(RNAi)* hermaphrodites exhibit sterility and an increased number of apoptotic cells in the germ line (Green et al., 2011). However, eight of the 22 *C. elegans* T-box genes still have no reported phenotypes, suggesting they have redundant or restricted functions that have not yet been recognized. A clearer understanding of the spatial expression patterns of these genes will inform genetic screens to identify subtle tissue-specific defects, and ongoing lineage analyses of reporter gene expression are beginning to provide this crucial data (Murray et al., 2012; Sarov et al., 2012).

5. POSTTRANSLATIONAL MECHANISMS AFFECTING T-BOX FACTOR ACTIVITY

T-box genes are frequently expressed in dynamic, tissue-specific patterns that play an important role in regulating their activity. However posttranslational mechanisms also strongly affect T-box factor activity. T-box factors interact with a variety of other DNA-binding transcription factors and chromatin regulatory proteins (reviewed in Boogerd, Moorman, & Barnett, 2009), and they can also be posttranslationally modified by SUMOylation, ubiquitination, and phosphorylation (Abrahams, Mowla, Parker, Goding, & Prince, 2008; Andreou et al., 2007; Chen et al., 2009; Huber et al., 2013; Jang, Park, Hong, & Hwang, 2013; Peres, Mowla, & Prince, 2015; Roy Chowdhuri et al., 2006). Understanding the in vivo function of these interactions and modifications is necessary to unravel the molecular mechanisms by which T-box factors function, and the strengths of *C. elegans* for genetic analyses facilitate this understanding.

Activity of the conserved T-box factors MLS-1 and TBX-2 depends on interaction with the Groucho-family transcriptional corepressor UNC-37 (Huber, Crum, & Okkema, 2016; Miller & Okkema, 2011). Reducing UNC-37/Groucho activity produces phenotypes similar to those found in *mls-1* and *tbx-2* mutants, including a transformation of uterine muscles to a vulval muscle fate and loss of ABa-derived pharyngeal muscles. MLS-1 and TBX-2 interact with UNC-37/Groucho through engrailed homology 1 (eh1) motifs located outside of their T-boxes. eh1 motifs are frequently found in metazoan T-box factors (Copley, 2005), suggesting that interaction with Groucho-family corepressors is a conserved mechanism for T-box repressor activity. Indeed, Groucho has been implicated in the function of several vertebrate and Drosophila T-box factors (Farin et al., 2007; Formaz-Preston, Ryu, Svendsen, & Brook, 2012; Kaltenbrun et al., 2013).

Likewise, posttranslational SUMOylation strongly affects activity of *C. elegans* TBX-2 (Huber et al., 2013; Milton & Okkema, 2015; Roy Chowdhuri et al., 2006). Decreased SUMOylation results in phenotypes similar to those observed in *tbx-2* mutants and strongly enhances pharyngeal defects and lethality in a *tbx-2* hypomorphic mutant, indicating that SUMOylation is necessary for full TBX-2 function in vivo. TBX-2 is

SUMOylated in mammalian cell assays at lysine residues in two consensus SUMOylation sites (Huber et al., 2013). One of these sites is located near the TBX-2 C-terminus (V**K**$_{400}$KE), while the second site is in a highly conserved region of the T-box DNA-binding domain (L**K**$_{231}$IE; Fig. 3). Proteolysis of the human T-bet T-box factor is regulated by ubiquitination of this conserved K_{231} (Jang et al., 2013), indicating this residue is accessible and a hotspot for modification. Human TBX-22 and mouse Tbx-5 have also been shown to be SUMOylated, although these proteins are modified at different sites than those in *C. elegans* TBX-2 (Andreou et al., 2007; Beketaev et al., 2014).

Finally, nuclear localization of *C. elegans* TBX-2 may also be regulated. TBX-2 protein has been detected in both the nucleus and cytoplasm using TBX-2::GFP fusion proteins and antibodies targeting endogenous TBX-2 protein (Miyahara et al., 2004; Smith & Mango, 2007). The functional relevance of this differential localization has not been demonstrated; however, similar cytoplasmic and nuclear localization has been observed for several vertebrate T-box factors, suggesting that this may also be an important conserved mechanism for regulating T-box factor activity (Bimber, Dettman, & Simon, 2007; McLane et al., 2013; Peres et al., 2015).

6. CONCLUDING REMARKS

The T-box factor family in *C. elegans* is exceptionally diverse, containing members of conserved subfamilies, as well as a clade of divergent factors that are unlike those found in other organisms. This diversity provides unique and exciting opportunities to increase our understanding of both the mechanistic and developmental functions of T-box factors. Current studies indicate that the *C. elegans* T-box factors bind a wider variety of DNA sequences than most characterized T-box factors, and indeed, the high level of sequence divergence among these factors in the C-terminus of the T-box raises the interesting hypothesis that some members may not bind DNA at all. Systematic studies of how these factors interact with DNA and other factors will increase our mechanistic knowledge of how T-box factors function. Likewise, genetic analyses have revealed roles for *C. elegans* T-box genes in processes such as sex determination and neural development, which are not typically thought of as T-box-dependent processes, and it is likely that characterization of *C. elegans* T-box genes will continue to reveal novel functions. No phenotypes have currently been associated with knockdown of eight of the 22 *C. elegans* T-box genes (Table 1), and we have no understanding of the developmental function of these factors. The lack of phenotypes could result from genetic redundancy, as has been observed between related T-box factors and with transcription factors in other families, and specialized screens may be necessary to identify their function. Finally, the diversity among the *C. elegans* T-box genes provides a context to examine the mechanisms and constraints that allowed this gene family to evolve differently than T-box genes in other organisms.

ACKNOWLEDGMENTS

I am indebted to Teresa Orenic, Ankur Saxena, Jennifer Schmidt, and Paul Huber for comments on this manuscript. Work in my laboratory is supported by grants from the National Institutes of Health.

REFERENCES

Abrahams, A., Mowla, S., Parker, M. I., Goding, C. R., & Prince, S. (2008). UV-mediated regulation of the anti-senescence factor Tbx2. *The Journal of Biological Chemistry, 283*, 2223–2230.

Ackerman, D., & Gems, D. (2012). Insulin/IGF-1 and hypoxia signaling act in concert to regulate iron homeostasis in *Caenorhabditis elegans*. *PLoS Genetics, 8*, e1002498.

Agulnik, S. I., Garvey, N., Hancock, S., Ruvinsky, I., Chapman, D. L., Agulnik, I., et al. (1996). Evolution of mouse T-box genes by tandem duplication and cluster dispersion. *Genetics, 144*, 249–254.

Agulnik, S. I., Ruvinsky, I., & Silver, L. M. (1997). Three novel T-box genes in *Caenorhabditis elegans*. *Genome, 40*, 458–464.

Ahringer, J. (1996). Posterior patterning by the *Caenorhabditis elegans even-skipped* homolog *vab-7*. *Genes & Development, 10*, 1120–1130.

Albertson, D. G., & Thomson, J. N. (1976). The pharynx of *Caenorhabditis elegans*. *Philosophical Transactions of the Royal Society B, 275*, 299–325.

Altun, Z. F., & Hall, D. H. (2009a). Epithelial system, hypodermis. *WormAtlas*. http://dx.doi.org/10.3908/wormatlas.1.13.

Altun, Z. F., & Hall, D. H. (2009b). Muscle system, somatic muscle. *WormAtlas*. http://dx.doi.org/10.3908/wormatlas.1.7.

Andachi, Y. (2004). *Caenorhabditis elegans* T-box genes *tbx-9* and *tbx-8* are required for formation of hypodermis and body-wall muscle in embryogenesis. *Genes to Cells, 9*, 331–344.

Andreou, A. M., Pauws, E., Jones, M. C., Singh, M. K., Bussen, M., Doudney, K., et al. (2007). *TBX22* missense mutations found in patients with X-linked cleft palate affect DNA binding, sumoylation, and transcriptional repression. *American Journal of Human Genetics, 81*, 700–712.

Appleford, P. J., Gravato-Nobre, M., Braun, T., & Woollard, A. (2008). Identification of cis-regulatory elements from the *C. elegans* T-box gene *mab-9* reveals a novel role for *mab-9* in hypodermal function. *Developmental Biology, 317*, 695–704.

Araya, C. L., Kawli, T., Kundaje, A., Jiang, L., Wu, B., Vafeados, D., et al. (2014). Regulatory analysis of the *C. elegans* genome with spatiotemporal resolution. *Nature, 512*, 400–405.

Bargmann, C. I. (2006). Chemosensation in *C. elegans*. *WormBook: The online review of C. elegans biology*. (pp. 1–29). http://dx.doi.org/10.1895/wormbook.1.123.1. http://www.wormbook.org.

Beketaev, I., Kim, E. Y., Zhang, Y., Yu, W., Qian, L., & Wang, J. (2014). Potentiation of Tbx5-mediated transactivation by SUMO conjugation and protein inhibitor of activated STAT 1 (PIAS1). *The International Journal of Biochemistry & Cell Biology*, *50*, 82–92.

Bimber, B., Dettman, R. W., & Simon, H. G. (2007). Differential regulation of Tbx5 protein expression and sub-cellular localization during heart development. *Developmental Biology*, *302*, 230–242.

Blaxter, M. (1998). *Caenorhabditis elegans* is a nematode. *Science*, *282*, 2041–2046.

Boogerd, C. J., Moorman, A. F., & Barnett, P. (2009). Protein interactions at the heart of cardiac chamber formation. *Annals of Anatomy*, *191*, 505–517.

Bowerman, B., Draper, B. W., Mello, C. C., & Priess, J. R. (1993). The maternal gene *skn-1* encodes a protein that is distributed unequally in early *C. elegans* embryos. *Cell*, *74*, 443–452.

Braybrook, C., Doudney, K., Marcano, A. C., Arnason, A., Bjornsson, A., Patton, M. A., et al. (2001). The T-box transcription factor gene *TBX22* is mutated in X-linked cleft palate and ankyloglossia. *Nature Genetics*, *29*, 179–183.

Brenner, S. (2003). Nobel lecture. Nature's gift to science. *Bioscience Reports*, *23*, 225–237.

Broitman-Maduro, G., Lin, K. T., Hung, W. W., & Maduro, M. F. (2006). Specification of the *C. elegans* MS blastomere by the T-box factor TBX-35. *Development*, *133*, 3097–3106.

Broitman-Maduro, G., Owraghi, M., Hung, W. W., Kuntz, S., Sternberg, P. W., & Maduro, M. F. (2009). The NK-2 class homeodomain factor CEH-51 and the T-box factor TBX-35 have overlapping function in *C. elegans* mesoderm development. *Development*, *136*, 2735–2746.

Carmi, I., Kopczynski, J. B., & Meyer, B. J. (1998). The nuclear hormone receptor SEX-1 is an X-chromosome signal that determines nematode sex. *Nature*, *396*, 168–173.

Castellanos, R., Xie, Q., Zheng, D., Cvekl, A., & Morrow, B. E. (2014). Mammalian TBX1 preferentially binds and regulates downstream targets via a tandem T-site repeat. *PloS One*, *9*, e95151.

Chen, Y. L., Liu, B., Zhou, Z. N., Hu, R. Y., Fei, C., Xie, Z. H., et al. (2009). Smad6 inhibits the transcriptional activity of Tbx6 by mediating its degradation. *The Journal of Biological Chemistry*, *284*, 23481–23490.

Chisholm, A. D., & Hodgkin, J. (1989). The *mab-9* gene controls the fate of B, the major male-specific blast cell in the tail region of *Caenorhabditis elegans*. *Genes & Development*, *3*, 1413–1423.

Colbert, H. A., & Bargmann, C. I. (1995). Odorant-specific adaptation pathways generate olfactory plasticity in *C. elegans*. *Neuron*, *14*, 803–812.

Coll, M., Seidman, J. G., & Muller, C. W. (2002). Structure of the DNA-bound T-box domain of human TBX3, a transcription factor responsible for ulnar-mammary syndrome. *Structure*, *10*, 343–356.

Conlon, F. L., Fairclough, L., Price, B. M., Casey, E. S., & Smith, J. C. (2001). Determinants of T box protein specificity. *Development*, *128*, 3749–3758.

Consortium. (1998). Genome sequence of the nematode *C. elegans*: A platform for investigating biology. The *C. elegans* Sequencing Consortium. *Science*, *282*, 2012–2018.

Copley, R. R. (2005). The EH1 motif in metazoan transcription factors. *BMC Genomics*, *6*, 169.

Corsi, A. K., Wightman, B., & Chalfie, M. (2015). A transparent window into biology: A primer on *Caenorhabditis elegans*. *Genetics*, *200*, 387–407.

Diogo, R., Kelly, R. G., Christiaen, L., Levine, M., Ziermann, J. M., Molnar, J. L., et al. (2015). A new heart for a new head in vertebrate cardiopharyngeal evolution. *Nature*, *520*, 466–473.

El Omari, K., De Mesmaeker, J., Karia, D., Ginn, H., Bhattacharya, S., & Mancini, E. J. (2012). Structure of the DNA-bound T-box domain of human TBX1, a transcription factor associated with the DiGeorge syndrome. *Proteins*, *80*, 655–660.

Farboud, B., Nix, P., Jow, M. M., Gladden, J. M., & Meyer, B. J. (2013). Molecular antagonism between X-chromosome and autosome signals determines nematode sex. *Genes & Development*, *27*, 1159–1178.

Farin, H. F., Bussen, M., Schmidt, M. K., Singh, M. K., Schuster-Gossler, K., & Kispert, A. (2007). Transcriptional repression by the T-box proteins Tbx18 and Tbx15 depends on Groucho corepressors. *The Journal of Biological Chemistry*, *282*, 25748–25759.

Formaz-Preston, A., Ryu, J. R., Svendsen, P. C., & Brook, W. J. (2012). The Tbx20 homolog Midline represses *wingless* in conjunction with Groucho during the maintenance of segment polarity. *Developmental Biology*, *369*, 319–329.

Gerstein, M. B., Lu, Z. J., Van Nostrand, E. L., Cheng, C., Arshinoff, B. I., Liu, T., et al. (2010). Integrative analysis of the *Caenorhabditis elegans* genome by the modENCODE project. *Science*, *330*, 1775–1787.

Gladden, J. M., & Meyer, B. J. (2007). A ONECUT homeodomain protein communicates X chromosome dose to specify *Caenorhabditis elegans* sexual fate by repressing a sex switch gene. *Genetics*, *177*, 1621–1637.

Good, K., Ciosk, R., Nance, J., Neves, A., Hill, R. J., & Priess, J. R. (2004). The T-box transcription factors TBX-37 and TBX-38 link GLP-1/Notch signaling to mesoderm induction in *C. elegans* embryos. *Development*, *131*, 1967–1978.

Green, R. A., Kao, H. L., Audhya, A., Arur, S., Mayers, J. R., Fridolfsson, H. N., et al. (2011). A high-resolution *C. elegans* essential gene network based on phenotypic profiling of a complex tissue. *Cell*, *145*, 470–482.

Greulich, F., Rudat, C., & Kispert, A. (2011). Mechanisms of T-box gene function in the developing heart. *Cardiovascular Research*, *91*, 212–222.

Hashimshony, T., Feder, M., Levin, M., Hall, B. K., & Yanai, I. (2015). Spatiotemporal transcriptomics reveals the evolutionary history of the endoderm germ layer. *Nature*, *519*, 219–222.

Haun, C., Alexander, J., Stainier, D. Y., & Okkema, P. G. (1998). Rescue of *Caenorhabditis elegans* pharyngeal development by a vertebrate heart specification gene. *Proceedings of the National Academy of Sciences of the United States of America*, *95*, 5072–5075.

Hillier, L. W., Reinke, V., Green, P., Hirst, M., Marra, M. A., & Waterston, R. H. (2009). Massively parallel sequencing of the polyadenylated transcriptome of *C. elegans*. *Genome Research*, *19*, 657–666.

Hodgkin, J. (1983). Male phenotypes and mating efficiency in *Caenorhabditis elegans*. *Genetics*, *103*, 43–64.

Hodgkin, J., Zellan, J. D., & Albertson, D. G. (1994). Identification of a candidate primary sex determination locus, fox-1, on the X chromosome of *Caenorhabditis elegans*. *Development*, *120*, 3681–3689.

Horner, M. A., Quintin, S., Domeier, M. E., Kimble, J., Labouesse, M., & Mango, S. E. (1998). *pha-4*, an HNF-3 homolog, specifies pharyngeal organ identity in *Caenorhabditis elegans*. *Genes & Development*, *12*, 1947–1952.

Huber, P., Crum, T., Clary, L. M., Ronan, T., Packard, A. V., & Okkema, P. G. (2013). Function of the *C. elegans* T-box factor TBX-2 depends on SUMOylation. *Cellular and Molecular Life Sciences*, *70*, 4157–4168.

Huber, P., Crum, T., & Okkema, P. G. (2016). Function of the *C. elegans* T-box factor TBX-2 depends on interaction with the UNC-37/Groucho corepressor. *Developmental Biology*, *416*, 266–276.

Jang, E. J., Park, H. R., Hong, J. H., & Hwang, E. S. (2013). Lysine 313 of T-box is crucial for modulation of protein stability, DNA binding, and threonine phosphorylation of T-bet. *The Journal of Immunology*, *190*, 5764–5770.

Jolma, A., Yan, J., Whitington, T., Toivonen, J., Nitta, K. R., Rastas, P., et al. (2013). DNA-binding specificities of human transcription factors. *Cell*, *152*, 327–339.

Kalb, J. M., Lau, K. K., Goszczynski, B., Fukushige, T., Moons, D., Okkema, P. G., et al. (1998). *pha-4* is *Ce-fkh-1*, a fork head/HNF-3alpha, beta, gamma homolog that functions in organogenesis of the *C. elegans* pharynx. *Development, 125*, 2171–2180.

Kaltenbrun, E., Greco, T. M., Slagle, C. E., Kennedy, L. M., Li, T., Cristea, I. M., et al. (2013). A Gro/TLE-NuRD corepressor complex facilitates Tbx20-dependent transcriptional repression. *Journal of Proteome Research, 12*, 5395–5409.

Kamath, R. S., Fraser, A. G., Dong, Y., Poulin, G., Durbin, R., Gotta, M., et al. (2003). Systematic functional analysis of the *Caenorhabditis elegans* genome using RNAi. *Nature, 421*, 231–237.

Kiontke, K., & Fitch, D. H. (2005). The phylogenetic relationships of Caenorhabditis and other rhabditids. *WormBook: The online review of C. elegans biology* (pp. 1–11). http://dx.doi.org/10.1895/wormbook.1.11.1. http://www.wormbook.org.

Kiontke, K., & Fitch, D. H. (2013). Nematodes. *Current Biology: CB, 23*, R862–R864.

Kispert, A., & Herrmann, B. G. (1993). The *Brachyury* gene encodes a novel DNA binding protein. *The EMBO Journal, 12*, 3211–3220.

Kohara, Y. (2001). Systematic analysis of gene expression of the *C. elegans* genome. *Tanpakushitsu Kakusan Koso, 46*, 2425–2431.

Kostas, S. A., & Fire, A. (2002). The T-box factor MLS-1 acts as a molecular switch during specification of nonstriated muscle in *C. elegans*. *Genes & Development, 16*, 257–269.

Levin, M., Hashimshony, T., Wagner, F., & Yanai, I. (2012). Developmental milestones punctuate gene expression in the *Caenorhabditis embryo*. *Developmental Cell, 22*, 1101–1108.

Lin, R., Thompson, S., & Priess, J. R. (1995). pop-1 encodes an HMG box protein required for the specification of a mesoderm precursor in early *C. elegans* embryos. *Cell, 83*, 599–609.

Lints, R., & Hall, D. H. (2009a). Male, introduction. *WormAtlas*. http://dx.doi.org/10.3908/wormatlas.2.1.

Lints, R., & Hall, D. H. (2009b). Reproductive system, egglaying apparatus. *WormAtlas*. http://dx.doi.org/10.3908/wormatlas.1.24.

Madl, J. E., & Herman, R. K. (1979). Polyploids and sex determination in *Caenorhabditis elegans*. *Genetics, 93*, 393–402.

Maduro, M. F. (2006). Endomesoderm specification in *Caenorhabditis elegans* and other nematodes. *Bioessays, 28*, 1010–1022.

Maduro, M. F., Lin, R., & Rothman, J. H. (2002). Dynamics of a developmental switch: Recursive intracellular and intranuclear redistribution of *Caenorhabditis elegans* POP-1 parallels Wnt-inhibited transcriptional repression. *Developmental Biology, 248*, 128–142.

Maduro, M. F., Meneghini, M. D., Bowerman, B., Broitman-Maduro, G., & Rothman, J. H. (2001). Restriction of mesendoderm to a single blastomere by the combined action of SKN-1 and a GSK-3beta homolog is mediated by MED-1 and -2 in *C. elegans*. *Molecular Cell, 7*, 475–485.

Mango, S. E. (2007). The *C. elegans* pharynx: A model for organogenesis. *WormBook: The online review of C. elegans biology* (pp. 1–26). http://dx.doi.org/10.1895/wormbook.1.129.1. http://www.wormbook.org.

Mango, S. E., Lambie, E. J., & Kimble, J. (1994). The *pha-4* gene is required to generate the pharyngeal primordium of *Caenorhabditis elegans*. *Development, 120*, 3019–3031.

McLane, L. M., Banerjee, P. P., Cosma, G. L., Makedonas, G., Wherry, E. J., Orange, J. S., et al. (2013). Differential localization of T-bet and Eomes in CD8 T cell memory populations. *The Journal of Immunology, 190*, 3207–3215.

Mello, C. C., Draper, B. W., & Priess, J. R. (1994). The maternal genes *apx-1* and *glp-1* and establishment of dorsal-ventral polarity in the early *C. elegans* embryo. *Cell, 77*, 95–106.

Meyer, B. J. (2010). Targeting X chromosomes for repression. *Current Opinion in Genetics & Development, 20,* 179–189.

Miller, R. R., & Okkema, P. G. (2011). The *Caenorhabditis elegans* T-box factor MLS-1 requires Groucho co-repressor interaction for uterine muscle specification. *PLoS Genetics, 7,* e1002210.

Milton, A. C., & Okkema, P. G. (2015). *Caenorhabditis elegans* TBX-2 directly regulates its own expression in a negative autoregulatory loop. *G3: Genes, Genomes Genetics, 5,* 1177–1186.

Miyahara, K., Suzuki, N., Ishihara, T., Tsuchiya, E., & Katsura, I. (2004). TBX2/TBX3 transcriptional factor homologue controls olfactory adaptation in *Caenorhabditis elegans. Journal of Neurobiology, 58,* 392–402.

Muller, C. W., & Herrmann, B. G. (1997). Crystallographic structure of the T domain-DNA complex of the Brachyury transcription factor. *Nature, 389,* 884–888.

Murray, J. I., Boyle, T. J., Preston, E., Vafeados, D., Mericle, B., Weisdepp, P., et al. (2012). Multidimensional regulation of gene expression in the *C. elegans* embryo. *Genome Research, 22,* 1282–1294.

Narasimhan, K., Lambert, S. A., Yang, A. W., Riddell, J., Mnaimneh, S., Zheng, H., et al. (2015). Mapping and analysis of *Caenorhabditis elegans* transcription factor sequence specificities. *eLife, 4,* e06967.

Neves, A., & Priess, J. R. (2005). The REF-1 family of bHLH transcription factors pattern *C. elegans* embryos through Notch-dependent and Notch-independent pathways. *Developmental Cell, 8,* 867–879.

Nicoll, M., Akerib, C. C., & Meyer, B. J. (1997). X-chromosome-counting mechanisms that determine nematode sex. *Nature, 388,* 200–204.

Notredame, C., Higgins, D. G., & Heringa, J. (2000). T-coffee: A novel method for fast and accurate multiple sequence alignment. *Journal of Molecular Biology, 302,* 205–217.

Okkema, P. G., & Fire, A. (1994). The *Caenorhabditis elegans* NK-2 class homeoprotein CEH-22 is involved in combinatorial activation of gene expression in pharyngeal muscle. *Development, 120,* 2175–2186.

Okkema, P. G., Ha, E., Haun, C., Chen, W., & Fire, A. (1997). The *Caenorhabditis elegans* NK-2 homeobox gene *ceh-22* activates pharyngeal muscle gene expression in combination with *pha-1* and is required for normal pharyngeal development. *Development, 124,* 3965–3973.

Papaioannou, V. E. (2001). T-box genes in development: From hydra to humans. *International Review of Cytology, 207,* 1–70.

Papaioannou, V. E. (2014). The T-box gene family: Emerging roles in development, stem cells and cancer. *Development, 141,* 3819–3833.

Peres, J., Mowla, S., & Prince, S. (2015). The T-box transcription factor, TBX3, is a key substrate of AKT3 in melanomagenesis. *Oncotarget, 6,* 1821–1833.

Pocock, R., Ahringer, J., Mitsch, M., Maxwell, S., & Woollard, A. (2004). A regulatory network of T-box genes and the *even-skipped* homologue *vab-7* controls patterning and morphogenesis in *C. elegans. Development, 131,* 2373–2385.

Pocock, R., Mione, M., Hussain, S., Maxwell, S., Pontecorvi, M., Aslam, S., et al. (2008). Neuronal function of Tbx20 conserved from nematodes to vertebrates. *Developmental Biology, 317,* 671–685.

Powell, J. R., Jow, M. M., & Meyer, B. J. (2005). The T-box transcription factor SEA-1 is an autosomal element of the X: A signal that determines *C. elegans* sex. *Development Cell, 9,* 339–349.

Priess, J. R., Schnabel, H., & Schnabel, R. (1987). The *glp-1* locus and cellular interactions in early *C. elegans* embryos. *Cell, 51,* 601–611.

Priess, J. R., & Thomson, J. N. (1987). Cellular interactions in early *C. elegans* embryos. *Cell, 48,* 241–250.

Reinke, V., Krause, M., & Okkema, P. (2013). Transcriptional regulation of gene expression in *C. elegans*. *WormBook: The online review of C. elegans biology* (pp. 1–34). http://dx.doi.org/10.1895/wormbook.1.45.2. http://www.wormbook.org.

Roy Chowdhuri, S., Crum, T., Woollard, A., Aslam, S., & Okkema, P. G. (2006). The T-box factor TBX-2 and the SUMO conjugating enzyme UBC-9 are required for ABa-derived pharyngeal muscle in *C. elegans*. *Developmental Biology*, *295*, 664–677.

Sarov, M., Murray, J. I., Schanze, K., Pozniakovski, A., Niu, W., Angermann, K., et al. (2012). A genome-scale resource for in vivo tag-based protein function exploration in *C. elegans*. *Cell*, *150*, 855–866.

Sawyer, J. M., Glass, S., Li, T., Shemer, G., White, N. D., Starostina, N. G., et al. (2011). Overcoming redundancy: An RNAi enhancer screen for morphogenesis genes in *Caenorhabditis elegans*. *Genetics*, *188*, 549–564.

Sebe-Pedros, A., Ariza-Cosano, A., Weirauch, M. T., Leininger, S., Yang, A., Torruella, G., et al. (2013). Early evolution of the T-box transcription factor family. *Proceedings of the National Academy of Sciences of the United States of America*, *110*, 16050–16055.

Shaye, D. D., & Greenwald, I. (2011). OrthoList: A compendium of *C. elegans* genes with human orthologs. *PloS One*, *6*, e20085.

Simmer, F., Moorman, C., Van Der Linden, A. M., Kuijk, E., Van Den Berghe, P. V., Kamath, R., et al. (2003). Genome-wide RNAi of *C. elegans* using the hypersensitive *rrf-3* strain reveals novel gene functions. *PLoS Biology*, *1*, E12.

Singhvi, A., Frank, C. A., & Garriga, G. (2008). The T-box gene *tbx-2*, the homeobox gene *egl-5* and the asymmetric cell division gene *ham-1* specify neural fate in the HSN/PHB lineage. *Genetics*, *179*, 887–898.

Skipper, M., Milne, C. A., & Hodgkin, J. (1999). Genetic and molecular analysis of *fox-1*, a numerator element involved in *Caenorhabditis elegans* primary sex determination. *Genetics*, *151*, 617–631.

Smith, P. A., & Mango, S. E. (2007). Role of T-box gene *tbx-2* for anterior foregut muscle development in *C. elegans*. *Developmental Biology*, *302*, 25–39.

Song, M. R., Shirasaki, R., Cai, C. L., Ruiz, E. C., Evans, S. M., Lee, S. K., et al. (2006). T-Box transcription factor Tbx20 regulates a genetic program for cranial motor neuron cell body migration. *Development*, *133*, 4945–4955.

Sonnichsen, B., Koski, L. B., Walsh, A., Marschall, P., Neumann, B., Brehm, M., et al. (2005). Full-genome RNAi profiling of early embryogenesis in *Caenorhabditis elegans*. *Nature*, *434*, 462–469.

Stennard, F. A., Costa, M. W., Lai, D., Biben, C., Furtado, M. B., Solloway, M. J., et al. (2005). Murine T-box transcription factor Tbx20 acts as a repressor during heart development, and is essential for adult heart integrity, function and adaptation. *Development*, *132*, 2451–2462.

Stirnimann, C. U., Ptchelkine, D., Grimm, C., & Muller, C. W. (2010). Structural basis of TBX5-DNA recognition: The T-box domain in its DNA-bound and -unbound form. *Journal of Molecular Biology*, *400*, 71–81.

Sulston, J. E., Albertson, D. G., & Thomson, J. N. (1980). The *Caenorhabditis elegans* male: Postembryonic development of nongonadal structures. *Developmental Biology*, *78*, 542–576.

Sulston, J. E., & Horvitz, H. R. (1977). Post-embryonic cell lineages of the nematode, *Caenorhabditis elegans*. *Developmental Biology*, *56*, 110–156.

Sulston, J. E., Schierenberg, E., White, J. G., & Thomson, J. N. (1983). The embryonic cell lineage of the nematode *Caenorhabditis elegans*. *Developmental Biology*, *100*, 64–119.

Woollard, A., & Hodgkin, J. (2000). The *Caenorhabditis elegans* fate-determining gene *mab-9* encodes a T-box protein required to pattern the posterior hindgut. *Genes & Development*, *14*, 596–603.

CHAPTER THREE

T-Box Genes and Developmental Gene Regulatory Networks in Ascidians

A. Di Gregorio[1]
New York University College of Dentistry, New York, NY, United States
[1]Corresponding author: e-mail address: adg13@nyu.edu

Contents

1. Introduction 57
2. Brachyury and the Temporal Control of Notochord Specification, Development, and Differentiation 59
3. Tbx2/3: An Essential Mediator of Brachyury Function in the Notochord 66
4. The Lineage-Specific Duplication of *Tbx6* Genes in Ascidians 68
5. Tbx6a 69
6. Tbx6b, Tbx6c, and the Interpretation of the Maternally Inherited Muscle Determinant 70
7. Tbx6b and Tbx6c in Heart Development 73
8. Tbx1/10 and the Evolutionary Origins of the Pharyngeal Muscles 74
9. Tbx15/18/22 (VegTR) and the Establishment of Muscle Cell Identity 75
10. Maternal *T* and Tail Development 76
11. Tbx20: In or Out of the Ascidian Heart? 77
12. Transcriptional Regulation of T-Box Genes Expression 78
    12.1 Brachyury: A Balancing Act Between Multiple Activators and a Repressor 79
    12.2 Tbx2/3: A Composite Expression Pattern Orchestrated by Additive, Partly Redundant CRMs 81
    12.3 Tbx6b and Tbx6c: Early and Late Elements, Autoregulation, and Crossregulation 82
13. Concluding Remarks 83
Acknowledgments 84
References 84

Abstract

Ascidians are invertebrate chordates with a biphasic life cycle characterized by a dual body plan that displays simplified versions of chordate structures, such as a premetamorphic 40-cell notochord topped by a dorsal nerve cord and post-metamorphic pharyngeal slits. These relatively simple chordates are characterized by

Current Topics in Developmental Biology, Volume 122
ISSN 0070-2153
http://dx.doi.org/10.1016/bs.ctdb.2016.08.007

rapid development, compact genomes and ease of transgenesis, and thus provide the opportunity to rapidly characterize the genomic organization, developmental function, and transcriptional regulation of evolutionarily conserved gene families. This review summarizes the current knowledge on members of the T-box family of transcription factors in *Ciona* and other ascidians. In both chordate and nonchordate animals, these genes control a variety of morphogenetic processes, and their mutations are responsible for malformations and developmental defects in organisms ranging from flies to humans. In ascidians, T-box transcription factors are required for the formation and specialization of essential structures, including notochord, muscle, heart, and differentiated neurons. In recent years, the experimental advantages offered by ascidian embryos have allowed the rapid accumulation of a wealth of information on the molecular mechanisms that regulate the expression of T-box genes. These studies have also elucidated the strategies employed by these transcription factors to orchestrate the appropriate spatial and temporal deployment of the numerous target genes that they control.

ABBREVIATIONS

AS atrial siphon
ASM atrial siphon muscle
bp base pair(s)
cDNA complementary DNA
ChIP chromatin immunoprecipitation
chrom. chromosome
CRM *cis*-regulatory module
En endostyle
ESN(s) epidermal sensory neuron(s)
EST(s) expressed sequence tag(s)
hr(s) hour(s)
GFP green fluorescent protein
GST glutathione *S*-transferase
Ht heart
kb kilobase(s), or 1000 base pairs
Int intestine
LoM longitudinal muscle
NC neural complex
OS oral siphon
OSM oral siphon muscle
PGS pharyngeal gill slits
St stomach
Tbx T-box
TLC trunk lateral cell
TVC trunk ventral cell

1. INTRODUCTION

The phylum Chordata comprises animals characterized by the presence of the notochord, an axial structure of mesodermal origin necessary for their development (e.g., Stemple, 2005). Most chordates develop a vertebral column during their early life, and therefore are grouped in the subphylum Vertebrata, which includes humans. Animals that develop a notochord but not a vertebral column, defined as invertebrate chordates, are grouped into two subphyla of marine organisms, Cephalochordata and Urochordata (or tunicates). The main representative of extant cephalochordates is amphioxus, commonly known as lancelet, while tunicates are grouped into three main classes, Ascidiacea (ascidians), Thaliacea (thaliaceans), and Larvacea (larvaceans or appendicularians). Contrary to what had been believed for several decades, recent molecular phylogenies have suggested that tunicates are more closely related to vertebrates than cephalochordates (Delsuc, Brinkmann, Chourrout, & Philippe, 2006). Nevertheless, after tunicates branched off from the main chordate lineage, ~500 million years ago, their genomes have undergone several rearrangements, which have led to clade-specific gene duplications and gene losses, and consequent adaptations and divergence.

Among tunicates, ascidians are the largest and most-studied class. Numerous ascidian genomes are publicly available in a searchable format, and several studies have been carried out on evolutionarily conserved gene families. In particular, solitary ascidians of the genera *Ciona* and *Halocynthia* are widely used for studies of gene expression, regulation, and function (e.g., Kourakis & Smith, 2015; Lemaire, 2009). One of the main reasons for the continued interest in these model organisms is the similarity of their embryonic body plan to that of vertebrates (Fig. 1A and B). In addition to sharing the notochord with developing vertebrates, ascidian embryos also display a dorsal tubular nervous system, ventrally located endodermal derivatives, and a simple "brain," comprising the sensory vesicle and visceral ganglion, able to receive and process sensory inputs and to coordinate the swimming movements of paraxially located, nonsegmental muscle cells (e.g., Passamaneck & Di Gregorio, 2005). Recent studies have identified ascidian structures related to neurogenic placodes (Abitua et al., 2015; Manni et al., 2004) as well as migratory cells that share key molecular markers and the ability to differentiate into various cell types with vertebrate neural crest

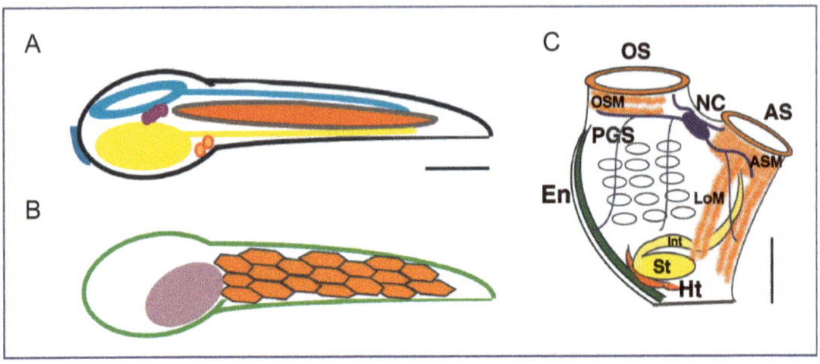

Fig. 1 Main ascidian tissues, before and after metamorphosis. Schematic illustrations of the *Ciona* embryonic (A, B) and adult tissues (C). (A) Side view of a tailbud showing its inner tissues, color-coded as follows: *blue*, nervous system; *dark blue*, presumptive adhesive organ (palps); *red*, notochord; *yellow*, endoderm; *purple*, trunk lateral cells (TLCs); *orange*, trunk ventral cells (TVCs). (B) Side view of a tailbud, showing its outermost tissues: the muscle cells flanking the notochord on each side (*orange*), mesenchyme (*light purple*), and a sectional view of the epidermis (*green*) that covers the entire body. (C) Simplified view of the adult body plan. Abbreviations: *AS*, atrial siphon; *ASM*, atrial siphon muscle; *En*, endostyle; *Ht*, heart; *Int*, intestine; *LoM*, longitudinal muscle; *NC*, neural complex; *OS*, oral siphon; *OSM*, oral siphon muscle; *PGS*, pharyngeal gill slits; *St*, stomach. For simplicity, neither the gonad nor the gonoducts are shown. Scale bars: in (A), ~100 μm; in (B), ~2 cm.

cells (Jeffery et al., 2008; Stolfi, Ryan, Meinertzhagen, & Christiaen, 2015). Ascidian larvae are unable to feed themselves, hence they swim for an interval ranging between several minutes and a few days, depending upon their species, and then settle to begin metamorphosis. At metamorphosis, the body plan is massively rearranged; both notochord and embryonic muscle are eliminated, while part of the cells of the sensory vesicle and anterior nerve cord act as primordial stem cells in the formation of the adult neural complex (NC, Fig. 1) (Horie et al., 2011). The newly formed body plan contains a different set of chordate features, such as pharyngeal gill slits (PGS, Fig. 1), a thyroid-like structure, called endostyle (En, Fig. 1), and a tubular heart (Ht, Fig. 1) (Karaiskou, Swalla, Sasakura, & Chambon, 2015).

Comparative genomic studies suggest that tunicates have undergone gene losses that have affected various ancestral gene families as well as individual genes present in nonchordate invertebrates (Berná & Alvarez-Valin, 2014). For example, *tailless*, a gene encoding a nuclear receptor, and genes controlling the circadian rhythms, are missing in *Ciona* but are present in *Drosophila* (Dehal et al., 2002). In particular, a complete set of *Hox* genes was likely present in a common chordate ancestor, as suggested by the presence of a complete *Hox* cluster in the amphioxus *Branchiostoma floridae* (Garcia-Fernández & Holland, 1994), while a few *Hox* genes are absent in *Ciona* (Spagnuolo et al., 2003). A similar scenario can be envisioned for members of another evolutionarily conserved family of transcription factors, the T-box family. *Brachyury*-related and *Tbx7/8*-related genes have been identified in amoebas, and in sponges the T-box gene family expands to include *Tbr1*-, *Tbx1*-, and *Tbx2*-related genes (Holstien et al., 2010; Papaioannou, 2014; Sebé-Pedrós, de Mendoza, Lang, Degnan, & Ruiz-Trillo, 2011). The genome of *Branchiostoma floridae* contains representatives of all the T-box genes subfamilies found in vertebrates, including a *Tbx4/5* ortholog that seems to have been lost from the ascidian genomes (Paps, Holland, & Shimeld, 2012; Ruvinsky, Silver, & Gibson-Brown, 2000). On the other hand, in addition to gene losses, ascidian genomes also display lineage-specific gene duplications, as in the case of the multiple *Tbx6*-related genes found in *Ciona* (order Phlebobranchia) (Takatori et al., 2004) and *Halocynthia* (order Stolidobranchia) (Stolfi, Sasakura, et al., 2015).

In sum, the favorable phylogenetic position, the simplified yet conserved body plan organization, and the presence of representative T-box genes render ascidians informative model organisms and a valuable reference for comparative studies of T-box genes. This review mainly focuses on the T-box genes identified in the ascidian *Ciona intestinalis*, their expression patterns and the mechanisms controlling their expression, and on cognate genes identified thus far from other ascidian species. The developmental roles of these genes and their positions within the simplified gene regulatory networks of ascidians are summarized and compared to those of their vertebrate counterparts.

2. BRACHYURY AND THE TEMPORAL CONTROL OF NOTOCHORD SPECIFICATION, DEVELOPMENT, AND DIFFERENTIATION

Brachyury (Greek for "short tail"), also known as "*T*," for "tail," was the first gene of the T-box family to be identified (Herrmann, Labeit, Poustka, King, & Lehrach, 1990). The gene was isolated in mouse by positional cloning and it was shown to encode a sequence-specific transcription factor with a novel DNA-binding domain, the T-domain (encoded by a nucleotide sequence named T-box) (Kispert & Herrmann, 1993). The identification of mouse *Brachyury* is an example of forward genetic approach. The *Brachyury* mutation had been first reported in 1927, and the gene was named after the short-tail phenotype that characterizes heterozygous mouse mutants (Dobrovolskaïa-Zavadskaïa, 1927). Mouse embryos homozygous for the *Brachyury* mutation die in utero by gestation day 11, display defects in the formation of the allantois, hindgut, and posterior mesoderm, and, in particular, fail to form a notochord (Gluecksohn-Schoenheimer, 1940). The notochord is the main feature of the phylum Chordata and is an axial structure of mesodermal origin required not only for support but also for the regionalization of the neural tube dorsal to it and the patterning of its flanking somites (Stemple, 2005). The molecular mechanisms underlying the appearance of the notochord during the evolutionary history of multicellular animals are still under investigation (Satoh, Tagawa, & Takahashi, 2012). While *Brachyury* orthologous genes are represented in the genomes of nonchordate animals, and are expressed in different structures and tissues (Swalla, 2006), none of the chordate animals investigated thus far is able to properly develop a notochord if the function of this gene is impaired. *Brachyury*$^{-/-}$mutants have been identified or generated via reverse genetic approaches in different chordates, and all mutants analyzed thus far display severely impaired notochord formation (Chiba, Jiang, Satoh, & Smith, 2009; Nibu, José-Edwards, & Di Gregorio, 2013; Smith, 1999). Ascidian *T* (*As-T*), a *Brachyury*-related gene from *Halocynthia roretzi*, was the first T-box gene identified in ascidians (Yasuo & Satoh, 1993), followed by the *Ciona Brachyury* gene (*Ci-Bra*; Table 1) (Corbo, Levine, & Zeller, 1997). Unlike their vertebrate counterparts, these genes were exclusively expressed in notochord cells and in their lineage-restricted precursors (Fig. 2). Differently from *Ciona* and *Halocynthia*, two *Brachyury* orthologs have been identified in amphioxus, and the expression patterns obtained for them include the paraxial mesoderm in addition to the notochord (Holland, Koschorz, Holland, & Herrmann, 1995). All ascidian species sampled thus far seem to contain a single copy of *Brachyury*. Of note, *Brachyury* is present in the genomes of the tailless ascidians *Molgula occulta* and *Molgula tectiformis* and is originally expressed in the notochord precursors (Takada, Satoh, & Swalla, 2002). However, in *M. occulta Brachyury* expression disappears prematurely and only 20 notochord cells are formed, instead of 40, and these cells fail to undergo convergent extension, the main morphogenetic process that ensures tail elongation. In *M. tectiformis*, the notochord precursors stop dividing even earlier, and as a consequence only 10 notochord cells are formed and convergent extension is not achieved (Takada et al., 2002). The lack of convergent extension and tail elongation is also observed in *N*-ethyl-nitrosourea-induced *Ci-Bra* mutants, which contain a premature

Table 1 T-Box Genes in the Ascidian *Ciona intestinalis*

T-Box Subfamily	*Ciona intestinalis* Gene Name	Gene Model	Genomic Location	Consensus Binding Sequence
Brachyury (T)	*Ci-Bra*	KH.S1404.1	Scaffold 1404	TNRCACYT[a] TCACACCTAGGTGTCA[b]
Tbx2, Tbx3	*Ci-Tbx2/3*	KH.L96.87	Scaffold L96	N/D
Tbx6	*Ci-Tbx6a*	KH.L8.11	Scaffold L8	N/D
	Ci-Tbx6b	KH.S654.3	Scaffold 654	GWTCACACCT[c]
	Ci-Tbx6c	KH.S654.2	Scaffold 654	GWTCACACCT[c]
	Ci-Tbx6d?	KH.S654.1?	Scaffold 654	N/D
Tbx1, Tbx10	*Ci-Tbx1/10*	KH.C7.628	Chrom. 7	N/D
Tbx15/18/22	*Ci-VegTR*	KH.S589.4	Scaffold 589	TNNCAC[d]
Tbr/Eomesodermin/ T-bet/Tbx21	*Ci-mT*	KH.C3.773	Chrom. 3	N/D
Tbx20	*Ci-Tbx20*	KH.C1.224	Chrom. 1	N/D

[a]Katikala et al. (2013).
[b]Kubo et al. (2010).
[c]Yagi et al. (2005).
[d]Erives and Levine (2000).

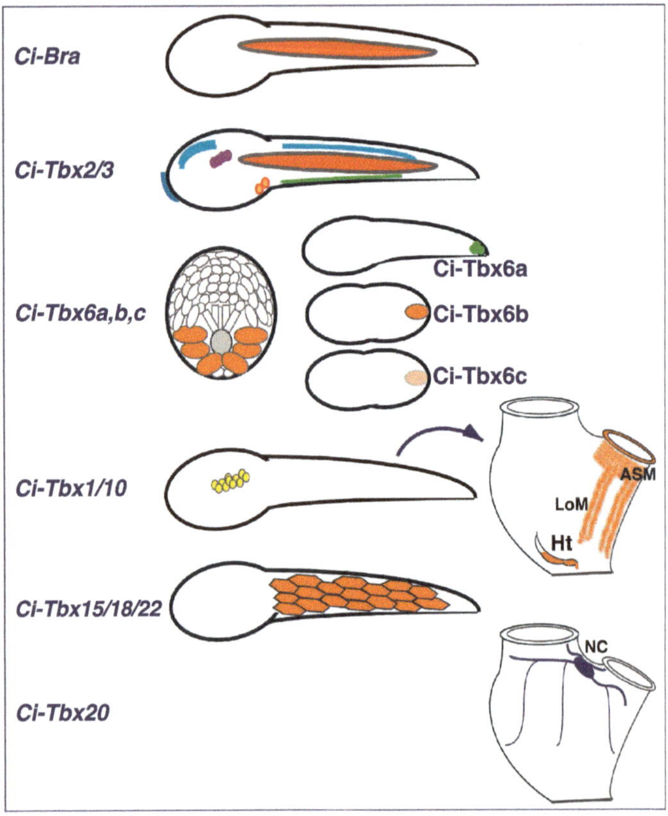

Fig. 2 Expression patterns reported for T-box genes in *Ciona intestinalis*. Schematic summary of the expression patterns reported thus far for *C. intestinalis* T-box genes. The ubiquitously expressed *Ci-mT* is not depicted. Abbreviations and color-coding are as in Fig. 1. Epidermis is colored in *green*. Expression of *Ci-Tbx2/3* is detected only in part of the sensory vesicle and part of the nerve cord (*blue*). For the *Tbx6*-related genes, the main stage of expression, late gastrula/neural plate, is illustrated, and muscle precursors are colored in *orange*. In the case of *Ci-Tbx6a*, expression is still detected in the epidermis at the tip of the tail (*green*) at the early-tailbud stage, while *Ci-Tbx6b* and *Ci-Tbx6c* are reportedly expressed, in late neurulae, only in a few muscle cells at the tip of the tail (*orange*; *light orange* indicates fading hybridization signal in muscle). In the case of *Ci-Tbx1/10*, trunk endodermal cells are individually represented, to indicate that the expression of this gene is limited to a cohort of endodermal cells. Embryos and juveniles are not drawn to scale.

stop codon in the *Ci-Bra* coding region that causes the synthesis of a severely truncated protein composed of only the first 48 amino acid residues of wild-type Ci-Bra (Chiba et al., 2009). In the appendicularian *Oikopleura dioica*, a representative of another class of urochordates, double-stranded RNA interference experiments have shown that impairments in the function of *Brachyury* cause notochord malformations and occasionally loss of an identifiable notochord (Omotezako, Nishino, Onuma, & Nishida, 2013). Together, these findings suggest that the main function of Brachyury in urochordates is the control of notochord morphogenesis.

The identification of *Ci-Bra* was accompanied by the first promoter/enhancer analysis in *Ciona*, through the development of a straightforward electroporation protocol that has since enabled the identification of hundreds of regulatory regions (Corbo et al., 1997; Di Gregorio & Levine, 2002; Irvine, 2013). In addition to that, electroporation rendered misexpression experiments feasible and economical. In particular, the misexpression of Ci-Bra in endodermal and neural precursors allowed the identification, through a subtraction screen, of numerous candidate Ci-Bra-downstream genes, many of which turned out to be expressed in the notochord (Di Gregorio & Levine, 1999; Hotta et al., 2000; Hotta, Takahashi, Erives, Levine, & Satoh, 1999; Hotta, Takahashi, Satoh, & Gojobori, 2008; Takahashi et al., 1999), including the direct Ci-Bra targets *Ci-tropomyosin* (Di Gregorio & Levine, 1999) and *Ci-leprecan* (Dunn & Di Gregorio, 2009). Additional Ci-Bra target genes, including a few evolutionarily conserved transcription factors, were identified in later studies (José-Edwards et al., 2011; José-Edwards, Oda-Ishii, Nibu, & Di Gregorio, 2013; Kugler, Passamaneck, Feldman, Regnier, & Di Gregorio, 2008; Thompson & Di Gregorio, 2015). Through electrophoretic mobility assays, an in vitro-synthesized glutathione *S*-transferase (GST)-Ci-Bra fusion protein was found to bind both the palindromic mouse Brachyury consensus binding sequence (Kispert, Koschorz, & Herrmann, 1995) and half-sites with the generic sequence TNNCAC (Di Gregorio & Levine, 1999). An extended consensus binding sequence for Ci-Bra, TNRCACYT, was later determined on the basis of the functional relevance of Ci-Bra-binding sites within notochord CRMs (Katikala et al., 2013) (Table 1). For a genome-wide screen chromatin immunoprecipitation (ChIP-chip assay) for Ci-Bra targets, a Ci-Bra-green fluorescent protein (GFP) fusion protein was expressed in the notochord precursors of *Ciona,* under the control of the *Ci-Bra* promoter. After immunoprecipitation with a GFP antibody, the purified DNA was hybridized to whole-genome *Ciona* microarrays, in parallel with control DNA from wild-type embryos. The Ci-Bra-GFP fusion protein was shown to bind 2924 genomic loci in early *Ciona* embryos, corresponding to ~2092 individual genes, including 194 transcription factors (Kubo et al., 2010). The number of Brachyury-bound regions in *Ciona* is higher than that of the loci reportedly bound by the corresponding proteins in mouse (1942 bound targets; Lolas, Valenzuela, Tjian, & Liu, 2014) and *Xenopus* (~1400 bound targets; Gentsch et al., 2013). However, it has been shown that not all the binding events detected through immunoprecipitation experiments are necessarily associated with detectable changes in the expression of genes whose loci are occupied by a transcription factor (Fisher et al., 2012). Consistent with these findings, of the ~1400 genes occupied by Xbra, 81 have been confirmed to strictly depend upon this factor for their expression (Gentsch et al., 2013; Lolas et al., 2014). It seems therefore likely that the number of genes whose expression is affected by Ci-Bra might be lower than the number of binding events detected through ChIP experiments. On the other hand, the number of indirect targets of Ci-Bra is on the rise, as more genes targeted by other notochord transcription factors positioned downstream of Ci-Bra are being identified (José-Edwards et al., 2013).

Notochord formation and differentiation in ascidians occur within approximately 18–20 h after fertilization, depending upon the incubation temperature, and involve precisely timed morphogenetic movements, such as intercalation, elongation, and tubulogenesis (Denker & Jiang, 2012). This observation indicated that the proper temporal sequence of these movements was ensured by the sequential deployment of notochord genes. Accordingly, the close observation of the expression patterns of 39 bona fide Ci-Bra notochord targets underscored their sequential onset of expression, and suggested that it could be the result of different mechanisms of temporal regulation of notochord gene expression by Ci-Bra (Hotta et al., 1999). To shed light on these mechanisms, notochord *cis*-regulatory modules (CRMs, or enhancers) were isolated and characterized from notochord genes representative of the early-, middle-, and late-onset Ci-Bra targets. This analysis revealed that notochord CRMs associated with early-onset Ci-Bra targets, i.e., genes that become detectable in notochord precursors around early gastrulation, rely for their activity upon multiple functional Ci-Bra-binding sites that act cooperatively; middle-onset Ci-Bra targets, i.e., genes that become detectable in notochord precursors around the neural plate stage, rely upon individual functional Ci-Bra-binding sites; last, notochord CRMs associated with late-onset Ci-Bra targets, i.e., genes that become detectable in notochord by neurulation, are indirectly controlled by Ci-Bra through transcriptional intermediaries (Katikala et al., 2013). These findings propose a mechanistic explanation for the accurate temporal deployment of Ci-Bra target genes observed during notochord development in *Ciona* embryos (Fig. 3).

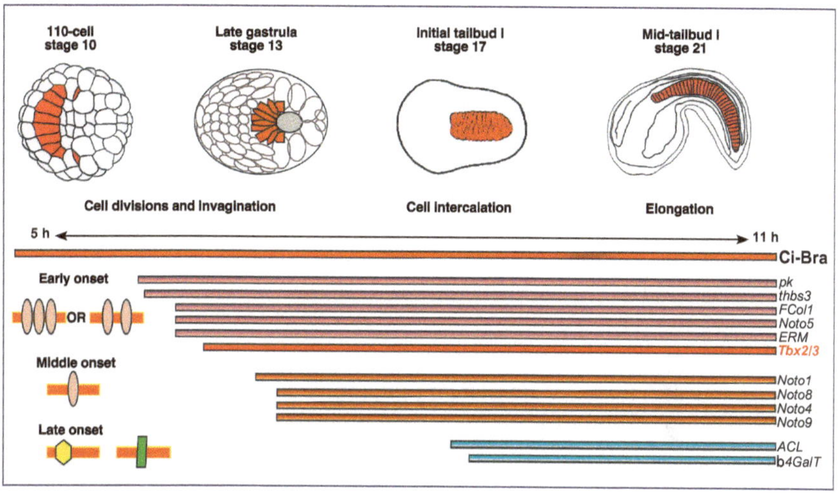

Fig. 3 Different *cis*-regulatory mechanisms ensure temporal regulation of notochord gene expression by Ci-Bra. (*Top*) Schematic drawings of *Ciona* embryos at the stages indicated above them, according to Hotta et al. (2007). The notochord precursors and the definitive 40 notochord cells are colored in *red*. The stages of notochord development related to each stage are indicated below each drawing (Denker & Jiang, 2012). These morphogenetic events take place between 5 and 11 h after fertilization at the incubation temperature of 18°C. (*Bottom*) The window of expression of *Ci-Bra* and of notochord genes whose transcription is controlled by Ci-Bra are symbolized by *horizontal bars* underneath the stages of notochord development. Expression of *Ci-Bra* (*red bar*) is detected in notochord precursors beginning at the 64-cell stage (not shown; Corbo et al., 1997). Expression windows of early-onset Ci-Bra target genes are colored in *pink*; expression of *Ci-Tbx2/3* is highlighted in *dark pink*. Expression windows of middle-onset Ci-Bra target genes are colored in amber, and those of late-onset Ci-Bra target genes are colored in *blue*. Genes with different onset of expression are associated with notochord CRMs (schematics on the *left side*) controlled by Ci-Bra through multiple, presumably cooperative binding sites (early onset), through individual binding sites (middle onset) or indirectly, via transcriptional intermediaries (schematics on the *left*) (Katikala et al., 2013). CRMs and their binding sites are not drawn to scale. Gene names are abbreviated as follows: *pk, Ci-prickle; thbs3, Ci-thrombospondin 3; FCol1, fibrillar collagen 1; Noto1–Noto9, Ciona notochord gene 1–9; ERM, ezrin–radixin–moesin; ACL, Ci-ATP citrate lyase; b4galT, beta-1,4-galactosyltransferase* (Hotta et al., 2000; Katikala et al., 2013). *Symbols*: *Pink ovals*: functional Ci-Bra-binding sites. *Yellow hexagon* and *green vertical bar*: binding sites for transcription factors that act as intermediaries of Ci-Bra.

The analysis of *Ciona* notochord CRMs has also provided the first examples of the long-sought minimal notochord regulatory regions controlled synergistically by Ci-Bra and Ci-FoxA-a and dependent upon binding sites for both these transcription factors (José-Edwards et al., 2015; Passamaneck et al., 2009). Moreover, the notochord CRM associated with another putative Ci-Bra target gene, *Ci-Ephrin3*, has been shown to require a functional Ci-Bra-binding site closely associated with an $(AC)_6$ microsatellite sequence (José-Edwards et al., 2015). Remarkably, ChIP-chip studies of Brachyury occupancy in differentiating mouse embryonic stem cells show that mouse Brachyury often binds $(AC)_{\geq 6}$ microsatellite repeats (Evans et al., 2012), thus suggesting a possible evolutionary conservation of this mechanism of regulation of gene expression.

3. Tbx2/3: AN ESSENTIAL MEDIATOR OF BRACHYURY FUNCTION IN THE NOTOCHORD

Ci-Tbx2/3 is the only T-box gene reportedly expressed in the *Ciona* notochord in addition to *Ci-Bra* (Imai, Hino, Yagi, Satoh, & Satou, 2004; Takatori et al., 2004; Table 1). Differently from the notochord-specific *Ci-Bra*, *Ci-Tbx2/3* is expressed in additional territories, including ventral epidermis and the epidermal sensory neurons (ESNs) interspersed in it, presumptive palps, trunk lateral cells (TLCs), and nerve cord (Imai et al., 2004; Takatori et al., 2004) (Fig. 2). Expression of *Ci-Tbx2/3* commences early, approximately at the 32-cell stage, in B6.2 blastomeres, the precursors of muscle, mesenchyme cells, and secondary notochord, and after one round of cell divisions is observed in B7.7 mesenchyme precursors as well as in an additional cell pair, A7.6, the TLCs precursors (Imai et al., 2004) (for a summary of the nomenclature used for ascidian blastomeres, see Meinertzhagen & Okamura, 2001). By neurulation, expression expands to a region of the trunk epidermis and to the presumptive adhesive organ, which will form palps containing peripheral sensory neurons. In the ventral epidermis, Ci-Tbx2/3 is required for the expression of *Ci-Msx* (formerly *Ci-msxb*; Aniello et al., 1999) in ESNs (Waki, Imai, & Satou, 2015).

By the early-tailbud stage, a narrow region expressing *Ci-Tbx2/3* becomes distinguishable in the nerve cord; by the mid-tailbud stage, this territory of expression increases in length and eventually spans two-thirds of the length of the nerve cord (José-Edwards et al., 2013). Notochord expression is first detected at the neural plate stage and persists in this structure past the late-tailbud stage. In embryos homozygous for a mutation in the *Ci-Bra* coding region (Chiba et al., 2009), *Ci-Tbx2/3* remains expressed at normal levels in sensory vesicle, presumptive palps, TLCs and spinal cord, but is no longer detectable in the notochord (José-Edwards et al., 2013). Conversely, other notochord genes, such as *Ci-Lmx-like*, remain expressed in the notochord of *Ci-Bra*$^{-/-}$embryos (José-Edwards et al., 2011). These findings suggested that *Ci-Tbx2/3*, unlike other notochord transcription factor genes, depends upon Ci-Bra for its notochord expression. The developmental function of Ci-Tbx2/3 has been investigated through morpholino oligonucleotide injection. These experiments revealed a possible autoregulatory feedback of this factor on its own transcription (Imai, Levine, Satoh, & Satou, 2006) and highlighted the strong effects of a global knockdown of this gene on both tail extension and trunk development (José-Edwards et al., 2013). The specific effects of the perturbation in the levels and function of Ci-Tbx2/3 in the notochord have been investigated by expressing in notochord precursors either a "passive" repressor form of Ci-Tbx2/3, i.e., a truncated version of the protein that contained only its DNA-binding domain, or a Ci-Tbx2/3::VP16 fusion protein, which was expected to act as a strong activator of gene expression (José-Edwards et al., 2013). Both these mutant versions of Ci-Tbx2/3 were expressed in the developing *Ciona* notochord via the *Ci-Bra* promoter and were instrumental for the identification of candidate Ci-Tbx2/3 target genes through microarray screens. These experiments identified 81 potential Ci-Tbx2/3-downstream genes, 70 of which yielded detectable signals in whole-mount in situ hybridizations and were expressed in all the *Ci-Tbx2/3* domains described earlier. In particular, 20 of these genes (~29%) were expressed in the notochord, including a previously uncharacterized thrombospondin type 1-domain containing protein, a perspective mediator of cell–cell and cell–matrix interactions in the extracellular matrix (José-Edwards et al., 2013). Importantly, one of the target genes of Ci-Tbx2/3 was found to be *Ci-Noto4*, which encodes a phosphotyrosine-binding domain that had been previously shown to be required for notochord intercalation and for the acquisition of the characteristic notochord cell shape (Yamada, Ueno, Satoh, & Takahashi, 2011). Interestingly, *Ci-Noto4* and other genes identified through the experiments described earlier had also been previously described as genes activated by Ci-Bra (Takahashi et al., 1999), which suggested that Ci-Bra controlled them through Ci-Tbx2/3. However, other genes, such as *dual oxidase-c* (*Ci-Duox-c*) appear to be repressed by Ci-Tbx2/3 (José-Edwards et al., 2013). In conclusion, the results of the microarray screens suggest that Ci-Tbx2/3 can operate as both an activator and a repressor in a context-dependent fashion, despite the lack of recognizable activation or repression domains in the Ci-Tbx2/3 protein sequence. Similarly, in vertebrates, Tbx2 and Tbx3 are thought to act as both activators and repressors in a context-dependent manner (Washkowitz, Gavrilov, Begum, & Papaioannou, 2012). These observations suggest that Ci-Tbx2/3 strengthens and refines the Ci-Bra-downstream gene regulatory network during critical stages of notochord formation.

4. THE LINEAGE-SPECIFIC DUPLICATION OF *Tbx6* GENES IN ASCIDIANS

Transcription factors of the Tbx6 subfamily are responsible for the specification of paraxial mesoderm (Wardle & Papaioannou, 2008). *Tbx6* genes are typically absent from the notochord and are expressed in the paraxial mesoderm, which is progressively subdivided into metameric somites as morphogenesis proceeds. Mutant mice lacking Tbx6 function form irregular somites in the neck region of the embryos and have their posterior somites replaced by two ectopic neural tube-like structures on the sides of the axial neural tube found in wild-type embryos (Chapman & Papaioannou, 1998). Similar expression in the paraxial mesoderm has been reported for *Tbx6* orthologs in other chordates, such as zebrafish (Wardle & Papaioannou, 2008). Within the two early chordate subphyla, the cephalochordate amphioxus develops paraxially located somites, while tunicates lack a segmented paraxial mesoderm, although they develop paraxially located, nonsyncitial muscle cells (Passamaneck, Hadjantonakis, & Di Gregorio, 2007).

The *C. intestinalis* genome contains three bona fide *Tbx6*-related genes, *Ci-Tbx6a*, *Ci-Tbx6b*, and *Ci-Tbx6c* (Takatori et al., 2004). *Ci-Tbx6b* and *Ci-Tbx6c* are clustered on scaffold 654 in a region spanning ~11 kb (Brozovic et al., 2016). An additional *Tbx6*-related gene, *Ci-Tbx6d*, was originally described (Dehal et al., 2002), but was later reported to be the result of an error in the genome assembly (Takatori et al., 2004) and no additional information is presently available about it.

Ci-Tbx6a, *Ci-Tbx6b*, and *Ci-Tbx6c* are believed to have originated through two rounds of lineage-specific duplications, the first one consisting in the duplication of a single ancestral gene into *Ci-Tbx6a* and *Ci-Tbx6b/c* and the second involving the duplication of *Ci-Tbx6b/c* and the consequent divergence of the resulting two copies of this gene into *Ci-Tbx6b* and *Ci-Tbx6c* (Takatori et al., 2004).

In *H. roretzi*, a T-box gene, *As-T2*, is expressed in muscle cells in a pattern very similar to that of the *Ciona Tbx6* genes, and remains expressed in epidermal cells located at the tip of the tail (Yasuo, Kobayashi, Shimauchi, & Satoh, 1996). This narrow site of epidermal expression is more closely reminiscent of the expression of *Ci-Tbx6a* (Fig. 2) (see later). Overexpression of *As-T2* induces ectopic expression of myosin heavy chain and actin, and partially suppresses expression of an epidermal marker, *HrEpiC* (Mitani, Takahashi, & Satoh, 1999). Differently from most other ascidian T-box genes, which are usually underrepresented in the endoderm, *As-T2* is very transiently expressed in the endodermal precursors of *Halocynthia* around the 32-/44-cell stage, although it is rapidly downregulated in this lineage and does not seem to affect endodermal development (Mitani, Takahashi, & Satoh, 2001).

The analysis of the draft genomes of three recently sequenced *Molgula* species, *Molgula occidentalis*, *M. oculata*, and *M. occulta*, identified only two *Tbx6*-related genes in each species (Stolfi et al., 2014). In situ hybridizations carried out for the two *Tbx6*-related genes of *M. occidentalis* indicate that both genes are expressed in a broad posterior territory in pregastrula embryos, similarly to the *Tbx6* genes of *Ciona* (Stolfi et al., 2014).

5. Tbx6a

Ci-Tbx6a (Table 1) is first detected at the 16-cell stage specifically in B5.1 blastomeres, and one cell division later in some of their descendants, namely the B6.2 blastomeres, which will give rise to muscle, mesenchyme cells and secondary notochord; at the same time, *Ci-Tbx6a* expression begins in the B6.4 blastomeres, which will originate muscle and mesenchyme cells (Nishida, 1987). At the gastrula and neural plate stages, *Ci-Tbx6a* expression remains detectable in muscle and mesenchyme precursors, while at neurulation it is downregulated in muscle cells and becomes confined to a small group of epidermal cells at the tip of the developing tail (Imai et al., 2004; Takatori et al., 2004) (Fig. 2). This narrow region of the embryo is also the site of expression of *CiHox12* (Keys et al., 2005) a member of the dispersed *Ciona Hox* cluster (Di Gregorio et al., 1995; Ikuta, Yoshida, Satoh, & Saiga, 2004) and of *Ci-Wnt5* (Imai et al., 2004). Expression of *Ci-Tbx6a* is dependent upon Ci-macho1, the muscle determinant described later, as shown by its downregulation in Ci-macho1 morphant embryos (Yagi, Satoh, & Satou, 2004). Although morpholino-mediated knockdown of Ci-Tbx6a was attempted (Imai et al., 2006), the specific function of this gene in muscle development remains to be elucidated; it seems conceivable that it might be at least in part overlapping with the functions of the other two *Tbx6*-related genes. Hence, the presumed role of Ci-Tbx6a in mediating the function of Ci-macho1 in muscle development remains uncharacterized.

6. Tbx6b, Tbx6c, AND THE INTERPRETATION OF THE MATERNALLY INHERITED MUSCLE DETERMINANT

Ci-Tbx6b and *Ci-Tbx6c* are the most closely related *Tbx6* genes found in *Ciona* (Table 1); nevertheless, their expression patterns are similar but not identical, as are their functions in embryonic development. At the 16-cell stage, both genes are expressed exclusively in the B5.1 blastomeres, similar to *Ci-Tbx6a*. However, differently from *Ci-Tbx6a*, neither *Ci-Tbx6b* nor *Ci-Tbx6c* is expressed in the B6.2; both genes remain confined to the B6.4 pair and its progeny, and thus appear to be strictly limited to muscle and mesenchyme precursors (Takatori et al., 2004). Interestingly, in late gastrulae, expression of *Ci-Tbx6c* expands to encompass all muscle precursors and becomes indistinguishable from that of *Ci-Tbx6b* (Fig. 2). In late neurulae, expression of both genes is downregulated in most cells of the developing muscle and is restricted to the posterior-most muscle precursors (Takatori et al., 2004) (Fig. 2). Neither *Ci-Tbx6b* nor *Ci-Tbx6c* is detected in tailbud embryos (Imai et al., 2004), which suggests that their function is limited to early muscle development. In ascidians, muscle formation relies upon the quintessential mosaic development, whereby the asymmetric distribution of a maternal "determinant" to select blastomeres causes the early specification, and invariance, of lineage(s) (Conklin, 1905; Reverberi & Ortolani, 1955). The molecular nature of the muscle determinant was first elucidated in *H. roretzi*, where it was shown, through blastomere dissections followed by subtractive hybridization, that the maternally stored mRNA for the zinc-finger transcription factor macho-1 is necessary and sufficient to determine the primary muscle cell fate (Nishida & Sawada, 2001). Subsequent studies showed that the *macho1* counterpart in *C. intestinalis*, *Ci-macho1*, possesses similar properties (Satou et al., 2002). Morpholino oligonucleotide translational knockdown of Ci-macho1 induced the downregulation of numerous muscle genes, including *Ci-Tbx6a*, *Ci-Tbx6b*, and *Ci-Tbx6c* (Yagi, Satoh, et al., 2004; Yagi, Satou, & Satoh, 2004). Further studies determined that among all the putative Ci-macho1-downstream genes, only *Ci-Tbx6b* and *Ci-Tbx6c* were able to induce ectopic expression of structural muscle genes in non-muscle precursors, and revealed that both genes are mediators of Ci-macho1 function in muscle determination (Yagi, Takatori, Satou, & Satoh, 2005). Gene knock-down experiments carried out using morpholino oligonucleotides directed against both *Ci-Tbx6b* and *Ci-Tbx6c* placed *Ciona MyoD* (aka *Ci-MRF* or *Ci-MDF*; Meedel, Chang, & Yasuo, 2007) downstream of both these transcription factors, showing that its expression is downregulated in B8.7 and B8.8 muscle progenitors in Ci-Tbx6b/c morphant embryos (Imai et al., 2006). By regulating expression of *Ci-MyoD* through Ci-Tbx6b and/or Ci-Tbx6c, Ci-macho1 indirectly controls genes required for muscle differentiation, including the transcription factor genes *Ci-Otp* and *Ci-Mox* (Imai et al., 2006) and several structural genes, such as genes encoding for myosin light and heavy chains, troponin subunits, and *Ciona muscle actin* (Yagi et al., 2005). Noticeably, the activation of *Ci-MyoD* expression requires the cooperation of the zinc-finger protein Ci-ZicL (Imai et al., 2006; Yagi et al., 2005), in addition to the Ci-Tbx6-related transcription factors.

ChIP-chip experiments uncovered additional branches of the macho1/Tbx6-dependent muscle gene regulatory network by identifying genes controlled by Ci-Tbx6b. A genome-wide search for Ci-Tbxb targets was carried out by expressing the GFP-tagged complementary DNA (cDNA) for this gene in *Ciona* embryos, under the control of its own promoter region (Kubo et al., 2010). These experiments identified 864 genes whose loci were occupied by Ci-Tbx6b during early development, 126 of which encode for transcription factors (Kubo et al., 2010). One of the most significant interactions uncovered through this approach is the binding of Ci-Tbx6b to the promoter regions of *Ci-MRF* and *Ci-ZicL* (Kubo et al., 2010). Among the structural muscle genes bound in vivo by Ci-Tbx6b are muscle actins, alpha actinin, myosin heavy chains, sarcoplasmic reticulum calcium ATPase, and several other genes required for muscle differentiation (Kubo et al., 2010), some of which had been previously identified as putative targets of Ci-Tbx6 proteins because of the presence of Ci-Tbx6b/c consensus binding sequence in their upstream regions (Yagi et al., 2005).

The ChIP-chip experiments described earlier also uncovered 155 genes bound by both Ci-Tbx6b and Ci-MRF, at least 65 of which are expressed in muscle cells; these results suggest that these two transcription factors form a feed-forward loop that prolongs the maternal input of Ci-Macho1 to later developmental stages (Kubo et al., 2010). Binding sites for Ci-Tbx6b and Ci-MRF were found in the promoter regions of over 75% of the known muscle genes identified through this approach. In further support of the

Ci-Tbx6/Ci-MRF synergy, binding sites for Ci-Tbx6b/c and Ci-MRF were identified and functionally characterized in the *cis*-regulatory regions of a group of coregulated muscle genes in *C. intestinalis*, and proven to be required, although to a highly variable extent, for the muscle activity of these regions in vivo (Brown, Johnson, & Sidow, 2007). Furthermore, it was found that their respective arrangements and functional relevance were quite variable when different muscle *cis*-regulatory regions from the same species were compared; however, when the evolutionary conservation of the respective *cis*-regulatory architectural requirements was assessed in the corresponding regions of *Ciona savignyi*, it became apparent that these requirements, once established, were highly constrained (Brown et al., 2007).

In addition to cooperating as mediators of Ci-macho1 function, Ci-Tbx6b and Ci-Tbx6c regulate each other. Misexpression experiments unveiled the existence of a cross-regulatory positive feedback between *Ci-Tbx6b* and *Ci-Tbx6c*, as the ectopic expression of each of these genes in notochord cells causes the ectopic expression of the other in this territory (Kugler et al., 2010).

In *H. roretzi* (Stolidobranchia), Hr-macho1 similarly controls the *Tbx6*-related gene(s) (Sawada, Fukushima, & Nishida, 2005). Studies in this species have shown that in addition to promoting muscle cell fate by acting downstream of Hr-macho1, Hr-Tbx6 (named As-T2 in Yasuo et al., 1996) actively suppresses the mesenchymal fate, and is antagonized by FGF (Kumano, Negoro, & Nishida, 2014). Further investigations revealed that the suppression of mesenchymal fate is achieved by Hr-Tbx6 through its ability to control not only cell differentiation but also the number of cell divisions (Kuwajima, Kumano, & Nishida, 2014). This scenario is likely to be more complex, as recent studies suggest that the recently assembled *H. roretzi* genome contains four copies of *Tbx6*-related genes (Stolfi, Sasakura, et al., 2015).

In the ascidians of the genus *Molgula* (order Stolidobranchia), *Tbx6* orthologs are present in two very different species, the tailed species *M. oculata* and the tailless species *M. tectiformis*. In the tailed species, *Tbx6*-related genes are required to activate and/or maintain muscle gene expression. In the tailless species, which lacks the tail and its musculature, *Tbx6*-related genes are expressed in muscle precursors, which suggests that the activators responsible for the expression of muscle genes are still functional (Gyoja, 2006; Takada et al., 2002). The draft sequences of three *Molgula* genomes, *M. oculata* and *M. occidentalis*, both tailed, and *M. occulta*, tailless, revealed the presence of two copies of *Tbx6*-related genes in a head-to-head configuration in each of these genomes (Stolfi et al., 2014). Last, the genome of another solitary ascidian, *Phallusia mammillata*, which, like *Ciona*, belongs to the order Phlebobranchia, also contains three copies of *Tbx6*-related genes (Stolfi, Sasakura, et al., 2015). Together, these data suggest that the ascidian-specific duplication of *Tbx6* genes occurred prior to the separation of the orders Phlebobranchia and Stolidobranchia.

7. Tbx6b AND Tbx6c IN HEART DEVELOPMENT

Similarly to its vertebrate counterparts, ascidian Mesp, a transcription factor of the bHLH family, is required for the specification of the heart precursors and for heart formation (Bondue & Blanpain, 2010; Satou, Imai, & Satoh, 2004). The ascidian heart precursors originate from a subset of larval mesenchyme cells, the trunk ventral cells (TVCs; Fig. 1A), and through asymmetric divisions eventually give rise to the juvenile heart, a V-shaped tube surrounded by a fluid-filled pericardium (Cota, Segade, & Davidson, 2014; Davidson & Christiaen, 2006; Davidson & Levine, 2003; Pope & Rowley, 2002). In *C. intestinalis*, Ci-Mesp controls heart formation by activating expression of evolutionarily conserved components of the heart gene regulatory network, such as *Hand*, *Gata4/5/6*, *NK4*, and *FoxF* (Imai et al., 2006). The identification of Tbx6-binding sites within an enhancer associated with *Ci-Mesp*, and the specific expression of *Ci-Tbx6c* in B7.5 blastomeres (Takatori et al., 2004), which are precursors of both muscle and heart (Satou et al., 2004), originally suggested that Ci-Tbx6c was the only T-domain transcription factor required for expression of *Ci-Mesp*. However, more recent studies indicated that the ectopic expression of *Ci-Tbx6b*, unlike that of *Ci-Tbx6c*, was sufficient to ectopically express this crucial component of the heart-specific gene cascade (Christiaen, Stolfi, Davidson, & Levine, 2009). In light of these findings, it seemed more plausible to assume that both Ci-Tbx6b and Ci-Tbx6c might affect *Ci-Mesp* expression, and to hypothesize that the activation of the *Ci-Mesp* cardiac enhancer required not only the widespread Tbx6-related transcription factors but also a more localized activator (Christiaen et al., 2009). Experimental evidence confirmed this hypothesis and revealed that this additional factor is a homeodomain activator, Ci-Lhx3, a mediator of the beta-catenin pathway (Wada, Hamada, Kobayashi, & Satoh, 2008). Expression of *Ci-Mesp* is therefore restricted to the only blastomere pair in which Ci-Tbx6b/c factors and Ci-Lhx3 are simultaneously expressed (Christiaen et al., 2009).

Interestingly, in *M. occidentalis*, Tbx6-r.b, one of the Tbx6-related genes, also regulates *Mo-Mesp* expression, however, it does not require Mo-Lhx3 and most likely works synergistically with a different transcription factor, which is yet to be identified (Stolfi et al., 2014).

8. Tbx1/10 AND THE EVOLUTIONARY ORIGINS OF THE PHARYNGEAL MUSCLES

Ci-Tbx1/10 is equally related to both the *Tbx1* and the *Tbx10* subfamilies of T-box genes (Takatori et al., 2004; Table 1). Its expression was first reported in mid-tailbud embryos, in a small group of endodermal cells located ventrolaterally to the sensory vesicle, and it expanded slightly in hatched larvae (Takatori et al., 2004) (Fig. 2). It was later reported that *Ci-Tbx1/10* is also expressed in the large lateral secondary TVCs of late tailbuds, and, after these cells divide, in the longitudinal muscle (LoM) precursors of juveniles, at the time when these precursors separate from the precursors of the atrial siphon muscle (ASM; Fig. 2) (Stolfi et al., 2010; Wang, Razy-Krajka, Siu, Ketcham, & Christiaen, 2013). Of note, transcription of *Ci-Tbx1/10* is specifically repressed in the secondary heart precursors by Ci-NK4 and is maintained only in the LoM precursors, which are considered the evolutionary equivalent of vertebrate pharyngeal muscles (Wang et al., 2013). Interestingly, recent findings indicate that in addition to being expressed in the ASM, *Ci-Tbx1/10* is expressed in the oral siphon muscle (OSM, Fig. 1) as well; in both siphons, this transcription factor ultimately induces muscle gene expression by activating Ci-Mrf, although through different modalities (Tolkin & Christiaen, 2016). In parallel, in the secondary heart precursors, Ci-Tbx1/10 inhibits *Ci-GATAa*, a main component of the heart-specific gene regulatory network (Ragkousi, Beh, Sweeney, Starobinska, & Davidson, 2011) and in the ASM precursors it activates expression of *Ci-Collier/OLF/EBF* (*Ci-COE*), which promotes pharyngeal muscle fate (Razy-Krajka et al., 2014), thus inducing these cells to give rise to pharyngeal muscle rather than becoming part of the heart (Wang et al., 2013). Recent studies have determined that the expression of Tbx1/10 in secondary TVCs reported for *Ciona* is conserved in *M. occidentalis* (Stolfi et al., 2014).

Together, these findings suggest a role for *Ci-Tbx1/10* in specification and development of the siphon musculature that resembles the function of human TBX1 in pharyngeal development. In fact, in humans, mutations in the *TBX1* locus cause DiGeorge/velocardiofacial syndrome, which is characterized by faulty development of the pharyngeal pouches, in addition to a variety of other phenotypic manifestations (Choe & Crump, 2014).

9. Tbx15/18/22 (VegTR) AND THE ESTABLISHMENT OF MUSCLE CELL IDENTITY

Ciona VegT-Related (*Ci-VegTR*) was identified as a maternally expressed gene whose transcripts are localized to the vegetal cytoplasm of zygotes and are later segregated to the muscle lineage precursors deriving from the B4.1 blastomeres (Erives & Levine, 2000; Table 1). The gene was later renamed *Ci-Tbx15/18/22* on the basis of its phylogenetic relationship with members of these subfamilies, and it was shown to be expressed at low levels also in muscle cells from neurula to late-tailbud stage (Takatori et al., 2004) (Fig. 2). An in vitro-synthesized GST-fusion protein containing the T-domain of Ci-VegTR was shown to bind the TCCCAC and TGGCAC core sequences found in the *Ci-snail* muscle enhancer, which are required for its muscle activity (Erives & Levine, 2000). *Ci-snail* encodes an evolutionarily conserved zinc-finger transcriptional repressor expressed in muscle cells, in the lateral ependymal cells of the larval nerve cord, and in a group of cells of the sensory vesicle, the descendants of blastomere pair A9.30, which give rise to cholinergic neurons of the visceral ganglion, the motor center of the larva (Erives, Corbo, & Levine, 1998; Stolfi & Levine, 2011). In addition to contributing to the regionalization of nerve cord (Di Gregorio, Corbo, & Levine, 2001) and sensory vesicle (Imai, Stolfi, Levine, & Satou, 2009), transcriptional repression by Ci-Snail is required to prevent expression of *Ci-Bra* in muscle cells, and to establish a crucial developmental boundary between notochord and muscle cells (Corbo et al., 1997; Fujiwara, Corbo, & Levine, 1998) (Fig. 4A). The presumed Ci-VegTR-dependent early activation of *Ci-snail* expression in muscle precursors would suggest that Ci-VegTR might contribute to the early establishment of muscle identity by indirectly inducing the repression of *Ci-Bra* expression in these cells. Another hypothesis on the function of Ci-Tbx15/18/22 stems from the reported functions of vertebrate Tbx15 and Tbx18, which in mouse act as Groucho-dependent transcriptional corepressors (Farin et al., 2007). In particular, Tbx18 has been reported to downregulate the Tbx6-dependent activation of Delta-like 1 expression in somitic mesoderm (Farin et al., 2007). Interestingly, the expression of all three *Tbx6* genes in muscle precursors is abruptly downregulated approximately at the time when zygotic transcripts for *Ci-VegTR* become detectable

(Takatori et al., 2004). This suggests the possibility that Ci-VegTR might also act as a transcriptional repressor during later developmental stages, and be responsible for the rapid downregulation of the *Ci-Tbx6* genes in muscle cells.

In amphioxus, *Tbx15/18/22* is expressed in the developing somites, which are located bilaterally to the notochord (Beaster-Jones, Horton, Gibson-Brown, Holland, & Holland, 2006). Although the specific role of *AmphiTbx15/18/22* is still undetermined, it has been suggested on the basis of its expression pattern that the ancestral function of *Tbx15/18/22* in the last common chordate ancestor was to pattern somites (Beaster-Jones et al., 2006). In support of this hypothesis, zebrafish *tbx18* is expressed in presomitic mesoderm and newly developed somites (Begemann, Gibert, Meyer, & Ingham, 2002); similarly, in mouse, *Tbx18* is expressed in the anterior compartments of developing somites (Kraus, Haenig, & Kispert, 2001), while *Tbx22* shows a complementary pattern in the posterior compartments (Bush, Lan, Maltby, & Jiang, 2002). Hence, the ancestral function of *Tbx18* and *Tbx22* would have been maintained in amphioxus, which develops distinguishable somites and a proper paraxial mesoderm, but was presumably lost in ascidians, which display unsegmented, nonsyncitial muscle in the embryonic territory that is occupied by somites in amphioxus. In turn, this would imply that the loss of segmented paraxial mesoderm is a derived feature of ascidians.

10. MATERNAL *T* AND TAIL DEVELOPMENT

Ci-mT is a member of the Tbr/Eomesodermin/T-bet/Tbx21 subfamily (Takatori et al., 2004; Table 1). It is strongly expressed in zygotes and remains ubiquitously expressed in early embryos; at the 110-cell stage, the expression of this gene, while still ubiquitous, intensifies in a pair of vegetal blastomeres, reportedly the A7.6 pair (Takatori et al., 2004). These blastomeres are the precursors of the TLCs, traditionally known as the blood cell precursors (Kawaminani & Nishida, 1997). According to more recent studies, in hatched larvae TLCs derivatives also give rise to a population of migrating mesenchymal cells that will become part of the tunic of the juvenile (Tokuoka, Imai, Satou, & Satoh, 2004). In addition, studies in *C. savignyi* indicate that TLCs originate components of the ciliary epithelium of the gill slits in the branchial basket of the juvenile, as well as blood cells, tunic cells, tunic granules and a subpopulation of cells that will become part of the gut, while in *H. roretzi* they reportedly give rise to part of the longitudinal mantle muscle (Bishop, Hall, & Bates, 2010 and references therein). In both species, TLC descendants form the OSM (Hirano & Nishida, 1997; Wang et al., 2013; Fig. 1). More recently, TLCs have been indicated as the evolutionary precursors of neural crest cells (Jeffery et al., 2008), although other studies suggest otherwise (Abitua et al., 2015). Around gastrulation, the hybridization signal for *Ci-mT* becomes undetectable (Imai et al., 2004). In situ expression data are not available for this gene in juveniles, however expressed sequence tag (EST) counts from a cDNA library derived from *Ciona* heart cells detected one *Ci-mT* cDNA clone out of 13,243 (0.0076% representation) (Satou, Kawashima, Shoguchi, Nakayama, & Satoh, 2005).

Ci-mT was the subject of pioneering gene inactivation experiments, which introduced the transposon-mediated knockdown of maternally expressed genes in ascidians (Iitsuka et al., 2014). This technique allowed the reduction of *Ci-mT* mRNA to 4.1% of the quantity detected in wild-type eggs, and caused severe defects in tail development in embryos derived from *Ci-mT*-depleted eggs (Iitsuka et al., 2014). This phenotype, which is characterized by the presence of bents in the tail, could not be rescued by the injection of in vitro transcribed *Ci-mT* mRNA; this finding has been interpreted as a requirement of the maternally stored Ci-mT transcription factor, synthesized during oogenesis, for the proper completion of tail morphogenesis (Iitsuka et al., 2014).

The sequence of a putative ortholog of *Ci-mT* from *H. roretzi*, *As-mT*, has been deposited in GenBank (accession number: AB013619.1; Takada et al., unpublished) but its function is still unknown. No clear orthologs of *Ci-mT* could be readily identified in vertebrates.

11. Tbx20: IN OR OUT OF THE ASCIDIAN HEART?

The expression of *Ci-Tbx20* (Table 1) was unsuccessfully investigated through extensive WMISH experiments on embryos at various developmental stages preceding metamorphosis (Takatori et al., 2004). EST counts indicated the presence of a low-abundance transcript (0.0034% representation) in a cDNA library prepared from blood cells (Ghost database; Satou et al., 2005), suggesting the hypothesis that this gene might be expressed in at least one of the several types of blood cells found in the adult ascidian (Ogasawara et al., 2006; Rowley, 1982; Terajima et al., 2003). In vertebrate and in *Drosophila* embryos, transcription factors of the Tbx20 family are required for heart and dorsal vessel development, respectively, as well as for vasculogenesis (Cai et al., 2013; Miskolczi-McCallum, Scavetta, Svendsen, Soanes, & Brook, 2005). However, no transcripts have been detected in a cDNA library prepared from the *Ciona* heart (Ghost database; Satou et al., 2005) and expression of *Ci-Tbx20* in the developing ascidian heart and its precursors is yet to be reported. These preliminary conclusions appear surprising, given the widespread conservation of *Tbx20* orthologs expression in several invertebrates, vertebrates, and in amphioxus, where *Tbx20* is expressed in the presumed precursors of the myocardium (Belgacem, Escande, Escriva, & Bertrand, 2011).

Nevertheless, subsequent studies showed that *Ci-Tbx20* is expressed in the motor neurons of the NC (Fig. 1), the central nervous system of the postmetamorphic ascidian, along with the homeobox genes *Ci-Phox2* and *CiHox1* (Dufour et al., 2006). These neurons innervate the muscle of the branchial basket, the filtering apparatus that occupies most of the adult body and is responsible for the capture of food particles contained in the seawater, through the mucus secreted by the endostyle (Holley, 1986) (Fig. 1). In contrast, expression of *Ci-Tbx20* is not detected in the motor neurons of the larval (premetamorphic) visceral ganglion, which are responsible for the swimming movements of the hatched larvae (Stolfi & Levine, 2011), consistent with the finding that expression of vertebrate *Tbx20* orthologs is largely confined to fully differentiated neurons (Kraus et al., 2001). These findings suggest that the motor neurons of the ascidian postmetamorphic NC are homologous to the cranial motor neurons of vertebrates, and directly relate this ascidian structure to the vertebrate hindbrain (Dufour et al., 2006).

12. TRANSCRIPTIONAL REGULATION OF T-BOX GENES EXPRESSION

Over the past two decades, the fortunate combination of experimental advantages offered by *Ciona* has rendered this simple chordate a prime model for the identification and characterization of CRMs, and of the molecular mechanisms controlling their function. In particular, the *cis*-regulatory sequences associated with T-box genes exemplify alternative strategies of control of gene expression found in this organism. Currently, detailed information is available on *cis*-regulatory regions associated with *C. intestinalis Brachyury*, *Tbx2/3*, *Tbx6b*, and *Tbx6c*, and with *H. roretzi As-T* and *As-T2*.

12.1 Brachyury: A Balancing Act Between Multiple Activators and a Repressor

The first *cis*-regulatory region to be identified and validated in vivo in *Ciona* was a ~3.5-kb genomic fragment encompassing the *Ci-Bra* promoter and a notochord enhancer directly upstream of it (Corbo et al., 1997). Serial 5′-truncations and site-directed mutations of this 3.5-kb region were tested through electroporation into zygotes, and a minimal 557-bp notochord enhancer, including a 5′-located 434-bp region, the putative TATA box, and the first 17 codons of the protein-coding region, was identified. The minimal notochord enhancer contained, in particular, putative binding sites for the transcription factor Suppressor of Hairless, Su(H), and 3 E-boxes (binding sites for bHLH transcription factors) (Corbo et al., 1997). Subsequent studies clarified that the Su(H) binding sites were required for notochord activity of this enhancer, particularly in the secondary notochord, the posterior-most part of the notochord that derives from B-line blastomeres and includes only eight cells (Corbo, Fujiwara, Levine, & Di Gregorio, 1998; Yagi, Satoh, et al., 2004; Yagi, Satou, et al., 2004). It was also shown that the acquisition of secondary notochord cell fate was induced in the B-line blastomeres first by Nodal signaling, and later by the activation of the Notch/Su(H) pathway by means of A-line derived Delta2 (Hudson & Yasuo, 2006) (Fig. 4A). This seems a noteworthy similarity with the transcriptional regulation of the expression of *no tail*, one of the two zebrafish *Brachyury* genes, which is controlled by two separate enhancers, one responsive to Nodal signaling, the other to Wnt and BMP signaling (Harvey, Tümpel, Dubrulle, Schier, & Smith, 2010).

Differently from what is seen in the secondary notochord, Ci-Bra expression in the primary notochord, consisting in the anterior-most 32 cells (A-line), requires the zinc-finger transcription factor ZicL, whose expression in notochord precursors is activated, in turn, by another transcription factor, FoxD (Yagi, Satoh, et al., 2004; Yagi, Satou, et al., 2004). This latter research also elucidated that activation of *Ci-Bra* in notochord precursors requires ab initio maternally provided beta-catenin, which is necessary for the specification of the endodermal precursors adjacent to the notochord precursors (Yagi, Satoh, et al., 2004; Yagi, Satou, et al., 2004). Of note, the mutation of binding sites for the Ci-Snail transcriptional repressor within the minimal *Ci-Bra* notochord enhancer causes its ectopic expression in muscle (Fujiwara et al., 1998) (Fig. 4A).

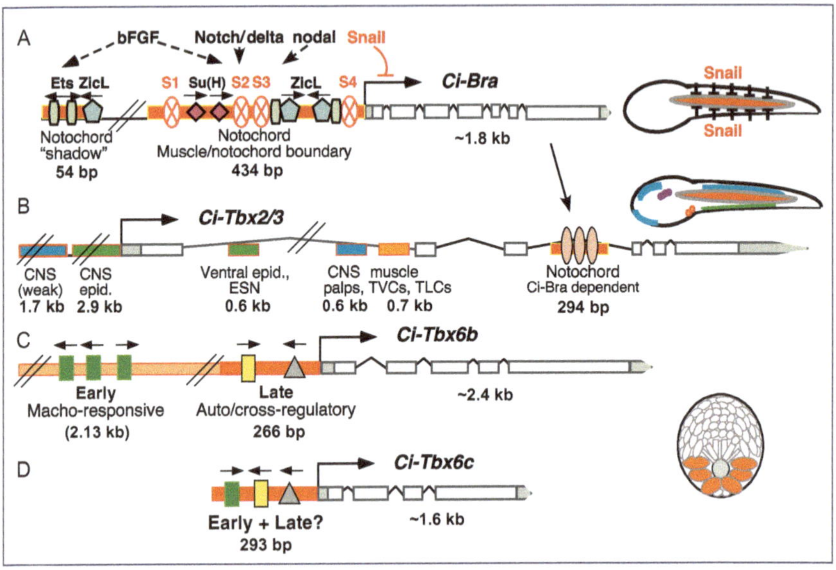

Fig. 4 Organization, regulatory inputs, and activity of *cis*-regulatory modules associated with *Ciona* T-box genes. (A) Schematic representation of the *Ci-Bra* genomic locus, with its proximal 434-bp notochord enhancer adjacent to the transcription start site (*black arrow*) (Corbo et al., 1997; Yagi, Satoh, & Satou, 2004; Yagi, Satou, & Satoh, 2004) and the distal minimal 54-bp "shadow" notochord enhancer (Farley, Olson, Zhang, Rokhsar, & Levine, 2016). For simplicity, the E-boxes in the 434-bp enhancer are not shown. Symbols: *vertical light green bars*, Ets-binding sites; *light blue pentagons*, ZicL-binding sites; *purple diamonds*, Suppressor of Hairless, Su(H), binding sites. The binding sites for the Ci-Snail repressor, which is expressed in muscle cells, are shown as *crossed red ovals* (S1–S4). The schematic of the embryo on the right side indicates that the presence of Ci-Snail prevents expression of *Ci-Bra* in muscle cells. The pathways that influence these enhancers are listed, and are connected to their target enhancers by *black dashed arrows*. (B) Schematic representation of the *Ci-Tbx2/3* genomic locus and the multiple CRMs that it contains. Some of the CRMs are partly redundant and direct expression in partially overlapping territories (José-Edwards et al., 2013). The notochord CRM is depicted as a *red rectangle* containing three Ci-Bra-binding sites required for notochord activity (*pink ovals*). (C) Schematic representation of the *Ci-Tbx6b* genomic locus. This locus contains an early, distally located Macho1-responsive region spanning 2.13 kb. Proximal to the promoter region is a late-onset muscle CRM that contains Tbx6-binding sites and can presumably mediate both the autoregulation by Ci-Tbx6b itself and the cross-regulation of this gene by Ci-Tbx6c and/or other Tbx6-related transcription factors. *Green vertical bars* indicate Macho1-binding sites; the *yellow vertical bar* represents a Tbx6-binding site required for muscle activity and the *gray triangle* indicates an E-box sequence whose contribution to muscle activity appears dispensable (Kugler et al., 2010). (D) Schematic representation of the *Ci-Tbx6c* locus and the muscle CRM located proximal to the transcription start site. The *Ci-Tbx6c* muscle CRM contains a Macho1-binding site, a Tbx6-binding site, and an E-box sequence related to the sites found in the *Ci-Tbx6b* early and late muscle elements. These sequences guided the identification of this muscle CRM, and their respective roles in its activity remain to be determined (*question mark*). Symbols are as in (C). CRMs and coding regions are not drawn to scale. *Oblique lines* indicate parts of the genomic sequences that have been omitted due to space constraints.

In addition to this main, well-characterized notochord enhancer located directly upstream of the putative TATA box, the *Ci-Bra* genomic locus had also been shown to contain another notochord enhancer, located further upstream of the transcription start site (Dr. Shigeki Fujiwara, personal communication). Within this second enhancer, a minimal 54-bp fragment active in the notochord was recently identified and shown to contain two binding sites for Ets transcription factor(s) as well as by a putative ZicL-binding site; reversal of the orientation of one of the Ets-binding sites resulted in a reduction of the notochord activity (Farley et al., 2016) (Fig. 4A). Putative Ets-binding sites are also found in the proximal *Ci-Bra* notochord enhancer, and potentially they might be providing an additional site of input for the bFGF signaling pathway (Fig. 4A).

The regulation of *Brachyury* expression has also been analyzed in another ascidian species, *H. roretzi*. A 900-bp region upstream of the *H. roretzi Brachyury* gene, *Hr-Bra* (aka *As-T*) was tested in vivo and was found to require direct activation by Ets and Zic-binding sites for its notochord activity (Matsumoto, Kumano, & Nishida, 2007), consistent with previous reports of the ability of bFGF to induce expression of *As-T* (Nakatani, Yasuo, Satoh, & Nishida, 1996).

12.2 Tbx2/3: A Composite Expression Pattern Orchestrated by Additive, Partly Redundant CRMs

The broad expression pattern of *Ci-Tbx2/3*, described earlier (Fig. 2), is the composite output of a set of CRMs scattered throughout its ~15-kb long genomic locus (Fig. 4B). These CRMs were identified through the systematic dissection and in vivo analysis of genomic fragments covering most of the *Ci-Tbx2/3* coding region, along with 4.6 kb upstream of it (José-Edwards et al., 2013). This analysis revealed the presence of a 1.7-kb CRM weakly active in sensory vesicle and mesenchyme located ~4 kb upstream of the transcription start site, a 2.9-kb CRM directly upstream of the transcription start site, active in sensory vesicle, muscle and trunk epidermis, and four intronic CRMs, three of which found in the 9-kb first intron of *Ci-Tbx2/3* (José-Edwards et al., 2013) (Fig. 4B). Among them, a 0.6-kb CRM able to direct reporter gene expression in sensory vesicle and strongly active in the ventral midline epidermis and ESNs, which could inform future studies aimed at the identification of the direct activators of *Ci-Tbx2/3* expression in this BMP-dependent lineage (Waki et al., 2015). Two additional, partially redundant CRMs active in TVCs, anterior sensory vesicle and presumptive palps, were isolated from the first intron as well, and a 1.9-kb notochord CRM was found by testing a genomic fragment straddling the second and third intron. This notochord CRM was reduced to a 294-bp minimal region, which turned out to be dependent upon cooperative Ci-Bra-binding sites for its activity (José-Edwards et al., 2013).

12.3 Tbx6b and Tbx6c: Early and Late Elements, Autoregulation, and Crossregulation

The *cis*-regulatory region upstream of *Ci-Tbx6b* provided a first example of two distinct "early" and "late" CRMs associated with the same muscle gene in *Ciona* (Kugler et al., 2010). In fact, the 2.4-kb *Ci-Tbx6b* upstream region contains three Macho1-binding sites and is active by the 32-cell stage in muscle precursors. When the Macho1-binding sites are deleted or mutated, a delayed activation of the *cis*-regulatory region is observed, whereby reporter gene transcription is not observed before the early neurula stage. In addition to this early, Macho1-dependent region, another muscle CRM, spanning 266 bp upstream of the transcription start site, was found to possess late-onset muscle activity, starting from the early neurula stage (Kugler et al., 2010) (Fig. 4C). This muscle *cis*-regulatory element contains a putative, dispensable CREB-binding site, a dispensable E-box, and a T-box-binding site, which is effectively bound in vitro by both Ci-Tbx6b and Ci-Tbx6c proteins. Hence, this T-box-binding site might be mediating both the reported autoregulation of Ci-Tbx6b and the cross-regulatory interaction with Ci-Tbx6c.

A 293-bp muscle CRM was identified upstream of *Ci-Tbx6c* by searching for the binding sites identified in the *Ci-Tbx6b* muscle CRM; differently from the *Ci-Tbx6b* muscle CRM, this *Ci-Tbx6c* muscle CRM contains a single Macho1-binding site located upstream of a T-box-binding site and an E-box. Hence, the presumed binding sites for early and late activators are directly juxtaposed within a compact element (Kugler et al., 2010) (Fig. 4D).

In *Halocynthia*, the genomic locus of the *Tbx6*-related gene *As-T2* contains a ~410-bp distal CRM that is likely responsible for the expression in epidermal cells at the tip of the tail, and a ~270-bp proximal CRM that recapitulates the expression of *As-T2* in muscle cells (Mitani et al., 2001). Neighboring the *As-T2* muscle CRM there are bona fide autoregulatory T-box-binding sites, with the conserved sequence TTCACACTT, which is also found in the *muscle actin 4* (*HrMA4*) and *myosin heavy chain* (*HrMHC*) promoter regions, and is bound by As-T2 in vitro (Mitani et al., 2001).

13. CONCLUDING REMARKS

In summary, in chordates Brachyury appears to have maintained an evolutionarily conserved, constrained role in notochord formation, which is consistent with its function in the axial mesoderm of vertebrates, and its expression in the ascidian notochord is safeguarded by two separable notochord CRMs. In *Ciona*, Brachyury is required for the notochord expression of another T-box transcription factor, Ci-Tbx2/3, which in turn controls several notochord genes. Studies in zebrafish suggest that *Tbx2*-related genes, such as zebrafish *Tbx-c*, are expressed in the vertebrate notochord, similarly to *Ci-Tbx2/3* (Dheen et al., 1999), although their relationship with Brachyury in this and other vertebrates remains to be determined. Similarly to their vertebrate counterparts, ascidian *Tbx6*-related genes are expressed in the mesoderm flanking the notochord. At least two of these genes, Tbx6b and Tbx6c, have been recruited by the ascidian-specific maternal muscle determinant Macho1 to ensure the completion of the late steps of the muscle differentiation program by ensuring activation and/or maintenance of myosin and actin genes, and by suppressing the mesenchymal fate in the common precursors of muscle and mesenchyme. On the other hand, the function of Tbx6-related transcription factors in the activation of *Mesp* expression and the consequent specification of heart precursors is preserved in ascidians and in other organisms, as is the role of *Ciona* Tbx1/10 in pharyngeal mesoderm specification and development, which closely resembles that of vertebrate members of the Tbx1 subfamily. Tbx15/18/22 (VegTR) is expressed in the mesoderm flanking the notochord, similarly to its counterpart in amphioxus; however, the absence of paraxial segmentation and somites in ascidians renders the determination of functional homology between *Tbx15/18/22* and its amphioxus and vertebrate orthologs particularly challenging.

In addition to the evolutionarily conserved members of the T-box family, at least two ascidian genomes contain a maternally expressed T-box gene, *maternal-T*, which despite being related to members of the *Tbr/Eomesodermin/T-bet/Tbx21* subfamily, appears to lack identifiable orthologs in other animals and therefore could be specific to ascidians, where it is required for proper tail extension. The presumed lack of *Tbx20* expression in heart precursors and cardiac tissues in the tunicate lineage would represent a surprising derived loss of a function that has otherwise been observed in a broad range of chordates and arthropods. To address these important points, future studies should be aimed at elucidating the post-metamorphic expression patterns of ascidian T-box genes that are still either fragmentary or unknown, as well as the expression of these crucial transcription factors in other tunicates.

ACKNOWLEDGMENTS

Thanks to all present and past lab members and collaborators, and in particular to Dr. Shigeki Fujiwara (Kochi University, Japan) for sharing his unpublished results, and to Drs. Alberto Stolfi and Lionel Christiaen (New York University, USA) for comments on the manuscript. Research in the lab is supported by Grant R01GM100466 from the National Institute of General Medical Sciences of the National Institutes of Health.

REFERENCES

Abitua, P. B., Gainous, T. B., Kaczmarczyk, A. N., Winchell, C. J., Hudson, C., Kamata, K., et al. (2015). The pre-vertebrate origins of neurogenic placodes. *Nature*, *524*, 462–465.

Aniello, F., Locascio, A., Villani, M. G., Di Gregorio, A., Fucci, L., & Branno, M. (1999). Identification and developmental expression of *Ci-msh-b*: A novel homologue of *Drosophila msh* gene in *Ciona intestinalis*. *Mechanisms of Development*, *88*, 123–126.

Beaster-Jones, L., Horton, A. C., Gibson-Brown, J. J., Holland, N. D., & Holland, L. Z. (2006). The amphioxus gene, *AmphiTbx15/18/22*, illuminates the origins of chordate segmentation. *Evolution & Development*, *8*, 119–129.

Begemann, G., Gibert, Y., Meyer, A., & Ingham, P. W. (2002). Cloning of zebrafish T-box genes *tbx15* and *tbx18* and their expression during embryonic development. *Mechanisms of Development*, *114*, 137–141.

Belgacem, M. R., Escande, M. L., Escriva, H., & Bertrand, S. (2011). Amphioxus Tbx6/16 and Tbx20 embryonic expression patterns reveal ancestral functions in chordates. *Gene Expression Patterns*, *11*, 239–243.

Berná, L., & Alvarez-Valin, F. (2014). Evolutionary genomics of fast evolving tunicates. *Genome Biology and Evolution*, *6*, 1724–1738.

Bishop, C. D., Hall, B. K., & Bates, W. R. (2010). HSP90 expression in two migratory cell types during ascidian development: Test cells deposit HSP90 on the larval tunic. *International Journal of Developmental Biology*, *54*, 1337–1346.

Bondue, A., & Blanpain, C. (2010). Mesp1: A key regulator of cardiovascular lineage commitment. *Circulation Research*, *107*, 1414–1427.

Brown, C. D., Johnson, D. S., & Sidow, A. (2007). Functional architecture and evolution of transcriptional elements that drive gene coexpression. *Science*, *317*, 1557–1560.

Brozovic, M., Martin, C., Dantec, C., Dauga, D., Mendez, M., Simion, P., et al. (2016). ANISEED 2015: A digital framework for the comparative developmental biology of ascidians. *Nucleic Acids Research*, *44*(D1), D808–D818.

Bush, J. O., Lan, Y., Maltby, K. M., & Jiang, R. (2002). Isolation and developmental expression analysis of *Tbx22*, the mouse homolog of the human X-linked cleft palate gene. *Developmental Dynamics*, *225*, 322–326.

Cai, X., Zhang, W., Hu, J., Zhang, L., Sultana, N., Wu, B., et al. (2013). Tbx20 acts upstream of Wnt signaling to regulate endocardial cushion formation and valve remodeling during mouse cardiogenesis. *Development*, *140*, 3176–3187.

Chapman, D. L., & Papaioannou, V. E. (1998). Three neural tubes in mouse embryos with mutations in the T-box gene Tbx6. *Nature*, *391*, 695–697.

Chiba, S., Jiang, D., Satoh, N., & Smith, W. C. (2009). *Brachyury* null mutant-induced defects in juvenile ascidian endodermal organs. *Development, 136*, 35–39.
Choe, C. P., & Crump, J. G. (2014). Tbx1 controls the morphogenesis of pharyngeal pouch epithelia through mesodermal Wnt11r and Fgf8a. *Development, 141*, 3583–3593.
Christiaen, L., Stolfi, A., Davidson, B., & Levine, M. (2009). Spatio-temporal intersection of Lhx3 and Tbx6 defines the cardiac field through synergistic activation of *Mesp*. *Developmental Biology, 328*, 552–560.
Conklin, E. (1905). The organization and cell lineage of the ascidian egg. *Journal of the Academy of Natural Sciences of Philadelphia, 13*, 1–119.
Corbo, J. C., Fujiwara, S., Levine, M., & Di Gregorio, A. (1998). Suppressor of hairless activates *Brachyury* expression in the *Ciona* embryo. *Developmental Biology, 203*, 358–368.
Corbo, J. C., Levine, M., & Zeller, R. W. (1997). Characterization of a notochord-specific enhancer from the *Brachyury* promoter region of the ascidian, *Ciona intestinalis*. *Development, 124*, 589–602.
Cota, C. D., Segade, F., & Davidson, B. (2014). Heart genetics in a small package, exploiting the condensed genome of *Ciona intestinalis*. *Briefings in Functional Genomics, 13*, 3–14.
Davidson, B., & Christiaen, L. (2006). Linking chordate gene networks to cellular behavior in ascidians. *Cell, 124*, 247–250.
Davidson, B., & Levine, M. (2003). Evolutionary origins of the vertebrate heart: Specification of the cardiac lineage in *Ciona intestinalis*. *Proceedings of the National Academy of Sciences of the United States of America, 100*, 11469–11473.
Dehal, P., Satou, Y., Campbell, R. K., Chapman, J., Degnan, B., De Tomaso, A., et al. (2002). The draft genome of *Ciona intestinalis*: Insights into chordate and vertebrate origins. *Science, 298*, 2157–2167.
Delsuc, F., Brinkmann, H., Chourrout, D., & Philippe, H. (2006). Tunicates and not cephalochordates are the closest living relatives of vertebrates. *Nature, 439*, 965–968.
Denker, E., & Jiang, D. (2012). *Ciona intestinalis* notochord as a new model to investigate the cellular and molecular mechanisms of tubulogenesis. *Seminars in Cell & Developmental Biology, 23*, 308–319.
Dheen, T., Sleptsova-Friedrich, I., Xu, Y., Clark, M., Lehrach, H., Gong, Z., et al. (1999). Zebrafish tbx-c functions during formation of midline structures. *Development, 126*, 2703–2713.
Di Gregorio, A., Corbo, J. C., & Levine, M. (2001). The regulation of *forkhead/HNF-3beta* expression in the *Ciona* embryo. *Developmental Biology, 229*, 31–43.
Di Gregorio, A., & Levine, M. (1999). Regulation of *Ci-tropomyosin-like*, a brachyury target gene in the ascidian, *Ciona intestinalis*. *Development, 126*, 5599–5609.
Di Gregorio, A., & Levine, M. (2002). Analyzing gene regulation in ascidian embryos: New tools for new perspectives. *Differentiation, 70*, 132–139.
Di Gregorio, A., Spagnuolo, A., Ristoratore, F., Pischetola, M., Aniello, F., Branno, M., et al. (1995). Cloning of ascidian homeobox genes provides evidence for a primordial chordate cluster. *Gene, 156*, 253–257.
Dobrovolskaïa-Zavadskaïa, N. (1927). Sur la mortification spontanée de la queue chez la souris nouveau-nee et sur l'existence d'un caractere hereditaire "non-viable". *Comptes Rendus des Séances de la Société de Biologie, Paris, 97*, 114–116.
Dufour, H. D., Chettouh, Z., Deyts, C., de Rosa, R., Goridis, C., Joly, J. S., et al. (2006). Precraniate origin of cranial motoneurons. *Proceedings of the National Academy of Sciences of the United States of America, 103*, 8727–8732.
Dunn, M. P., & Di Gregorio, A. (2009). The evolutionarily conserved *leprecan* gene: Its regulation by Brachyury and its role in the developing *Ciona* notochord. *Developmental Biology, 328*, 561–574.

Erives, A., Corbo, J. C., & Levine, M. (1998). Lineage-specific regulation of the *Ciona snail* gene in the embryonic mesoderm and neuroectoderm. *Developmental Biology, 194*, 213–225.

Erives, A., & Levine, M. (2000). Characterization of a maternal T-box gene in *Ciona intestinalis*. *Developmental Biology, 225*, 169–178.

Evans, A. L., Faial, T., Gilchrist, M. J., Down, T., Vallier, L., Pedersen, R. A., et al. (2012). Genomic targets of brachyury (T) in differentiating mouse embryonic stem cells. *PloS One*, 7(3), e33346.

Farin, H. F., Bussen, M., Schmidt, M. K., Singh, M. K., Schuster-Gossler, K., & Kispert, A. (2007). Transcriptional repression by the T-box proteins Tbx18 and Tbx15 depends on Groucho corepressors. *The Journal of Biological Chemistry, 282*, 25748–25759.

Farley, E. K., Olson, K. M., Zhang, W., Rokhsar, D. S., & Levine, M. S. (2016). Syntax compensates for poor binding sites to encode tissue specificity of developmental enhancers. *Proceedings of the National Academy of Sciences of the United States of America, 113*, 6508–6513.

Fisher, W. W., Li, J. J., Hammonds, A. S., Brown, J. B., Pfeiffer, B. D., Weiszmann, R., et al. (2012). DNA regions bound at low occupancy by transcription factors do not drive patterned reporter gene expression in *Drosophila*. *Proceedings of the National Academy of Sciences of the United States of America, 109*, 21330–21335.

Fujiwara, S., Corbo, J. C., & Levine, M. (1998). The snail repressor establishes a muscle/notochord boundary in the *Ciona* embryo. *Development, 125*, 2511–2520.

Garcia-Fernàndez, J., & Holland, P. W. (1994). Archetypal organization of the amphioxus *Hox* gene cluster. *Nature, 370*, 563–566.

Gentsch, G. E., Owens, N. D., Martin, S. R., Piccinelli, P., Faial, T., Trotter, M. W., et al. (2013). In vivo T-box transcription factor profiling reveals joint regulation of embryonic neuromesodermal bipotency. *Cell Reports, 4*, 1185–1196.

Gluecksohn-Schoenheimer, S. (1940). The effect of an early lethal (t) in the house mouse. *Genetics, 25*, 391–400.

Gyoja, F. (2006). Expression of a muscle determinant gene, macho-1, in the anural ascidian *Molgula tectiformis*. *Development Genes and Evolution, 216*, 285–289.

Harvey, S. A., Tümpel, S., Dubrulle, J., Schier, A. F., & Smith, J. C. (2010). *No tail* integrates two modes of mesoderm induction. *Development, 137*, 1127–1135.

Herrmann, B. G., Labeit, S., Poustka, A., King, T. R., & Lehrach, H. (1990). Cloning of the *T* gene required in mesoderm formation in the mouse. *Nature, 343*, 617–622.

Hirano, T., & Nishida, H. (1997). Developmental fates of larval tissues after metamorphosis in ascidian *Halocynthia roretzi*. I. Origin of mesodermal tissues of the juvenile. *Developmental Biology, 192*, 199–210.

Holland, P. W., Koschorz, B., Holland, L. Z., & Herrmann, B. G. (1995). Conservation of brachyury (T) genes in amphioxus and vertebrates: Developmental and evolutionary implications. *Development, 121*, 4283–4291.

Holley, M. C. (1986). Cell shape, spatial patterns of cilia, and mucus-net construction in the ascidian endostyle. *Tissue and Cell, 18*, 667–684.

Holstien, K., Rivera, A., Windsor, P., Ding, S., Leys, S. P., Hill, M., et al. (2010). Expansion, diversification, and expression of T-box family genes in Porifera. *Development Genes and Evolution, 220*, 251–262.

Horie, T., Shinki, R., Ogura, Y., Kusakabe, T. G., Satoh, N., & Sasakura, Y. (2011). Ependymal cells of chordate larvae are stem-like cells that form the adult nervous system. *Nature, 469*, 525–528.

Hotta, K., Mitsuhara, K., Takahashi, H., Inaba, K., Oka, K., Gojobori, T., et al. (2007). A web-based interactive developmental table for the ascidian *Ciona intestinalis*, including 3D real-image embryo reconstructions: I. From fertilized egg to hatching larva. *Developmental Dynamics, 236*, 1790–1805.

Hotta, K., Takahashi, H., Asakura, T., Saitoh, B., Takatori, N., Satou, Y., et al. (2000). Characterization of brachyury-downstream notochord genes in the *Ciona intestinalis* embryo. *Developmental Biology, 224*, 69–80.

Hotta, K., Takahashi, H., Erives, A., Levine, M., & Satoh, N. (1999). Temporal expression patterns of 39 brachyury-downstream genes associated with notochord formation in the *Ciona intestinalis* embryo. *Development, Growth & Differentiation, 41*, 657–664.

Hotta, K., Takahashi, H., Satoh, N., & Gojobori, T. (2008). Brachyury-downstream gene sets in a chordate, *Ciona intestinalis*: Integrating notochord specification, morphogenesis and chordate evolution. *Evolution & Development, 10*, 37–51.

Hudson, C., & Yasuo, H. (2006). A signalling relay involving nodal and delta ligands acts during secondary notochord induction in *Ciona* embryos. *Development, 133*, 2855–2864.

Iitsuka, T., Mita, K., Hozumi, A., Hamada, M., Satoh, N., & Sasakura, Y. (2014). Transposon-mediated targeted and specific knockdown of maternally expressed transcripts in the ascidian *Ciona intestinalis*. *Scientific Reports, 4*, 5050.

Ikuta, T., Yoshida, N., Satoh, N., & Saiga, H. (2004). *Ciona intestinalis Hox* gene cluster: Its dispersed structure and residual colinear expression in development. *Proceedings of the National Academy of Sciences of the United States of America, 101*, 15118–15123.

Imai, K. S., Hino, K., Yagi, K., Satoh, N., & Satou, Y. (2004). Gene expression profiles of transcription factors and signaling molecules in the ascidian embryo: Towards a comprehensive understanding of gene networks. *Development, 131*, 4047–4058.

Imai, K. S., Levine, M., Satoh, N., & Satou, Y. (2006). Regulatory blueprint for a chordate embryo. *Science, 312*, 1183–1187.

Imai, K. S., Stolfi, A., Levine, M., & Satou, Y. (2009). Gene regulatory networks underlying the compartmentalization of the *Ciona* central nervous system. *Development, 136*, 285–293.

Irvine, S. Q. (2013). Study of *Cis*-regulatory elements in the ascidian *Ciona intestinalis*. *Current Genomics, 14*, 56–67.

Jeffery, W. R., Chiba, T., Krajka, F. R., Deyts, C., Satoh, N., & Joly, J. S. (2008). Trunk lateral cells are neural crest-like cells in the ascidian *Ciona intestinalis*: Insights into the ancestry and evolution of the neural crest. *Developmental Biology, 324*, 152–160.

José-Edwards, D. S., Kerner, P., Kugler, J. E., Deng, W., Jiang, D., & Di Gregorio, A. (2011). The identification of transcription factors expressed in the notochord of *Ciona intestinalis* adds new potential players to the brachyury gene regulatory network. *Developmental Dynamics, 240*, 1793–1805.

José-Edwards, D. S., Oda-Ishii, I., Kugler, J. E., Passamaneck, Y. J., Katikala, L., Nibu, Y., et al. (2015). Brachyury, Foxa2 and the cis-regulatory origins of the notochord. *PLOS Genetics, 11*(12), e1005730.

José-Edwards, D. S., Oda-Ishii, I., Nibu, Y., & Di Gregorio, A. (2013). Tbx2/3 is an essential mediator within the brachyury gene network during *Ciona* notochord development. *Development, 140*, 2422–2433.

Karaiskou, A., Swalla, B. J., Sasakura, Y., & Chambon, J. P. (2015). Metamorphosis in solitary ascidians. *Genesis, 53*, 34–47.

Katikala, L., Aihara, H., Passamaneck, Y. J., Gazdoiu, S., José-Edwards, D. S., Kugler, J. E., et al. (2013). Functional brachyury binding sites establish a temporal read-out of gene expression in the *Ciona* notochord. *PLoS Biology, 11*(10), e1001697.

Kawaminani, S., & Nishida, H. (1997). Induction of trunk lateral cells, the blood cell precursors, during ascidian embryogenesis. *Developmental Biology, 181*, 14–20.

Keys, D. N., Lee, B. I., Di Gregorio, A., Harafuji, N., Detter, J. C., Wang, M., et al. (2005). A saturation screen for cis-acting regulatory DNA in the *Hox* genes of *Ciona intestinalis*. *Proceedings of the National Academy of Sciences of the United States of America, 102*, 679–683.

Kispert, A., & Herrmann, B. G. (1993). The *Brachyury* gene encodes a novel DNA binding protein. *The EMBO Journal, 12*, 3211–3220.

Kispert, A., Koschorz, B., & Herrmann, B. G. (1995). The T protein encoded by brachyury is a tissue-specific transcription factor. *The EMBO Journal, 14*, 4763–4772.

Kourakis, M. J., & Smith, W. C. (2015). An organismal perspective on *C. intestinalis* development, origins and diversification. *eLife, 4*, e06024.

Kraus, F., Haenig, B., & Kispert, A. (2001). Cloning and expression analysis of the mouse T-box gene Tbx18. *Mechanisms of Development, 100*, 83–86.

Kubo, A., Suzuki, N., Yuan, X., Nakai, K., Satoh, N., Imai, K. S., et al. (2010). Genomic cis-regulatory networks in the early *Ciona intestinalis* embryo. *Development, 137*, 1613–1623.

Kugler, J. E., Gazdoiu, S., Oda-Ishii, I., Passamaneck, Y. J., Erives, A. J., & Di Gregorio, A. (2010). Temporal regulation of the muscle gene cascade by Macho1 and Tbx6 transcription factors in *Ciona intestinalis. Journal of Cell Science, 123*, 2453–2463.

Kugler, J. E., Passamaneck, Y. J., Feldman, T. G., Regnier, T. W., & Di Gregorio, A. (2008). Evolutionary conservation of vertebrate notochord genes in the ascidian *Ciona intestinalis. Genesis, 46*, 697–710.

Kumano, G., Negoro, N., & Nishida, H. (2014). Transcription factor Tbx6 plays a central role in fate determination between mesenchyme and muscle in embryos of the ascidian, *Halocynthia roretzi. Development, Growth & Differentiation, 56*, 310–322.

Kuwajima, M., Kumano, G., & Nishida, H. (2014). Regulation of the number of cell division rounds by tissue-specific transcription factors and Cdk inhibitor during ascidian embryogenesis. *PloS One, 9*(3), e90188.

Lemaire, P. (2009). Unfolding a chordate developmental program, one cell at a time: Invariant cell lineages, short-range inductions and evolutionary plasticity in ascidians. *Developmental Biology, 332*, 48–60.

Lolas, M., Valenzuela, P. D., Tjian, R., & Liu, Z. (2014). Charting Brachyury-mediated developmental pathways during early mouse embryogenesis. *Proceedings of the National Academy of Sciences of the United States of America, 111*, 4478–4483.

Manni, L., Lane, N. J., Joly, J. S., Gasparini, F., Tiozzo, S., Caicci, F., et al. (2004). Neurogenic and non-neurogenic placodes in ascidians. *Journal of Experimental Zoology. Part B, Molecular and Developmental Evolution, 302*, 483–504.

Matsumoto, J., Kumano, G., & Nishida, H. (2007). Direct activation by Ets and Zic is required for initial expression of the *Brachyury* gene in the ascidian notochord. *Developmental Biology, 306*, 870–882.

Meedel, T. H., Chang, P., & Yasuo, H. (2007). Muscle development in *Ciona intestinalis* requires the b-HLH myogenic regulatory factor gene Ci-MRF. *Developmental Biology, 302*, 333–344.

Meinertzhagen, I. A., & Okamura, Y. (2001). The larval ascidian nervous system: The chordate brain from its small beginnings. *Trends in Neurosciences, 24*, 401–410.

Miskolczi-McCallum, C. M., Scavetta, R. J., Svendsen, P. C., Soanes, K. H., & Brook, W. J. (2005). The *Drosophila melanogaster* T-box genes *midline* and *H15* are conserved regulators of heart development. *Developmental Biology, 278*, 459–472.

Mitani, Y., Takahashi, H., & Satoh, N. (1999). An ascidian T-box gene *As-T2* is related to the Tbx6 subfamily and is associated with embryonic muscle cell differentiation. *Developmental Dynamics, 215*, 62–68.

Mitani, Y., Takahashi, H., & Satoh, N. (2001). Regulation of the muscle-specific expression and function of an ascidian T-box gene, *As-T2. Development, 128*, 3717–3728.

Nakatani, Y., Yasuo, H., Satoh, N., & Nishida, H. (1996). Basic fibroblast growth factor induces notochord formation and the expression of As-T, a brachyury homolog, during ascidian embryogenesis. *Development, 122*, 2023–2031.

Nibu, Y., José-Edwards, D. S., & Di Gregorio, A. (2013). From notochord formation to hereditary chordomas: The many roles of brachyury. *BioMed Research International, 2013*, 826435.

Nishida, H. (1987). Cell lineage analysis in ascidian embryos by intracellular injection of a tracer enzyme. III. Up to the tissue restricted stage. *Developmental Biology, 121*, 526–541.

Nishida, H., & Sawada, K. (2001). macho-1 encodes a localized mRNA in ascidian eggs that specifies muscle fate during embryogenesis. *Nature, 409*, 724–729.

Ogasawara, M., Nakazawa, N., Azumi, K., Yamabe, E., Satoh, N., & Satake, M. (2006). Identification of thirty-four transcripts expressed specifically in hemocytes of *Ciona intestinalis* and their expression profiles throughout the life cycle. *DNA Research, 13*, 25–35.

Omotezako, T., Nishino, A., Onuma, T. A., & Nishida, H. (2013). RNA interference in the appendicularian *Oikopleura dioica* reveals the function of the *Brachyury* gene. *Development Genes and Evolution, 223*, 261–267.

Papaioannou, V. E. (2014). The T-box gene family: Emerging roles in development, stem cells and cancer. *Development, 141*, 3819–3833.

Paps, J., Holland, P. W., & Shimeld, S. M. (2012). A genome-wide view of transcription factor gene diversity in chordate evolution: Less gene loss in amphioxus? *Briefings in Functional Genomics, 11*, 177–186.

Passamaneck, Y. J., & Di Gregorio, A. (2005). *Ciona intestinalis*: Chordate development made simple. *Developmental Dynamics, 233*, 1–19.

Passamaneck, Y. J., Hadjantonakis, A. K., & Di Gregorio, A. (2007). Dynamic and polarized muscle cell behaviors accompany tail morphogenesis in the ascidian *Ciona intestinalis*. *PloS One, 2*(1), e714.

Passamaneck, Y. J., Katikala, L., Perrone, L., Dunn, M. P., Oda-Ishii, I., & Di Gregorio, A. (2009). Direct activation of a notochord cis-regulatory module by brachyury and FoxA in the ascidian *Ciona intestinalis*. *Development, 136*, 3679–3689.

Pope, E. C., & Rowley, A. F. (2002). The heart of *Ciona intestinalis*: Eicosanoid-generating capacity and the effects of precursor fatty acids and eicosanoids on heart rate. *Journal of Experimental Biology, 205*, 1577–1583.

Ragkousi, K., Beh, J., Sweeney, S., Starobinska, E., & Davidson, B. (2011). A single GATA factor plays discrete, lineage specific roles in ascidian heart development. *Developmental Biology, 352*, 154–163.

Razy-Krajka, F., Lam, K., Wang, W., Stolfi, A., Joly, M., Bonneau, R., et al. (2014). Collier/OLF/EBF-dependent transcriptional dynamics control pharyngeal muscle specification from primed cardiopharyngeal progenitors. *Developmental Cell, 29*, 263–276.

Reverberi, G., & Ortolani, G. (1962). Twin larvae from halves of the same egg in ascidians. *Developmental Biology, 5*, 84–100.

Rowley, A. F. (1982). Ultrastructural and cytochemical studies on the blood cells of the sea squirt, *Ciona intestinalis*. I. Stem cells and amoebocytes. *Cell and Tissue Research, 223*, 403–414.

Ruvinsky, I., Silver, L. M., & Gibson-Brown, J. J. (2000). Phylogenetic analysis of T-Box genes demonstrates the importance of amphioxus for understanding evolution of the vertebrate genome. *Genetics, 156*, 1249–1257.

Satoh, N., Tagawa, K., & Takahashi, H. (2012). How was the notochord born? *Evolution & Development, 14*, 56–75.

Satou, Y., Imai, K. S., & Satoh, N. (2004). The ascidian Mesp gene specifies heart precursor cells. *Development, 131*, 2533–2541.

Satou, Y., Kawashima, T., Shoguchi, E., Nakayama, A., & Satoh, N. (2005). An integrated database of the ascidian, *Ciona intestinalis*: Towards functional genomics. *Zoological Science, 22*, 837–843.

Satou, Y., Yagi, K., Imai, K. S., Yamada, L., Nishida, H., & Satoh, N. (2002). macho-1-related genes in *Ciona* embryos. *Development Genes and Evolution, 212*, 87–92.

Sawada, K., Fukushima, Y., & Nishida, H. (2005). Macho-1 functions as transcriptional activator for muscle formation in embryos of the ascidian *Halocynthia roretzi. Gene Expression Patterns, 5*, 429–437.

Sebé-Pedrós, A., de Mendoza, A., Lang, B. F., Degnan, B. M., & Ruiz-Trillo, I. (2011). Unexpected repertoire of metazoan transcription factors in the unicellular holozoan *Capsaspora owczarzaki. Molecular Biology and Evolution, 28*, 1241–1254.

Smith, J. (1999). T-box genes: What they do and how they do it. *Trends in Genetics, 15*, 154–158.

Spagnuolo, A., Ristoratore, F., Di Gregorio, A., Aniello, F., Branno, M., & Di Lauro, R. (2003). Unusual number and genomic organization of *Hox* genes in the tunicate *Ciona intestinalis. Gene, 309*, 71–79.

Stemple, D. L. (2005). Structure and function of the notochord: An essential organ for chordate development. *Development, 132*, 2503–2512.

Stolfi, A., Gainous, T. B., Young, J. J., Mori, A., Levine, M., & Christiaen, L. (2010). Early chordate origins of the vertebrate second heart field. *Science, 329*, 565–568.

Stolfi, A., & Levine, M. (2011). Neuronal subtype specification in the spinal cord of a protovertebrate. *Development, 138*, 995–1004.

Stolfi, A., Lowe, E. K., Racioppi, C., Ristoratore, F., Brown, C. T., Swalla, B. J., et al. (2014). Divergent mechanisms regulate conserved cardiopharyngeal development and gene expression in distantly related ascidians. *eLife, 3*, e03728.

Stolfi, A., Ryan, K., Meinertzhagen, I. A., & Christiaen, L. (2015). Migratory neuronal progenitors arise from the neural plate borders in tunicates. *Nature, 527*, 371–374.

Stolfi, A., Sasakura, Y., Chalopin, D., Satou, Y., Christiaen, L., Dantec, C., et al. (2015). Guidelines for the nomenclature of genetic elements in tunicate genomes. *Genesis, 53*, 1–14.

Swalla, B. J. (2006). Building divergent body plans with similar genetic pathways. *Heredity, 97*, 235–243.

Takada, N., Satoh, N., & Swalla, B. J. (2002). Expression of *Tbx6*, a muscle lineage T-box gene, in the tailless embryo of the ascidian *Molgula tectiformis. Development Genes and Evolution, 212*, 354–356.

Takahashi, H., Hotta, K., Erives, A., Di Gregorio, A., Zeller, R. W., Levine, M., et al. (1999). Brachyury downstream notochord differentiation in the ascidian embryo. *Genes & Development, 13*, 1519–1523.

Takatori, N., Hotta, K., Mochizuki, Y., Satoh, G., Mitani, Y., Satoh, N., et al. (2004). T-box genes in the ascidian *Ciona intestinalis*: Characterization of cDNAs and spatial expression. *Developmental Dynamics, 230*, 743–753.

Terajima, D., Yamada, S., Uchino, R., Ikawa, S., Ikeda, M., Shida, K., et al. (2003). Identification and sequence of seventy-nine new transcripts expressed in hemocytes of *Ciona intestinalis*, three of which may be involved in characteristic cell-cell communication. *DNA Research, 10*, 203–212.

Thompson, J. M., & Di Gregorio, A. (2015). *Insulin-like* genes in ascidians: Findings in *Ciona* and hypotheses on the evolutionary origins of the pancreas. *Genesis, 53*, 82–104.

Tokuoka, M., Imai, K. S., Satou, Y., & Satoh, N. (2004). Three distinct lineages of mesenchymal cells in *Ciona intestinalis* embryos demonstrated by specific gene expression. *Developmental Biology, 274*, 211–224.

Tolkin, T., & Christiaen, L. (2016). Rewiring of an ancestral Tbx1/10-Ebf-Mrf network for pharyngeal muscle specification in distinct embryonic lineages. *BiorXiv*. http://dx.doi.org/10.1101/039289.

Wada, S., Hamada, M., Kobayashi, K., & Satoh, N. (2008). Novel genes involved in canonical Wnt/beta-catenin signaling pathway in early *Ciona intestinalis* embryos. *Development, Growth & Differentiation, 50*, 215–227.

Waki, K., Imai, K. S., & Satou, Y. (2015). Genetic pathways for differentiation of the peripheral nervous system in ascidians. *Nature Communications*, *6*, 8719.
Wang, W., Razy-Krajka, F., Siu, E., Ketcham, A., & Christiaen, L. (2013). NK4 antagonizes Tbx1/10 to promote cardiac versus pharyngeal muscle fate in the ascidian second heart field. *PLoS Biology*, *11*(12), e1001725.
Wardle, F. C., & Papaioannou, V. E. (2008). Teasing out T-box targets in early mesoderm. *Current Opinions in Genetics & Development*, *18*, 418–425.
Washkowitz, A. J., Gavrilov, S., Begum, S., & Papaioannou, V. E. (2012). Diverse functional networks of *Tbx3* in development and disease. *WIREs Systems Biology and Medicine*, *4*, 273–283.
Yagi, K., Satoh, N., & Satou, Y. (2004). Identification of downstream genes of the ascidian muscle determinant gene Ci-macho1. *Developmental Biology*, *274*, 478–489.
Yagi, K., Satou, Y., & Satoh, N. (2004). A zinc finger transcription factor, ZicL, is a direct activator of *Brachyury* in the notochord specification of *Ciona intestinalis*. *Development*, *131*, 1279–1288.
Yagi, K., Takatori, N., Satou, Y., & Satoh, N. (2005). Ci-Tbx6b and Ci-Tbx6c are key mediators of the maternal effect gene Ci-macho1 in muscle cell differentiation in *Ciona intestinalis* embryos. *Developmental Biology*, *282*, 535–549.
Yamada, S., Ueno, N., Satoh, N., & Takahashi, H. (2011). *Ciona intestinalis* Noto4 contains a phosphotyrosine interaction domain and is involved in the midline intercalation of notochord cells. *International Journal of Developmental Biology*, *55*, 11–18.
Yasuo, H., Kobayashi, M., Shimauchi, Y., & Satoh, N. (1996). The ascidian genome contains another T-domain gene that is expressed in differentiating muscle and the tip of the tail of the embryo. *Developmental Biology*, *180*, 773–779.
Yasuo, H., & Satoh, N. (1993). Function of vertebrate *T* gene. *Nature*, *364*, 582–583.

CHAPTER FOUR

Eomesodermin—At Dawn of Cell Fate Decisions During Early Embryogenesis

S. Probst*,†, S.J. Arnold*,†,1

*Institute of Experimental and Clinical Pharmacology and Toxicology, Faculty of Medicine, University of Freiburg, Freiburg, Germany
†BIOSS Centre for Biological Signalling Studies, Albert-Ludwigs-University of Freiburg, Freiburg, Germany
[1]Corresponding author: e-mail address: sebastian.arnold@uniklinik-freiburg.de

Contents

1. Introduction 94
2. Early *Eomes* Functions in the Trophectoderm 95
3. *Eomes* in the Visceral Endoderm Promotes Formation of the Embryonic AP Axis 98
4. *Eomes* Functions During the Formation of Germ Layers 99
5. *Eomes* Transcriptional Activities During Germ Layer Formation 106
6. Control of *Eomes* Expression During Gastrulation Onset 109
Acknowledgments 110
References 111

Abstract

Proteins of the large family of T-box transcription factors are implicated in a broad spectrum of developmental processes. Loss-of-function mutations of T-box(Tbx) factors frequently cause severe embryonic phenotypes, often resulting from defects in cell fate specification and lineage differentiation. This review summarizes current knowledge on the functions of the T-box transcription factor Eomesodermin (*Eomes*) from postfertilization development until gastrulation stages of vertebrate embryos. *Eomes* exhibits evolutionary conserved functions in cell lineage specification and morphogenesis during gastrulation in all studied vertebrate model systems. In addition, during mammalian embryogenesis, *Eomes* is crucially required in extraembryonic tissues that are specific for intrauterine development.

This chapter mainly focuses on mammalian development of mouse; however, common functions shared among other vertebrate model system, such as embryos of zebrafish and *Xenopus laevis*, will be compared in the context of specification cell lineages during gastrulation.

Furthermore, this review recapitulates the current understanding of the molecular functions and transcriptional targets of *Eomes* as component of transcriptional complexes that guide cell-type specification and morphogenesis of early vertebrate embryos.

Current Topics in Developmental Biology, Volume 122
ISSN 0070-2153
http://dx.doi.org/10.1016/bs.ctdb.2016.09.001

1. INTRODUCTION

The first ortholog of the *Eomesodermin* (short *Eomes*) gene was discovered and cloned from *Xenopus laevis* in the lab of John Gurdon (Ryan, Garrett, Mitchell, & Gurdon, 1996). Here, *Eomes* was identified as one of the earliest expressed genes after onset of zygotic gene transcription in mesoderm-forming cells. This first report demonstrated that *Eomes* is sufficient and also required for the induction of mesoderm-specific gene activation. The name *Eomesodermin* was chosen according to the immediate and early expression at the onset of mesoderm formation (Eoσ, Greek for dawn). In this first analysis of *Eomes* function, Ryan and colleagues had already speculated that *Eomes* might also be expressed and functionally required in endoderm progenitors, which was demonstrated later in zebrafish (Bjornson et al., 2005) and mouse (Arnold, Hofmann, Bikoff, & Robertson, 2008). Soon after the first discovery of *Eomes* in *Xenopus*, studies in other species followed, describing additional expression domains, such as specific expression in neuronal progenitors of the developing cerebral cortex (Ciruna & Rossant, 1999; Hancock, Agulnik, Silver, & Papaioannou, 1999). Due to close homology of *Eomes* to another cortically expressed Tbx-factor, *Tbr1* (T-box brain gene 1), the name *Tbr2* is commonly used in the field of neurobiology despite the official gene nomenclature of *Eomes*. Following studies identified additional expression domains in different cell types of the adaptive and innate immune system of adult mice, such as central memory $CD8^+$ T cells and natural killer cells, respectively (Gordon et al., 2012; Pearce et al., 2003; Tayade et al., 2005).

The generation of the first loss-of-function allele in mouse revealed that *Eomes* is expressed and functionally required in trophectoderm (TE) cells (Russ et al., 2000). The TE is specific to intrauterine development of mammalian embryos and gives rise to the embryonic portions of the fetomaternal connection that forms the placenta. Loss of *Eomes* function leads to embryonic lethality shortly after implantation due to compromised TE development. Only after the generation of a conditional loss-of-function allele, it became feasible to analyze *Eomes* functions in more detail during later stages and at specific expression sites (Arnold et al., 2008; Russ et al., 2000). It was then shown that *Eomes* has crucial functions in another extraembryonic tissue of the mouse embryo, the visceral endoderm (VE), that is required for establishing the embryonic anterior–posterior (AP) axis (Nowotschin et al., 2013). Probably the most intensively studied function of *Eomes* during early development is the specification of the germ layers at gastrulation. Here, *Eomes* is crucially required for the formation of endoderm and mesoderm, which is a conserved function found in all studied vertebrate model systems, mouse, zebrafish, and *Xenopus*.

Remarkably, loss of *Eomes* function causes severe phenotypes at any given expression site, despite overlapping expression of other T-box factors. Interestingly, experiments in differentiating embryonic stem (ES) cells have suggested considerable functional redundancy between T-box factors, e.g., *Eomes*, *Tbx3*, *Brachyury*, and *Tbx6*. Thus, there is room for future studies to resolve in more detail the specific and nonredundant molecular functions of these early expressed T-box factors during development.

This chapter aims to present an overview of early embryonic *Eomes* functions until gastrulation stages. We discuss how *Eomes* acts at each expression site, following the chronological order of expression domains in mouse and include and compare the numerous studies from other vertebrate model systems and stem cell-based approaches. The overview of the embryonic functions is followed by paragraphs summarizing the current understanding of the control mechanisms for *Eomes* expression, and its molecular functions as a component of transcriptional complexes.

2. EARLY *EOMES* FUNCTIONS IN THE TROPHECTODERM

In the mouse embryo, first expression of *Eomes* is found in the mammalian-specific cell lineage of the TE during early preimplantation stages. Intrauterine development depends on the proper establishment of the fetomaternal connection, which is generated from cells of the TE on the embryonic and the decidua on the maternal side. The separation of the TE from the inner cell mass (ICM), which gives rise to all embryonic cell types and the yolk sac, is the first definitive cell lineage decision that takes place in the mammalian embryo at late morula to early blastocyst stage around embryonic days 2.5–3.5 (E2.5–3.5) in mouse. This lineage separation is initiated by activities of the transcription factors TEA domain family member4 (*Tead4*) that induces *Caudal-related homeobox2* (*Cdx2*) in prospective TE cells (Nishioka et al., 2009; Yagi et al., 2007). *Cdx2* suppresses ICM fate through counteracting activities of the pluripotency factor *Oct3/4*, thus enforcing the TE-ICM lineage determination (Niwa et al., 2005; Strumpf et al., 2005). Shortly after the TE lineage is established, *Eomes* expression initiates in cells of the mural TE, the cells lining the blastocyst cavity, and in polar TE, the cells located closest to the ICM (Hancock et al., 1999; Ralston & Rossant, 2008; Russ et al., 2000). While expression is lost early in the mural TE, *Eomes* is maintained in polar TE, which contains trophoblast stem cells (TSCs) that are the cellular source for all TE-derived cells giving rise to the embryonic part of the placenta. *Eomes* specifically marks and is functionally required for the maintenance of the population of undifferentiated TSCs from early time points until this placental stem cell pool ceases around E9.5 in mouse (Ciruna & Rossant, 1999; Kuales, Weiss, Sedelmeier, Pfeifer, & Arnold, 2015; Kwon & Hadjantonakis, 2007).

Targeted deletion of *Eomes* gene function leads to early embryonic lethality around E5.5 due to a failure of TE development (Russ et al., 2000). *Eomes*-deficient embryos form morphologically normal blastocysts that are proficient to implant (Strumpf et al., 2005); however, they arrest shortly after implantation and do not develop significantly beyond the late blastocyst stage. The TE is the only *Eomes*-expressing tissue in the peri-implantation stage embryo. Thus, it was concluded that *Eomes* is required for the maintenance and early differentiation of TSCs (Russ et al., 2000; Strumpf et al., 2005). Similar to ES cells, TSCs can be derived from blastocysts and cultured indefinitely as an in vitro correlate representing stem cells of the trophoblast in the embryo (Tanaka, Kunath, Hadjantonakis, Nagy, & Rossant, 1998). Supporting the view that TSCs depend on *Eomes* function, it is not possible to isolate TSCs from *Eomes*-deficient blastocyst outgrowths (Russ et al., 2000; Strumpf et al., 2005), and the inducible deletion of *Eomes* gene function in TSCs impairs self-renewal (Kidder & Palmer, 2010; Kuales et al., 2015). In conclusion, *Eomes* seems not to be primarily required for the early lineage determination of TE cells, but is necessary for subsequent maintenance of TSCs and potentially also for early differentiation steps (Strumpf et al., 2005).

In addition to *Eomes* a handful of other transcription factors act in a regulatory network that specifies and maintains TSC fate. Among those are *Tead4*, *Cdx2*, *Tcfap2c*, *Gata3*, *Elf5*, and *Ets2* (for an overview, see Latos & Hemberger, 2014). Interestingly, the embryonic loss-of-function phenotypes of these transcription factors vary considerably. Loss of *Tead4* and *Cdx2* gene function leads to early lethality before implantation (Nishioka et al., 2008; Strumpf et al., 2005), *Elf5*, *Ets2*, and *Tcfap2* mutations generate relatively late phenotypes after gastrulation onset (Donnison et al., 2005; Georgiades & Rossant, 2006; Werling & Schorle, 2002; Yamamoto et al., 1998), and *Eomes*-deficient embryos arrest after implantation (Russ et al., 2000; Strumpf et al., 2005). Despite these variations of embryonic phenotypes, the analysis of target genes and chromatin binding of these transcriptional regulators in TE cells have revealed a high degree of co-regulation and co-occupancy of target genes. Also self-reinforcement of the key TSC factors by auto- and cross-regulation has been demonstrated. For example, EOMES binds its own promoter and those of *Tcfap2c*, *Gata3*, and *Ets2* (Kidder & Palmer, 2010). In addition, considerable overlap of EOMES-binding sites is seen with those of CDX2, ELF5 (Chuong, Rumi, Soares, & Baker, 2013), and TCFAP2C (Kidder & Palmer, 2010). This significant overlap in chromatin binding suggests the existence of a core TSC transcriptional complex, similar to the pluripotency maintaining complex containing the factors OCT3/4, SOX2, and NANOG (Boyer et al., 2005). However, to date the composition of the core TSC-maintaining transcriptional complexes including details of the direct physical interaction of participating factors remains to be studied in closer detail.

The key transcription factors for TSCs are not only required for TSC maintenance, but they are also sufficient for the induced transdifferentiation of ES cells to TSCs when overexpressed in ES cells. This remarkable reprogramming capacity was first described for *Cdx2* and *Eomes* (Niwa et al., 2005) and later shown for other factors, such as *Tead4*, *Tcfap2*, *Gata3*, and *Elf5* (Kuckenberg et al., 2010; Ng et al., 2008; Nishioka et al., 2009; Ralston et al., 2010). Interestingly, resulting TSCs are not equally capable to self-renew. *Tead4*, *Cdx2*, and *Tcfap2* transdifferentiated cells undergo self-renewal similar to TSCs isolated from embryos. In contrast, *Eomes*, *Elf5*, and *Gata3*-induced cells are not sustainable over longer culture periods and are biased toward differentiation. While these differences during reprogramming by various factors are not entirely understood, they reflect some of the experimental observations of distinct target gene induction. For example, *Cdx2* acts upstream of *Eomes* and suppresses the ES cell-specific pluripotency program. In contrast, *Eomes* does not activate *Cdx2* expression and is not able to restrict and suppress the *Oct3/4*-driven pluripotency transcriptional network (Niwa et al., 2005). Thus, *Eomes* as a single factor induces TSC characteristics in ES cells, but is not sufficient to promote the entire TSC program.

Two recent studies reported on the identification of reprogramming factors to directly convert murine embryonic and adult-derived fibroblast to so-called induced TSCs (iTSCs), which resemble embryo-derived TSCs by morphology, marker expression, epigenetic signature, and function (Benchetrit et al., 2015; Kubaczka et al., 2015). Both reports transfected fibroblasts with a cocktail of candidate TSC reprogramming factors and revealed sets of four transcription factors, which induce iTSC lineage conversion. Remarkably, three of the four factors, namely *Eomes*, *Tfap2c*, and

Gata3, were found by both independent approaches (Benchetrit et al., 2015; Kubaczka et al., 2015). Of note the two other key TSC transcription factors, *Cdx2* and *Elf5*, are not required for the conversion to iTSCs. This places *Eomes* into the center of transcriptional regulators for the TSCs cell lineage.

3. *EOMES* IN THE VISCERAL ENDODERM PROMOTES FORMATION OF THE EMBRYONIC AP AXIS

The second expression domain of *Eomes* in mouse is found in the VE overlying the epiblast of the cup-shaped postimplantation embryos starting from E5.0 (Fig. 2, I) (Arnold, Sugnaseelan, Groszer, Srinivas, & Robertson, 2009; Chazaud & Rossant, 2006; Nowotschin et al., 2013). Here, *Eomes* expression is detected in a few of the VE cells, first without an obvious pattern, later with a bias toward expression cells on the anterior side of the embryo. This expression domain was not detected in every study when using in situ hybridization analysis or a lacZ reporter that was inserted into the *Eomes* gene locus, possibly indicating a dynamic expression in the VE (Arnold et al., 2008; Ciruna & Rossant, 1999; Russ et al., 2000). Following the generation of antibodies directed against EOMES and using an $Eomes^{GFP}$ reporter allele, the robust expression of *Eomes* in the VE became more obvious (Arnold et al., 2009; Nowotschin et al., 2013).

In the postimplantation embryo, one prominent function of the VE is the establishment of the embryonic AP axis by reciprocal tissue interactions of the VE and the epiblast. A specialized group of VE cells is induced at the distal tip of the embryo under the influence of epiblast-derived *Nodal* signals, namely the distal visceral endoderm (DVE) (Rivera-Pérez & Hadjantonakis, 2015; Stower & Srinivas, 2014). This group of DVE cells, together with newly recruited cells with similar expression signatures and functions, forms an important organizing center, the anterior visceral endoderm (AVE) that is located at the anterior embryonic–extraembryonic border (Takaoka, Yamamoto, & Hamada, 2011). The AVE fulfills crucial patterning functions for the induction of the AP axis by restricting signaling activities to the posterior side of the embryo by the secretion of Nodal, Wnt, and BMP inhibitors (Rivera-Pérez & Hadjantonakis, 2015; Stower & Srinivas, 2014). First indications that *Eomes* might act in the VE of the early postimplantation embryo came from studies describing a genetic interaction of *Eomes* with the TGFβ signaling molecule *Nodal* by the generation of embryos that were heterozygous for loss-of-function alleles of *Nodal* and *Eomes*. Heterozygosity of either allele alone does not lead to embryonic phenotypes. In contrast, one-third of double heterozygous embryos exhibit an early failure to establish the AP axis (Arnold et al., 2008). This genetic interaction of *Nodal* and *Eomes* can be explained in two ways as was demonstrated in other studies: First, by the activation of *Eomes* expression in the VE by SMAD2-dependent *Nodal* signaling from the epiblast (Nowotschin et al., 2013), and second, by cooperative activation of target genes by EOMES, together with SMAD2 in the VE (Nowotschin et al., 2013).

To investigate more thoroughly the functions of *Eomes* in the VE, a conditional gene-deletion approach was chosen to specifically delete *Eomes* within the VE using *Ttr*-Cre (*Eomes*$^{\Delta VE}$) (Kwon & Hadjantonakis, 2009; Nowotschin et al., 2013). In *Eomes*$^{\Delta VE}$ embryos, DVE cells are first established and can be detected by expression of a Hex-GFP reporter. However, DVE cells are reduced in number and fail to be maintained. Additionally, no further cells are recruited to the anteriorly migrating cells to form the fully functional AVE. Thus, *Eomes*$^{\Delta VE}$ embryos do not establish a proper AP axis and exhibit grossly increased Nodal expression in the epiblast. At least part of this phenotype can be explained by the absence of *Lim homeobox protein 1* (*Lhx1*) expression which is directly activated by EOMES in the VE and required for correct formation and migration of the AVE (Fig. 2, I; Perea-Gómez, Shawlot, Sasaki, Behringer, & Ang, 1999; Shawlot & Behringer, 1995). Collectively, it was shown that *Eomes* is an important component of the Nodal-driven regulatory network involved in specification and migration of the AVE and therefore the AP axis of the mouse embryo (Nowotschin et al., 2013).

4. *EOMES* FUNCTIONS DURING THE FORMATION OF GERM LAYERS

Evolutionary conserved roles of *Eomes* for lineage specification at gastrulation have been explored in different vertebrate model organisms and are the most widely studied early embryonic functions of *Eomes*. First described in *Xenopus*, but later also confirmed in other model systems and during stem cell differentiation, the *Eomes* gene is now recognized as one of the earliest functional indicators for gastrulation onset and is required for the specification of mesoderm subtypes and definitive endoderm.

In *Xenopus*, first expression of *Eomes* is found around onset of zygotic gene transcription in an equatorial band in all mesoderm-forming cells (Fig. 1A and B). Expression precedes that of other mesodermal markers and *Eomes* is more broadly expressed than *Brachyury*, which is considered

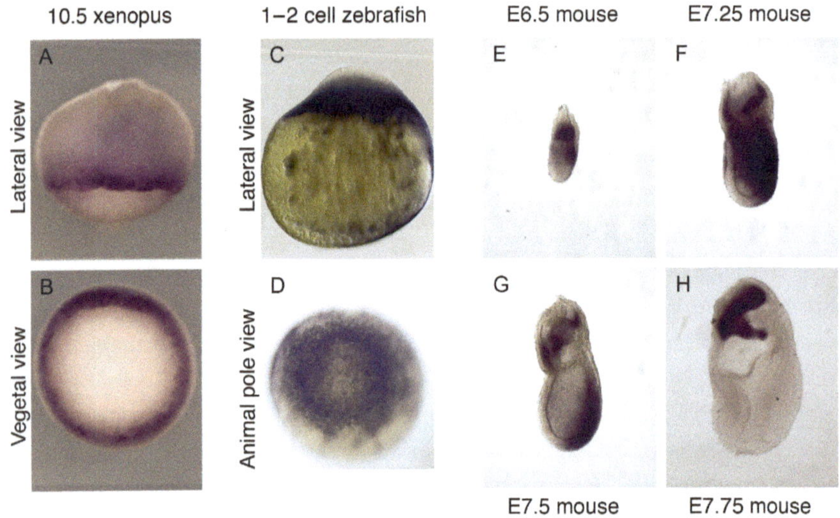

Fig. 1 *Eomes* expression in different species. In (A) (*lateral view*) and (B) (*vegetal view*) the expression of *Eomes* in a *Xenopus* embryo at stage 10.5 (gastrula) is depicted. (C) Shows the lateral and (D) the animal pole view of a 1–2 cell zebrafish embryo with localized *eomesa* expression at the vegetal margin of the blastodisc. In the mouse embryo (E–H) *Eomes* is expressed in the extraembryonic ectoderm and the posterior epiblast at E6.5 (E). One day later the extraembryonic pattern is restricted to the developing chorion, where it stays expressed until after E7.75 (F–H). In the embryonic part, *Eomes* expression in the epiblast and primitive streak extends toward the distal tip of the embryo and is detected in the mesodermal wings at E7.25 (F). Expression in the primitive streak and epiblast is lost after E7.5 and no further *Eomes* transcripts are detected in the embryonic part of the embryo (G–H). Mouse embryos are oriented with anterior to the left and posterior to the right. *Images of zebrafish and* Xenopus *embryos were kindly provided by D. Kimelman and M. Asashima, respectively.*

a panmesoderm marker (Ryan et al., 1996). When ectopically expressed *Eomes* induces a range of mesodermal genes in a concentration-dependent manner. In reverse, the inhibition of *Eomes* functions by RNA injection of a dominant negative EOMES–ENGRAILED fusion protein generates severe gastrulation phenotypes and subsets of mesodermal marker genes fail to be expressed. Thus, in *Xenopus*, *Eomes* is necessary and sufficient for mesoderm formation (Ryan et al., 1996). Of note, to date no detailed analysis and no formal proof of the functional importance of *Eomes* for endoderm development in *Xenopus* was performed. However, already the initial report by Ryan and colleagues and a following global analysis of gene regulatory networks for endoderm formation suggested expression in endoderm-generating cells in *Xenopus* (Ryan et al., 1996; Sinner et al., 2006). Since

Xenopus endoderm development is initiated by the maternally expressed T-box factor *VegT* (Horb & Thomsen, 1997), it will be interesting to learn if *Eomes* also exhibits functions during formation of endoderm as seen in other vertebrate model organisms (described below).

The zebrafish genome contains two *Eomes* homologs, *eomesa* and *eomesb*. However, only *eomesa* seems to be expressed during early developmental stages (Nelson et al., 2014). In contrast to *Xenopus*, *eomesa* has a strong maternal contribution in addition to early zygotic expression following the midblastula transition (Bruce et al., 2003). *Eomesa* mRNA is detected throughout the oocyte and becomes localized to a ring-like domain at the vegetal edge of blastodisc around the first cleavage (Fig. 1C and D). Expression is maintained in a vegetal ring at later stages until *Eomes* mRNA becomes undetectable after 3.5 h postfertilization (Bjornson et al., 2005). *Eomesa*-expressing blastomeres at the margins of the blastodisc give rise to a mixed population of mesoderm and endoderm progenitors, which are separated after onset of gastrulation, when cells involute to form the germ layers. Similar to *Xenopus*, also in zebrafish the forced expression of *eomesa* activates a subset of mesodermal genes and is able to induce organizer genes, such as *goosecoid* (*gsc*), *chordin* (*chd*), and *floating head* (*flh*), and leads to the formation of a secondary body axis (Bruce et al., 2003). Several following reports have identified the transcriptional targets of *eomesa* to induce early mesendoderm in zebrafish. Here, *eomesa* acts in combination with nodal-related factors that are also cross-regulated by *eomesa* (Nelson et al., 2014; Slagle, Aoki, & Burdine, 2011; Xu et al., 2014).

A very thorough study by Bjornson and colleagues demonstrated the interaction of *eomesa* with transcription factors of the regulatory gene network that initiates and controls endoderm formation (Bjornson et al., 2005). Here, Eomes cooperates with the transcription factors Gata5 and Bon downstream of Nodal signaling to activate endoderm-specific genes, including the key endoderm determinant in zebrafish, the sox transcription factor *casanova* (*cas/sox32*) in marginal blastomeres. Eomes physically interacts with Gata5 and Bon to form transcriptional complexes that directly bind and activate target gene transcription (Bjornson et al., 2005). However, the regulation of *cas/sox32* is not entirely eliminated in maternal-zygotic *eomesa* (MZ*eomesa)* mutants and remaining *cas/sox32* activities seem to be sufficient for the formation of endoderm and a gut tube (Du, Draper, Mione, Moens, & Bruce, 2012). More generally, MZ*eomesa* zebrafish mutants show only surprisingly mild endoderm and mesoderm phenotypes, especially when compared to loss-of-function mutations in other species, e.g., the mouse embryo as described later. This might be explained by suggested overlapping functions of *eomesa* and *foxh1* specific to zebrafish mesendoderm development, which could compensate for *eomesa* deficiency (Nelson et al., 2014; Slagle et al., 2011).

In the mouse embryo, *Eomes* expression is initiated at the posterior proximal pole of the mouse epiblast, the site of gastrulation onset around E6.5 (Figs. 1E–H and 2, II). Here, under the influence of increasing signaling levels of the TGFβ/Nodal-, Wnt-, and FGF pathways, cells of the posterior epiblast form the primitive streak (PS), which is the structure at which mesoderm and definitive endoderm (DE) cells are fate specified. At the PS, cells undergo an epithelial-to-mesenchymal transition (EMT) and delaminate from the epithelial epiblast, thus generating the mesoderm germ layer between the remaining epiblast (later referred to as ectoderm) and the VE layer (reviewed in Arnold & Robertson, 2009; Fig. 2, II, III). The nascent mesoderm layer initially contains genuine mesoderm cells and prospective DE cells that will successively integrate into the VE layer, thereby replacing most VE cells (Kwon, Viotti, & Hadjantonakis, 2008; Viotti, Nowotschin, & Hadjantonakis, 2014). Different subtypes of mesoderm and endoderm are specified consecutively in a spatiotemporal pattern during cell ingression at the PS. The first cells to leave the PS are the extraembryonic and anterior mesoderm (head mesenchyme and cardiogenic mesoderm), followed by DE and other anterior PS derivatives and finally paraxial and intermediate mesoderm (reviewed in Tam & Behringer, 1997). The generation of different cell types is dependent on dynamic changes of signaling levels and the transcriptional activities when cells ingress through the streak. Patterning of the PS is best understood for SMAD2/3/4-mediated Nodal signaling which is acting in an anterior-to-posterior graded fashion to generate different cell types (Chu, Dunn, Anderson, Oxburgh, & Robertson, 2004; Dunn, Vincent, Oxburgh, Robertson, & Bikoff, 2004; Vincent, Dunn, Hayashi, Norris, & Robertson, 2003).

Eomes mRNA expression is first seen before the formation of a discernable PS and represents one of the earliest genes to mark gastrulation onset. mRNA expression in the epiblast and the anterior portion of the PS is maintained until formation of the embryonic node at E7.5 when expression is abruptly lost (Fig. 1E–H; Ciruna & Rossant, 1999; Russ et al., 2000). Genetic lineage-tracing of *Eomes*-expressing cells using an $Eomes^{Cre}$ knock-in allele in combination with a Cre-inducible reporter ($ROSA26^{R}$; Soriano, 1999) revealed that *Eomes*-expressing cells give rise to anterior mesoderm derivatives, such as head mesenchyme and cardiovascular

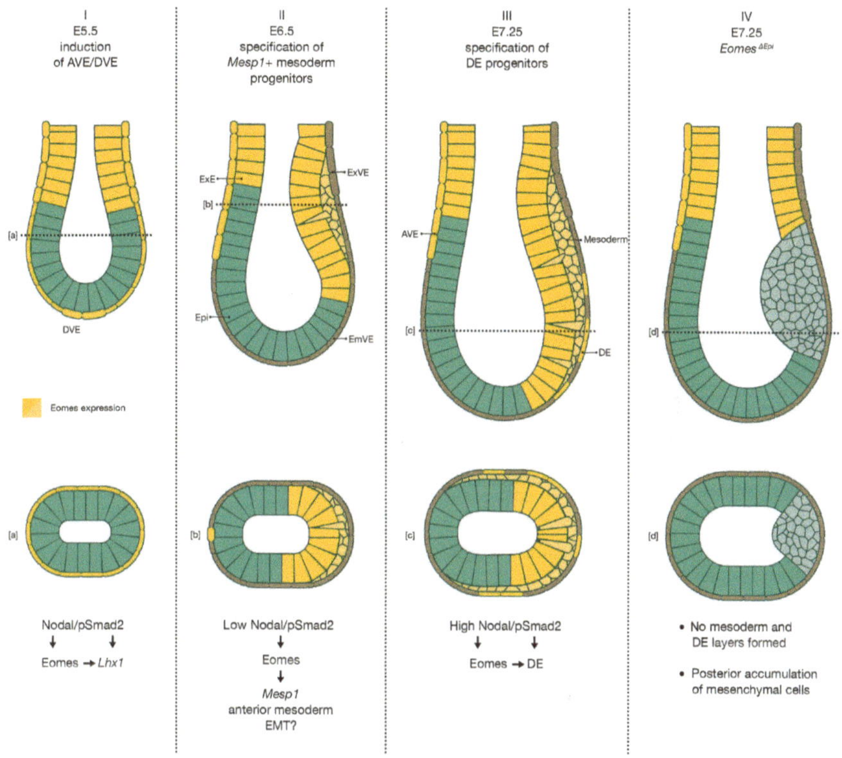

Fig. 2 Functions of *Eomes* during AVE/DVE induction and gastrulation. The roles of *Eomes* in normal development from E5.5 to E7.25 are depicted in I to III. IV shows an *Eomes*$^{\Delta Epi}$ embryo at E7.25. Embryos are shown in sagittal sections (*upper drawings*) and transversal sections (*lower drawings a–d*). The transversal sectional planes shown in a–d are indicated on the sagittal views by *dotted lines*. *Eomes* expression is indicated by *yellow color*. All embryos are oriented with anterior to the left, posterior to the right, proximal to the top, and distal to the bottom. (I) At E5.5 *Eomes* is expressed in the visceral endoderm (VE). Here, *Eomes* is necessary for the formation of the AVE. *Eomes* is induced by Nodal/pSMAD2 signaling and activates together with Nodal/pSMAD2 the expression of *Lhx1*, a transcription factor important in specification and migration of the AVE. (II) One day later at E6.5 *Eomes* is required for the initiation of gastrulation. *Eomes* is expressed in the posterior epiblast and is crucial for delamination of mesendodermal cells from the primitive streak (PS). *Eomes* directly activates expression of the transcription factor *Mesp1* in the early PS in the presence of low Nodal/pSMAD2 signaling levels. *Mesp1* specifies the early lineage of anterior mesoderm. (III) Slightly later *Eomes* expression extends further distally and is a major regulator of the specification of the definitive endoderm (DE). DE cells leave the PS together with mesoderm and integrate into the VE cell layer. DE cells are visible as *Eomes*-expressing (*yellow*) cells in the VE layer at this stage. *Eomes* in concert with high Nodal/pSMAD2 signaling regulates the expression of a set of core DE regulators. (IV) In epiblast-specific *Eomes* mutants, cells in the PS acquire mesenchymal morphology, but fail to delaminate

(Continued)

progenitors, the extraembryonic mesoderm, and cells of the DE-derived primary gut tube (Costello et al., 2011). To study TE/VE-independent *Eomes* functions within the epiblast and PS, a conditional gene-deletion approach was applied by *Eomes* gene inactivation with the epiblast-specific Sox2.Cre deleter strain (*Eomes*$^{\Delta Epi}$) (Arnold et al., 2008). *Eomes*$^{\Delta Epi}$ embryos are morphologically severely disturbed and development is blocked at early gastrulation stage. No mesoderm layer is formed and instead mesenchymal cells accumulate in the PS region. Marker analysis reveals that general mesoderm and PS markers, such as *Brachyury*, *Wnt3*, or *Ffg8*, are robustly expressed in *Eomes*$^{\Delta Epi}$ embryos (Arnold et al., 2008). However, early anterior mesoderm and cardiac marker genes, including the key regulator for cardiomyocyte specification *Mesp1*, are absent in *Eomes*$^{\Delta Epi}$ embryos, indicating that the specification of this specific subset of anterior mesoderm depends on *Eomes* (Costello et al., 2011). Furthermore, marker analysis of key genes for DE specification, such as *Hex1* and *Cer1*, shows a complete lack of expression in the anterior PS region. In addition, VE cells overlying the epiblast are not replaced by epiblast-derived DE cells in *Eomes*$^{\Delta Epi}$ embryos (Fig. 2, II–IV). Experiments using chimeric embryos consisting of wild-type and *Eomes*-deficient (*Eomes*$^{-/-}$) cells demonstrate that the failure to specify both anterior mesoderm derivatives and the DE cell lineage is a cell autonomous function, since *Eomes*-deficient ES cells are entirely excluded from the head mesenchyme, the heart, and the gut tube (Arnold et al., 2008; Costello et al., 2011). In contrast, *Eomes* null cells can efficiently contribute to structures of posterior paraxial and lateral plate mesoderm. Thus, the generation of more posterior mesoderm is independent of *Eomes* as expression of more widely expressed mesoderm markers, such as *Brachyury*, *Wnt3*, and *Fgf8*, in *Eomes*$^{\Delta Epi}$ embryos also suggested (Arnold et al., 2008).

The SMAD2/3/4-mediated Nodal signaling pathway has been shown to play crucial roles in patterning of the PS. The specification of anterior PS derivatives, such as prechordal plate, the node, notochord, and including the DE, requires highest levels of SMAD2/3 signaling activities (Chu et al., 2004; Dunn et al., 2004; Hoodless et al., 2001; Tremblay, Hoodless, Bikoff, & Robertson, 2000; Vincent et al., 2003). A genetic interaction of *Eomes* with

Fig. 2—Cont'd from the PS. This leads to the posterior accumulation of cells in the PS region. The mesoderm layer is absent, and the anterior mesoderm and DE lineages are not specified. *AVE*, anterior visceral endoderm; *DE*, definitive endoderm; *DVE*, distal visceral endoderm; *EmVE*, embryonic visceral endoderm; *Epi*, epiblast; *ExE*, extraembryonic ectoderm; *ExVE*, extraembryonic visceral endoderm.

Nodal signaling as observed for establishing the AP axis by the AVE is also found during the specification of DE. Embryos that heterozygously carry mutant alleles of *Eomes* and *Nodal* show in about 30% of cases a failure to properly form DE, as demonstrated by marker gene expression and phenotypic observation (Arnold et al., 2008). On a molecular level *Eomes* and Nodal signaling are tightly linked by reciprocal cross-regulation as demonstrated in several studies. *Eomes* expression is induced by Activin/Nodal/SMAD signals (see below), and in zebrafish zygotic expression of Nodal-related factors is regulated by *eomes* (Xu et al., 2014). Importantly, the defects in mesoderm and DE specification in *Eomes*$^{\Delta Epi}$ embryos are most likely not primarily mediated by compromised Nodal signaling, or other signaling molecules, since *Eomes*$^{-/-}$ cells fail to form anterior mesoderm and DE in a strictly cell autonomous fashion, while neighboring wild-type cells contribute normally to all forming tissues (Arnold et al., 2008; Costello et al., 2011).

Collectively, gain- and loss-of-function studies in different vertebrate model systems have demonstrated that *Eomes* is a central component of the gene regulatory network that establishes mesoderm subtypes and endoderm. *Eomes* acts closely together with SMAD2/3-mediated Nodal signals that have prominent patterning functions during germ layer formation. The role of *Eomes* in endoderm formation seems grossly conserved between species. In contrast, the specification of mesoderm downstream of *Eomes* requires further analysis in *Xenopus* and zebrafish, to determine if in these species only subsets of mesoderm depend on *Eomes*, as observed in mouse, or if indeed formation of the broad mesendoderm population is compromised by loss of *Eomes*. For example, it was suggested that regulation of zygotic VegT for paraxial mesoderm development is also dependent on *Eomes* in *Xenopus* (Fukuda et al., 2010).

In addition to the lack of marker gene expression for anterior mesoderm and DE, the most prominent and morphologically visible phenotype in *Eomes*$^{\Delta Epi}$ mouse embryos is a pronounced thickening of the PS (Arnold et al., 2008; Fig. 2, IV). *Eomes*-deficient cells fail to delaminate from the PS region, and thus neither the mesodermal cell layer is forming, nor are epiblast-derived cells integrating into the VE layer. Instead, cells with mesenchymal morphology and marker expression accumulate along the posterior side of the embryo thus causing developmental arrest. Cells in the PS normally undergo an EMT, including loss of E-Cadherin (*Cdh1*) mRNA and protein expression and induction of N-Cadherin (*Cdh2*) in nascent mesoderm (Carver, Jiang, Lan, Oram, & Gridley, 2001; Zohn et al., 2006). In *Eomes*$^{\Delta Epi}$ embryos, a persistence of E-Cadherin transcripts and protein is observed in accumulating PS cells, while the upstream regulators of E-Cadherin suppression, *Fgf8* and *Snail*, are robustly expressed (Arnold et al., 2008; Carver et al., 2001; Sun, Meyers, Lewandoski, & Martin, 1999). Interestingly, *Eomes*-deficient cells from dissected mutant PS show grossly normal migratory behavior when observed in culture (Arnold et al., 2008). Thus, PS cells in *Eomes*$^{\Delta Epi}$ embryos undergo a partial EMT, obtain some mesenchymal characteristics, while they also maintain an epithelial signature. However, they are not able to leave the streak and form mesoderm.

Specification of cell fate and morphogenesis of the mesodermal cell layer may not be separate functions of *Eomes*. Interestingly, the phenotype of *Mesp1/2* double knockout embryos is remarkably similar to the accumulating cells in the streak of *Eomes*$^{\Delta Epi}$ embryos (Kitajima, Takagi, Inoue, & Saga, 2000). Expression of *Mesp1* and *Mesp2* is mostly absent in *Eomes*$^{\Delta Epi}$ embryos (Costello et al., 2011), indicating that the observed phenotype might be mediated by lack of *Mesp1/2* gene induction. Furthermore, this indicates that anterior mesoderm is required for the initial steps of mesoderm formation, and that posterior mesoderm derivatives, indicated by expression of mesodermal genes like *T*, *Wnt3*, and *Fgf8* in *Eomes*$^{\Delta Epi}$ mutants, are unable to initiate formation of the mesoderm layer. However, to date it remains unclear what are *Mesp1/2*- and/or *Eomes*-regulated factors that mediate the initial steps of mesoderm formation.

Interestingly, in zebrafish maternally expressed *eomesa* is also implicated in early cell migration during epiboly (Bruce, Howley, Dixon Fox, & Ho, 2005; Du et al., 2012). During epiboly cells of the blastoderm layer migrate and spread over the yolk so that it is entirely engulfed by the end of gastrulation. MZ*eomesa* embryos exhibit a delay in epiboly initiation, possibly mediated by altered microtubule dynamics of yolk cells and changed adhesive properties of deep blastoderm cells (Du et al., 2012). However, the precise mechanisms and transcriptional targets of *eomesa* leading to the observed epiboly phenotype require further analysis (Du et al., 2012). It will be interesting to learn if similar cellular processes during cell migration are regulated by *Eomes* in zebrafish and mouse.

5. *EOMES* TRANSCRIPTIONAL ACTIVITIES DURING GERM LAYER FORMATION

T-box (Tbx) factors frequently act at developmentally critical transition steps during cell lineage decisions and often are potent inducers of cell fate changes when experimentally induced. *Eomes* shares with other Tbx factors the common, defining feature of the DNA-binding T-box domain, which binds to the conserved consensus motif of T-half sites (reviewed in Papaioannou, 2014). To date relatively little is known about the regulation of protein modifications, stability, localization, and transcriptional activities of Tbx-proteins, including EOMES. It is mostly assumed that transcriptional activities of T-box factors are regulated at the level of region-specific expression and their assembly with different protein interaction partners to form variable transcriptional complexes.

During gastrulation EOMES activities are intrinsically connected to Nodal/Smad2/3 signaling. *Eomes* expression itself is controlled by SMAD2/3 signaling (discussed below), and also transcriptional functions are guided by SMAD2/3 signaling by the cooperative regulation of many target genes by EOMES and SMAD2/3 (Arnold et al., 2008; Bjornson et al., 2005; Brown et al., 2011; Nelson et al., 2014; Slagle et al., 2011; Teo et al., 2011). Genome-wide comparisons of binding sites of EOMES and SMAD2/3 in differentiating human and murine ES cells as well as in zebrafish and *Xenopus* embryos show that they share binding to many evolutionary conserved target genes (Gentsch et al., 2013; Kartikasari et al., 2013; Nelson et al., 2014; Teo et al., 2011). Among the EOMES/SMAD2/3 bound gene loci are key genes for DE specification and differentiation, including *Sox17*, *Gsc*, and *Foxa2*, which can be precipitated in as shown by successive tandem chromatin immunoprecipitation experiments (tandem ChIP) (Teo et al., 2011). Importantly, EOMES and SMAD2/3 binding to key DE genes are likely of functional importance, as target gene expression is lost in *Eomes*- or *Smad2*-deficient mouse embryos (Arnold et al., 2008; Brennan et al., 2001), ES cell (Brown et al., 2011), and in zebrafish embryos (Bjornson et al., 2005; Nelson et al., 2014). In *Xenopus*, a direct interaction of Eomes and Smad2 proteins has been demonstrated previously (Picozzi, Wang, Cronk, & Ryan, 2009). Studies in zebrafish further suggest that *foxh1*, another mediator of Nodal signals, shares additional combinatorial activities with *eomesa*, possibly involving a complex of all three factors *eomesa*, *foxh1*, and *smad2* (Nelson et al., 2014; Slagle et al., 2011). Also during establishment of the AP axis of the mouse embryo *Eomes* functions downstream of Nodal/Smad2/3 signaling by direct binding to regulatory sequences and activation of *Lhx1* transcription required for proper formation and migration of the AVE (Nowotschin et al., 2013).

In addition, EOMES also activates target genes independently of Nodal signaling. Importantly, SMAD2/3-independent target genes are acting outside of the DE specification program to direct cells to the anterior mesoderm lineage. This was, for example, shown for the direct, EOMES-dependent and SMAD2/3-independent regulation of the key cardiogenic transcription factor *Mesp1* during mouse gastrulation and in stem cell differentiation (Costello et al., 2011; van den Ameele et al., 2012). *Mesp1* is more efficiently activated in differentiating ES cells in the presence of low Nodal/Smad2/3 signaling levels, shown by undetectable signaling active (phosphorylated) pSmad2, when compared to cells that have elevated pSMAD2/3 signaling levels after stimulation with Activin A (Costello et al., 2011; van den Ameele et al., 2012).

EOMES is generally considered to mainly function as a transcriptional activator, as demonstrated by injection of *eomes-VP16* transactivating constructs into zebrafish embryos, which mimics forced *eomesa* expression. In contrast, *eomes-eng* repressor constructs mimic *eomesa* loss of function (Bjornson et al., 2005; Bruce et al., 2003). However, genome-wide analysis of EOMES chromatin binding in combination with transcriptional profiling has challenged this view. EOMES binding is found in close proximity to genes that are positively regulated in the absence of EOMES and there is almost the same numbers of up- and downregulated candidate *Eomes*-target genes after EOMES depletion in differentiating ES cells (Teo et al., 2011). Identified, negatively regulated *Eomes*-target genes include pluripotency factors, posterior mesodermal genes, and ectodermal genes, which are normally suppressed in *Eomes*-expressing cells during gastrulation. This suggests that *Eomes* might also be able to act as negative regulator of gene expression (Nelson et al., 2014; Teo et al., 2011).

The modes of transcriptional activation and repression by EOMES are not fully explored yet. EOMES binds very broadly to chromatin and about 11,000 candidate target genes were revealed by ChIP analysis in human ES cells during differentiation to endoderm. Among those 30% are differentially expressed following shRNA-mediated EOMES ablation, including key endoderm determining factors (Teo et al., 2011). Some reports indicate that Tbx factors, such as TBX21/T-bet, which is closely related to EOMES, exhibit separable transactivating and histone-modifying activities (Lewis, Miller, Miazgowicz, Beima, & Weinmann, 2007; Miller, Huang, Miazgowicz, Brassil, & Weinmann, 2008). While not formally shown for key DE genes, one report also suggests for EOMES that it not only activates target gene transcription but additionally recruits histone-modifying enzyme activity. Accordingly, EOMES interacts with the H3K27-demethylase JMJD3, thereby removing repressive H3K27me3 marks at target gene enhancers, including its own enhancer (Kartikasari et al., 2013). Additionally, EOMES and other Tbx-proteins interact with RbBp5, a component of H3K4–methltransferase complexes that establish permissive H3K4me3 chromatin marks (Miller et al., 2008). Such a dual mechanism of direct transactivation of target genes and remodeling of chromatin state might act to ensure robust target gene activation as required for stable fate commitment during cell lineage determination.

6. CONTROL OF *EOMES* EXPRESSION DURING GASTRULATION ONSET

The *Eomes* gene in mouse is located on chromosome 9 and in humans on chromosome 3 in relatively gene-poor genomic regions. Different studies using chromatin immunoprecipitation data and in silico genome analyses suggested several, nonoverlapping enhancer elements. These seem to regulate the murine *Eomes* gene locus during gastrulation onset (Kartikasari et al., 2013; Teo et al., 2011), in the TE lineage (Ng et al., 2008), and in the murine retina (Mao et al., 2008). However, the functional significance of these enhancers for *Eomes* expression at gastrulation stage awaits experimental validation. Another study by Wolf and colleagues used interspecies genomic alignments and generated transgenic reporter lines to identify a 4.0 kb genomic element–20 kb 5′ of the transcriptional start site of *Eomes* that is sufficient to recapitulate endogenous expression using a GFP reporter. This putative enhancer contains several potential transcription factor binding sites, but their functional relevance has not been tested experimentally (Wolf, Klein, Garcia, Hyttel, & Serup, 2012).

The most widely studied upstream inducers of *Eomes* are SMAD2/3-mediated Nodal/Activin/TGFβ-signals. In Nodal-deficient mouse embryos, no *Eomes* expression can be observed (Brennan et al., 2001). Similarly, regulation of *Eomes* by Activin has also been shown in *Xenopus* and a *Xenopus* Activin-responsive genomic element has been described 5 kb upstream of the transcriptional start (Ryan, Garrett, Bourillot, Stennard, & Gurdon, 2000; Ryan et al., 1996).

Eomes is among the early developmental genes that in ES cells are found in a bivalent chromatin state (Kartikasari et al., 2013). This poised chromatin state is characterized by co-occurrence of permissive H3K4me3 and repressive H3K27me3 histone marks and is a chromatin profile often associated with regulators of lineage decisions (Bernstein et al., 2006). A thorough study by Teo et al. in human ES cells demonstrates that transcription factors of the pluripotency network, OCT3/4, NANOG, and SOX2, bind to *Eomes* genomic elements at pluripotent state, thus suppressing expression (Teo et al., 2011). With onset of differentiation to the endoderm lineage, SOX2 is not maintained and NANOG in combination with SMAD2/3, possibly also involving OCT3/4, is positively regulating *Eomes* transcription. Kartikasari and colleagues suggested that TBX3 and the H3K27me3 demethylase JMJD3 are also recruited to the *Eomes* locus at differentiation onset, thus further removing repressive histone marks (Kartikasari et al., 2013). After first activation of *Eomes*, transcription levels are further amplified by enhanced Nodal/Activin signaling, and possibly by a positive feed forward of *Eomes* on its own promoter (Kartikasari et al., 2013; Teo et al., 2011). The view of positive self-regulation of *Eomes* is further supported by the ectopic overexpression of *eomes* in zebrafish embryos, which also leads to self-induction (Bruce et al., 2003). In contrast, in mouse embryos loss-of-function mutations of *Eomes* lead to upregulated expression from the *Eomes* locus rather suggesting negative regulation of EOMES on itself (Russ et al., 2000).

In summary, during recent years studies on *Eomes* revealed its central and conserved role as a lineage specifying transcription factor during gastrulation of vertebrate embryos. In mouse, *Eomes* is important for the generation of endoderm and subtypes of anterior mesoderm. These functions are grossly conserved in nonmammalian embryos, but further studies will be required to unravel the precise contribution of *Eomes* function to different cell lineages, e.g., different mesodermal subtypes. In mammalian embryos, *Eomes* obtained additional functions in extraembryonic tissues, TE and VE, which are unique for intrauterine development. This acquisition of additional gene functions involved in the generation of evolutionary younger tissue types might represent an interesting example of gene evolution.

ACKNOWLEDGMENTS

We thank the members of the Arnold lab for suggestions and comments on the manuscript. We thank David Kimelman and Makoto Asashima for their kind contribution of images and Chasper Würmli for his help with illustrations. The authors research was supported by a research grant from the Novartis Foundation for Medical-Biological Research, Switzerland, to S.P. S.J.A. was supported by the Emmy Noether Program (AR732/1-1) and the Collaborative Research Centre 850 (CRC/SFB850, project B03) of the German Research Foundation (DFG).

REFERENCES

Arnold, S. J., Hofmann, U. K., Bikoff, E. K., & Robertson, E. J. (2008). Pivotal roles for eomesodermin during axis formation, epithelium-to-mesenchyme transition and endoderm specification in the mouse. *Development, 135*(3), 501–511. http://doi.org/10.1242/dev.014357.

Arnold, S. J., & Robertson, E. J. (2009). Making a commitment: Cell lineage allocation and axis patterning in the early mouse embryo. *Nature Reviews. Molecular Cell Biology, 10*(2), 91–103. http://doi.org/10.1038/nrm2618.

Arnold, S. J., Sugnaseelan, J., Groszer, M., Srinivas, S., & Robertson, E. J. (2009). Generation and analysis of a mouse line harboring GFP in the Eomes/Tbr2 locus. *Genesis, 47*(11), 775–781. http://doi.org/10.1002/dvg.20562.

Benchetrit, H., Herman, S., van Wietmarschen, N., Wu, T., Makedonski, K., Maoz, N., et al. (2015). Extensive nuclear reprogramming underlies lineage conversion into functional trophoblast stem-like cells. *Cell Stem Cell, 17*(5), 543–556. http://doi.org/10.1016/j.stem.2015.08.006.

Bernstein, B. E., Mikkelsen, T. S., Xie, X., Kamal, M., Huebert, D. J., Cuff, J., et al. (2006). A bivalent chromatin structure marks key developmental genes in embryonic stem cells. *Cell, 125*(2), 315–326. http://doi.org/10.1016/j.cell.2006.02.041.

Bjornson, C. R. R., Griffin, K. J. P., Farr, G. H., III, Terashima, A., Himeda, C., Kikuchi, Y., et al. (2005). Eomesodermin is a localized maternal determinant required for endoderm induction in zebrafish. *Developmental Cell, 9*(4), 523–533. http://doi.org/10.1016/j.devcel.2005.08.010.

Boyer, L. A., Lee, T. I., Cole, M. F., Johnstone, S. E., Levine, S. S., Zucker, J. P., et al. (2005). Core transcriptional regulatory circuitry in human embryonic stem cells. *Cell, 122*(6), 947–956. http://doi.org/10.1016/j.cell.2005.08.020.

Brennan, J., Lu, C. C., Norris, D. P., Rodriguez, T. A., Beddington, R. S., & Robertson, E. J. (2001). Nodal signalling in the epiblast patterns the early mouse embryo. *Nature, 411*(6840), 965–969. http://doi.org/10.1038/35082103.

Brown, S., Teo, A., Pauklin, S., Hannan, N., Cho, C. H.-H., Lim, B., et al. (2011). Activin/nodal signaling controls divergent transcriptional networks in human embryonic stem cells and in endoderm progenitors. *Stem Cells, 29*(8), 1176–1185. http://doi.org/10.1002/stem.666.

Bruce, A. E. E., Howley, C., Dixon Fox, M., & Ho, R. K. (2005). T-box gene eomesodermin and the homeobox-containing Mix/Bix gene mtx2 regulate epiboly movements in the zebrafish. *Developmental Dynamics, 233*(1), 105–114. http://doi.org/10.1002/dvdy.20305.

Bruce, A. E. E., Howley, C., Zhou, Y., Vickers, S. L., Silver, L. M., King, M. L., et al. (2003). The maternally expressed zebrafish T-box gene eomesodermin regulates organizer formation. *Development, 130*(22), 5503–5517. http://doi.org/10.1242/dev.00763.

Carver, E. A., Jiang, R., Lan, Y., Oram, K. F., & Gridley, T. (2001). The mouse snail gene encodes a key regulator of the epithelial-mesenchymal transition. *Molecular and Cellular Biology, 21*(23), 8184–8188. http://doi.org/10.1128/MCB.21.23.8184-8188.2001.

Chazaud, C., & Rossant, J. (2006). Disruption of early proximodistal patterning and AVE formation in Apc mutants. *Development, 133*(17), 3379–3387. http://doi.org/10.1242/dev.02523.

Chu, G. C., Dunn, N. R., Anderson, D. C., Oxburgh, L., & Robertson, E. J. (2004). Differential requirements for Smad4 in TGFbeta-dependent patterning of the early mouse embryo. *Development, 131*(15), 3501–3512. http://doi.org/10.1242/dev.01248.

Chuong, E. B., Rumi, M. A. K., Soares, M. J., & Baker, J. C. (2013). Endogenous retroviruses function as species-specific enhancer elements in the placenta. *Nature Genetics, 45*(3), 325–329. http://doi.org/10.1038/ng.2553.

Ciruna, B. G., & Rossant, J. (1999). Expression of the T-box gene Eomesodermin during early mouse development. *Mechanisms of Development*, *81*(1-2), 199–203.

Costello, I., Pimeisl, I.-M., Dräger, S., Bikoff, E. K., Robertson, E. J., & Arnold, S. J. (2011). The T-box transcription factor eomesodermin acts upstream of Mesp1 to specify cardiac mesoderm during mouse gastrulation. *Nature Cell Biology*, *13*(9), 1084–1091. http://doi.org/10.1038/ncb2304.

Donnison, M., Beaton, A., Davey, H. W., Broadhurst, R., L'Huillier, P., & Pfeffer, P. L. (2005). Loss of the extraembryonic ectoderm in Elf5 mutants leads to defects in embryonic patterning. *Development*, *132*(10), 2299–2308. http://doi.org/10.1242/dev.01819.

Du, S., Draper, B. W., Mione, M., Moens, C. B., & Bruce, A. (2012). Differential regulation of epiboly initiation and progression by zebrafish Eomesodermin A. *Developmental Biology*, *362*(1), 11–23. http://doi.org/10.1016/j.ydbio.2011.10.036.

Dunn, N. R., Vincent, S. D., Oxburgh, L., Robertson, E. J., & Bikoff, E. K. (2004). Combinatorial activities of Smad2 and Smad3 regulate mesoderm formation and patterning in the mouse embryo. *Development*, *131*(8), 1717–1728. http://doi.org/10.1242/dev.01072.

Fukuda, M., Takahashi, S., Haramoto, Y., Onuma, Y., Kim, Y.-J., Yeo, C.-Y., et al. (2010). Zygotic VegT is required for Xenopus paraxial mesoderm formation and is regulated by nodal signaling and eomesodermin. *The International Journal of Developmental Biology*, *54*(1), 81–92. http://doi.org/10.1387/ijdb.082837mf.

Gentsch, G. E., Owens, N. D. L., Martin, S. R., Piccinelli, P., Faial, T., Trotter, M. W. B., et al. (2013). In vivo T-box transcription factor profiling reveals joint regulation of embryonic neuromesodermal bipotency. *Cell Reports*, *4*(6), 1185–1196. http://doi.org/10.1016/j.celrep.2013.08.012.

Georgiades, P., & Rossant, J. (2006). Ets2 is necessary in trophoblast for normal embryonic anteroposterior axis development. *Development*, *133*(6), 1059–1068. http://doi.org/10.1242/dev.02277.

Gordon, S. M., Chaix, J., Rupp, L. J., Wu, J., Madera, S., Sun, J. C., et al. (2012). The transcription factors T-bet and eomes control key checkpoints of natural killer cell maturation. *Immunity*, *36*(1), 55–67. http://doi.org/10.1016/j.immuni.2011.11.016.

Hancock, S. N., Agulnik, S. I., Silver, L. M., & Papaioannou, V. E. (1999). Mapping and expression analysis of the mouse ortholog of Xenopus eomesodermin. *Mechanisms of Development*, *81*(1-2), 205–208.

Hoodless, P. A., Pye, M., Chazaud, C., Labbé, E., Attisano, L., Rossant, J., et al. (2001). FoxH1 (Fast) functions to specify the anterior primitive streak in the mouse. *Genes & Development*, *15*(10), 1257–1271. http://doi.org/10.1101/gad.881501.

Horb, M. E., & Thomsen, G. H. (1997). A vegetally localized T-box transcription factor in Xenopus eggs specifies mesoderm and endoderm and is essential for embryonic mesoderm formation. *Development*, *124*(9), 1689–1698.

Kartikasari, A. E. R., Zhou, J. X., Kanji, M. S., Chan, D. N., Sinha, A., Grapin-Botton, A., et al. (2013). The histone demethylase Jmjd3 sequentially associates with the transcription factors Tbx3 and Eomes to drive endoderm differentiation. *The EMBO Journal*, *32*(10), 1393–1408. http://doi.org/10.1038/emboj.2013.78.

Kidder, B. L., & Palmer, S. (2010). Examination of transcriptional networks reveals an important role for TCFAP2C, SMARCA4, and EOMES in trophoblast stem cell maintenance. *Genome Research*, *20*(4), 458–472. http://doi.org/10.1101/gr.101469.109.

Kitajima, S., Takagi, A., Inoue, T., & Saga, Y. (2000). MesP1 and MesP2 are essential for the development of cardiac mesoderm. *Development*, *127*(15), 3215–3226.

Kuales, G., Weiss, M., Sedelmeier, O., Pfeifer, D., & Arnold, S. J. (2015). A resource for the transcriptional signature of bona fide trophoblast stem cells and analysis of their embryonic persistence. *Stem Cells International*, *2015*(16), 13. http://doi.org/10.1155/2015/218518. Article ID 218518.

Kubaczka, C., Senner, C. E., Cierlitza, M., Araúzo-Bravo, M. J., Kuckenberg, P., Peitz, M., et al. (2015). Direct induction of trophoblast stem cells from murine fibroblasts. *Cell Stem Cell*, *17*(5), 557–568. http://doi.org/10.1016/j.stem.2015.08.005.

Kuckenberg, P., Buhl, S., Woynecki, T., van Fürden, B., Tolkunova, E., Seiffe, F., et al. (2010). The transcription factor TCFAP2C/AP-2gamma cooperates with CDX2 to maintain trophectoderm formation. *Molecular and Cellular Biology*, *30*(13), 3310–3320. http://doi.org/10.1128/MCB.01215-09.

Kwon, G. S., & Hadjantonakis, A.-K. (2007). Eomes::GFP—A tool for live imaging cells of the trophoblast, primitive streak, and telencephalon in the mouse embryo. *Genesis*, *45*(4), 208–217. http://doi.org/10.1002/dvg.20293.

Kwon, G. S., & Hadjantonakis, A.-K. (2009). Transthyretin mouse transgenes direct RFP expression or Cre-mediated recombination throughout the visceral endoderm. *Genesis*, *47*(7), 447–455. http://doi.org/10.1002/dvg.20522.

Kwon, G. S., Viotti, M., & Hadjantonakis, A.-K. (2008). The endoderm of the mouse embryo arises by dynamic widespread intercalation of embryonic and extraembryonic lineages. *Developmental Cell*, *15*(4), 509–520. http://doi.org/10.1016/j.devcel.2008.07.017.

Latos, P. A., & Hemberger, M. (2014). Review: The transcriptional and signalling networks of mouse trophoblast stem cells. *Placenta*, *35*(Suppl.), S81–S85. http://doi.org/10.1016/j.placenta.2013.10.013.

Lewis, M. D., Miller, S. A., Miazgowicz, M. M., Beima, K. M., & Weinmann, A. S. (2007). T-bet's ability to regulate individual target genes requires the conserved T-box domain to recruit histone methyltransferase activity and a separate family member-specific transactivation domain. *Molecular and Cellular Biology*, *27*(24), 8510–8521. http://doi.org/10.1128/MCB.01615-07.

Mao, C.-A., Kiyama, T., Pan, P., Furuta, Y., Hadjantonakis, A.-K., & Klein, W. H. (2008). Eomesodermin, a target gene of Pou4f2, is required for retinal ganglion cell and optic nerve development in the mouse. *Development*, *135*(2), 271–280. http://doi.org/10.1242/dev.009688.

Miller, S. A., Huang, A. C., Miazgowicz, M. M., Brassil, M. M., & Weinmann, A. S. (2008). Coordinated but physically separable interaction with H3K27-demethylase and H3K4-methyltransferase activities are required for T-box protein-mediated activation of developmental gene expression. *Genes & Development*, *22*(21), 2980–2993. http://doi.org/10.1101/gad.1689708.

Nelson, A. C., Cutty, S. J., Niini, M., Stemple, D. L., Flicek, P., Houart, C., et al. (2014). Global identification of Smad2 and Eomesodermin targets in zebrafish identifies a conserved transcriptional network in mesendoderm and a novel role for eomesodermin in repression of ectodermal gene expression. *BMC Biology*, *12*(1), 81. http://doi.org/10.1186/s12915-014-0081-5.

Ng, R. K., Dean, W., Dawson, C., Lucifero, D., Madeja, Z., Reik, W., et al. (2008). Epigenetic restriction of embryonic cell lineage fate by methylation of Elf5. *Nature Cell Biology*, *10*(11), 1280–1290. http://doi.org/10.1038/ncb1786.

Nishioka, N., Inoue, K.-I., Adachi, K., Kiyonari, H., Ota, M., Ralston, A., et al. (2009). The hippo signaling pathway components Lats and Yap pattern Tead4 activity to distinguish mouse trophectoderm from inner cell mass. *Developmental Cell*, *16*(3), 398–410. http://doi.org/10.1016/j.devcel.2009.02.003.

Nishioka, N., Yamamoto, S., Kiyonari, H., Sato, H., Sawada, A., Ota, M., et al. (2008). Tead4 is required for specification of trophectoderm in pre-implantation mouse embryos. *Mechanisms of Development*, *125*(3-4), 270–283. http://doi.org/10.1016/j.mod.2007.11.002.

Niwa, H., Toyooka, Y., Shimosato, D., Strumpf, D., Takahashi, K., Yagi, R., et al. (2005). Interaction between Oct3/4 and Cdx2 determines trophectoderm differentiation. *Cell*, *123*(5), 917–929. http://doi.org/10.1016/j.cell.2005.08.040.

Nowotschin, S., Costello, I., Piliszek, A., Kwon, G. S., Mao, C.-A., Klein, W. H., et al. (2013). The T-box transcription factor eomesodermin is essential for AVE induction in the mouse embryo. *Genes & Development, 27*(9), 997–1002. http://doi.org/10.1101/gad.215152.113.

Papaioannou, V. E. (2014). The T-box gene family: Emerging roles in development, stem cells and cancer. *Development, 141*(20), 3819–3833. http://doi.org/10.1242/dev.104471.

Pearce, E. L., Mullen, A. C., Martins, G. A., Krawczyk, C. M., Hutchins, A. S., Zediak, V. P., et al. (2003). Control of effector CD8+ T cell function by the transcription factor Eomesodermin. *Science, 302*(5647), 1041–1043. http://doi.org/10.1126/science.1090148.

Perea-Gómez, A., Shawlot, W., Sasaki, H., Behringer, R. R., & Ang, S. (1999). HNF3beta and Lim1 interact in the visceral endoderm to regulate primitive streak formation and anterior-posterior polarity in the mouse embryo. *Development, 126*(20), 4499–4511.

Picozzi, P., Wang, F., Cronk, K., & Ryan, K. (2009). Eomesodermin requires transforming growth factor-beta/activin signaling and binds Smad2 to activate mesodermal genes. *The Journal of Biological Chemistry, 284*(4), 2397–2408. http://doi.org/10.1074/jbc.M808704200.

Ralston, A., Cox, B. J., Nishioka, N., Sasaki, H., Chea, E., Rugg-Gunn, P., et al. (2010). Gata3 regulates trophoblast development downstream of Tead4 and in parallel to Cdx2. *Development, 137*(3), 395–403. http://doi.org/10.1242/dev.038828.

Ralston, A., & Rossant, J. (2008). Cdx2 acts downstream of cell polarization to cell-autonomously promote trophectoderm fate in the early mouse embryo. *Developmental Biology, 313*(2), 614–629. http://doi.org/10.1016/j.ydbio.2007.10.054.

Rivera-Pérez, J. A., & Hadjantonakis, A.-K. (2015). The dynamics of morphogenesis in the early mouse embryo. *Cold Spring Harbor Perspectives in Biology*, 7(11), a015867. http://doi.org/10.1101/cshperspect.a015867.

Russ, A. P., Wattler, S., Colledge, W. H., Aparicio, S. A., Carlton, M. B., Pearce, J. J., et al. (2000). Eomesodermin is required for mouse trophoblast development and mesoderm formation. *Nature, 404*(6773), 95–99. http://doi.org/10.1038/35003601.

Ryan, K., Garrett, N., Bourillot, P., Stennard, F., & Gurdon, J. B. (2000). The Xenopus eomesodermin promoter and its concentration-dependent response to activin. *Mechanisms of Development, 94*(1-2), 133–146.

Ryan, K., Garrett, N., Mitchell, A., & Gurdon, J. B. (1996). Eomesodermin, a key early gene in Xenopus mesoderm differentiation. *Cell, 87*(6), 989–1000.

Shawlot, W., & Behringer, R. R. (1995). Requirement for Lim1 in head-organizer function. *Nature, 374*(6521), 425–430. http://doi.org/10.1038/374425a0.

Sinner, D., Kirilenko, P., Rankin, S., Wei, E., Howard, L., Kofron, M., et al. (2006). Global analysis of the transcriptional network controlling xenopus endoderm formation. *Development, 133*(10), 1955–1966. http://doi.org/10.1242/dev.02358.

Slagle, C. E., Aoki, T., & Burdine, R. D. (2011). Nodal-dependent mesendoderm specification requires the combinatorial activities of FoxH1 and eomesodermin. *PLoS Genetics*, 7(5), e1002072. http://doi.org/10.1371/journal.pgen.1002072.

Soriano, P. (1999). Generalized lacZ expression with the ROSA26 Cre reporter strain. *Nature Genetics, 21*(1), 70–71. http://doi.org/10.1038/5007.

Stower, M. J., & Srinivas, S. (2014). Heading forwards: Anterior visceral endoderm migration in patterning the mouse embryo. *Philosophical Transactions of the Royal Society of London. Series B, Biological Sciences, 369*(1657), 20130546. http://doi.org/10.1098/rstb.2013.0546.

Strumpf, D., Mao, C.-A., Yamanaka, Y., Ralston, A., Chawengsaksophak, K., Beck, F., et al. (2005). Cdx2 is required for correct cell fate specification and differentiation of trophectoderm in the mouse blastocyst. *Development, 132*(9), 2093–2102. http://doi.org/10.1242/dev.01801.

Sun, X., Meyers, E. N., Lewandoski, M., & Martin, G. R. (1999). Targeted disruption of Fgf8 causes failure of cell migration in the gastrulating mouse embryo. *Genes & Development, 13*(14), 1834–1846.

Takaoka, K., Yamamoto, M., & Hamada, H. (2011). Origin and role of distal visceral endoderm, a group of cells that determines anterior–posterior polarity of the mouse embryo. *Nature Cell Biology, 13*(7), 743–752. http://doi.org/10.1038/ncb2251.

Tam, P. P., & Behringer, R. R. (1997). Mouse gastrulation: The formation of a mammalian body plan. *Mechanisms of Development, 68*(1-2), 3–25.

Tanaka, S., Kunath, T., Hadjantonakis, A. K., Nagy, A., & Rossant, J. (1998). Promotion of trophoblast stem cell proliferation by FGF4. *Science, 282*(5396), 2072–2075.

Tayade, C., Fang, Y., Black, G. P., V A, P., Erlebacher, A., & Croy, B. A. (2005). Differential transcription of Eomes and T-bet during maturation of mouse uterine natural killer cells. *Journal of Leukocyte Biology, 78*(6), 1347–1355. http://doi.org/10.1189/jlb.0305142.

Teo, A. K. K., Arnold, S. J., Trotter, M. W. B., Brown, S., Ang, L. T., Chng, Z., et al. (2011). Pluripotency factors regulate definitive endoderm specification through eomesodermin. *Genes & Development, 25*(3), 238–250. http://doi.org/10.1101/gad.607311.

Tremblay, K. D., Hoodless, P. A., Bikoff, E. K., & Robertson, E. J. (2000). Formation of the definitive endoderm in mouse is a Smad2-dependent process. *Development, 127*(14), 3079–3090.

van den Ameele, J., Tiberi, L., Bondue, A., Paulissen, C., Herpoel, A., Iacovino, M., et al. (2012). Eomesodermin induces Mesp1 expression and cardiac differentiation from embryonic stem cells in the absence of activin. *EMBO Reports, 13*(4), 355–362. http://doi.org/10.1038/embor.2012.23.

Vincent, S. D., Dunn, N. R., Hayashi, S., Norris, D. P., & Robertson, E. J. (2003). Cell fate decisions within the mouse organizer are governed by graded Nodal signals. *Genes & Development, 17*(13), 1646–1662. http://doi.org/10.1101/gad.1100503.

Viotti, M., Nowotschin, S., & Hadjantonakis, A. K. (2014). SOX17 links gut endoderm morphogenesis and germ layer segregation. *Nature Cell Biology, 16*(12), 1146–1156. http://doi.org/10.1038/ncb3070. Epub 2014 Nov 24. PMID: 25419850.

Werling, U., & Schorle, H. (2002). Transcription factor gene AP-2 gamma essential for early murine development. *Molecular and Cellular Biology, 22*(9), 3149–3156. http://doi.org/10.1128/MCB.22.9.3149-3156.2002.

Wolf, X. A. K., Klein, T., Garcia, R., Hyttel, P., & Serup, P. (2012). Identification of a conserved cis-acting region driving expression of mouse eomesodermin to the primitive streak, node, and definitive endoderm. *Gene Expression Patterns, 12*(1-2), 85–93. http://doi.org/10.1016/j.gep.2011.06.003.

Xu, P., Zhu, G., Wang, Y., Sun, J., Liu, X., Chen, Y.-G., et al. (2014). Maternal eomesodermin regulates zygotic nodal gene expression for mesendoderm induction in zebrafish embryos. *Journal of Molecular Cell Biology, 6*(4), 272–285. http://doi.org/10.1093/jmcb/mju028.

Yagi, R., Kohn, M. J., Karavanova, I., Kaneko, K. J., Vullhorst, D., DePamphilis, M. L., et al. (2007). Transcription factor TEAD4 specifies the trophectoderm lineage at the beginning of mammalian development. *Development, 134*(21), 3827–3836. http://doi.org/10.1242/dev.010223.

Yamamoto, H., Flannery, M. L., Kupriyanov, S., Pearce, J., McKercher, S. R., Henkel, G. W., et al. (1998). Defective trophoblast function in mice with a targeted mutation of Ets2. *Genes & Development, 12*(9), 1315–1326.

Zohn, I. E., Li, Y., Skolnik, E. Y., Anderson, K. V., Han, J., & Niswander, L. (2006). p38 and a p38-interacting protein are critical for downregulation of E-cadherin during mouse gastrulation. *Cell, 125*(5), 957–969. http://doi.org/10.1016/j.cell.2006.03.048.

CHAPTER FIVE

Cooperation Between T-Box Factors Regulates the Continuous Segregation of Germ Layers During Vertebrate Embryogenesis

G.E. Gentsch[1], R.S. Monteiro, J.C. Smith
Developmental Biology Laboratory, The Francis Crick Institute, Mill Hill Laboratory, London, United Kingdom
[1]Corresponding author: e-mail address: george.gentsch@crick.ac.uk

Contents

1. Controversy of Germ Layer Segregation: Holmdahl's Blastema vs Vogt's Predetermination Model 118
2. Multiple Functions of T-Box Factors in Development and Disease 120
   2.1 Evolution of T-Box Function to Specify Germ Layers 121
   2.2 How Do T-Box Factors Regulate Transcription? 122
3. Early Embryonic Roles of T-Box Factors in Segregating and Differentiating Germ Layers in Vertebrates 126
   3.1 T-Box Factors Are Consecutively Expressed in Zones of Germ Layer Segregation 126
   3.2 T-Box Factors Control the Emergence of Distinct Germ Layer Subtypes 131
4. Implications for Reverse Engineering ES Cells 141
5. Summary 143
Acknowledgments 143
References 143

Abstract

A wild-type vertebrate embryo first generates its head and then extends its main body axis by successively appending trunk and tail. This rostro-caudal (head-to-tail) development is initiated by a set of morphogenetic movements known as gastrulation that recruits multipotent cells into one of the three morphologically distinct germ layers: ectoderm, endoderm, and mesoderm. These primordial tissues go on to form complementary sets of connective tissues and organs to build the head and at least some of the trunk. In contrast, the tail appears to be formed without clear germ layer segregation from a terminally located growth zone (the tailbud). Recent research shows that the tailbud retains some pregastrulation multipotency to generate derivatives of different germ layers such as spinal cord (ectoderm) and skeletal muscle (mesoderm). This review

Current Topics in Developmental Biology, Volume 122
ISSN 0070-2153
http://dx.doi.org/10.1016/bs.ctdb.2016.07.012

discusses the emerging role of T-box transcription factors in this process and their intricate relationship with other genetic components to regulate multipotency and germ layer segregation during axial elongation—from gastrulation to the termination of tail growth.

1. CONTROVERSY OF GERM LAYER SEGREGATION: HOLMDAHL'S BLASTEMA VS VOGT'S PREDETERMINATION MODEL

The idea that vertebrate organs originate from three germ layers was introduced in the early 19th century. Dissections of the early chick embryo revealed the cell movements of gastrulation, during which a rather uniform tissue, the blastoderm, delaminates into three discrete germ layers: ectoderm, containing both surface and neural ectoderm; endoderm; and mesoderm (Pander, 1817; Remak, 1855). However, the notion that gastrulation is the primary branching point in organogenesis does not apply to all body compartments along the anteroposterior axis. In line with the cephalocaudal "law," the head, trunk, and tail compartments are formed successively in all vertebrate embryos (Kingsbury, 1932). While the head and at least some of the trunk are shaped by gastrulation, the tail appears to develop without germ layer segregation from a terminally located population of mesenchymal cells (Griffith, Wiley, & Sanders, 1992). As early as 1884, histological evidence indicated that the posterior nervous system (spinal cord) derives from mesodermal progenitors rather than from ectoderm, thus contradicting the germ layer theory (Kölliker, 1884).

With the different modes of development at the rostral and caudal ends of the embryo in mind, axial elongation can be viewed separately as that driven by gastrulation (primary elongation) and that driven by blastema or limb bud-like outgrowth (secondary elongation), respectively (Grüneberg, 1956; Holmdahl, 1925). Similarly, morphogenesis of the neural tube in amniotes switches at the lumbosacral level from the fusion of neural folds in the head and trunk (primary neurulation) to the cavitation of the medullary cord in the tail (secondary neurulation) (Beck & Slack, 1999; Kingsbury, 1932; Schoenwolf, 1984; Schoenwolf & Delongo, 1980). The grafting, extirpation, and in vitro culturing of tailbud tissue showed that it can generate secondary neural tubes and other axial structures of the other two germ layers (Criley, 1969; Griffith & Sanders, 1991; Schoenwolf, 1978; Seevers, 1932).

These observations notwithstanding, tail formation has also been regarded as a continuation of gastrulation, mainly because of the seamless transition from trunk tissue to tail (Vogt, 1926). According to Vogt, the tailbud contains a mosaic population of cells predetermined during gastrulation. In fact, specific regions within the amphibian tailbud, the chordoneural hinge, and the posterior wall of the neurenteric canal, both of which give rise to mesodermal and neural structures, could be traced back to the dorsal blastopore lip of gastrula embryos (Gont, Steinbeisser, Blumberg, & De Robertis, 1993; Pasteels, 1939, 1942). Similarly derived cell populations were subsequently found in other vertebrate species (Cambray & Wilson, 2007; Catala, Teillet, & Le Douarin, 1995; Nakao & Ishizawa, 1984). Close inspection of the tailbud using chimeric grafts or fluorescently labeled cell clusters also revealed ongoing gastrulation-like cell movements such as ingression and divergence (Catala et al., 1995; Kanki & Ho, 1997; Knezevic, De Santo, & Mackem, 1998).

Holmdahl's idea of a homogeneous growth center was also challenged because the mitotic indices of cells within the tailbud were similar to, or even lower than, those of their derivatives or those of primary germ layers (Gaertner, 1949; Kanki & Ho, 1997; Mills & Bellairs, 1989; Pasteels, 1943; Schoenwolf, 1977). Indeed, quiescence seems to be essential for mesodermal progenitors to undergo normal muscle differentiation (Bouldin, Snelson, Farr, & Kimelman, 2014). In addition, tailbud and chordoneural hinge tissue both retain the inductive capacity of the gastrula organizer (Spemann & Mangold, 1924), although they consistently induce uncommitted host cells to form tail rather than head structures (Bytinski-Salz, 1931; Gont et al., 1993; Knezevic et al., 1998).

More recent studies reconcile the opposing models of Holmdahl and Vogt, suggesting that the tailbud is neither uniformly multipotent nor a mosaic of fixed lineages. On the one hand, lineage tracing of single or small clusters of cells confirmed that the tailbud contains independent populations of progenitor cells with distinct potencies, rather like the zone of germ layer segregation during gastrulation (Davis & Kirschner, 2000; Lawson, Meneses, & Pedersen, 1991; Row, Tsotras, Goto, & Martin, 2016; Selleck & Stern, 1991; Tzouanacou, Wegener, Wymeersch, Wilson, & Nicolas, 2009; Wymeersch et al., 2016). On the other hand, heterotopic grafts showed that potency is defined by anatomical position rather than intrinsic properties of the progenitor cell. For example, fields that impose neuromesodermal bipotency are found at the posterior border of the organizer (eg, node–streak border in amniotes) or in the chordoneural hinge

(Cambray & Wilson, 2002, 2007; McGrew et al., 2008; Row et al., 2016; Wymeersch et al., 2016).

The continuous and spatially restricted segregation of germ layers is reflected by specific gene expression profiles in and around populations of multipotent progenitor cells. Various studies report genes that retain their expression patterns beyond gastrulation to the tailbud stages (Beck & Slack, 1998; Beck, Whitman, & Slack, 2001; Gont et al., 1993; Joly et al., 1992; Knezevic et al., 1998; Wilson & Beddington, 1997). This review deals with genes of the T-box transcription factor family that are induced and maintained to regulate primary cell fate decisions in vertebrates. We place most emphasis on the *Xenopus* embryo and point out any features that seem to be unique to this species. However, since the mechanisms of cephalocaudal development are highly conserved among vertebrates, most observations have proved to be applicable to other vertebrate embryos as well. We will discuss (1) how T-box factors have evolved ultimately to specify germ layers in vertebrates; (2) how they control other gene networks; (3) how they and their up- and downstream regulators affect multipotency and differentiation; and (4) what conclusions can be drawn to reverse engineer embryonic stem (ES) cells in vitro.

2. MULTIPLE FUNCTIONS OF T-BOX FACTORS IN DEVELOPMENT AND DISEASE

The T-box family of transcription factors is required, in a concentration-dependent manner, for various events during vertebrate development, including stem cell survival, differentiation, proliferation, epithelial-to-mesenchymal transition (EMT), and left–right asymmetry (Papaioannou, 2014). Bearing in mind the observed dose sensitivity, it is not surprising that some family members have been associated with familial and sporadic cancers including chordomas, mammary carcinoma, and adenocarcinoma (Ciriello et al., 2015; Jacobs et al., 2000; Kelley et al., 2014; Park et al., 2008; Yang et al., 2009). T-box mutations have also been linked to several human congenital disorders (Papaioannou, 2014; Ghosh, Brook, & Wilsdon, 2017). For example, missense mutations of the T-box gene *Brachyury* have been detected in patients with neural tube defects including caudal regression syndrome and *spina bifida occulta* (Ghebranious et al., 2008; Postma et al., 2014; Shaheen et al., 2015). *Brachyury* (from the Greek *brakhus oura* meaning short tail), the founding member of the T-box family, was named after its originally reported semidominant heterozygous mutation that produced a short-tailed mouse (Dobrovolskaia-Zavadskaia, 1927).

2.1 Evolution of T-Box Function to Specify Germ Layers

The most ancient T-box factor *Brachyury* has been tracked back in evolution as far as the last common ancestor of the metazoan/fungal (opisthokont) clade (Sebé-Pedrós et al., 2013; Sebé-Pedrós & Ruiz-Trillo, 2017). Subsequently, the T-box family expanded with the emergence of true metazoan animals through the duplication of preexisting genes or genomes to as many as 32 T-box genes in the bilaterian lineage (Sebé-Pedrós et al., 2013). The vertebrate T-box factors have been divided into five subfamilies and eight groups based on the phylogenetic analysis of a prototypic set of T-box domains predating vertebrate genome duplication and functional redundancy: *Brachyury/T* (*T/Tbx19*), *Tbx1* (*Tbx1/10*, *Tbx15/18/22*, *Tbx20*), *Tbx2* (*Tbx2/3*, *Tbx4/5*), *Tbx6* (*Tbx6/Tbx16/Mga*), and *Eomesodermin* (*Tbr1/Eomesodermin/Tbx21*) (Ruvinsky, Silver, & Gibson-Brown, 2000; Takatori et al., 2004; Windner, Bird, Patterson, Doris, & Devoto, 2012). The last two subfamilies (*Tbx6* and *Eomes*) have so far only been found in bilaterians.

Spatial expression data obtained from lower metazoans such as the sponge (Adell & Müller, 2005) and the jellyfish (Scholz & Technau, 2003; Yamada, Pang, Martindale, & Tochinai, 2007) suggest that the original function of T-box factors was the control of cell identity and morphogenetic movements. The diversification of the T-box family then created new embryonic domains of gene expression. In particular, orthologues of the *Brachyury*, *Eomesodermin* (*Eomes*), and *Tbx6* subfamilies were co-opted to pattern lineage-specific germ layers in larval development of invertebrates. *Brachyury* transcription was consistently deployed during gastrulation to create terminal structures such as the endoderm-derived hindgut (Arendt, Technau, & Wittbrodt, 2001; Croce, Lhomond, & Gache, 2001; Gross & McClay, 2001; Harada, Yasuo, & Satoh, 1995; Kispert, Herrmann, Leptin, & Reuter, 1994; Peterson, Cameron, Tagawa, Satoh, & Davidson, 1999; Peterson, Harada, Cameron, & Davidson, 1999; Rast, Cameron, Poustka, & Davidson, 2002; Singer, Harbecke, Kusch, Reuter, & Lengyel, 1996; Tagawa, Humphreys, & Satoh, 1998). *Eomes* expression was often transiently detected at the onset of gastrulation to specify both mesoderm and endoderm (Fuchikami et al., 2002; Shoguchi, Satoh, & Maruyama, 1999, 2000; Tagawa, Humphreys, & Satoh, 2000).

Sustained expression of *Brachyury* at the posterior end of the embryo promoted chordate evolution by setting up the phylum-defining notochord, a mesoderm-derived structure supporting and patterning the main body axis (Bassham & Postlethwait, 2000; Chiba, Jiang, Satoh, & Smith, 2009; Holland, Koschorz, Holland, & Herrmann, 1995; José-Edwards et al., 2015; Yasuo & Satoh, 1993). Meanwhile, *Tbx6* homologues have been introduced into the bilaterian lineage to specify both rostral and caudal mesoderm-derived structures such as the heart and muscular systems, respectively (Andrikou, Iovene, Rizzo, Oliveri, & Arnone, 2013; Reim & Frasch, 2005; Reim, Lee, & Frasch, 2003; Yasuo, Kobayashi, Shimauchi, & Satoh, 1996).

2.2 How Do T-Box Factors Regulate Transcription?

Like most families of transcription factors that coordinate gene regulatory networks, a highly conserved DNA-binding domain is present in all members of the T-box family. The so-called T-box domain of about 180 amino acids binds double-stranded DNA in a sequence-specific manner. A high-affinity DNA-binding motif was first identified in vitro through the systematic evolution of ligands by exponential enrichment (SELEX) (Kispert & Herrmann, 1993). Brachyury showed preferential binding to a near-palindromic DNA sequence. X-ray crystallography confirmed that a truncated version of *Xenopus* Brachyury (Xbra) binds this motif as a homodimer (Müller & Herrmann, 1997). Further analysis using both fewer rounds of SELEX and electrophoretic mobility shift assays suggested that a half-site of the palindromic motif is sufficient for Brachyury to bind DNA as a monomer (Casey, O'Reilly, Conlon, & Smith, 1998; Casey et al., 1999; Conlon, Fairclough, Price, Casey, & Smith, 2001; Tada, Casey, Fairclough, & Smith, 1998). Indeed, genome-wide chromatin profiling indicates that in vivo T-box factors of the Brachyury, Eomes, and Tbx6 subfamily most frequently recognize variants of an eight to nine base pair core with the consensus sequence (T)TVRCACHT (Fig. 1; Faial et al., 2015; Gentsch et al., 2013; Lolas, Valenzuela, Tjian, & Liu, 2014; Morley et al., 2009; Nelson et al., 2014; Teo et al., 2011; Tsankov et al., 2015; Windner et al., 2015). Some motif variants were detected with a strong preference for adenines 5′ to the core sequence, which is common to transcription factors requiring an enlarged minor groove for DNA binding (Gentsch et al., 2013; Jolma et al., 2013). Surface plasmon resonance estimated a high affinity for the most stringently conserved motif TTCACACCT, with half the

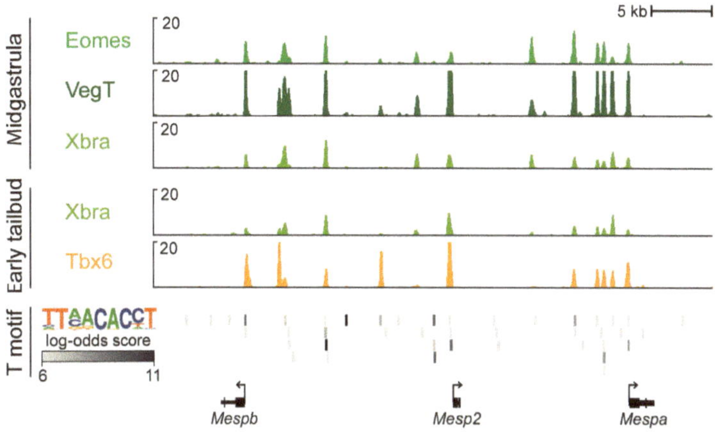

Fig. 1 Chromatin recruitment of Eomes, VegT, Xbra, and Tbx6 to the *Mesp* gene cluster (*Mespb*, *Mesp2*, and *Mespa*) at mid-gastrula and early tailbud stages. These T-box factors most frequently recognize a consensus sequence that resembles the canonical T-motif. Genomic sites (*enlarged boxes* underneath the chromatin profiles) that agree with the position–weight matrix of the motif are shown in *shades of gray* representing log-odds score. Higher score indicates a better match. ChIP-Seq profiles are normalized to 10 million mapped reads. *Data from Gentsch, G. E., Owens, N. D. L., Martin, S. R., Piccinelli, P., Faial, T., Trotter, et al. (2013). In vivo T-box transcription factor profiling reveals joint regulation of embryonic neuromesodermal bipotency.* Cell Reports, 4*(6), 1185–1196. http://doi.org/10.1016/j.celrep.2013.08.012 and unpublished (G.E.G. and J.C.S.) work.*

sites occupied by Xbra (defined as the dissociation constant K_d) at a concentration of ~14 nM. The most strongly conserved base pairs at the second and fifth position are critical, as they provide major contact points in the minor and major groove with the Xbra T-domain (Gentsch et al., 2013; Müller & Herrmann, 1997). However, at a high nuclear concentration, Xbra may use its homodimerization capacity in vivo to bind sites with near-palindromic sequences that are more degenerate and of lower affinity than single motifs (Conlon et al., 2001; Gentsch et al., 2013).

Steric hindrance may prevent Eomes and VegT, a Tbx16 orthologue, from binding these sites efficiently in *Xenopus* embryos (Gentsch et al., 2013; G.E.G. & J.C.S., unpublished). Similarly, both Tbx1 and Tbx2 (representing two other T-box subfamilies) were reported to bind perfect palindromic sites in vitro, but only Tbx1 did so as a homodimer (Sinha, Abraham, Gronostajski, & Campbell, 2000). The crystal structure of the DNA-bound Tbx3 (the closest orthologue to Tbx2) confirmed that at least some T-box factors can bind the canonical DNA-binding motif as a monomer (Coll, Seidman, & Müller, 2002). More recent high-throughput SELEX has concluded that different orientations and gaps between two canonical core motifs are critical for all groups of T-box factors to homodimerize (Jolma et al., 2013). However, such perfect palindromic DNA sequences have not evolved in vivo as they are rarely found in vertebrate genomes. Furthermore, most transcription factors occupy chromatin in order of (i) degree of accessibility; (ii) posttranslational modifications; and (iii) intrinsic DNA affinities (Arvey, Agius, Noble, & Leslie, 2012; Li et al., 2011; Pique-Regi et al., 2011).

How the recruitment of a T-box factor to chromatin regulates tissue-specific gene transcription is still a matter of debate. Several *cis* and *trans* qualities have been suggested to change target gene activity. Number, affinity, spacing, orientation, and promoter proximity of T-box binding motifs can influence *cis*-regulation of target genes. For example, inverted repeats of the canonical binding motif yielded the highest response to Brachyury when separated by 24 base pairs in a heterologous reporter assay (Kispert, Koschorz, & Herrmann, 1995). The number of functional T-box binding sites was also correlated with the temporal readout of gene induction (Katikala et al., 2013). The notochord-specific activity of wild-type and mutant *cis*-regulatory modules was transiently tested in ascidian embryos. This study suggested that Brachyury regulates early-onset target genes via two to three synergistically acting binding sites. Reducing the number of these sites delayed gene expression to the time of onset of later-induced target genes driven by only one T-box binding site.

A T-box factor may also modulate target gene activity based on its nuclear concentration, posttranslational modifications, specific protein interactions in and outside the T-domain, and developmental context such as previous and present expression of other factors. Nuclear localization signal and nuclear export signal (NES) have both been found in most T-box factors, where they regulate shuttling between nucleus and cytoplasm. In particular, the evolutionarily conserved NES within the T-box domain interacts with the nuclear export receptor CRM1, a member of the importin-β superfamily (Kulisz & Simon, 2008). Further deletion analysis of mouse Brachyury revealed that the C-terminal half (nonconserved among T-box factors) contains two autonomous transactivation domains both of which are damped by adjacent repression domains (Kispert, Koschorz, et al., 1995). Thus, the intrinsic capacity of any nuclear protein interacting with one of these domains or with the T-box domain itself determines to what degree a T-box factor acts as a transcriptional activator or repressor. Given the right developmental context such as naïve ectodermal tissue of the *Xenopus* embryo (the so-called animal cap), Xbra, VegT, or Eomes are all capable of directly activating the same mesoderm-specific genes. However, some target genes are preferentially induced by one or the other T-box factor, such as *Xbra3* (*Xbra* paralogue) by Xbra or the endoderm-specific gene *Sox17β* by VegT (Gentsch et al., 2013). Distinct mesodermal derivatives are generated depending on the level of Xbra and the expression of other factors (Cunliffe & Smith, 1992, 1994; O'Reilly, Smith, & Cunliffe, 1995). For example, Xbra requires Foxa4 in a synergistic and dose-dependent manner to form notochord tissue (O'Reilly et al., 1995). Indeed, the DNA occupancy levels and regulatory roles of Brachyury are cell lineage dependent as shown for both endoderm and posterior mesoderm derived from human ES cells by stimulating the Nodal and BMP pathway, respectively (Faial et al., 2015). The SMAD signaling mediators and Brachyury interact with each other probably through their respective MH2 and N-terminal domains, as shown for Smad1, to synergistically promote these lineage-specific gene regulatory networks (Faial et al., 2015; Messenger et al., 2005). Another stable interaction was detected between the homeobox and T-box factor DNA-binding domains. In zebrafish and *Xenopus*, T-box and Mix homeobox proteins form transcriptional regulatory complexes to specify endoderm (Bjornson et al., 2005; Pereira et al., 2011).

While most members of the T-box family have been demonstrated to facilitate gene activation, direct evidence for T-box factor-mediated repression is less abundant. T-box factors often regulate dual cell fate decisions. Thus, their functional disruption promotes one cell fate at the expense of another. However, the mechanism of cell fate suppression has not been explicitly investigated in the context of germ layer formation. In melanoma cells, the binding of tumor suppressor protein retinoblastoma was reported to enhance the intrinsic capacity of Tbx2 to repress the cell cycle regulator *Cdkn1a* (Vance, Shaw, Rodriguez, Ott, & Goding, 2010).

Despite increased understanding of the T-box regulatory code, more mechanistic insights are required to predict functional *cis*-regulatory modules and how they work in vivo. Recent advances in genome engineering such as the type II CRISPR system make it possible now to readily examine *cis*-regulation (Cong et al., 2013).

3. EARLY EMBRYONIC ROLES OF T-BOX FACTORS IN SEGREGATING AND DIFFERENTIATING GERM LAYERS IN VERTEBRATES

The following sections focus on the T-box factors *Eomes*, *VegT*, *Brachyury*, and *Tbx6*, because these contribute to vertebrate germ layer segregation and differentiation during axial elongation—from gastrulation to the termination of tail growth. Other reviews give a broader account of T-box biology and focus less on amphibians (Naiche, Harrelson, Kelly, & Papaioannou, 2005; Papaioannou, 2014; Takashima & Suzuki, 2013; Wardle & Papaioannou, 2008).

3.1 T-Box Factors Are Consecutively Expressed in Zones of Germ Layer Segregation

Unlike other vertebrates, amphibian oocytes possess a polarized cylindrical symmetry along the so-called animal–vegetal axis, where several maternal transcripts, such as a specific isoform of *VegT* (m*VegT*), are translocated to the vegetal cortex during oogenesis (Fukuda et al., 2010; Horb & Thomsen, 1997; Kloc & Etkin, 1995; Stennard, Carnac, & Gurdon, 1996; Stennard, Zorn, Ryan, Garrett, & Gurdon, 1999; Zhang & King, 1996). As a consequence of this asymmetry, the bulk of the vegetal hemisphere goes on to form only derivatives of the endodermal germ layer (Bauer, Huang, & Moody, 1994; Dale & Slack, 1987; Wylie, Snape, Heasman, & Smith, 1987). As in other vertebrate species, many maternal transcripts such as m*VegT* start to be degraded after the mid-blastula transition (stage 8.5 of Nieuwkoop and Faber) (Duval et al., 1990; Owens et al., 2016; Stennard et al., 1999).

In *Xenopus*, three T-box genes are sequentially activated in an equatorial position between the animal and vegetal poles during the few hours between the mid-blastula and early gastrula stages (Figs. 2 and 3): *Eomes* at stage 8.5, another isoform of *VegT* (z*VegT*) at stage 9, and *Xbra* at stage 9.5 (Fukuda et al., 2010; Horb & Thomsen, 1997; Owens et al., 2016; Ryan, Garrett, Mitchell, & Gurdon, 1996; Smith, Price, Green, Weigel, & Herrmann, 1991; Stennard et al., 1999; Zhang & King, 1996). The gene expression domains mark the so-called marginal zone, which contains both involuting (superficial, epithelial endoderm, and deep nonepithelial mesoderm) and noninvoluting (surface and neural ectoderm) progenitor cells (Keller, 1991; Keller, Danilchik, Gimlich, & Shih, 1985). The expression of each

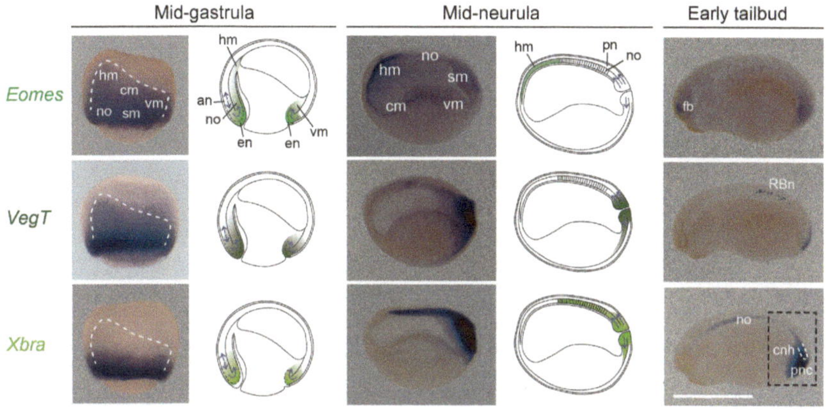

Fig. 2 Dynamic expression patterns of the T-box factors *Eomes*, *VegT*, and *Xbra* in *Xenopus* embryos from mid-gastrula to early tailbud stage. Schematic illustrations of sagittal sections are shown to the right of corresponding whole-mount in situ hybridizations (WMISH). *Arrows* show the approximate direction of cell movements. Orientations are as follows: mid-gastrula embryos, dorsal to the right and future posterior end to the bottom; mid-neurula and early tailbud, dorsal to the top and posterior to the left. *Dashed line* in WMISH of mid-gastrula embryos marks the leading edge of the mesodermal mantle. Schematic diagram of the tailbud (*boxed area*) is shown in Fig. 5. Abbreviations of germ layer subtypes, progenitor fields, and other embryonic structures: *an*, anterior neural plate; *cm*, cardiovascular mesoderm; *cnh*, chordoneural hinge (marked by *dashed line*); *en*, suprablastoporal endoderm; *fb*, forebrain; *hm*, head mesoderm; *no*, notochord; *pn*, posterior neuroectoderm; *pnc*, posterior wall of the neurenteric canal; *RBn*, Rohon-Beard neurons, *sm*, somitic mesoderm; *vm*, ventral mesoderm (e.g., forms ventral blood island, see Fig. 4). Scale bar, 0.5 mm. *Modified from Gentsch, G. E., Owens, N. D. L., Martin, S. R., Piccinelli, P., Faial, T., Trotter, et al. (2013). In vivo T-box transcription factor profiling reveals joint regulation of embryonic neuromesodermal bipotency.* Cell Reports, 4*(6), 1185–1196. http://doi.org/10.1016/j.celrep.2013.08.012.*

gene starts on the dorsal side of the embryo before spreading ventrally. In *Xenopus*, the dorsoventral axis is specified by sperm-induced cortical rotation and the nuclear translocation of stabilized β-catenin on the future dorsal side of the 16-cell embryo (Larabell et al., 1997; Rowning et al., 1997; Schneider, Steinbeisser, Warga, & Hausen, 1996).

Eomes and z*VegT* transcript levels peak early during gastrulation (stage 10) when the dorsal blastopore lip becomes visible due to the involution and anterior migration of deep mesodermal cells across the blastocoel roof (Keller, 1981; Owens et al., 2016). By this stage, the expression patterns of *Eomes*, z*VegT*, and *Xbra* overlap substantially around the blastopore (Figs. 2 and 3). As development proceeds, expression persists around the blastopore but declines in involuted mesoderm, albeit at different rates for

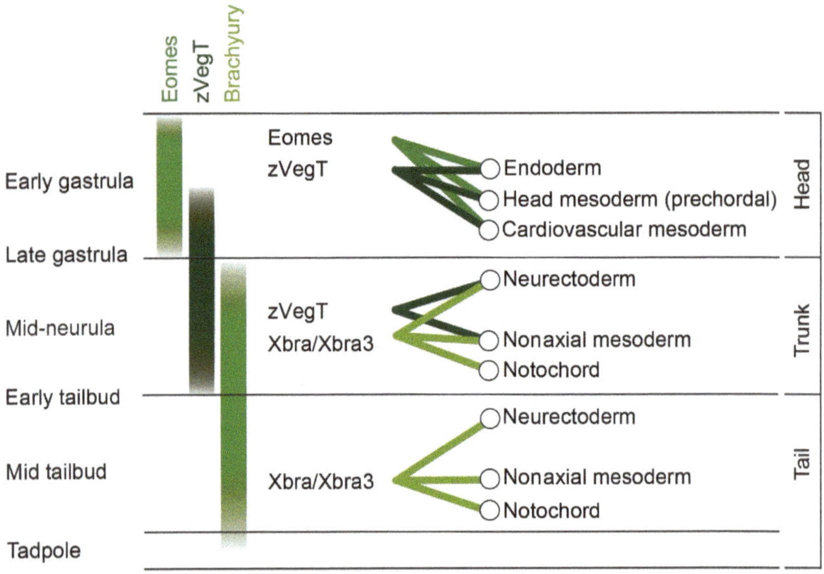

Fig. 3 Progenitor cells expressing Eomes, zVegT, and/or Xbra/Xbra3 are allocated to germ layer subtypes to form separate or continuous embryonic structures throughout the head, trunk, and tail. Based on spatiotemporal expression analysis (temporal order shown to the *left*), cell lineage tracing, and loss-of-function experiments, the contribution is illustrated by lines between T-box factor and germ layer subtype. Xbra/Xbra3 expression terminates in tadpoles when axial elongation is completed. While Eomes and VegT-positive cells are known to significantly contribute to endoderm early, later contribution is minor or uncertain (not shown). *Modified from Gentsch, G. E., Owens, N. D. L., Martin, S. R., Piccinelli, P., Faial, T., Trotter, et al. (2013). In vivo T-box transcription factor profiling reveals joint regulation of embryonic neuromesodermal bipotency.* Cell Reports, *4(6), 1185–1196. http://doi.org/10.1016/j.celrep.2013.08.012.*

different genes. For example, *Eomes* transcripts persist quite significantly in the mesoderm mantle, while expression of *Xbra* persists only in the developing notochord throughout axial elongation (Fig. 2). Eventually, as animal cells continue to engulf the vegetal hemisphere, the T-box factor-expressing blastopore collar approaches the future posterior end (Fukuda et al., 2010; Gentsch et al., 2013; Keller, 1978; Moosmann et al., 2013). In the wake of T-box gene induction and morphogenetic movements, germ layer segregation occurs first dorsally and yields head structures like the forebrain or prechordal plate (head mesoderm) from the noninvoluting and involuting marginal zone, respectively (Figs. 2 and 3). As gastrulation proceeds, germ layer derivatives of the trunk are generated. Prospective mesoderm that is formed from involuting cells around the ventrolateral blastopore lip produces the cardiovascular system and blood.

Around mid-gastrulation (stage 10.5), two more T-box genes, *Xbra3* and *Tbx6*, are activated as part of the next wave of zygotic transcription (Collart et al., 2014; Gentsch et al., 2013). *Xbra3* is a synexpressed paralogue of *Xbra* (Hayata, Eisaki, Kuroda, & Asashima, 1999). *Tbx6* is expressed predominantly in cells committed to presomitic mesoderm (Uchiyama, Kobayashi, Yamashita, Ohno, & Yabe, 2001). While *Eomes* transcription declines toward the end of gastrulation, z*VegT*, *Xbra/Xbra3*, and *Tbx6* remain strongly expressed in cells of the circumblastoporal collar, although z*VegT* and *Tbx6* are excluded from the dorsal side that later develops into the chordoneural hinge. z*VegT* expression disappears toward the end of primary neurulation as defined by the closure of the posterior neuropore (Figs. 2 and 3). The *Brachyury* paralogues and *Tbx6* continue to be expressed in separate and overlapping domains within the tailbud. The formation of the neurenteric canal connecting the lumen of the spinal cord with the anus divides the tailbud into the anterior and posterior wall of the neurenteric canal. While both walls contain populations of *Xbra/Xbra3*-positive cells, *Tbx6* transcripts are restricted to cells of the posterior wall and presomitic mesoderm (Gont et al., 1993; Uchiyama et al., 2001). The anterior wall is known in most vertebrates as the chordoneural hinge (Figs. 2 and 5). Both walls are at least bipotent in that they both produce different derivatives of mainly two germ layers: mesoderm and neuroectoderm. Most cells of the chordoneural hinge contribute to the notochord, the floor plate, and lateral horns of the spinal cord, whereas cells of the posterior wall give rise to somitic muscle and the roof plate of the spinal cord (Teillet, Lapointe, & Le Douarin, 1998). According to studies with other vertebrate embryos, the caudal populations of *Xbra/Xbra3*- and/or *Tbx6*-expressing cells are likely to shrink gradually over time at least in part because somitogenesis occurs more quickly than tail elongation (Gomez et al., 2008; Olivera-Martinez, Harada, Halley, & Storey, 2012; Sanders, Khare, Ooi, & Bellairs, 1986).

All these *Xenopus* T-box factors have been retained in other vertebrate clades except for *VegT* (also known as *Tbx16*, *spadetail*, or *Tbx6L*) and *Tbx6*, which were secondarily lost during mammalian and bird evolution, respectively (Ahn, You, & Kim, 2012; Kuraku & Kuratani, 2011). Meanwhile, gene-specific duplications have been found for some of these T-box factors in other vertebrates species, such as the *Brachyury* paralogues *ntl* (*no tail*) and *bra* (*brachyury*) in zebrafish (Martin & Kimelman, 2008).

T-box gene homologues also resemble each other with respect to their spatiotemporal gene regulation. During gastrulation, a collective of *Eomes*, *Tbx16*, and *Brachyury* genes is dynamically expressed within the early zone of germ layer segregation that is the primitive streak in birds, reptiles, and mammals or the germ ring in fish. Later, members of the *Brachyury* and *Tbx6* subfamily remain the only T-box factors strongly expressed at the posterior end of the embryo, while *Eomes* downregulates its expression toward primary neurulation. The expression pattern of *Tbx6* subfamily often extends further than *Brachyury* to the anterior limit of the presomitic mesoderm. This stereotypical order of T-box gene expression in time and space is an early molecular readout of the allocation of multipotent cells to specific germ layer lineages in the vertebrate embryo. Fate maps of the early mouse embryo show that the anteroposterior pattern of mesodermal subtypes (cranial, cardiac, paraxial, intermediate, and lateral plate mesoderm) is set up according to the temporal order of cell allocation to germ layers, as the primitive streak elongates toward the distal tip of the embryo. In the vicinity of anterior (cranial and cardiac) mesoderm, definitive endoderm segregates early from surface ectoderm in the most anterior region of the primitive streak (Kinder et al., 1999; Lawson et al., 1991; Tam, Parameswaran, Kinder, & Weinberger, 1997; Tzouanacou et al., 2009).

In this way, the up- and downregulation of *Eomes*, the earliest-expressed T-box gene in the primitive streak, provides a marker of definitive endoderm and anterior mesoderm (Costello et al., 2011). By contrast, *Brachyury*-expressing cells predominantly contribute to later developing structures of the trunk and tail such as posterior neuroectoderm and several mesodermal subtypes (Anderson et al., 2013; Garriock et al., 2015; Imuta, Kiyonari, Jang, Behringer, & Sasaki, 2013; Perantoni et al., 2005; Tzouanacou et al., 2009). Whether *Eomes*, *Brachyury*, or *Tbx6*, the expression of these T-box genes is usually transient in germ layer subpopulations that emerge and differentiate. The exceptions are the primordial fields forming the chordoneural hinge and the notochord, which show persistent *Brachyury* expression, perhaps to maintain uniform cell identity along the rostro-caudal axis (Chapman, Agulnik, Hancock, Silver, & Papaioannou, 1996; Ciruna & Rossant, 1999; Griffin, Amacher, Kimmel, & Kimelman, 1998; Hancock, Agulnik, Silver, & Papaioannou, 1999; Kispert & Herrmann, 1994; Kispert, Ortner, Cooke, & Herrmann, 1995; Knezevic & De Santo, 1997; Nikaido et al., 2002; Russ et al., 2000; Schulte-Merker, Ho, Herrmann, & Nüsslein-Volhard, 1992;

Schulte-Merker, van Eeden, Halpern, Kimmel, & Nusslein-Volhard, 1994; Wilkinson, Bhatt, & Herrmann, 1990).

3.2 T-Box Factors Control the Emergence of Distinct Germ Layer Subtypes

In this section, we outline the importance of correctly regulating genes of the *Brachyury*, *Eomes*, and *Tbx6* subfamily to segregate and differentiate the germ layers in various vertebrate species. Different and common regulatory inputs are used to express these T-box factors as a collective to sustain triploblastic development throughout embryogenesis. We compile studies of single loss-of-function experiments before synthesizing a more coherent view of T-box-mediated gene regulation.

3.2.1 Eomes Controls Anterior Fate Restrictions

The function of *Eomes* has not been extensively investigated in *Xenopus* (Fukuda et al., 2010; Gentsch et al., 2013). Initial attempts to disrupt *Eomes* activity in vivo made use of a dominant-negative version of Eomes that carried the repressor domain of the *Drosophila* engrailed protein. This severely disrupted gastrulation when overexpressed (Ryan et al., 1996). However, the fusion construct probably interferes with other factors showing similar chromatin recruitments (Gentsch et al., 2013). In contrast to these original observations, the *Eomes* phenotype is expected to be rather weak, because of its functional redundancy with co-expressed z*VegT* (Gentsch et al., 2013). In zebrafish, the *eomes* paralogue a (*eomesa*) is maternally expressed and is functionally more reminiscent of *Xenopus* m*VegT*. Both maternal factors contribute to the normal expression of endoderm-related Mix-like and Sox factors, but functional redundancies with other maternal factors make them nonessential for endoderm development (Du, Draper, Mione, Moens, & Bruce, 2012; G.E.G., T. Spruce, & J.C.S., unpublished; Nelson et al., 2014).

The functional redundancy observed for *Xenopus Eomes* is less marked in mice because of the loss of *Tbx16* genes in placental mammals. However, *Eomes* was co-opted to regulate the differentiation and proliferation of the extra-embryonic trophoblast lineage in mouse, which is essential for periimplantation development (Russ et al., 2000; Strumpf et al., 2005). Hence, understanding the embryonic role of *Eomes* required generating epiblast-specific mutants either by aggregating null ES cells with tetraploid host embryos or by conditionally knocking out the gene (Arnold, Hofmann, Bikoff, & Robertson, 2008; Russ et al., 2000). First, the *Eomes*-deficient epiblast fails to relax cell adhesions by downregulating *E-cadherin*.

Consequently, EMT is blocked and germ layers cannot delaminate from the primitive streak. Second, critical T-box target genes for the specification of definitive endoderm (*Cerberus*, *Foxa1*, *Foxa2*, *Gata6*, *Gsc*, *Hex*, *Lhx1*, and *Sox17*) or cardiac mesoderm (*Mesp1* and *Mesp2*) are not induced, preventing the epiblast from producing these germ layer derivatives. In line with this phenotype, and with *Eomes* having a cell-autonomous role, *Eomes* null ES cells rarely contribute to definitive endoderm or anterior mesoderm when injected into wild-type blastocysts (chimera) (Arnold et al., 2008; Costello et al., 2011). A more global analysis using in vitro specified human ES cells provides some evidence that Eomes regulates definitive endoderm by both feed-forward and repressive activity to control differentiation and block alternative cell fates such as pluripotency or mesoderm, respectively (Teo et al., 2011).

3.2.2 Brachyury Controls Posterior Fate Restrictions

Brachyury is a well-studied developmental gene. Several reports underline its importance in regulating germ layer segregation and differentiation during and, in particular, after gastrulation. In mice, different alleles have been reported for *Brachyury* (*T*), and these produce a range of phenotypes in both hetero- and homozygote embryos. The more severe alleles $T^{Wisconsin}$ (T^{Wis}) and $T^{Curtailed}$ (T^{C}), both of which probably act in a dominant-negative (antimorph) fashion, affect more anterior structures than the original *T* null allele (Dobrovolskaia-Zavadskaia, 1927; Searle, 1966; Shedlovsky, King, & Dove, 1988). The introduction of a single wild-type copy of *T* into the *T/+* or $T^{C}/+$ mouse genome alleviates the phenotype in a dosage-dependent way (Stott, Kispert, & Herrmann, 1993).

The loss-of-function phenotype is quite uniform across vertebrate species due to the well-conserved expression pattern of *Brachyury*. While mice rely on just one *Brachyury* gene, two *Brachyury* paralogues, *Xbra/Xbra3* and *ntl/bra*, work synergistically in *Xenopus* and zebrafish (Gentsch et al., 2013; Herrmann, Labeit, Poustka, King, & Lehrach, 1990; Martin & Kimelman, 2008). In doing so, *Xbra* and *ntl* predominate *Xbra3* and *bra*. In contrast to *Xenopus* and zebrafish, homozygous *T* mice die midway through somitogenesis at around E10.5. This is probably because of nutritional deprivation caused the malformation of the mesoderm-derived allantois which normally establishes a connection (umbilical cord) to the maternal circulation (Gluecksohn-Schoenheimer, 1944; Inman & Downs, 2006).

The complete loss of *Brachyury* activity truncates the posterior end of the embryo while leaving anterior structures largely intact. The migration and balance of germ layer progenitors are perturbed such that more neuroectoderm is formed at the expense of mesoderm. The notochord fails to develop as its primordial plate degenerates. Both production and segmentation of paraxial mesoderm stop after generating the first 8–12 somites. However, these remaining somites are markedly misshapen, reflecting a deteriorating process of mesoderm fate restriction toward the posterior. The excess production of neuroectodermal progenitors yields a convoluted spinal cord and ectopic neural tube-like structures in anatomical positions that are normally occupied by somites (Chesley, 1935; Gentsch et al., 2013; Grüneberg, 1958; Halpern, Ho, Walker, & Kimmel, 1993; Martin & Kimelman, 2008; Schulte-Merker et al., 1994; Yamaguchi, Takada, Yoshikawa, Wu, & McMahon, 1999; Yanagisawa, Fujimoto, & Urushihara, 1981). In addition, disruption of *Brachyury* activity in *T* or T^{Wis} homozygotes leads to an irregular node with fewer or disorganized monocilia (Concepcion & Papaioannou, 2014; Fujimoto & Yanagisawa, 1984; Herrmann, 1991). Consequently, beating cilia cannot generate the directional fluid flow that is required for left-sided Nodal signaling around the node and in the lateral plate mesoderm. This randomizes left–right asymmetry, which means that the heart is frequently inverted (*situs inversus*) and features ventrally displaced ventricular loops (Concepcion & Papaioannou, 2014; King, Beddington, & Brown, 1998). Similarly, zebrafish embryos require *ntl* and *tbx16* expression in the dorsal organizer (dorsal forerunner cells) to break left–right symmetry (Amack, Wang, & Yost, 2007; Amack & Yost, 2004; Essner, Amack, Nyholm, Harris, & Yost, 2005). Other tissue-specific knockdowns reveal that continuous *T* expression in the notochord is required to sustain its cell fate and function. Notochord cells without T protein adopt a more proliferative, neural fate and stop producing sonic hedgehog (shh). The absence of shh signaling along the rostro-caudal axis causes the loss or malformation of all except cervical (neck) vertebrae (Zhu, Kwan, & Mackem, 2016).

Some of these Brachyury-dependent embryonic events were identified by exploiting the pluripotency of ES cells in chimeric embryos and in vitro. Among wild-type cells in chimeric mouse embryos, *T* null ES cells fail to contribute to posterior mesoderm suggesting that Brachyury acts cell-autonomously. Instead, they predominantly populate anterior mesoderm including the first seven somites, definitive endoderm, neuroectoderm, or the zones of germ layer segregation (Beddington, Rashbass, & Wilson,

1992; Wilson & Beddington, 1997; Wilson, Manson, Skarnes, & Beddington, 1995). It appears that *T* null cells that are determined to transit the primitive streak late, or to emerge from the tailbud, lose migratory capacity. Time-lapse recordings determined that *T* null cells isolated from the late primitive streak move significantly slower on extracellular matrix than do equivalent wild-type cells (Hashimoto, Fujimoto, & Nakatsuji, 1987). Studies with primitive streak-like subpopulations derived from human ES cells in vitro also concluded that Brachyury prevents allocation to an endodermal cell lineage by activating essential genes for posterior mesoderm development (Bernardo et al., 2011; Faial et al., 2015; Mendjan et al., 2014).

3.2.3 Tbx6 Subfamily Shows Subfunctionalization to Control Both Fate Restrictions and Somitogenesis

In vertebrates, *Tbx6* and *Tbx16* have complementary roles in the specification of germ layer subtypes such as presomitic mesoderm. In *Xenopus*, z*VegT* supports mesoderm fate restriction during trunk elongation (Fukuda et al., 2010; Gentsch et al., 2013), whereas *Tbx6* allows paraxial mesoderm to be polarized along the main body axis (G.E.G. & J.C.S., unpublished; Tazumi, Yabe, & Uchiyama, 2010; Tazumi, Yabe, Yokoyama, Aihara, & Uchiyama, 2008). This presomitic patterning is essential for generating somite boundaries and regulating skeletal myogenesis. Similarly, zebrafish *spadetail* mutants show reduced formation of trunk mesoderm while leaving head, tail, and notochord relatively normal (Griffin et al., 1998; Kimmel, Kane, Walker, Warga, & Rothman, 1989). The loss of functional *Tbx6* (*fused somites*) also completely disrupts somitogenesis and myogenesis in zebrafish embryos (Nikaido et al., 2002; Windner et al., 2012, 2015). These observations indicate that these members of the *Tbx6* subfamily collaborate to generate and segment paraxial mesoderm and allow it to differentiate. In a manner that resembles the shared roles of *Eomes* and *Brachyury*, *Tbx16* promotes the segregation of mesoderm and modulates adhesion cell-autonomously to facilitate the convergent cell movements of mesodermal subtypes such as paraxial trunk mesoderm (Ho & Kane, 1990; Yamamoto et al., 1998).

Since placental mammals have lost *Tbx16*, mouse *Tbx6* better conforms to the *Tbx6* prototype by both specifying mesoderm and regulating somitogenesis (Chapman & Papaioannou, 1998; Watabe-Rudolph, Schlautmann, Papaioannou, & Gossler, 2002). In contrast to *Xenopus* and zebrafish that show subfunctionalization for *Tbx6*, mouse embryos lacking functional *Tbx6* fail to maintain paraxial mesoderm after the seventh somite and generate instead two ectopic neural tubes in situ (Chapman & Papaioannou, 1998). Accordingly, the developmental potential of *Tbx6* null ES cells is restricted as they fail to colonize posterior somites in chimeric embryos (Chapman, Cooper-Morgan, Harrelson, & Papaioannou, 2003). Germ layer conversion from mesoderm to neuroectoderm is reminiscent of the *T* phenotype and reflects the prevalent neuromesodermal bipotency of primitive streak or tailbud cells (Tzouanacou et al., 2009; Yamaguchi et al., 1999). Tbx6-mediated repression of neural identity occurs indirectly via a distinct *Sox2* enhancer (N1), which is normally active in the primitive streak and in caudal lateral epiblast containing long-term neuromesodermal progenitor cells (Takemoto et al., 2011; Yoshida et al., 2014). Furthermore, despite not being detectable in the node pit, *Tbx6*, like *T*, regulates left–right asymmetry (Hadjantonakis, Pisano, & Papaioannou, 2008).

3.2.4 Cooperation Between T-Box Factors Controls Fate Restrictions of all Mesodermal Subtypes

Eomes, Brachyury, and Tbx6/Tbx16 show similar chromatin binding profiles (Fig. 1) and similar abilities to promote germ layer segregation and differentiation in different regions of the embryo, so it seems likely that these T-box factors collaborate to control multipotency during axial elongation (Fig. 3). In *Xenopus*, the combined knockdown of *Eomes*, *zVegT*, and *Xbra/Xbra3* causes the loss of all mesoderm subtypes and the formation of an oversized neural tube that expands into the vacant anatomical positions of axial and paraxial mesoderm (Fig. 4; Gentsch et al., 2013). Double mutants for *spadetail* and *no tail* lack trunk and tail mesoderm, suggesting that *Tbx16* and *Brachyury* also act synergistically in zebrafish (Amacher, Draper, Summers, & Kimmel, 2002; Griffin et al., 1998). The head mesoderm of this double mutant remains intact, perhaps because of the early expression of maternal *Eomes* (Du et al., 2012; Nelson et al., 2014; Slagle, Aoki, & Burdine, 2011). Although these combinatorial loss-of-function experiments have not yet been carried out in mice, we anticipate similar semiredundancies between *Eomes*, *T*, and *Tbx6*.

Analysis of genome-wide chromatin profiles (Fig. 1) and differential expression under loss- and gain-of-function conditions both in vivo (Fig. 4) and in vitro suggests that T-box transcription factors control germ layer segregation in vertebrates in the following manner:

(1) *Induction and maintenance of germ layer segregation.* Early embryonic cells employ an extended network of pluripotency factors and paracrine

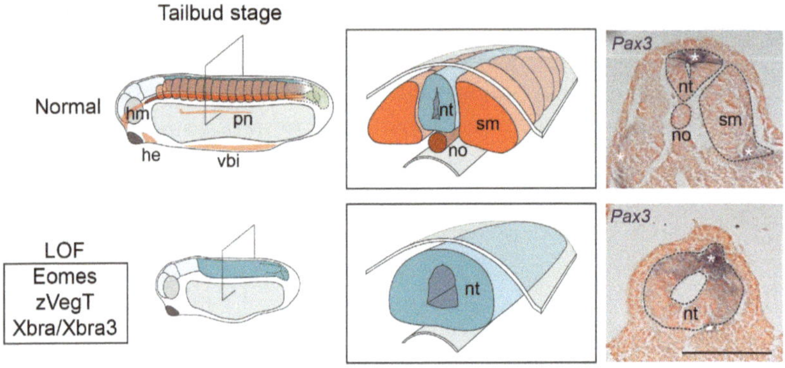

Fig. 4 *Xenopus* embryos deficient (LOF, loss of function) for *Eomes*, *zVegT*, and *Xbra/Xbra3* fail to form mesodermal derivatives such as head mesoderm (hm), heart (he), ventral blood island (vbi), and pronephros (pn). Cross sections through the trunk show an oversized neural tube (nt) that expanded at the expense of axial (*no*, notochord) and paraxial mesoderm (*sm*, skeletal muscle) at least in part due to the premature expression of pro-neural *Sox2/Sox3* and pleiotropic *Pax3* (transcribed in the roof plate of the neural tube and hypaxial muscle, marked by *asterisks*). Scale bar, 0.2 mm. *Modified from Gentsch, G. E., Owens, N. D. L., Martin, S. R., Piccinelli, P., Faial, T., Trotter, et al. (2013). In vivo T-box transcription factor profiling reveals joint regulation of embryonic neuromesodermal bipotency.* Cell Reports, *4(6), 1185–1196. http://doi.org/10.1016/j.celrep.2013.08.012.*

TGF-β signals to induce *Eomes*, *Tbx16*, and *Brachyury* in the zone of germ layer segregation (Agius, Oelgeschlager, Wessely, Kemp, & De Robertis, 2000; Brennan et al., 2001; Conlon et al., 1994; Feldman et al., 1998; Lee et al., 2013; Luxardi, Marchal, Thomé, & Kodjabachian, 2010; Skirkanich, Luxardi, Yang, Kodjabachian, & Klein, 2011; Yoon, Wills, Chuong, Gupta, & Baker, 2011). Most of these gene inductions are also sensitive to FGF signals and the accumulation of nuclear β-catenin (Ciruna & Rossant, 2001; Demagny, Araki, & De Robertis, 2014; Harvey, Tümpel, Dubrulle, Schier, & Smith, 2010; Rudloff & Kemler, 2012; Valenta et al., 2011; Yamaguchi et al., 1999). However, we note that z*VegT* does not respond to FGF (Stennard et al., 1999), which may explain its earlier transcriptional termination than *Xbra* at the posterior end of the embryo. The collective of Eomes, Brachyury, and Tbx16 maintains germ layer progenitors throughout axial elongation by maintaining the expression of several FGF (*fgf4*, *fgf8*, *fgf20*) and Wnt (*wnt3a*, *wnt8*) ligands (Fig. 5B and C). Canonical Wnt and FGF trigger an autoregulatory loop that is necessary for the continuous expression of

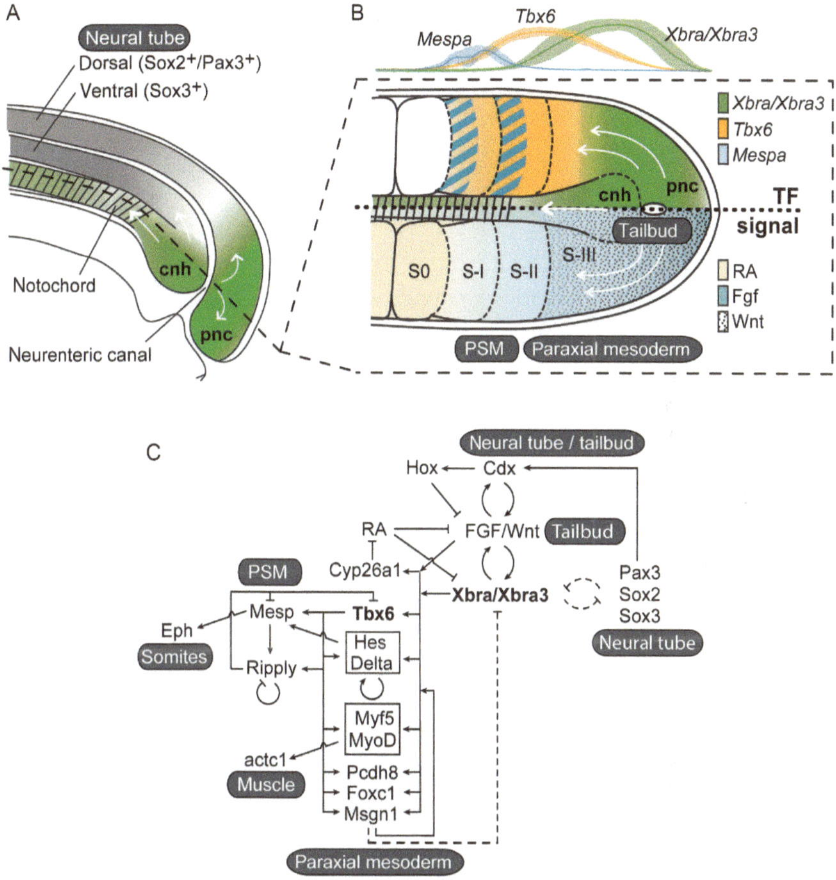

Fig. 5 Model of a T-box factor-dependent genetic regulatory network controlling neuromesodermal trajectories within the terminal growth zone of *Xenopus* embryo at early tailbud stage. Schematic diagram of a sagittal section (A) and a horizontal section (B) through the posterior region of an early tailbud embryo (see *boxed area* in Fig. 2) illustrating the expression of *Xbra*/*Xbra3*, *Tbx6*, *Mespa*, *Sox2/3*, and *Pax3*, signaling activity of FGF, Wnt, and retinoic acid (RA) and the recruitment of mesodermal and neural cells (*white arrows*) from the chordoneural hinge (cnh) and posterior wall of the neurenteric canal (pnc). Most cells of the cnh give rise to the notochord and the ventrolateral horns of the neural tube, whereas cells in the pnc contribute to paraxial mesoderm and the dorsal roof of the neural tube. *S0*, newly forming somite; *S-I/-II/-III*, presomitic mesoderm. *Graph above the horizontal section* shows semiquantitative measurements of transcript levels within the tailbud (G.E.G. & J.C.S., unpublished). (C) Genetic regulatory inputs and feedback loops with several functional nodes being active in tailbud (consisting of cnh and pnc), neural tube, paraxial mesoderm, presomitic mesoderm (PSM): maintenance of progenitor cells; specification of paraxial mesoderm (also by inhibiting alternative fates); myogenic differentiation; patterning of presomitic mesoderm; protection from neuralization; and determination of regional (Hox) identity. *Dashed lines* represent uncertain genetic interactions causing repression.

at least *Brachyury* and some of the other T-box target genes (Boulet & Capecchi, 2012; Casey et al., 1998; Isaacs, Pownall, & Slack, 1994; Martin & Kimelman, 2008; Naiche, Holder, & Lewandoski, 2011; Schulte-Merker & Smith, 1995; Thorpe, Weidinger, & Moon, 2005). In zebrafish, chick, mice, and human cells, the expression level of Brachyury increases with the level of Wnt signaling. In particular, bipotential neuromesodermal progenitors expressing both pro-neural Sox2 and Brachyury protein at lower levels are found in the dorsal section of the posterior wall of the tailbud that produces intermediate levels of Wnt activity (Bouldin et al., 2015; Olivera-Martinez et al., 2012; Wymeersch et al., 2016). Tailbud outgrowth also relies on Brachyury activating *Cyp26a1* to catabolize retinoic acid emerging from adjacent somites (Fig. 5B and C). High levels of retinoic acid cause termination of the Brachyury-FGF/Wnt feedback loop and apoptosis, both of which also naturally occur by the end of axial elongation (Gentsch et al., 2013; Martin & Kimelman, 2010; Olivera-Martinez et al., 2012; Shum et al., 1999).

(2) *Germ layer subtype restriction.* As well as maintaining a signaling environment for multipotent progenitors, T-box factors also set up gene regulatory networks that define distinct germ layer subtypes of the endoderm, mesoderm, and ectoderm. The transcriptional output of these networks depends on the activities of extracellular signals that change dynamically across different anatomical domains and developmental stages. Thus, transplanted cells within the zone of germ layer segregation adopt the fate of the new embryonic position. However, this plasticity means that germ layer subtypes have tight regulatory mechanisms in place to promote their identity while repressing alternative fates. That means that the failure to specify endoderm or mesoderm derepresses the alternative, neuroectodermal cell fate. The disruption of *Eomes*, *Brachyury*, or *Tbx16/Tbx6*, or any mesoderm-promoting signal transduction therefore causes neurulation. Indeed, the loss of Nodal, FGF, and Wnt, or gain of retinoic acid signaling, within the zone of germ layer segregation yields ectopic neural tissue of corresponding regional identity in mice (Abu-Abed et al., 2003; Camus, Perea-Gomez, Moreau, & Collignon, 2006; Ciruna, Schwartz, Harpal, Yamaguchi, & Rossant, 1997; Deng et al., 1994; Sakai et al., 2001; Yamaguchi et al., 1999; Yoshikawa, Fujimori, McMahon, & Takada, 1997). Neurulation of tail tip tissue also marks the normal end of axial elongation (Olivera-Martinez et al., 2012). In *Xenopus*, T-box factor-deficient tailbuds specifically overexpress the pro-neural genes *Sox3* and *Sox2/Pax3* in anatomical positions of the chordoneural hinge and the posterior wall of the neurenteric canal, respectively (Fig. 5A). They represent the disturbed niches of distinct neuromesodermal cell lineages producing ventrolateral and dorsal neural tube tissue at the expense of axial and nonaxial mesoderm (Gentsch et al., 2013; G.E.G. & J.C.S., unpublished). In zebrafish, repressed canonical Wnt signaling causes more progenitors being recruited to the neural instead of mesodermal cell lineage (Martin & Kimelman, 2012; Row et al., 2016).

The mechanism by which T-box factors antagonize neural regulators and vice versa is not well understood. Mutual repression is a possibility since both SoxB1 and T-box factors occupy the distal elements of each other (Gentsch et al., 2013; G.E.G. & J.C.S., unpublished). However, the SoxB1/Pax3 and T-box factors that are involved are mainly known as activators and probably require the recruitment of repressive cofactors (Fig. 5C).

Further fate restrictions concern the separation of adjacent mesodermal subtypes. *Tbx16* expression is excluded from the dorsal midline through the repressive activity of not, a notochord homeobox factor segregating axial from paraxial mesoderm (Amacher & Kimmel, 1998; Gont, Fainsod, Kim, & De Robertis, 1996; Yasuo & Lemaire, 2001). In a reciprocal manner, Mesogenin 1 (Msgn1) diverts midline progenitors to paraxial mesoderm by repressing *Brachyury* and promoting critical T-box target genes for mesoderm maturation (Fig. 5C; Chalamalasetty et al., 2014; Row et al., 2016). Similarly, the T-box target genes of the forkhead family *Foxc1* and *Foxc2* regulate the allocation of mesodermal progenitors to rather paraxial than intermediate mesoderm (Fig. 5C; Wilm, James, Schultheiss, & Hogan, 2004).

(3) *Germ layer specification.* The T-box factors control the maturation of endodermal and mesodermal germ layer subtypes by binding most of their *cis*-regulatory elements simultaneously, but activating them in time and space according to extrinsic signals such as Wnt, FGF, BMP, Notch/Delta, and retinoic acid (Gentsch et al., 2013). In the case of paraxial mesoderm specification, *Tbx6*, *Msgn1*, *Foxc1*, and *Protocadherin 8* (*Pcdh8*) are induced at an early stage with high FGF/Wnt signaling and maintained in a coherent (type I; Alon, 2007) feed-forward loop (via Tbx6 and Msgn1) to restrict cell fate, remodel cell adhesions, convey positional information for somitogenesis, and establish directional cell migration from the niche of progenitor cells to presomitic mesoderm (Fig. 5C; Bajard et al., 2014; Chalamalasetty et al., 2014; Chapman & Papaioannou, 1998; Fior et al., 2012; Lawton et al., 2013; Manning & Kimelman, 2015; Wilm et al., 2004; Wittler et al., 2007; Yamamoto et al., 1998; Yoon & Wold, 2000). This T-box-dependent maturation process also involves bHLH-driven myogenic determination (*Myf5*, *MyoD*) and activation of Notch-related oscillations in the expression of genes encoding the Notch ligand *DeltaC* and many *Hes* family bHLH factors (Buckingham & Rigby, 2014; Gentsch et al., 2013; Jen, Gawantka, Pollet, Niehrs, & Kintner, 1999; Jen, Wettstein, Turner, Chitnis, & Kintner, 1997). Probably under the influence of retinoic acid, Tbx6 and the oscillating gene regulatory network activate *Mesp* (*mesoderm posterior*) bHLH factors and corepressors of the *Ripply* family in the anterior half of the future somite (Kawamura et al., 2005; Moreno & Kintner, 2004; Sparrow et al., 1998; Yasuhiko et al., 2006). Mesp establishes the boundary to the previous somite in part by activating Eph/Ephrin signaling, and it patterns somites along the rostro-caudal axis, while Ripply factors conclude presomitic patterning by repressing *Tbx6* expression (Fig. 5C; Durbin et al., 1998; Nakajima, Morimoto, Takahashi, Koseki, & Saga, 2006; Takahashi et al., 2000; Windner et al., 2015).

The entire specification of paraxial mesoderm or probably any other germ layer subtype also relies on correctly resolving the conflict of DNA replication and genome activity. Lifting spatially restricted cell divisions causes T-box factors to be ectopically expressed, which severely impairs morphogenesis during axial elongation (Bouldin et al., 2014; Murakami, Moody, Daar, & Morrison, 2004).

(4) *Determination of regional identity*. The *caudal-like homeobox* (*Cdx*) factors *Cdx1*, *Cdx2*, and *Cdx4*, normally induced and maintained by canonical Wnt, sequentially activate *Hox* genes at the posterior end of the embryo (Fig. 5C). Colinear *Hox* expression assigns distinct regional identities to embryonic structures along the main body axis. In a manner resembling the *Brachyury* phenotype, but more severe than it, *Cdx* null mice abrogate axial elongation after five somites due to the depletion of germ layer progenitors (van Rooijen et al., 2012). Since both *Brachyury* and *Cdx* expression depend on canonical Wnt signaling, T-box and Cdx factors are also suggested to regulate each other. However, despite T-box factors being recruited to and capable of inducing *Cdx* genes, embryos lacking a Brachyury/Wnt autoregulatory loop may rescue

Cdx expression by ectopically expressing *Pax3* in the tailbud. Pax3 is normally required for *Cdx2* expression in the roof plate of the forming spinal cord (Gentsch et al., 2013; Zhao et al., 2014). Thus, *Brachyury*-deficient embryos retain some *Hox*-specific regional identities. *Cdx* null embryos produce anterior mesoderm and express *Brachyury* until mesodermal progenitors are lost due to an inactive primitive streak (van Rooijen et al., 2012). Interestingly, among the Hox factors the most posteriorly expressed show the greatest potential to repress FGF and Wnt, which coordinates the acquisition of sacral identity with the termination of axial elongation (Fig. 5C; Denans, Iimura, & Pourquié, 2015). Taken together, T-box and Cdx factors use the same autoregulatory loops to maintain independent gene regulatory networks for segregating and conveying regional identity to germ layers (Gouti et al., 2014).

4. IMPLICATIONS FOR REVERSE ENGINEERING ES CELLS

Nieuwkoop's two-step "activation–transformation" model for neural induction was based on classical embryological experiments. In this model, the activation process defines a neural field which is later transformed from an anterior state to a more posterior one (Eyal-Giladi, 1954; Nieuwkoop, 1952). This model was consistent with the expression patterns of signaling molecules in vertebrate embryos. Thus, the organizer-derived inhibition of BMP signals promotes neural induction, while activation of the retinoic acid, FGF, or canonical Wnt pathway causes some degree of posteriorization. However, attempts to replicate this model in vitro, to generate posterior spinal cord tissue, have not succeeded well. Rather, neuroectoderm proves to be derived from dual-fated progenitors that also produce mesoderm rather than surface ectoderm during axial elongation (Tzouanacou et al., 2009).

This discovery has inspired new attempts to drive the fate of ES cells, all of which have reversed Nieuwkoop's model: First, ES cells are posteriorized by stimulating the FGF/Wnt-mediated autoregulatory loops of Brachyury and Cdx1/2/4. Second, pro-neural trajectories are activated with retinoic acid to produce neurons with posterior identity (Gouti et al., 2014; Lippmann et al., 2015; Tsakiridis et al., 2014; Turner et al., 2014). Modulating FGF and canonical Wnt activity to moderate levels can stabilize cells with characteristics of long-term neuromesodermal progenitors. First, they co-express Sox2 and Brachyury protein at low levels. These factors show toggle switch behavior, meaning that at high concentrations Brachyury and Sox2 are mutually exclusive (Fig. 6). Second, they can be cultured in these conditions quite homogeneously for up to 7 days (Lippmann et al., 2015). Third, they can also be caused to differentiate as mesodermal derivatives as shown for skeletal muscle by the continually stabilizing β-catenin in the absence of retinoic acid (Gouti et al., 2014). Fourth, in a manner reminiscent of its phenotype, *Brachyury* null ES cells fail to adopt mesodermal fate, but instead generate spinal cord neurons. This also underlines that the acquisition of regional identity is Brachyury independent (Gouti et al., 2014). Fifth, when grafted into the node/streak border, they contribute to both mesodermal and neural derivatives in vivo (Gouti et al., 2014). Sixth, the addition of retinoic acid at any point over 7 days stops progressive colinear *Hox* activation resulting in distinct neurons ranging from hindbrain to sacral spinal cord identity (Lippmann et al., 2015).

In conclusion, reverse engineering of ES cells can faithfully recapitulate germ layer fate decisions, which opens up new avenues for investigating T-box-related mechanisms like the toggle switch between Brachyury and Sox2 at single cell level or the progressive priming of chromatin landscapes for signal-induced transcription in homogeneous cell populations.

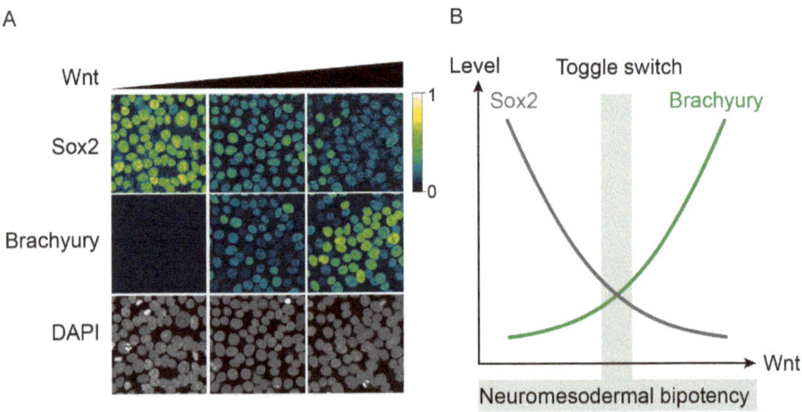

Fig. 6 Toggle switch between Sox2 and Brachyury. (A) Moderate activation of the canonical Wnt pathway (together with FGF) converts human ES cells into neuromesodermal progenitor-like cells. They co-express low levels of Brachyury and proneural Sox2 protein (relative levels shown, normalized to highest expression levels detected without or strong Wnt activation). (B) These factors show toggle switch behavior, meaning that at high concentrations Brachyury and Sox2 are mutually exclusive. *Data from G.E.G., A.S. Bernardo, and J.C.S. (unpublished).*

5. SUMMARY

Building on the network of pluripotency factors, the semiredundant collaboration of Eomes, Tbx6/16, and Brachyury sets up genes that ensure the continuous production and specification of dual-fated germ layer progenitors that reflect ongoing primary cell fate decisions throughout axial elongation to form head, trunk, and tail (Figs. 1–3).

Influenced by local FGF/Wnt-mediated autoregulatory loops, these T-box factors restrict progenitors to mesoderm at high levels, while lower levels also permit them to acquire a neural fate (Fig. 5). Thus, the low expression level of both Brachyury and pro-neural Sox2 protein is a key signature of neuromesodermal progenitors (Fig. 6). The establishment of *Hox*-related regional identity can progress independently of T-box factors via the Cdx factors (Fig. 5).

Different signaling environments employ a T-box factor defined gene regulatory network to drive distinct feed-forward loops that specify different mesodermal subtypes (for instance, *Mesp* expression in paraxial mesoderm, Figs. 1 and 5). All T-box factors show similar patterns of chromatin recruitment that are influenced by the developmental context of pluripotency and the conserved T-box binding motif.

Taken together, both dose and combination of T-box factors are essential to generate the correct ratio of germ layer derivatives to form a well-proportionate functional embryo (Fig. 4).

ACKNOWLEDGMENTS

G.E.G. and J.C.S. were supported by the Francis Crick Institute (Grant No. FCI01), which receives its core funding from Cancer Research UK, the UK Medical Research Council, and the Wellcome Trust. R.S.M. is supported by DevCom, a European Initial Training Network for Developmental and Computational Biology (Project Reference 607142).

REFERENCES

Abu-Abed, S., Dollé, P., Metzger, D., Wood, C., MacLean, G., Chambon, P., et al. (2003). Developing with lethal RA levels: Genetic ablation of Rarg can restore the viability of mice lacking Cyp26a1. *Development, 130*(7), 1449–1459.

Adell, T., & Müller, W. E. G. (2005). Expression pattern of the Brachyury and Tbx2 homologues from the sponge Suberites domuncula. *Biology of the Cell/Under the Auspices of the European Cell Biology Organization, 97*(8), 641–650. http://doi.org/10.1042/BC20040135.

Agius, E., Oelgeschlager, M., Wessely, O., Kemp, C., & De Robertis, E. M. (2000). Endodermal nodal-related signals and mesoderm induction in Xenopus. *Development, 127*(6), 1173–1183.

Ahn, D., You, K.-H., & Kim, C.-H. (2012). Evolution of the Tbx6/16 subfamily genes in vertebrates: Insights from Zebrafish. *Molecular Biology and Evolution*, *29*(12), 3959–3983. http://doi.org/10.1093/molbev/mss199.

Alon, U. (2007). Network motifs: Theory and experimental approaches. *Nature Reviews. Genetics*, *8*(6), 450–461. http://doi.org/10.1038/nrg2102.

Amacher, S. L., Draper, B. W., Summers, B. R., & Kimmel, C. B. (2002). The zebrafish T-box genes no tail and spadetail are required for development of trunk and tail mesoderm and medial floor plate. *Development*, *129*(14), 3311–3323.

Amacher, S. L., & Kimmel, C. B. (1998). Promoting notochord fate and repressing muscle development in zebrafish axial mesoderm. *Development*, *125*(8), 1397–1406.

Amack, J. D., Wang, X., & Yost, H. J. (2007). Two T-box genes play independent and cooperative roles to regulate morphogenesis of ciliated Kupffer's vesicle in zebrafish. *Developmental Biology*, *310*(2), 196–210. http://doi.org/10.1016/j.ydbio.2007.05.039.

Amack, J., & Yost, H. (2004). The T box transcription factor no tail in ciliated cells controls zebrafish left-right asymmetry. *Current Biology*, *14*(8), 685–690.

Anderson, M. J., Naiche, L. A., Wilson, C. P., Elder, C., Swing, D. A., & Lewandoski, M. (2013). TCreERT2, a transgenic mouse line for temporal control of Cre-mediated recombination in lineages emerging from the primitive streak or tail bud. *PLoS One*, *8*(4), e62479. http://doi.org/10.1371/journal.pone.0062479.

Andrikou, C., Iovene, E., Rizzo, F., Oliveri, P., & Arnone, M. I. (2013). Myogenesis in the sea urchin embryo: The molecular fingerprint of the myoblast precursors. *EvoDevo*, *4*(1), 1. http://doi.org/10.1186/2041-9139-4-33.

Arendt, D., Technau, U., & Wittbrodt, J. (2001). Evolution of the bilaterian larval foregut. *Nature*, *409*(6816), 81–85. http://doi.org/10.1038/35051075.

Arnold, S. J., Hofmann, U. K., Bikoff, E. K., & Robertson, E. J. (2008). Pivotal roles for eomesodermin during axis formation, epithelium-to-mesenchyme transition and endoderm specification in the mouse. *Development*, *135*(3), 501–511. http://doi.org/10.1242/dev.014357.

Arvey, A., Agius, P., Noble, W. S., & Leslie, C. (2012). Sequence and chromatin determinants of cell-type–specific transcription factor binding. *Genome Research*, *22*(9), 1723–1734. http://doi.org/10.1101/gr.127712.111.

Bajard, L., Morelli, L. G., Ares, S., Pecreaux, J., Jülicher, F., & Oates, A. C. (2014). Wnt-regulated dynamics of positional information in zebrafish somitogenesis. *Development*, *141*(6), 1381–1391. http://doi.org/10.1242/dev.093435.

Bassham, S., & Postlethwait, J. (2000). Brachyury (T) expression in embryos of a Larvacean Urochordate, Oikopleura dioica, and the ancestral role of T. *Developmental Biology*, *220*(2), 322–332. http://doi.org/10.1006/dbio.2000.9647.

Bauer, D. V., Huang, S., & Moody, S. A. (1994). The cleavage stage origin of Spemann's Organizer: Analysis of the movements of blastomere clones before and during gastrulation in Xenopus. *Development*, *120*(5), 1179–1189.

Beck, C. W., & Slack, J. M. (1998). Analysis of the developing Xenopus tail bud reveals separate phases of gene expression during determination and outgrowth. *Mechanisms of Development*, *72*(1–2), 41–52.

Beck, C. W., & Slack, J. M. (1999). A developmental pathway controlling outgrowth of the Xenopus tail bud. *Development*, *126*(8), 1611–1620.

Beck, C. W., Whitman, M., & Slack, J. M. (2001). The role of BMP signaling in outgrowth and patterning of the Xenopus tail bud. *Developmental Biology*, *238*(2), 303–314. http://doi.org/10.1006/dbio.2001.0407.

Beddington, R. S. P., Rashbass, P., & Wilson, V. (1992). Brachyury—A gene affecting mouse gastrulation and early organogenesis. *Development*, *116*(Suppl.), 157–165.

Bernardo, A. S., Faial, T., Gardner, L., Niakan, K. K., Ortmann, D., Senner, C. E., et al. (2011). BRACHYURY and CDX2 mediate BMP-induced differentiation of human and mouse pluripotent stem cells into embryonic and extraembryonic lineages. *Cell Stem Cell*, *9*(2), 144–155. http://doi.org/10.1016/j.stem.2011.06.015.

Bjornson, C. R. R., Griffin, K. J. P., Farr, G. H., Terashima, A., Himeda, C., Kikuchi, Y., et al. (2005). Eomesodermin is a localized maternal determinant required for endoderm induction in zebrafish. *Developmental Cell*, *9*(4), 523–533. http://doi.org/10.1016/j.devcel.2005.08.010.

Bouldin, C. M., Manning, A. J., Peng, Y.-H., Farr, G. H., Hung, K. L., Dong, A., et al. (2015). Wnt signaling and tbx16 form a bistable switch to commit bipotential progenitors to mesoderm. *Development*, *142*(14), 2499–2507. http://doi.org/10.1242/dev.124024.

Bouldin, C. M., Snelson, C. D., Farr, G. H., & Kimelman, D. (2014). Restricted expression of cdc25a in the tailbud is essential for formation of the zebrafish posterior body. *Genes & Development*, *28*(4), 384–395. http://doi.org/10.1101/gad.233577.113.

Boulet, A. M., & Capecchi, M. R. (2012). Signaling by FGF4 and FGF8 is required for axial elongation of the mouse embryo. *Developmental Biology*, *371*(2), 235–245. http://doi.org/10.1016/j.ydbio.2012.08.017.

Brennan, J., Lu, C. C., Norris, D. P., Rodriguez, T. A., Beddington, R. S. P., & Robertson, E. J. (2001). Nodal signalling in the epiblast patterns the early mouse embryo. *Nature*, *411*(6840), 965–969. http://doi.org/10.1038/35082103.

Buckingham, M., & Rigby, P. W. J. (2014). Gene regulatory networks and transcriptional mechanisms that control myogenesis. *Developmental Cell*, *28*(3), 225–238. http://doi.org/10.1016/j.devcel.2013.12.020.

Bytinski-Salz, H. (1931). Untersuchungen über die Induktionsfähigkeit der hinteren Medullarplattenbezirke. *Wilhelm Roux' Archiv für Entwicklungsmechanik der Organismen*, *123*(3–4), 518–564. http://doi.org/10.1007/BF01380647.

Cambray, N., & Wilson, V. (2002). Axial progenitors with extensive potency are localised to the mouse chordoneural hinge. *Development*, *129*(20), 4855–4866. http://doi.org/10.1016/S0925-4773(98)00015-X.

Cambray, N., & Wilson, V. (2007). Two distinct sources for a population of maturing axial progenitors. *Development*, *134*(15), 2829–2840. http://doi.org/10.1242/dev.02877.

Camus, A., Perea-Gomez, A., Moreau, A., & Collignon, J. (2006). Absence of nodal signaling promotes precocious neural differentiation in the mouse embryo. *Developmental Biology*, *295*(2), 743–755. http://doi.org/10.1016/j.ydbio.2006.03.047.

Casey, E. S., O'Reilly, M. A., Conlon, F. L., & Smith, J. C. (1998). The T-box transcription factor Brachyury regulates expression of eFGF through binding to a non-palindromic response element. *Development*, *125*(19), 3887–3894.

Casey, E., Tada, M., Fairclough, L., Wylie, C., Heasman, J., & Smith, J. (1999). Bix4 is activated directly by VegT and mediates endoderm formation in Xenopus development. *Development*, *126*(19), 4193–4200.

Catala, M., Teillet, M. A., & Le Douarin, N. M. (1995). Organization and development of the tail bud analyzed with the quail-chick chimaera system. *Mechanisms of Development*, *51*(1), 51–65.

Chalamalasetty, R. B., Garriock, R. J., Dunty, W. C., Kennedy, M. W., Jailwala, P., Si, H., et al. (2014). Mesogenin 1 is a master regulator of paraxial presomitic mesoderm differentiation. *Development*, *141*(22), 4285–4297. http://doi.org/10.1242/dev.110908.

Chapman, D. L., Agulnik, I., Hancock, S., Silver, L. M., & Papaioannou, V. E. (1996). Tbx6, a mouse T-box gene implicated in paraxial mesoderm formation at gastrulation. *Developmental Biology*, *180*(2), 534–542. http://doi.org/10.1006/dbio.1996.0326.

Chapman, D. L., Cooper-Morgan, A., Harrelson, Z., & Papaioannou, V. E. (2003). Critical role for Tbx6 in mesoderm specification in the mouse embryo. *Mechanisms of Development*, *120*(7), 837–847.

Chapman, D. L., & Papaioannou, V. E. (1998). Three neural tubes in mouse embryos with mutations in the T-box gene Tbx6. *Nature, 391*(6668), 695–697. http://doi.org/10.1038/35624.

Chesley, P. (1935). Development of the short-tailed mutant in the house mouse. *Journal of Experimental Zoology, 70*(3), 429–459. http://doi.org/10.1002/jez.1400700306.

Chiba, S., Jiang, D., Satoh, N., & Smith, W. C. (2009). Brachyury null mutant-induced defects in juvenile ascidian endodermal organs. *Development, 136*(1), 35–39. http://doi.org/10.1242/dev.030981.

Ciriello, G., Gatza, M. L., Beck, A. H., Wilkerson, M. D., Rhie, S. K., Pastore, A., et al. (2015). Comprehensive molecular portraits of invasive lobular breast cancer. *Cell, 163*(2), 506–519. http://doi.org/10.1016/j.cell.2015.09.033.

Ciruna, B. G., & Rossant, J. (1999). Expression of the T-box gene eomesodermin during early mouse development. *Mechanisms of Development, 81*(1–2), 199–203.

Ciruna, B., & Rossant, J. (2001). FGF signaling regulates mesoderm cell fate specification and morphogenetic movement at the primitive streak. *Developmental Cell, 1*(1), 37–49.

Ciruna, B. G., Schwartz, L., Harpal, K., Yamaguchi, T. P., & Rossant, J. (1997). Chimeric analysis of fibroblast growth factor receptor-1 (Fgfr1) function: A role for FGFR1 in morphogenetic movement through the primitive streak. *Development, 124*(14), 2829–2841.

Coll, M., Seidman, J. G., & Müller, C. W. (2002). Structure of the DNA-bound T-box domain of human TBX3, a transcription factor responsible for Ulnar-Mammary syndrome. *Structure, 10*(3), 343–356. http://doi.org/10.1016/S0969-2126(02)00722-0.

Collart, C., Owens, N. D. L., Bhaw-Rosun, L., Cooper, B., De Domenico, E., Patrushev, I., et al. (2014). High-resolution analysis of gene activity during the Xenopus mid-blastula transition. *Development, 141*(9), 1927–1939. http://doi.org/10.1242/dev.102012.

Concepcion, D., & Papaioannou, V. E. (2014). Nature and extent of left/right axis defects in T(Wis)/T(Wis) mutant mouse embryos. *Developmental Dynamics, 243*(8), 1046–1053. http://doi.org/10.1002/dvdy.24144.

Cong, L., Ran, F. A., Cox, D., Lin, S., Barretto, R., Habib, N., et al. (2013). Multiplex genome engineering using CRISPR/Cas systems. *Science, 339*(6121), 819–823. http://doi.org/10.1126/science.1231143.

Conlon, F., Fairclough, L., Price, B., Casey, E., & Smith, J. (2001). Determinants of T box protein specificity. *Development, 128*(19), 3749–3758.

Conlon, F. L., Lyons, K. M., Takaesu, N., Barth, K. S., Kispert, A., Herrmann, B., et al. (1994). A primary requirement for nodal in the formation and maintenance of the primitive streak in the mouse. *Development, 120*(7), 1919–1928.

Costello, I., Pimeisl, I.-M., Dräger, S., Bikoff, E. K., Robertson, E. J., & Arnold, S. J. (2011). The T-box transcription factor Eomesodermin acts upstream of Mesp1 to specify cardiac mesoderm during mouse gastrulation. *Nature Cell Biology, 13*(9), 1084–1091. http://doi.org/10.1038/ncb2304.

Criley, B. B. (1969). Analysis of the embryonic sources and mechanisms of development of posterior levels of chick neural tubes. *Journal of Morphology, 128*(4), 465–501. http://doi.org/10.1002/jmor.1051280406.

Croce, J., Lhomond, G., & Gache, C. (2001). Expression pattern of Brachyury in the embryo of the sea urchin *Paracentrotus lividus*. *Development Genes and Evolution, 211*(12), 617–619. http://doi.org/10.1007/s00427-001-0200-5.

Cunliffe, V., & Smith, J. (1992). Ectopic mesoderm formation in Xenopus embryos caused by widespread expression of a Brachyury homologue. *Nature, 358*(6385), 427–430.

Cunliffe, V., & Smith, J. (1994). Specification of mesodermal pattern in *Xenopus laevis* by interactions between Brachyury, noggin and Xwnt-8. *The EMBO Journal, 13*(2), 349–359.

Dale, L., & Slack, J. M. (1987). Fate map for the 32-cell stage of *Xenopus laevis*. *Development*, *99*(4), 527–551.

Davis, R. L., & Kirschner, M. W. (2000). The fate of cells in the tailbud of *Xenopus laevis*. *Development*, *127*(2), 255–267.

Demagny, H., Araki, T., & De Robertis, E. M. (2014). The tumor suppressor Smad4/DPC4 is regulated by phosphorylations that integrate FGF, Wnt, and TGF-β signaling. *Cell Reports*, *9*(2), 688–700. http://doi.org/10.1016/j.celrep.2014.09.020.

Denans, N., Iimura, T., & Pourquié, O. (2015). Hox genes control vertebrate body elongation by collinear Wnt repression. *eLife*, *4*, e04379. http://doi.org/10.7554/eLife.04379.

Deng, C. X., Wynshaw-Boris, A., Shen, M. M., Daugherty, C., Ornitz, D. M., & Leder, P. (1994). Murine FGFR-1 is required for early postimplantation growth and axial organization. *Genes & Development*, *8*(24), 3045–3057.

Dobrovolskaia-Zavadskaia, N. (1927). Sur la mortification spontanée de la queue chez la souris nouveau-née et sur l''existence d''un caractère (facteur) héréditaire "non viable". *Comptes Rendus Des Séances Et Mémoires De La Société De Biologie*, *97*, 114–116. http://doi.org/10.1002/jez.1400700306/abstract.

Du, S., Draper, B. W., Mione, M., Moens, C. B., & Bruce, A. (2012). Differential regulation of epiboly initiation and progression by zebrafish Eomesodermin A. *Developmental Biology*, *362*(1), 11–23. http://doi.org/10.1016/j.ydbio.2011.10.036.

Durbin, L., Brennan, C., Shiomi, K., Cooke, J., Barrios, A., Shanmugalingam, S., et al. (1998). Eph signaling is required for segmentation and differentiation of the somites. *Genes & Development*, *12*(19), 3096–3109.

Duval, C., Bouvet, P., Omilli, F., Roghi, C., Dorel, C., LeGuellec, R., et al. (1990). Stability of maternal mRNA in Xenopus embryos: Role of transcription and translation. *Molecular and Cellular Biology*, *10*(8), 4123–4129.

Essner, J. J., Amack, J. D., Nyholm, M. K., Harris, E. B., & Yost, H. J. (2005). Kupffer's vesicle is a ciliated organ of asymmetry in the zebrafish embryo that initiates left-right development of the brain, heart and gut. *Development*, *132*(6), 1247–1260. http://doi.org/10.1242/dev.01663.

Eyal-Giladi, H. (1954). Dynamic aspects of neural induction in amphibia. *Archives de Biologie (Liege)*, *65*(2), 179–259.

Faial, T., Bernardo, A. S., Mendjan, S., Diamanti, E., Ortmann, D., Gentsch, G. E., et al. (2015). Brachyury and SMAD signalling collaboratively orchestrate distinct mesoderm and endoderm gene regulatory networks in differentiating human embryonic stem cells. *Development*, *142*(12), 2121–2135. dev.117838, http://doi.org/10.1242/dev.117838.

Feldman, B., Gates, M. A., Egan, E. S., Dougan, S. T., Rennebeck, G., Sirotkin, H. I., et al. (1998). Zebrafish organizer development and germ-layer formation require nodal-related signals. *Nature*, *395*(6698), 181–185. http://doi.org/10.1038/26013.

Fior, R., Maxwell, A. A., Ma, T. P., Vezzaro, A., Moens, C. B., Amacher, S. L., et al. (2012). The differentiation and movement of presomitic mesoderm progenitor cells are controlled by Mesogenin 1. *Development*, *139*(24), 4656–4665. http://doi.org/10.1242/dev.078923.

Fuchikami, T., Mitsunaga-Nakatsubo, K., Amemiya, S., Hosomi, T., Watanabe, T., Kurokawa, D., et al. (2002). T-brain homologue (HpTb) is involved in the archenteron induction signals of micromere descendant cells in the sea urchin embryo. *Development*, *129*(22), 5205–5216. http://doi.org/10.1046/j.1440-169X.1997.t01-2-00013.x.

Fujimoto, H., & Yanagisawa, K. O. (1984). Defects in the archenteron of mouse embryos homozygous for the T-mutation. *Differentiation*, *25*(1), 44–47. http://doi.org/10.1111/j.1432-0436.1984.tb01336.x.

Fukuda, M., Takahashi, S., Haramoto, Y., Onuma, Y., Kim, Y.-J., Yeo, C.-Y., et al. (2010). Zygotic VegT is required for Xenopus paraxial mesoderm formation and is regulated by nodal signaling and Eomesodermin. *The International Journal of Developmental Biology*, *54*(1), 81–92. http://doi.org/10.1387/ijdb.082837mf.

Gaertner, R. A. (1949). Development of the posterior trunk and tail of the chick embryo. *Journal of Experimental Zoology*, *111*(2), 157–174. http://doi.org/10.1002/jez.1401110202.

Garriock, R. J., Chalamalasetty, R. B., Kennedy, M. W., Canizales, L. C., Lewandoski, M., & Yamaguchi, T. P. (2015). Lineage tracing of neuromesodermal progenitors reveals novel Wnt-dependent roles in trunk progenitor cell maintenance and differentiation. *Development*, *142*(9), 1628–1638. http://doi.org/10.1242/dev.111922.

Gentsch, G. E., Owens, N. D. L., Martin, S. R., Piccinelli, P., Faial, T., Trotter, M. W. B., et al. (2013). In vivo T-box transcription factor profiling reveals joint regulation of embryonic neuromesodermal bipotency. *Cell Reports*, *4*(6), 1185–1196. http://doi.org/10.1016/j.celrep.2013.08.012.

Ghebranious, N., Blank, R. D., Raggio, C. L., Staubli, J., McPherson, E., Ivacic, L., et al. (2008). A missense T(Brachyury) mutation contributes to vertebral malformations. *Journal of Bone and Mineral Research*, *23*(10), 1576–1583. http://doi.org/10.1359/jbmr.080503.

Ghosh, T. K., Brook, J. D., & Wilsdon, A. (2017). T-Box genes in human development and disease. *Current Topics in Developmental Biology*, *122*, 383–415.

Gluecksohn-Schoenheimer, S. (1944). The development of normal and homozygous Brachy (T/T) mouse embryos in the extraembryonic coelom of the chick. *Proceedings of the National Academy of Sciences of the United States of America*, *30*(6), 134.

Gomez, C., Ozbudak, E. M., Wunderlich, J., Baumann, D., Lewis, J., & Pourquié, O. (2008). Control of segment number in vertebrate embryos. *Nature*, *454*(7202), 335–339. http://doi.org/10.1038/nature07020.

Gont, L. K., Fainsod, A., Kim, S. H., & De Robertis, E. M. (1996). Overexpression of the homeobox gene Xnot-2 leads to notochord formation in Xenopus. *Developmental Biology*, *174*(1), 174–178. http://doi.org/10.1006/dbio.1996.0061.

Gont, L. K., Steinbeisser, H., Blumberg, B., & De Robertis, E. M. (1993). Tail formation as a continuation of gastrulation: The multiple cell populations of the Xenopus tailbud derive from the late blastopore lip. *Development*, *119*(4), 991–1004.

Gouti, M., Tsakiridis, A., Wymeersch, F. J., Huang, Y., Kleinjung, J., Wilson, V., et al. (2014). In vitro generation of neuromesodermal progenitors reveals distinct roles for wnt signalling in the specification of spinal cord and paraxial mesoderm identity. *PLoS Biology*, *12*(8), e1001937. http://doi.org/10.1371/journal.pbio.1001937.

Griffin, K. J., Amacher, S. L., Kimmel, C. B., & Kimelman, D. (1998). Molecular identification of spadetail: Regulation of zebrafish trunk and tail mesoderm formation by T-box genes. *Development*, *125*(17), 3379–3388.

Griffith, C. M., & Sanders, E. J. (1991). Effects of extracellular matrix components on the differentiation of chick embryo tail bud mesenchyme in culture. *Differentiation*, *47*(2), 61–68. http://doi.org/10.1111/j.1432-0436.1991.tb00223.x.

Griffith, C. M., Wiley, M. J., & Sanders, E. J. (1992). The vertebrate tail bud: Three germ layers from one tissue. *Anatomy and Embryology*, *185*(2), 101–113. http://doi.org/10.1007/BF00185911.

Gross, J. M., & McClay, D. R. (2001). The role of Brachyury (T) during gastrulation movements in the sea urchin *Lytechinus variegatus*. *Developmental Biology*, *239*(1), 132–147. http://doi.org/10.1006/dbio.2001.0426.

Grüneberg, H. (1956). A ventral ectodermal ridge of the tail in mouse embryos. *Nature*, *177*(4513), 787–788. http://doi.org/10.1038/177787b0.

Grüneberg, H. (1958). Genetical studies on the skeleton of the mouse. XXIII. The development of brachyury and anury. *Journal of Embryology and Experimental Morphology*, *6*(3), 424–443.

Hadjantonakis, A.-K., Pisano, E., & Papaioannou, V. E. (2008). Tbx6 regulates left/right patterning in mouse embryos through effects on nodal cilia and perinodal signaling. *PLoS One*, *3*(6), e2511. http://doi.org/10.1371/journal.pone.0002511.

Halpern, M., Ho, R., Walker, C., & Kimmel, C. (1993). Induction of muscle pioneers and floor plate is distinguished by the zebrafish no tail mutation. *Cell*, *75*(1), 99–111.

Hancock, S. N., Agulnik, S. I., Silver, L. M., & Papaioannou, V. E. (1999). Mapping and expression analysis of the mouse ortholog of Xenopus Eomesodermin. *Mechanisms of Development*, *81*(1–2), 205–208. http://doi.org/10.1016/S0925-4773(98)00244-5.

Harada, Y., Yasuo, H., & Satoh, N. (1995). A sea urchin homologue of the chordate Brachyury (T) gene is expressed in the secondary mesenchyme founder cells. *Development*, *121*(9), 2747–2754.

Harvey, S. A., Tümpel, S., Dubrulle, J., Schier, A. F., & Smith, J. C. (2010). no tail integrates two modes of mesoderm induction. *Development*, *137*(7), 1127–1135. http://doi.org/10.1242/dev.046318.

Hashimoto, K., Fujimoto, H., & Nakatsuji, N. (1987). An ECM substratum allows mouse mesodermal cells isolated from the primitive streak to exhibit motility similar to that inside the embryo and reveals a deficiency in the T/T mutant cells. *Development*, *100*(4), 587–598.

Hayata, T., Eisaki, A., Kuroda, H., & Asashima, M. (1999). Expression of Brachyury-like T-box transcription factor, Xbra3 in Xenopus embryo. *Development Genes and Evolution*, *209*(9), 560–563.

Herrmann, B. G. (1991). Expression pattern of the Brachyury gene in whole-mount TWis/TWis mutant embryos. *Development*, *113*(3), 913–917.

Herrmann, B., Labeit, S., Poustka, A., King, T., & Lehrach, H. (1990). Cloning of the T gene required in mesoderm formation in the mouse. *Nature*, *343*(6259), 617–622.

Ho, R. K., & Kane, D. A. (1990). Cell-autonomous action of zebrafish spt-1 mutation in specific mesodermal precursors. *Nature*, *348*(6303), 728–730. http://doi.org/10.1038/348728a0.

Holland, P. W., Koschorz, B., Holland, L. Z., & Herrmann, B. G. (1995). Conservation of Brachyury (T) genes in amphioxus and vertebrates: Developmental and evolutionary implications. *Development*, *121*(12), 4283–4291.

Holmdahl, D. E. (1925). Experimentelle Untersuchungen über die Lage der Grenze zwischen primärer und sekundärer Körperentwicklung beim Huhn. *Anatomischer Anzeiger*, *25*, 393–396.

Horb, M. E., & Thomsen, G. H. (1997). A vegetally localized T-box transcription factor in Xenopus eggs specifies mesoderm and endoderm and is essential for embryonic mesoderm formation. *Development*, *124*(9), 1689–1698.

Imuta, Y., Kiyonari, H., Jang, C.-W., Behringer, R. R., & Sasaki, H. (2013). Generation of knock-in mice that express nuclear enhanced green fluorescent protein and tamoxifen-inducible Cre recombinase in the notochord from Foxa2 and T loci. *Genesis*, *51*(3), 210–218. http://doi.org/10.1002/dvg.22376.

Inman, K. E., & Downs, K. M. (2006). Brachyury is required for elongation and vasculogenesis in the murine allantois. *Development*, *133*(15), 2947–2959. http://doi.org/10.1242/dev.02454.

Isaacs, H., Pownall, M., & Slack, J. (1994). eFGF regulates Xbra expression during Xenopus gastrulation. *The EMBO Journal*, *13*(19), 4469–4481.

Jacobs, J. J., Keblusek, P., Robanus-Maandag, E., Kristel, P., Lingbeek, M., Nederlof, P. M., et al. (2000). Senescence bypass screen identifies TBX2, which represses Cdkn2a (p19 (ARF)) and is amplified in a subset of human breast cancers. *Nature Genetics, 26*(3), 291–299. http://doi.org/10.1038/81583.

Jen, W. C., Gawantka, V., Pollet, N., Niehrs, C., & Kintner, C. (1999). Periodic repression of Notch pathway genes governs the segmentation of Xenopus embryos. *Genes & Development, 13*(11), 1486–1499.

Jen, W. C., Wettstein, D., Turner, D., Chitnis, A., & Kintner, C. (1997). The Notch ligand, X-Delta-2, mediates segmentation of the paraxial mesoderm in Xenopus embryos. *Development, 124*(6), 1169–1178.

Jolma, A., Yan, J., Whitington, T., Toivonen, J., Nitta, K. R., Rastas, P., et al. (2013). DNA-binding specificities of human transcription factors. *Cell, 152*(1–2), 327–339. http://doi.org/10.1016/j.cell.2012.12.009.

Joly, J. S., Maury, M., Joly, C., Duprey, P., Boulekbache, H., & Condamine, H. (1992). Expression of a zebrafish caudal homeobox gene correlates with the establishment of posterior cell lineages at gastrulation. *Differentiation, 50*(2), 75–87.

José-Edwards, D. S., Oda-Ishii, I., Kugler, J. E., Passamaneck, Y. J., Katikala, L., Nibu, Y., et al. (2015). Brachyury, Foxa2 and the cis-regulatory origins of the Notochord. *PLoS Genetics, 11*(12), e1005730. http://doi.org/10.1371/journal.pgen.1005730.

Kanki, J. P., & Ho, R. K. (1997). The development of the posterior body in zebrafish. *Development, 124*(4), 881–893.

Katikala, L., Aihara, H., Passamaneck, Y. J., Gazdoiu, S., José-Edwards, D. S., Kugler, J. E., et al. (2013). Functional Brachyury binding sites establish a temporal read-out of gene expression in the Ciona notochord. *PLoS Biology, 11*(10), e1001697. http://doi.org/10.1371/journal.pbio.1001697.

Kawamura, A., Koshida, S., Hijikata, H., Ohbayashi, A., Kondoh, H., & Takada, S. (2005). Groucho-associated transcriptional repressor ripply1 is required for proper transition from the presomitic mesoderm to somites. *Developmental Cell, 9*(6), 735–744. http://doi.org/10.1016/j.devcel.2005.09.021.

Keller, R. E. (1978). Time-lapse cinemicrographic analysis of superficial cell behavior during and prior to gastrulation in *Xenopus laevis*. *Journal of Morphology, 157*(2), 223–247. http://doi.org/10.1002/jmor.1051570209.

Keller, R. E. (1981). An experimental analysis of the role of bottle cells and the deep marginal zone in gastrulation of *Xenopus laevis*. *Journal of Experimental Zoology, 216*(1), 81–101. http://doi.org/10.1002/jez.1402160109.

Keller, R. (1991). Early embryonic development of *Xenopus laevis*. *Methods in Cell Biology, 36*, 61–113.

Keller, R. E., Danilchik, M., Gimlich, R., & Shih, J. (1985). The function and mechanism of convergent extension during gastrulation of *Xenopus laevis*. *Journal of Embryology and Experimental Morphology, 89*(Suppl.), 185–209.

Kelley, M. J., Shi, J., Ballew, B., Hyland, P. L., Li, W.-Q., Rotunno, M., et al. (2014). Characterization of T gene sequence variants and germline duplications in familial and sporadic chordoma. *Human Genetics, 133*(10), 1289–1297. http://doi.org/10.1007/s00439-014-1463-z.

Kimmel, C. B., Kane, D. A., Walker, C., Warga, R. M., & Rothman, M. B. (1989). A mutation that changes cell movement and cell fate in the zebrafish embryo. *Nature, 337*(6205), 358–362. http://doi.org/10.1038/337358a0.

Kinder, S. J., Tsang, T. E., Quinlan, G. A., Hadjantonakis, A. K., Nagy, A., & Tam, P. P. (1999). The orderly allocation of mesodermal cells to the extraembryonic structures and the anteroposterior axis during gastrulation of the mouse embryo. *Development, 126*(21), 4691–4701.

King, T., Beddington, R., & Brown, N. (1998). The role of the brachyury gene in heart development and left-right specification in the mouse. *Mechanisms of Development, 79*(1–2), 29–37.

Kingsbury, B. F. (1932). The "law" of cephalocaudal differential growth in its application to the nervous system. *Journal of Comparative Neurology, 56*(2), 431–463. http://doi.org/10.1002/cne.900560208.

Kispert, A., & Herrmann, B. G. (1993). The Brachyury gene encodes a novel DNA binding protein. *The EMBO Journal, 12*(8), 3211.

Kispert, A., & Herrmann, B. G. (1994). Immunohistochemical analysis of the Brachyury protein in wild-type and mutant mouse embryos. *Developmental Biology, 161*(1), 179–193. http://doi.org/10.1006/dbio.1994.1019.

Kispert, A., Herrmann, B. G., Leptin, M., & Reuter, R. (1994). Homologs of the mouse Brachyury gene are involved in the specification of posterior terminal structures in Drosophila, Tribolium, and Locusta. *Genes & Development, 8*(18), 2137–2150. http://doi.org/10.1101/gad.8.18.2137.

Kispert, A., Koschorz, B., & Herrmann, B. (1995). The T protein encoded by Brachyury is a tissue-specific transcription factor. *The EMBO Journal, 14*(19), 4763–4772.

Kispert, A., Ortner, H., Cooke, J., & Herrmann, B. G. (1995). The chick Brachyury gene: Developmental expression pattern and response to axial induction by localized activin. *Developmental Biology, 168*(2), 406–415. http://doi.org/10.1006/dbio.1995.1090.

Kloc, M., & Etkin, L. D. (1995). Two distinct pathways for the localization of RNAs at the vegetal cortex in Xenopus oocytes. *Development, 121*(2), 287–297.

Knezevic, V., & De Santo, R. (1997). Two novel chick T-box genes related to mouse Brachyury are expressed in different, non-overlapping mesodermal domains during gastrulation. *Development, 124*, 411–419.

Knezevic, V., De Santo, R., & Mackem, S. (1998). Continuing organizer function during chick tail development. *Development, 125*(10), 1791–1801.

Kölliker, A. (1884). Die embryonalen Keimblätter und die Gewebe. *Zeitschrift für Wissenschaftliche Zoologie, 40*, 179–213.

Kulisz, A., & Simon, H.-G. (2008). An evolutionarily conserved nuclear export signal facilitates cytoplasmic localization of the Tbx5 transcription factor. *Molecular and Cellular Biology, 28*(5), 1553–1564. http://doi.org/10.1128/MCB.00935-07.

Kuraku, S., & Kuratani, S. (2011). Genome-wide detection of gene extinction in early mammalian evolution. *Genome Biology and Evolution, 3*, 1449–1462. http://doi.org/10.1093/gbe/evr120.

Larabell, C. A., Torres, M., Rowning, B. A., Yost, C., Miller, J. R., Wu, M., et al. (1997). Establishment of the dorso-ventral axis in Xenopus embryos is presaged by early asymmetries in beta-catenin that are modulated by the Wnt signaling pathway. *The Journal of Cell Biology, 136*(5), 1123–1136.

Lawson, K. A., Meneses, J. J., & Pedersen, R. A. (1991). Clonal analysis of epiblast fate during germ layer formation in the mouse embryo. *Development, 113*, 891–911.

Lawton, A. K., Nandi, A., Stulberg, M. J., Dray, N., Sneddon, M. W., Pontius, W., et al. (2013). Regulated tissue fluidity steers zebrafish body elongation. *Development, 140*(3), 573–582. http://doi.org/10.1242/dev.090381.

Lee, M. T., Bonneau, A. R., Takacs, C. M., Bazzini, A. A., Divito, K. R., Fleming, E. S., et al. (2013). Nanog, Pou5f1 and SoxB1 activate zygotic gene expression during the maternal-to-zygotic transition. *Nature, 503*(7476), 360–364. http://doi.org/10.1038/nature12632.

Li, X.-Y., Thomas, S., Sabo, P. J., Eisen, M. B., Stamatoyannopoulos, J. A., & Biggin, M. D. (2011). The role of chromatin accessibility in directing the widespread, overlapping patterns of Drosophila transcription factor binding. *Genome Biology, 12*(4), R34. http://doi.org/10.1186/gb-2011-12-4-r34.

Lippmann, E. S., Williams, C. E., Ruhl, D. A., Estevez-Silva, M. C., Chapman, E. R., Coon, J. J., et al. (2015). Deterministic HOX patterning in human pluripotent stem cell-derived neuroectoderm. *Stem Cell Reports*, *4*(4), 632–644. http://doi.org/10.1016/j.stemcr.2015.02.018.

Lolas, M., Valenzuela, P. D. T., Tjian, R., & Liu, Z. (2014). Charting Brachyury-mediated developmental pathways during early mouse embryogenesis. *Proceedings of the National Academy of Sciences of the United States of America*, *111*(12), 4478–4483. http://doi.org/10.1073/pnas.1402612111.

Luxardi, G., Marchal, L., Thomé, V., & Kodjabachian, L. (2010). Distinct Xenopus Nodal ligands sequentially induce mesendoderm and control gastrulation movements in parallel to the Wnt/PCP pathway. *Development*, *137*(3), 417–426. http://doi.org/10.1242/dev.039735.

Manning, A. J., & Kimelman, D. (2015). Tbx16 and Msgn1 are required to establish directional cell migration of zebrafish mesodermal progenitors. *Developmental Biology*, *406*(2), 172–185. http://doi.org/10.1016/j.ydbio.2015.09.001.

Martin, B. L., & Kimelman, D. (2008). Regulation of canonical wnt signaling by brachyury is essential for posterior mesoderm formation. *Developmental Cell*, *15*(1), 121–133. http://doi.org/10.1016/j.devcel.2008.04.013.

Martin, B. L., & Kimelman, D. (2010). Brachyury establishes the embryonic mesodermal progenitor niche. *Genes & Development*, *24*(24), 2778–2783. http://doi.org/10.1101/gad.1962910.

Martin, B. L., & Kimelman, D. (2012). Canonical Wnt signaling dynamically controls multiple stem cell fate decisions during vertebrate body formation. *Developmental Cell*, *22*(1), 223–232. http://doi.org/10.1016/j.devcel.2011.11.001.

McGrew, M. J., Sherman, A., Lillico, S. G., Ellard, F. M., Radcliffe, P. A., Gilhooley, H. J., et al. (2008). Localised axial progenitor cell populations in the avian tail bud are not committed to a posterior Hox identity. *Development*, *135*(13), 2289–2299. http://doi.org/10.1242/dev.022020.

Mendjan, S., Mascetti, V. L., Ortmann, D., Ortiz, M., Karjosukarso, D. W., Ng, Y., et al. (2014). NANOG and CDX2 pattern distinct subtypes of human mesoderm during exit from pluripotency. *Cell Stem Cell*, *15*(3), 310–325. http://doi.org/10.1016/j.stem.2014.06.006.

Messenger, N. J., Kabitschke, C., Andrews, R., Grimmer, D., Núñez Miguel, R., Blundell, T. L., et al. (2005). Functional specificity of the Xenopus T-domain protein Brachyury is conferred by its ability to interact with Smad1. *Developmental Cell*, *8*(4), 599–610. http://doi.org/10.1016/j.devcel.2005.03.001.

Mills, C. L., & Bellairs, R. (1989). Mitosis and cell death in the tail of the chick embryo. *Anatomy and Embryology*, *180*(3), 301–308. http://doi.org/10.1007/BF00315888.

Moosmann, J., Ershov, A., Altapova, V., Baumbach, T., Prasad, M. S., LaBonne, C., et al. (2013). X-ray phase-contrast in vivo microtomography probes new aspects of Xenopus gastrulation. *Nature*, *497*(7449), 374–377. http://doi.org/10.1038/nature12116.

Moreno, T. A., & Kintner, C. (2004). Regulation of segmental patterning by retinoic acid signaling during Xenopus somitogenesis. *Developmental Cell*, *6*(2), 205–218.

Morley, R. H., Lachani, K., Keefe, D., Gilchrist, M. J., Flicek, P., Smith, J. C., et al. (2009). A gene regulatory network directed by zebrafish No tail accounts for its roles in mesoderm formation. *Proceedings of the National Academy of Sciences of the United States of America*, *106*(10), 3829–3834. http://doi.org/10.1073/pnas.0808382106.

Müller, C. W., & Herrmann, B. G. (1997). Crystallographic structure of the T domain-DNA complex of the Brachyury transcription factor. *Nature*, *389*(6653), 884–888. http://doi.org/10.1038/39929.

Murakami, M. S., Moody, S. A., Daar, I. O., & Morrison, D. K. (2004). Morphogenesis during Xenopus gastrulation requires Wee1-mediated inhibition of cell proliferation. *Development*, *131*(3), 571–580. http://doi.org/10.1242/dev.00971.

Naiche, L. A., Harrelson, Z., Kelly, R., & Papaioannou, V. (2005). T-box genes in vertebrate development. *Annual Review of Genetics*, *39*, 219–239.

Naiche, L. A., Holder, N., & Lewandoski, M. (2011). FGF4 and FGF8 comprise the wavefront activity that controls somitogenesis. *Proceedings of the National Academy of Sciences of the United States of America*, *108*(10), 4018–4023. http://doi.org/10.1073/pnas.1007417108.

Nakajima, Y., Morimoto, M., Takahashi, Y., Koseki, H., & Saga, Y. (2006). Identification of Epha4 enhancer required for segmental expression and the regulation by Mesp2. *Development*, *133*(13), 2517–2525. http://doi.org/10.1242/dev.02422.

Nakao, T., & Ishizawa, A. (1984). Light- and electron-microscopic observations of the tail bud of the larval lamprey (*Lampetra japonica*), with special reference to neural tube formation. *The American Journal of Anatomy*, *170*(1), 55–71. http://doi.org/10.1002/aja.1001700105.

Nelson, A. C., Cutty, S. J., Niini, M., Stemple, D. L., Flicek, P., Houart, C., et al. (2014). Global identification of Smad2 and Eomesodermin targets in zebrafish identifies a conserved transcriptional network in mesendoderm and a novel role for Eomesodermin in repression of ectodermal gene expression. *BMC Biology*, *12*(1), 81. http://doi.org/10.1186/s12915-014-0081-5.

Nieuwkoop, P. D. (1952). Activation and organization of the central nervous system in amphibians. Part III. Synthesis of a new working hypothesis. *The Journal of Experimental Zoology*, *120*(1), 83–108. http://doi.org/10.1002/jez.1401200104.

Nikaido, M., Kawakami, A., Sawada, A., Furutani-Seiki, M., Takeda, H., & Araki, K. (2002). Tbx24, encoding a T-box protein, is mutated in the zebrafish somite-segmentation mutant fused somites. *Nature Genetics*, *31*(2), 195–199. http://doi.org/10.1038/ng899.

Olivera-Martinez, I., Harada, H., Halley, P. A., & Storey, K. G. (2012). Loss of FGF-dependent mesoderm identity and rise of endogenous retinoid signalling determine cessation of body axis elongation. *PLoS Biology*, *10*(10), e1001415. http://doi.org/10.1371/journal.pbio.1001415.

O'Reilly, M., Smith, J., & Cunliffe, V. (1995). Patterning of the mesoderm in Xenopus: Dose-dependent and synergistic effects of Brachyury and Pintallavis. *Development*, *121*(5), 1351–1359.

Owens, N. D. L., Blitz, I. L., Lane, M. A., Patrushev, I., Overton, J. D., Gilchrist, M. J., et al. (2016). Measuring absolute RNA copy numbers at high temporal resolution reveals transcriptome kinetics in development. *Cell Reports*, *14*(3), 632–647. http://doi.org/10.1016/j.celrep.2015.12.050.

Pander, H. C. (1817). *Beiträge zur Entwickelungsgeschichte des Hühnchens im Eye*. Würzburg: H.L. Brönner.

Papaioannou, V. E. (2014). The T-box gene family: Emerging roles in development, stem cells and cancer. *Development*, *141*(20), 3819–3833. http://doi.org/10.1242/dev.104471.

Park, J. C., Chae, Y. K., Son, C. H., Kim, M. S., Lee, J., Ostrow, K., et al. (2008). Epigenetic silencing of human T (brachyury homologue) gene in non-small-cell lung cancer. *Biochemical and Biophysical Research Communications*, *365*(2), 221–226. http://doi.org/10.1016/j.bbrc.2007.10.144.

Pasteels, J. (1939). La formation de la queue chez les Vertébrés. *Annales De La Société Royale Zoologique De Belgique*, *70*, 33–51. http://doi.org/10.1111/dgd.12122/full.

Pasteels, J. (1942). New observations concerning the maps of presumptive areas of the young amphibian gastrula. (Amblystoma and Discoglossus). *Journal of Experimental Zoology*, *89*(2), 255–281. http://doi.org/10.1002/jez.1400890205.

Pasteels, J. (1943). Proliférations et croissance dans la gastrulation et la formation de la queue des Vertébrés. *Archives De Biologie Liege*, *54*, 1–55.

Perantoni, A. O., Timofeeva, O., Naillat, F., Richman, C., Pajni-Underwood, S., Wilson, C., et al. (2005). Inactivation of FGF8 in early mesoderm reveals an essential role in kidney development. *Development*, *132*(17), 3859–3871. http://doi.org/10.1242/dev.01945.

Pereira, L. A., Wong, M. S., Lim, S. M., Sides, A., Stanley, E. G., & Elefanty, A. G. (2011). Brachyury and related Tbx proteins interact with the Mixl1 homeodomain protein and negatively regulate Mixl1 transcriptional activity. *PLoS One*, *6*(12), e28394. http://doi.org/10.1371/journal.pone.0028394.

Peterson, K. J., Cameron, R. A., Tagawa, K., Satoh, N., & Davidson, E. H. (1999). A comparative molecular approach to mesodermal patterning in basal deuterostomes: The expression pattern of Brachyury in the enteropneust hemichordate *Ptychodera flava*. *Development*, *126*(1), 85–95.

Peterson, K. J., Harada, Y., Cameron, R. A., & Davidson, E. H. (1999). Expression pattern of Brachyury and not in the sea urchin: Comparative implications for the origins of mesoderm in the basal deuterostomes. *Developmental Biology*, *207*(2), 419–431. http://doi.org/10.1006/dbio.1998.9177.

Pique-Regi, R., Degner, J. F., Pai, A. A., Gaffney, D. J., Gilad, Y., & Pritchard, J. K. (2011). Accurate inference of transcription factor binding from DNA sequence and chromatin accessibility data. *Genome Research*, *21*(3), 447–455. http://doi.org/10.1101/gr.112623.110.

Postma, A. V., Alders, M., Sylva, M., Bilardo, C. M., Pajkrt, E., van Rijn, R. R., et al. (2014). Mutations in the T (brachyury) gene cause a novel syndrome consisting of sacral agenesis, abnormal ossification of the vertebral bodies and a persistent notochordal canal. *Journal of Medical Genetics*, *51*(2), 90–97. http://doi.org/10.1136/jmedgenet-2013-102001.

Rast, J. P., Cameron, R. A., Poustka, A. J., & Davidson, E. H. (2002). brachyury target genes in the early sea urchin embryo isolated by differential macroarray screening. *Developmental Biology*, *246*(1), 191–208. http://doi.org/10.1006/dbio.2002.0654.

Reim, I., & Frasch, M. (2005). The Dorsocross T-box genes are key components of the regulatory network controlling early cardiogenesis in Drosophila. *Development*, *132*(22), 4911–4925. http://doi.org/10.1242/dev.02077.

Reim, I., Lee, H.-H., & Frasch, M. (2003). The T-box-encoding Dorsocross genes function in amnioserosa development and the patterning of the dorsolateral germ band downstream of Dpp. *Development*, *130*(14), 3187–3204. http://doi.org/10.1242/dev.00548.

Remak, R. (1855). *Untersuchungen über die Entwickelung der Wirbelthiere*. Berlin: G. Reimer.

Row, R. H., Tsotras, S. R., Goto, H., & Martin, B. L. (2016). The zebrafish tailbud contains two independent populations of midline progenitor cells that maintain long-term germ layer plasticity and differentiate in response to local signaling cues. *Development*, *143*(2), 244–254. http://doi.org/10.1242/dev.129015.

Rowning, B. A., Wells, J., Wu, M., Gerhart, J. C., Moon, R. T., & Larabell, C. A. (1997). Microtubule-mediated transport of organelles and localization of beta-catenin to the future dorsal side of Xenopus eggs. *Proceedings of the National Academy of Sciences of the United States of America*, *94*(4), 1224–1229.

Rudloff, S., & Kemler, R. (2012). Differential requirements for β-catenin during mouse development. *Development*, *139*(20), 3711–3721. http://doi.org/10.1242/dev.085597.

Russ, A. P., Wattler, S., Colledge, W. H., Aparicio, S. A., Carlton, M. B., Pearce, J. J., et al. (2000). Eomesodermin is required for mouse trophoblast development and mesoderm formation. *Nature*, *404*(6773), 95–99. http://doi.org/10.1038/35003601.

Ruvinsky, I., Silver, L. M., & Gibson-Brown, J. J. (2000). Phylogenetic analysis of T-box genes demonstrates the importance of amphioxus for understanding evolution of the vertebrate genome. *Genetics*, *156*(3), 1249–1257.

Ryan, K., Garrett, N., Mitchell, A., & Gurdon, J. (1996). Eomesodermin, a key early gene in Xenopus mesoderm differentiation. *Cell*, *87*(6), 989–1000.

Sakai, Y., Meno, C., Fujii, H., Nishino, J., Shiratori, H., Saijoh, Y., et al. (2001). The retinoic acid-inactivating enzyme CYP26 is essential for establishing an uneven distribution of retinoic acid along the anterio-posterior axis within the mouse embryo. *Genes & Development*, *15*(2), 213–225. http://doi.org/10.1101/gad.851501.

Sanders, E. J., Khare, M. K., Ooi, V. C., & Bellairs, R. (1986). An experimental and morphological analysis of the tail bud mesenchyme of the chick embryo. *Anatomy and Embryology*, *174*(2), 179–185. http://doi.org/10.1007/BF00824333.

Schneider, S., Steinbeisser, H., Warga, R., & Hausen, P. (1996). Beta-catenin translocation into nuclei demarcates the dorsalizing centers in frog and fish embryos. *Mechanisms of Development*, *57*(2), 191–198.

Schoenwolf, G. C. (1977). Tail (end) bud contributions to the posterior region of the chick embryo. *Journal of Experimental Zoology*, *201*(2), 227–245. http://doi.org/10.1002/jez.1402010208.

Schoenwolf, G. C. (1978). Effects of complete tail bud extirpation on early development of the posterior region of the chick embryo. *The Anatomical Record*, *192*(2), 289–295. http://doi.org/10.1002/ar.1091920209.

Schoenwolf, G. C. (1984). Histological and ultrastructural studies of secondary neurulation in mouse embryos. *The American Journal of Anatomy*, *169*(4), 361–376. http://doi.org/10.1002/aja.1001690402.

Schoenwolf, G. C., & Delongo, J. (1980). Ultrastructure of secondary neurulation in the chick embryo. *The American Journal of Anatomy*, *158*(1), 43–63. http://doi.org/10.1002/aja.1001580106.

Scholz, C. B., & Technau, U. (2003). The ancestral role of Brachyury: Expression of NemBra1 in the basal cnidarian *Nematostella vectensis* (Anthozoa). *Development Genes and Evolution*, *212*(12), 563–570. http://doi.org/10.1007/s00427-002-0272-x.

Schulte-Merker, S., Ho, R. K., Herrmann, B. G., & Nüsslein-Volhard, C. (1992). The protein product of the zebrafish homologue of the mouse T gene is expressed in nuclei of the germ ring and the notochord of the early embryo. *Development*, *116*(4), 1021–1032.

Schulte-Merker, S., & Smith, J. (1995). Mesoderm formation in response to Brachyury requires FGF signalling. *Current Biology*, *5*(1), 62–67.

Schulte-Merker, S., van Eeden, F., Halpern, M., Kimmel, C., & Nusslein-Volhard, C. (1994). *no tail* (*ntl*) is the zebrafish homologue of the mouse T (*Brachyury*) gene. *Development*, *120*(4), 1009–1015.

Searle, A. G. (1966). Curtailed, a new dominant T-allele in the house mouse. *Genetical Research*, 7(1), 86–95. http://doi.org/10.1017/S0016672300009496.

Sebé-Pedrós, A., Ariza-Cosano, A., Weirauch, M. T., Leininger, S., Yang, A., Torruella, G., et al. (2013). Early evolution of the T-box transcription factor family. *Proceedings of the National Academy of Sciences of the United States of America*, *110*(40), 16050–16055. http://doi.org/10.1073/pnas.1309748110.

Sebé-Pedrós, A., & Ruiz-Trillo, I. (2017). Evolution and classification of the T-Box transcription factor family. *Current Topics in Developmental Biology*, *122*, 1–26.

Seevers, C. H. (1932). Potencies of the end bud and other caudal levels of the early chick embryo, with special reference to the origin of the metanephros. *The Anatomical Record*, *54*(2), 217–246. http://doi.org/10.1002/ar.1090540210.

Selleck, M. A., & Stern, C. D. (1991). Fate mapping and cell lineage analysis of Hensen's node in the chick embryo. *Development*, *112*(2), 615–626.

Shaheen, R., Alshail, E., Alaqeel, A., Ansari, S., Hindieh, F., & Alkuraya, F. S. (2015). T (brachyury) is linked to a Mendelian form of neural tube defects in humans. *Human Genetics*, *134*(10), 1139–1141. http://doi.org/10.1007/s00439-015-1589-7.

Shedlovsky, A., King, T. R., & Dove, W. F. (1988). Saturation germ line mutagenesis of the murine t region including a lethal allele at the quaking locus. *Proceedings of the National Academy of Sciences of the United States of America*, *85*(1), 180–184.

Shoguchi, E., Satoh, N., & Maruyama, Y. K. (1999). Pattern of Brachyury gene expression in starfish embryos resembles that of hemichordate embryos but not of sea urchin embryos. *Mechanisms of Development*, *82*(1–2), 185–189.

Shoguchi, E., Satoh, N., & Maruyama, Y. K. (2000). A starfish homolog of mouse T-brain-1 is expressed in the archenteron of Asterina pectinifera embryos: Possible involvement of two T-box genes in starfish gastrulation. *Development, Growth & Differentiation*, *42*(1), 61–68. http://doi.org/10.1046/j.1440-169x.2000.00481.x.

Shum, A. S., Poon, L. L., Tang, W. W., Koide, T., Chan, B. W., Leung, Y. C., et al. (1999). Retinoic acid induces down-regulation of Wnt-3a, apoptosis and diversion of tail bud cells to a neural fate in the mouse embryo. *Mechanisms of Development*, *84*(1–2), 17–30.

Singer, J. B., Harbecke, R., Kusch, T., Reuter, R., & Lengyel, J. A. (1996). Drosophila brachyenteron regulates gene activity and morphogenesis in the gut. *Development*, *122*(12), 3707–3718.

Sinha, S., Abraham, S., Gronostajski, R. M., & Campbell, C. E. (2000). Differential DNA binding and transcription modulation by three T-box proteins, T, TBX1 and TBX2. *Gene*, *258*(1–2), 15–29. http://doi.org/10.1016/S0378-1119(00)00417-0.

Skirkanich, J., Luxardi, G., Yang, J., Kodjabachian, L., & Klein, P. S. (2011). An essential role for transcription before the MBT in *Xenopus laevis*. *Developmental Biology*, *357*(2), 478–491. http://doi.org/10.1016/j.ydbio.2011.06.010.

Slagle, C. E., Aoki, T., & Burdine, R. D. (2011). Nodal-dependent mesendoderm specification requires the combinatorial activities of FoxH1 and Eomesodermin. *PLoS Genetics*, 7(5), e1002072. http://doi.org/10.1371/journal.pgen.1002072.

Smith, J., Price, B., Green, J., Weigel, D., & Herrmann, B. (1991). Expression of a Xenopus homolog of Brachyury (T) is an immediate-early response to mesoderm induction. *Cell*, *67*(1), 79–87.

Sparrow, D. B., Jen, W. C., Kotecha, S., Towers, N., Kintner, C., & Mohun, T. J. (1998). Thylacine 1 is expressed segmentally within the paraxial mesoderm of the Xenopus embryo and interacts with the Notch pathway. *Development*, *125*(11), 2041–2051.

Spemann, H., & Mangold, H. (1924). über Induktion von Embryonalanlagen durch Implantation artfremder Organisatoren. *Archiv für Mikroskopische Anatomie und Entwicklungsmechanik*, *100*(3–4), 599–638. http://doi.org/10.1007/BF02108133.

Stennard, F., Carnac, G., & Gurdon, J. B. (1996). The Xenopus T-box gene, Antipodean, encodes a vegetally localised maternal mRNA and can trigger mesoderm formation. *Development*, *122*(12), 4179–4188.

Stennard, F., Zorn, A. M., Ryan, K., Garrett, N., & Gurdon, J. B. (1999). Differential expression of VegT and Antipodean protein isoforms in Xenopus. *Mechanisms of Development*, *86*(1–2), 87–98.

Stott, D., Kispert, A., & Herrmann, B. G. (1993). Rescue of the tail defect of Brachyury mice. *Genes & Development*, 7(2), 197–203.

Strumpf, D., Mao, C.-A., Yamanaka, Y., Ralston, A., Chawengsaksophak, K., Beck, F., et al. (2005). Cdx2 is required for correct cell fate specification and differentiation of trophectoderm in the mouse blastocyst. *Development*, *132*(9), 2093–2102. http://doi.org/10.1242/dev.01801.

Tada, M., Casey, E., Fairclough, L., & Smith, J. (1998). Bix1, a direct target of Xenopus T-box genes, causes formation of ventral mesoderm and endoderm. *Development*, *125*(20), 3997–4006.

Tagawa, K., Humphreys, T., & Satoh, N. (1998). Novel pattern of Brachyury gene expression in hemichordate embryos. *Mechanisms of Development*, *75*(1–2), 139–143. http://doi.org/10.1016/S0925-4773(98)00078-1.

Tagawa, K., Humphreys, T., & Satoh, N. (2000). T-brain expression in the apical organ of hemichordate tornaria larvae suggests its evolutionary link to the vertebrate forebrain. *Journal of Experimental Zoology*, *288*(1), 23–31.

Takahashi, Y., Koizumi, K., Takagi, A., Kitajima, S., Inoue, T., Koseki, H., et al. (2000). Mesp2 initiates somite segmentation through the Notch signalling pathway. *Nature Genetics*, *25*(4), 390–396. http://doi.org/10.1038/78062.

Takashima, Y., & Suzuki, A. (2013). Regulation of organogenesis and stem cell properties by T-box transcription factors. *Cellular and Molecular Life Sciences: CMLS*, *70*(20), 3929–3945. http://doi.org/10.1007/s00018-013-1305-5.

Takatori, N., Hotta, K., Mochizuki, Y., Satoh, G., Mitani, Y., Satoh, N., et al. (2004). T-box genes in the ascidian *Ciona intestinalis*: Characterization of cDNAs and spatial expression. *Developmental Dynamics*, *230*(4), 743–753. http://doi.org/10.1002/dvdy.20082.

Takemoto, T., Uchikawa, M., Yoshida, M., Bell, D. M., Lovell-Badge, R., Papaioannou, V. E., et al. (2011). Tbx6-dependent Sox2 regulation determines neural or mesodermal fate in axial stem cells. *Nature*, *470*(7334), 394–398. http://doi.org/10.1038/nature09729.

Tam, P. P., Parameswaran, M., Kinder, S. J., & Weinberger, R. P. (1997). The allocation of epiblast cells to the embryonic heart and other mesodermal lineages: The role of ingression and tissue movement during gastrulation. *Development*, *124*(9), 1631–1642. http://doi.org/10.1016/0092-8674(94)90522-3.

Tazumi, S., Yabe, S., & Uchiyama, H. (2010). Paraxial T-box genes, Tbx6 and Tbx1, are required for cranial chondrogenesis and myogenesis. *Developmental Biology*, *346*(2), 170–180. http://doi.org/10.1016/j.ydbio.2010.07.028.

Tazumi, S., Yabe, S., Yokoyama, J., Aihara, Y., & Uchiyama, H. (2008). PMesogenin1 and 2 function directly downstream of Xtbx6 in Xenopus somitogenesis and myogenesis. *Developmental Dynamics*, *237*(12), 3749–3761. http://doi.org/10.1002/dvdy.21791.

Teillet, M. A., Lapointe, F., & Le Douarin, N. M. (1998). The relationships between notochord and floor plate in vertebrate development revisited. *Proceedings of the National Academy of Sciences of the United States of America*, *95*(20), 11733–11738. http://doi.org/10.1073/pnas.95.20.11733.

Teo, A. K. K., Arnold, S. J., Trotter, M. W. B., Brown, S., Ang, L. T., Chng, Z., et al. (2011). Pluripotency factors regulate definitive endoderm specification through eomesodermin. *Genes & Development*, *25*(3), 238–250. http://doi.org/10.1101/gad.607311.

Thorpe, C. J., Weidinger, G., & Moon, R. T. (2005). Wnt/beta-catenin regulation of the Sp1-related transcription factor sp5l promotes tail development in zebrafish. *Development*, *132*(8), 1763–1772. http://doi.org/10.1242/dev.01733.

Tsakiridis, A., Huang, Y., Blin, G., Skylaki, S., Wymeersch, F., Osorno, R., et al. (2014). Distinct Wnt-driven primitive streak-like populations reflect in vivo lineage precursors. *Development*, *141*(6), 1209–1221. http://doi.org/10.1242/dev.101014.

Tsankov, A. M., Gu, H., Akopian, V., Ziller, M. J., Donaghey, J., Amit, I., et al. (2015). Transcription factor binding dynamics during human ES cell differentiation. *Nature*, *518*(7539), 344–349. http://doi.org/10.1038/nature14233.

Turner, D. A., Hayward, P. C., Baillie-Johnson, P., Rué, P., Broome, R., Faunes, F., et al. (2014). Wnt/β-catenin and FGF signalling direct the specification and maintenance of a neuromesodermal axial progenitor in ensembles of mouse embryonic stem cells. *Development*, *141*(22), 4243–4253. http://doi.org/10.1242/dev.112979.

Tzouanacou, E., Wegener, A., Wymeersch, F. J., Wilson, V., & Nicolas, J.-F. (2009). Redefining the progression of lineage segregations during mammalian embryogenesis by clonal analysis. *Developmental Cell*, *17*(3), 365–376. http://doi.org/10.1016/j.devcel.2009.08.002.

Uchiyama, H., Kobayashi, T., Yamashita, A., Ohno, S., & Yabe, S. (2001). Cloning and characterization of the T-box gene Tbx6 in *Xenopus laevis*. *Development, Growth & Differentiation, 43*(6), 657–669.

Valenta, T., Gay, M., Steiner, S., Draganova, K., Zemke, M., Hoffmans, R., et al. (2011). Probing transcription-specific outputs of β-catenin in vivo. *Genes & Development, 25*(24), 2631–2643. http://doi.org/10.1101/gad.181289.111.

Vance, K. W., Shaw, H. M., Rodriguez, M., Ott, S., & Goding, C. R. (2010). The retinoblastoma protein modulates Tbx2 functional specificity. *Molecular Biology of the Cell, 21*(15), 2770–2779. http://doi.org/10.1091/mbc.E09-12-1029.

van Rooijen, C., Simmini, S., Bialecka, M., Neijts, R., van de Ven, C., Beck, F., et al. (2012). Evolutionarily conserved requirement of Cdx for post-occipital tissue emergence. *Development, 139*(14), 2576–2583. http://doi.org/10.1242/dev.079848.

Vogt, H. (1926). über Wachstum und Gestaltungsbewegungen am hinteren Körperende der Amphibien. *Anatomischer Anzeiger, 61*, 62–75.

Wardle, F. C., & Papaioannou, V. E. (2008). Teasing out T-box targets in early mesoderm. *Current Opinion in Genetics & Development, 18*(5), 418–425. http://doi.org/10.1016/j.gde.2008.07.017.

Watabe-Rudolph, M., Schlautmann, N., Papaioannou, V. E., & Gossler, A. (2002). The mouse rib-vertebrae mutation is a hypomorphic Tbx6 allele. *Mechanisms of Development, 119*(2), 251–256.

Wilkinson, D. G., Bhatt, S., & Herrmann, B. G. (1990). Expression pattern of the mouse T gene and its role in mesoderm formation. *Nature, 343*(6259), 657–659. http://doi.org/10.1038/343657a0.

Wilm, B., James, R. G., Schultheiss, T. M., & Hogan, B. L. M. (2004). The forkhead genes, Foxc1 and Foxc2, regulate paraxial versus intermediate mesoderm cell fate. *Developmental Biology, 271*(1), 176–189. http://doi.org/10.1016/j.ydbio.2004.03.034.

Wilson, V., & Beddington, R. (1997). Expression of T protein in the primitive streak is necessary and sufficient for posterior mesoderm movement and somite differentiation. *Developmental Biology, 192*(1), 45–58. http://doi.org/10.1006/dbio.1997.8701.

Wilson, V., Manson, L., Skarnes, W. C., & Beddington, R. S. (1995). The T gene is necessary for normal mesodermal morphogenetic cell movements during gastrulation. *Development, 121*(3), 877–886.

Windner, S. E., Bird, N. C., Patterson, S. E., Doris, R. A., & Devoto, S. H. (2012). Fss/Tbx6 is required for central dermomyotome cell fate in zebrafish. *Biology Open, 1*(8), 806–814. http://doi.org/10.1242/bio.20121958.

Windner, S. E., Doris, R. A., Ferguson, C. M., Nelson, A. C., Valentin, G., Tan, H., et al. (2015). Tbx6, Mesp-b and Ripply1 regulate the onset of skeletal myogenesis in zebrafish. *Development, 142*(6), 1159–1168. http://doi.org/10.1242/dev.113431.

Wittler, L., Shin, E.-H., Grote, P., Kispert, A., Beckers, A., Gossler, A., et al. (2007). Expression of Msgn1 in the presomitic mesoderm is controlled by synergism of WNT signalling and Tbx6. *EMBO Reports, 8*(8), 784–789. http://doi.org/10.1038/sj.embor.7401030.

Wylie, C. C., Snape, A., Heasman, J., & Smith, J. C. (1987). Vegetal pole cells and commitment to form endoderm in *Xenopus laevis*. *Developmental Biology, 119*(2), 496–502.

Wymeersch, F. J., Huang, Y., Blin, G., Cambray, N., Wilkie, R., Wong, F. C., et al. (2016). Position-dependent plasticity of distinct progenitor types in the primitive streak. *eLife, 5*, e10042. http://doi.org/10.7554/eLife.10042.

Yamada, A., Pang, K., Martindale, M. Q., & Tochinai, S. (2007). Surprisingly complex T-box gene complement in diploblastic metazoans. *Evolution & Development, 9*(3), 220–230. http://doi.org/10.1111/j.1525-142X.2007.00154.x.

Yamaguchi, T. P., Takada, S., Yoshikawa, Y., Wu, N., & McMahon, A. P. (1999). T (Brachyury) is a direct target of Wnt3a during paraxial mesoderm specification. *Genes & Development*, *13*(24), 3185–3190.

Yamamoto, A., Amacher, S. L., Kim, S. H., Geissert, D., Kimmel, C. B., & De Robertis, E. M. (1998). Zebrafish paraxial protocadherin is a downstream target of spadetail involved in morphogenesis of gastrula mesoderm. *Development*, *125*(17), 3389–3397.

Yanagisawa, K. O., Fujimoto, H., & Urushihara, H. (1981). Effects of the brachyury (T) mutation on morphogenetic movement in the mouse embryo. *Developmental Biology*, *87*(2), 242–248.

Yang, X. R., Ng, D., Alcorta, D. A., Liebsch, N. J., Sheridan, E., Li, S., et al. (2009). T (brachyury) gene duplication confers major susceptibility to familial chordoma. *Nature Genetics*, *41*(11), 1176–1178. http://doi.org/10.1038/ng.454.

Yasuhiko, Y., Haraguchi, S., Kitajima, S., Takahashi, Y., Kanno, J., & Saga, Y. (2006). Tbx6-mediated Notch signaling controls somite-specific Mesp2 expression. *Proceedings of the National Academy of Sciences of the United States of America*, *103*(10), 3651–3656. http://doi.org/10.1073/pnas.0508238103.

Yasuo, H., Kobayashi, M., Shimauchi, Y., & Satoh, N. (1996). The Ascidian genome contains another T-domain gene that is expressed in differentiating muscle and the tip of the tail of the embryo. *Developmental Biology*, *180*(2), 773–779.

Yasuo, H., & Lemaire, P. (2001). Role of Goosecoid, Xnot and Wnt antagonists in the maintenance of the notochord genetic programme in Xenopus gastrulae. *Development*, *128*(19), 3783–3793.

Yasuo, H., & Satoh, N. (1993). Function of vertebrate T gene. *Nature*, *364*(6438), 582–583. http://doi.org/10.1038/364582b0.

Yoon, S.-J., Wills, A. E., Chuong, E., Gupta, R., & Baker, J. C. (2011). HEB and E2A function as SMAD/FOXH1 cofactors. *Genes & Development*, *25*(15), 1654–1661. http://doi.org/10.1101/gad.16800511.

Yoon, J. K., & Wold, B. (2000). The bHLH regulator pMesogenin1 is required for maturation and segmentation of paraxial mesoderm. *Genes & Development*, *14*(24), 3204–3214.

Yoshida, M., Uchikawa, M., Rizzoti, K., Lovell-Badge, R., Takemoto, T., & Kondoh, H. (2014). Regulation of mesodermal precursor production by low-level expression of B1 Sox genes in the caudal lateral epiblast. *Mechanisms of Development*, *132*, 59–68. http://doi.org/10.1016/j.mod.2014.01.003.

Yoshikawa, Y., Fujimori, T., McMahon, A. P., & Takada, S. (1997). Evidence that absence of Wnt-3a signaling promotes neuralization instead of paraxial mesoderm development in the mouse. *Developmental Biology*, *183*(2), 234–242. http://doi.org/10.1006/dbio.1997.8502.

Zhang, J., & King, M. (1996). Xenopus VegT RNA is localized to the vegetal cortex during oogenesis and encodes a novel T-box transcription factor involved in mesodermal patterning. *Development*, *122*(12), 4119–4129.

Zhao, T., Gan, Q., Stokes, A., Lassiter, R. N. T., Wang, Y., Chan, J., et al. (2014). β-catenin regulates Pax3 and Cdx2 for caudal neural tube closure and elongation. *Development*, *141*(1), 148–157. http://doi.org/10.1242/dev.101550.

Zhu, J., Kwan, K. M., & Mackem, S. (2016). Putative oncogene Brachyury (T) is essential to specify cell fate but dispensable for notochord progenitor proliferation and EMT. *Proceedings of the National Academy of Sciences of the United States of America*, *113*(14), 3820–3825. http://doi.org/10.1073/pnas.1601252113.

CHAPTER SIX

T-Box Genes in *Drosophila* Mesoderm Development

I. Reim, M. Frasch[1], C. Schaub
Friedrich-Alexander University of Erlangen-Nürnberg, Erlangen, Germany
[1]Corresponding author: e-mail address: manfred.frasch@fau.de

Contents

1. Introduction 162
2. T-Box Genes in Visceral and Gonadal Mesoderm Development 163
   2.1 *Brachyenteron* and *Dorsocross* are Involved in Caudal Visceral Mesoderm Development 163
   2.2 Org-1 Functions in TVM Development 166
3. T-Box Genes in *Drosophila* Cardiogenesis 169
   3.1 Early Cardiac Specification Involves the *Dorsocross* T-Box Genes as Part of a Conserved Cardiogenic Network 172
   3.2 The *Dorsocross* and Tbx20-Related Genes Continue to Function in the Heart After Cardiac Specification 175
4. The Roles of the *Drosophila* Orthologs of *Tbx1* (*org-1*) and *Tbx20* (*mid* and *H15*) During Embryonic Somatic Muscle Assignments 177
   4.1 Org-1 is a Muscle Identity Gene in Specific Body Wall and Heart-Attached Muscles 178
   4.2 Midline Functions as a Muscle Identity Gene in Specific Lateral Body Wall Muscles 181
5. The Role of *Drosophila Tbx1* (*org-1*) in Lineage Plasticity, Reprogramming, and Commitment During Adult Myogenesis 182
6. Concluding Remarks and Outlook 185
Acknowledgments 187
References 187

Abstract

In *Drosophila* there are eight genes encoding transcription factors of the T-box family, which are known to exert a variety of crucial developmental functions during ectodermal patterning processes, neuronal cell specification, mesodermal tissue development, and the development of extraembryonic tissues. In this review, we focus on the prominent roles of *Drosophila* T-box genes in mesodermal tissues. First, we describe the contributions of *brachyenteron* (*byn*) and *optomotor-blind-related-gene-1* (*org-1*) to the development of the visceral mesoderm. Second, we provide an overview on the functions of the three *Dorsocross* paralogs (*Doc1–3*) and the two *Tbx20*-related paralogs

Current Topics in Developmental Biology, Volume 122
ISSN 0070-2153
http://dx.doi.org/10.1016/bs.ctdb.2016.06.003

(*midline* and *H15*) during *Drosophila* heart development. Third, we portray the roles of *org-1* and *midline/H15* in the specification of individual body wall and organ-attached muscles, including the function of *org-1* in the transdifferentiation of certain heart-attached muscles during metamorphosis. The functional analysis of these evolutionarily conserved T-box genes, along with their interactions with other types of transcription factors and various signaling pathways, has provided key insights into the regulation of *Drosophila* visceral mesoderm, muscle, and heart development.

1. INTRODUCTION

T-box gene-encoded transcription factors are among the chief regulators in the development of mesoderm-derived tissues in *Drosophila* as well as in vertebrates, and are of particular importance for the formation of the heart and other types of muscles. The *Drosophila* genome harbors a total of eight T-box genes (Table 1) that can be assigned to four classes, namely the Brachyury class, the Tbx1/15/20 class, the Tbx2/3 class, and the Tbx6 class (Reim, Lee, & Frasch, 2003; Sebé-Pedrós et al., 2013). These encompass the *T/Brachyury(Bra)*-ortholog *brachyenteron* (*byn*) (a.k.a. *Trg*; Kispert, Herrmann, Leptin, & Reuter, 1994); the *Tbx2* ortholog *optomotor-blind* (*omb*, a.k.a. *bifid*; Pflugfelder et al., 1990); the *Tbx1* ortholog *optomotor-blind-related-gene-1* (*org-1*) (Porsch et al., 1998); two paralogs related to *Tbx20*, *midline* (a.k.a. *H15-related* or *neuromancer2*) and *H15* (a.k.a. *neuromancer1*) (Buescher et al., 2004; Griffin et al., 2000); and the three paralogs of the *Dorsocross* locus, *Doc1*, *Doc2*, and *Doc3* (henceforth collectively called *Doc* due to their similar expression patterns and largely redundant functions). The *Doc* genes not only belong to the *Tbx6* class but also share some features with genes of the Tbx4/5 class (Hamaguchi, Yabe, Uchiyama, & Murakami, 2004; Reim & Frasch, 2010;

Table 1 *Drosophila* T-Box Genes

T-Box Genes *Drosophila melanogaster*	Full Name	Vertebrate Orthologs
byn	*brachyenteron*	*Brachyury/T*
Doc (*Doc1*, *Doc2*, *Doc3*)	*Dorsocross 1–3*	*Tbx6*
H15	– (*a.k.a., neuromancer1*)	*Tbx20*
mid	*midline* (*a.k.a. neuromancer2*)	*Tbx20*
omb	*optomotor-blind*	*Tbx2*
org-1	*optomotor-blind-related-gene-1*	*Tbx1*

Reim et al., 2003; Sebé-Pedrós & Ruiz-Trillo, 2017). With the exception of *omb*, all *Drosophila* T-box genes have been shown to be expressed in specific regions of the mesoderm or its derivatives. In this review, we summarize the contributions of T-box genes to various aspects of mesoderm and muscle development. In the first part, we will briefly describe the roles of T-box genes in visceral and gonadal mesoderm development. Thereafter, we will focus on the functions of the *Dorsocross*, *midline/H15*, and *org-1* genes during heart and skeletal muscle development.

2. T-BOX GENES IN VISCERAL AND GONADAL MESODERM DEVELOPMENT

2.1 *Brachyenteron* and *Dorsocross* are Involved in Caudal Visceral Mesoderm Development

In vertebrates, the orthologs of the most ancestral T-box gene, *T/Bra*, play an important role in the development of the posterior mesoderm and the notochord during early stages of embryogenesis (reviewed in Martin, 2016; Papaioannou, 2014). In the invertebrate *Drosophila* there is no equivalent of the notochord and the mesoderm is not generated successively; instead, it develops synchronously by invagination of mesodermal precursors along the anterior–posterior axis at the ventral side of the blastoderm embryo. Nevertheless, the *T/Bra* ortholog, *byn*, is expressed in posterior structures of the early embryo, which is reminiscent of the expression of its mammalian counterparts, although it plays only a limited role in mesodermal development. Analysis of *byn* mutants showed that the gene is required for the development of posterior tissues, namely the ectodermal hindgut and the caudal visceral mesoderm (CVM) (Kispert et al., 1994; Kusch & Reuter, 1999; Murakami et al., 1995; Singer, Harbecke, Kusch, Reuter, & Lengyel, 1996) (see Table 2 for list of abbreviations). The CVM represents the posterior-most portion of the mesoderm, and its precursors are located at the intersection of the striped *byn* expression domain with the ventral *twist* and *snail* expression domains in the posterior blastoderm (illustrated schematically in Fig. 1A; Kusch & Reuter, 1999; Ismat et al., 2010). The CVM gives rise to the founder cells (FCs) of the longitudinal visceral musculature (loVM), which migrate toward the anterior where they fuse with fusion-competent cells of the trunk visceral mesoderm (TVM) to form multinucleated longitudinal visceral muscle fibers (Fig. 1B–E). In addition to their contractile functions, these muscles also support midgut morphogenesis in conjunction with the TVM-derived circular gut musculature (Ismat et al., 2010; Lee, Zaffran, & Frasch, 2006).

Table 2 Abbreviations for Cell Types and Anatomical Terms

Cell Types and Anatomical Terms	
AM	Alary muscle
AMDC	Alary muscle-derived cell
AMP	Adult muscle precursor
CB	Cardioblast
ciVM	Circular visceral muscle/musculature
CM	Cardiogenic mesoderm
CVM	Caudal visceral mesoderm
FC	(Muscle) founder cell
FCM	Fusion-competent myoblast
gCB	Generic (Tin^{+}) cardioblasts
lAMP	Lateral adult muscle precursor
loVM	Longitudinal visceral muscle/musculature
oCB	Ostial (Tin^{-} Doc^{+}) cardioblasts
p/P	Progenitor
PC	Pericardial cell
pCMdSM	Presumptive cardiogenic/dorsal somatic mesoderm
SGP	Somatic gonadal precursor
SM	Somatic mesoderm/muscle
TARM	Thoracic alary-related muscle
TVM	Trunk visceral mesoderm
VLM	Ventral longitudinal muscles of the adult heart
VM	Visceral mesoderm/muscle/musculature

The combined activities of *byn* and other posterior/terminal genes together with ventral inputs from the mesoderm-intrinsic factor Snail result in activation of *HLH54F*, a bHLH transcription factor encoding gene that is crucial for the specification of CVM fate and subsequent migration of loVM progenitors (Ismat et al., 2010). Unlike *HLH54F*, which remains expressed throughout loVM development, *byn* expression ceases already during stage 11. *byn* and *HLH54F* mutants both display defects in the specification, migration, and survival of CVM cells, and consequently in the loVM-dependent morphogenesis

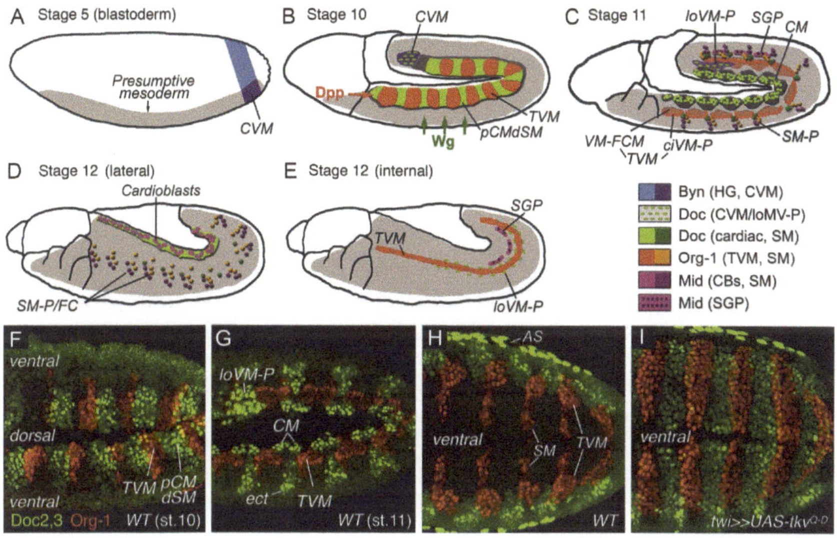

Fig. 1 Embryonic expression of *Drosophila* T-box genes in the mesoderm. (A–E) Schematic representation of the expression patterns of the T-box domain proteins Byn, Doc, Org-1, and Mid at indicated stages. The Mid paralog H15 is also expressed in cardioblasts, but only from stage 13 onward. With the exception of the early *byn* pattern, expression in nonmesodermal tissues as well as in early somatic mesoderm clusters at stage 10 has been omitted for clarity. (F–I) Immunostainings with antibodies against Doc and Org-1 (F, lateral view; H, ventral view) show the expression of these two T-box domain proteins in the wild type at stage 10. The alternating arrangements of the Doc^{+} cardiogenic/dorsal somatic mesoderm anlagen and the $Org\text{-}1^{+}$ trunk visceral mesoderm anlagen in the dorsal mesoderm can be discerned. (G) At stage 11, the different anlagen have been completely separated into different layers, with the TVM having moved inward. The expression of Org-1 is refined to a ventral row of ciVM founder cells and Doc to the cardiogenic mesoderm clusters near the dorsal margin, respectively. (H, I) Responsiveness of *org-1* and *Doc* expression to Dpp/BMP signaling is demonstrated by pan-mesodermal expression of constitutively active Thickveins (type I BMP receptor). Note the ventral expansion of the respective segmental expression domains in (I) as compared to (H). Abbreviations: *AS*, amnioserosa; *ciVM-P*, circular visceral muscle progenitors; *CM*, cardiogenic mesoderm; *CVM,* caudal visceral mesoderm; *ect*, ectoderm; *HG*, presumptive anlage of the hind gut; *loVM-P*, longitudinal visceral muscle progenitors; *pCMdSM*, presumptive cardiogenic/dorsal somatic mesoderm; *SGP*, somatic gonadal precursors; *SM*, somatic mesoderm; *SM-P/FC*, somatic muscle progenitors/founder cells; *TVM*, trunk visceral mesoderm; *VM-FCM*, visceral mesodermal fusion-competent myoblasts.

of the midgut at later stages (Ismat et al., 2010; Kusch & Reuter, 1999; Singer et al., 1996).

In addition to *byn*, the *Doc* genes are also expressed in portions of the hindgut and in migrating CVM/loVM progenitor cells, but *Doc* expression in the CVM initiates only after germ band extension when *byn* expression starts to fade (Fig. 1C–E) (Hamaguchi et al., 2012; Reim et al., 2003). Functional dissection of the *Doc* locus has led to the identification of an enhancer that drives *Doc* expression in migrating CVM/loVM progenitor cells, and its full activity was shown to require combined and direct inputs from HLH54F and the visceral mesoderm-specific FoxF transcription factor Biniou (Ismat et al., 2010; D. Schultheis, H. Jin, M. Frasch, & I. Reim, unpublished data; Zaffran, Küchler, Lee, & Frasch, 2001). While a role of the *Doc* genes in dorsal–ventral subdivision of the ectodermal hindgut has been demonstrated (Hamaguchi et al., 2012), their function in CVM development has not been studied in detail. Preliminary analysis of mutants lacking *Doc* expression driven by the earlier-described CVM/loVM progenitor enhancer suggests that these genes might be dispensable for CVM specification and loVM progenitor migration (M. Stenzel, D. Schultheis, & I. Reim, unpublished observation).

2.2 Org-1 Functions in TVM Development

Whereas the anlagen of the loVM are located in the CVM, the circular visceral musculature derives from populations of progenitor cells that are specified bilaterally in segmental blocks along the entire thoracic and abdominal mesoderm. These are the progenitors of the TVM. The TVM domain represents one of the domains delineated by intersecting segmental and dorsal–ventral patterning activities within each segment during subdivision of the trunk mesoderm at the extended germ band stage (stage 10) (illustrated in Fig. 1B; reviewed in Lee et al., 2006). At least two types of signals emanating from the overlaying ectoderm are critical for this subdivision. First, dorsally restricted Decapentaplegic (Dpp) activates BMP–Smad signaling and induces expression of the NK class homeobox gene *tinman* (*tin*, ortholog of *Nkx2-5*). Tin provides the cells of the dorsal mesoderm with the competence to acquire either TVM or dorsal somatic/cardiogenic mesoderm fates (Azpiazu & Frasch, 1993; Bodmer, 1993; Frasch, 1995; Xu, Yin, Hudson, Ferguson, & Frasch, 1998; Yin & Frasch, 1998). Second, segmentally repeated stripes of Wingless (Wg) in the ectoderm and canonical Wnt signaling in the underlying mesoderm block induction of visceral mesoderm fates while actively inducing somatic or cardiac mesoderm fates (Azpiazu, Lawrence, Vincent, & Frasch, 1996; Lee & Frasch, 2000). A critical target gene of Wnt signaling in this context is *sloppy paired* (*slp*), which encodes a forkhead domain transcriptional repressor. Slp directly represses expression of the TVM master regulator encoded by the NK class homeobox gene *bagpipe* (*bap*), which is activated by Tin and Smads (Azpiazu & Frasch, 1993; Azpiazu et al., 1996; Junion et al., 2012; Lee & Frasch, 2000, 2005; Riechmann, Irion, Wilson, Grosskortenhaus, & Leptin, 1997). Thus, Slp acts antagonistically toward TVM development, while in its absence, Dpp and Tin induce TVM.

Notably, the described patterning processes also lead to the appearance of alternating expression patterns of the *org-1* (*Drosophila Tbx1*) and *Dorsocross* T-box genes within the dorsal portion of the trunk mesoderm (Lee, Norris, Weiss, & Frasch, 2003; Reim & Frasch, 2005; Schaub & Frasch, 2013). The early expression domains of *org-1* coincide with *bap* and demark the presumptive TMV, whereas *Doc* expression within the dorsal mesoderm demarks the presumptive cardiogenic/dorsal somatic mesoderm (pCMdSM, discussed in more detail in Section 3.1) (Fig. 1B and F). In addition to *bap*, *org-1* is initially coexpressed with *tin*, the FoxF gene *biniou* (*bin*), and *Anaplastic lymphoma kinase* (*Alk*) in the segmented TVM (Azpiazu & Frasch, 1993; Englund et al., 2003; Lee et al., 2003, 2006; Schaub & Frasch, 2013; Zaffran et al., 2001). Expression of *org-1* is activated downstream of *tin* but independently of *bap* and *biniou*, and *dpp* provides the key signals from the dorsal ectoderm for its induction. This suggests a regulatory mechanism analogous to that of *bap*, such that the combined binding of Smads and Tin activates a Dpp-responsive *org-1* enhancer, whereas Wnt signal-activated Slp is required for its segmentally repeated repression (Schaub & Frasch, 2013; C. Schaub, H. Nagaso, & M. Frasch, unpublished data).

The binucleated circular visceral musculature arises upon fusion of circular visceral muscle founder cells (ciVM-FC), which are specified along the ventral margin of the TVM by secreted Jelly belly (Jeb) signals acting via the Alk receptor tyrosine kinase, and visceral fusion-competent myoblasts (VM-FCMs), which occupy dorsally adjacent areas of the TVM (Englund et al., 2003; Klapper et al., 2002; Lee et al., 2003, 2006; San-Martin, Ruiz-Gomez, Landgraf, & Bate, 2001). The circular visceral musculature together with the loVM forms the outer layer and the endodermal midgut epithelium forms the inner layer of the midgut tube. During later embryonic development, the midgut undergoes morphological diversifications along the anterior–posterior axis. From the anterior of the midgut, four blind protrusions (called gastric caeca) evaginate and upon the formation of three constrictions the midgut tube is subdivided along its length into four different compartments that ultimately morph into the gut loops. These morphological events are the readout of a regulatory patterning network that is first established in the FCs of the circular visceral musculature. Acting along the anterior–

posterior axis, it integrates the spatial activity of the homeotic selector genes and the secreted factors Dpp and Wg, and presumably leads to the activation of particular morphogenetic effectors at defined positions (Campos-Ortega & Hartenstein, 1997; Capovilla, Brandt, & Botas, 1994; Hursh, Padgett, & Gelbart, 1993; Lee et al., 2006; Mathies, Kerridge, & Scott, 1994; Müller, Thuringer, Biggin, Zust, & Bienz, 1989; Panganiban, Reuter, Scott, & Hoffmann, 1990; Reuter, Panganiban, Hoffmann, & Scott, 1990; Reuter & Scott, 1990; Sun, Hursh, Jackson, & Beachy, 1995; Tepass & Hartenstein, 1994).

During visceral mesoderm development the expression of *org-1* in all progenitors of the TVM is first narrowed down to the ciVM-FCs, and only after fusion it is reactivated in the nuclei from the fused VM-FCMs (Schaub & Frasch, 2013). Jeb-activated Alk signaling is required for *org-1* induction in the ciVM-FCs (Lee et al., 2003), whereas downregulation of *org-1* in the VM-FCMs depends on Notch signaling activity and the Gli zinc finger transcription factor *lameduck* (*lmd*) (Duan, Skeath, & Nguyen, 2001; Popichenko et al., 2013; Stute, Schimmelpfeng, Renkawitz-Pohl, Palmer, & Holz, 2004). Although *lmd* mRNA is expressed in both FCs and fusion-competent cells of the TVM, Lmd protein does not accumulate in the FCs. This is apparently due to its phosphorylation by activated Alk signaling components in ciVM-FCs, which triggers its nuclear export and subsequent degradation involving the Mind bomb 2 (Mib2) ubiquitin ligase (Carrasco-Rando & Ruiz-Gomez, 2008; Nguyen, Voza, Ezzeddine, & Frasch, 2007; Popichenko et al., 2013). Together with the localized expression of the Jeb ligand and the TVM-specific expression of the Alk receptor, this pathway makes a crucial contribution to the FC-specific expression of *org-1*. It can also explain the expansion of *org-1* expression into the nuclei of the former fusion-competent cells after myoblast fusion, when activated Alk signaling components spread into the binucleate syncytia and analogously promote the elimination of Lmd from the VM-FCM-derived nuclei and cytoplasm of these syncytia (Popichenko et al., 2013).

Whereas *org-1* is not required for the specification of the ciVM-FCs, it encodes a crucial factor required for proper patterning and morphogenesis of the midgut. The full activation and maintenance of ciVM-FC-specific *odd paired* (*opa*), *teashirt* (*tsh*), *Ultrabithorax* (*Ubx*), *dpp,* and *wg* expression in defined domains along the anterior–posterior axis strictly requires *org-1*. Consequently, the formation of midgut structures that depend on these gene activities, namely the gastric caeca and the anterior as well as the central midgut constrictions, are disrupted in *org-1* mutant embryos (Schaub & Frasch, 2013). Genetic analysis of known ciVM-FC-specific *cis*-regulatory modules (CRMs) of *wg* and *dpp* showed that their enhancer activity in localized domains of the TVM depends on Org-1 protein and on the presence of T-box domain-binding motifs in these CRMs. Together with demonstrated in vivo binding of Org-1 to these CRMs, these observations strongly indicate that Org-1 is a direct regulator of *dpp* and *wg* in the ciVM-FCs of the TVM and, in combination with the FoxF transcription factor Biniou (Bin), serves as an essential TVM-specific competence factor in the regional induction of these signaling genes. Org-1 and Bin act in concert with spatially restricted regulatory inputs such as Hox factors and signaling effectors during this process to achieve region-specific outputs in the TVM, which, in turn, differentially shape the midgut along its anterior–posterior axis (Schaub & Frasch, 2013; Zaffran et al., 2001).

The areas located immediately ventrally to the TVM give rise to other types of mesodermal derivatives and, depending on their location along the anterior–posterior axis and corresponding Hox gene expression, form either fat body or somatic gonadal precursors (SGPs) (van Doren, 2006). While there is no indication of any direct function of T-box genes during fat body development, the *Tbx20* ortholog *midline* (*mid*) is expressed in the clusters of the SGP-generating gonadal mesoderm (Reim, Mohler, & Frasch, 2005; Tripathy, Kunwar, Sano, & Renault, 2014) (Fig. 1C). Recently, a *mid* mutant was identified in a screen for genes that regulate early gonad development in the embryo (Tripathy et al., 2014). Mutants that lack functional Mid fail to form gonads and the germ cells are found scattered within the posterior embryo instead of being ensheathed by the SGPs. This study also showed that *mid* expression in SGPs requires the Nkx2-5 ortholog Tinman as transcriptional regulator, an intriguing parallel to cardiac development (see later). Notably, *mid* and its paralog *H15* become again important for germ cell-associated cells during later stages, when they contribute to the epidermal growth factor receptor mediated patterning of the ovarian follicle cell epithelium through a proposed negative feedback mechanism (Fregoso Lomas, Hails, Lachance, & Nilson, 2013).

3. T-BOX GENES IN *DROSOPHILA* CARDIOGENESIS

The study of cardiogenesis provides a good example of how the combined knowledge from different model systems can lead to major advances in the understanding of complex developmental processes. *Drosophila* heart research has contributed significantly to this understanding, mainly by identifying key regulatory factors of the basic cardiogenic network. Many aspects of cardiogenesis, such as the induction of bilateral cardiogenic fields and a core set of tissue-specific transcription factors, are conserved between *Drosophila* and vertebrates (Reim & Frasch, 2010; Zaffran & Frasch, 2002). *Drosophila* cardiogenesis proceeds through a series of gene regulatory and morphogenetic events that can be summarized as: (1) induction of the presumptive cardiogenic mesoderm (CM) as one of the regions emerging from mesoderm subdivision at about stage 10, (2) specification of the definitive CM involving a core set of conserved cardiogenic transcription factors at early stage 11, (3) specification and determination of progenitor cells including their diversification into different types of cardioblasts (CBs, precursors of the cardiomyocytes) or other heart-associated cell types (precursors of noncontractile pericardial cells/PCs and of the hematopoietic lymph gland) during stages 11–12, (4) cell movements including dorsal migration of all progenitor cells toward the dorsal midline and alignment of CBs into bilateral rows during stages 12–16, (5) terminal differentiation of cardiac cells, including heart tube morphogenesis, acquisition of contractility in cardiomyocytes, and formation of cardiac inflow tracts for the hemolymph (so-called ostia) at the final stage of embryogenesis (for overview, see Bodmer & Frasch, 2010; Reim & Frasch, 2010; Vogler & Bodmer, 2015). In addition, remodeling and partial histolysis of preexisting structures of the larval heart take place during metamorphosis (Molina & Cripps, 2001; Sellin, Albrecht, Kölsch, & Paululat, 2006). At the end of cardiogenesis, both the larval and the adult heart appear as simple linear tubes (in insects usually called "dorsal vessel") consisting of mononucleated contractile cardiomyocytes (with a subset of them forming the ostia) that are flanked by noncontractile nephrocyte-like PCs (Fig. 2A and B). In addition, specialized syncytial muscle fibers, the alary muscles (AMs), and—limited to the adult heart—the ventral longitudinal muscles (VLMs) are associated with the heart (for morphological details of the *Drosophila* heart and associated structures, see Lehmacher, Abeln, & Paululat, 2012; Rotstein & Paululat, 2016).

Here, we focus on the regulatory aspects of embryonic cardiogenesis as T-box genes, specifically the *Tbx6*-related *Dorsocross* genes (*Doc1-3*) and the *Tbx20* orthologs *mid* and *H15*, are key players in this process. The formation of heart-associated VLMs, which is regulated by the *Tbx1* ortholog *org-1*, will be discussed in the following section because the progenitors of these muscles originally arise from the somatic mesoderm, as do skeletal

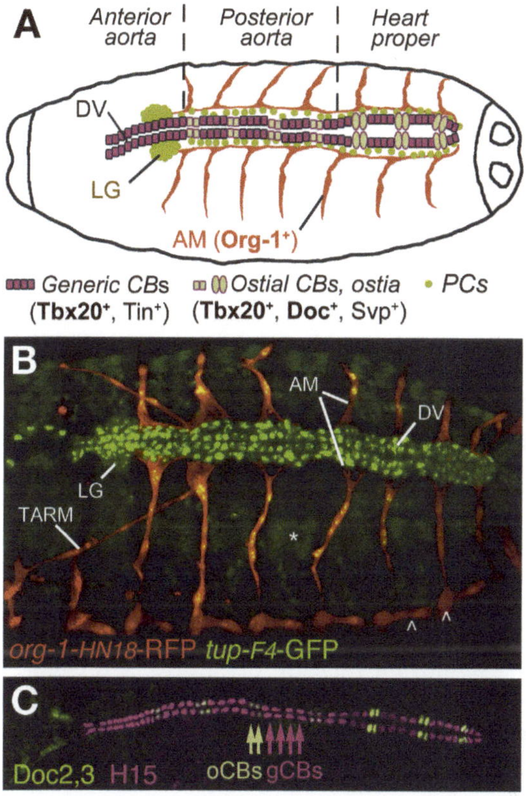

Fig. 2 Expression of T-box genes in the *Drosophila* heart and its associated tissues at the end of embryogenesis. (A) Schematic of the dorsal vessel (DV) with its cardioblast (CB) subtypes, and pericardial cells (PCs) flanked by seven pairs of associated Org-1$^+$ alary muscles (AMs). All CBs express both of the Tbx20 factors, Mid, and H15, while Doc expression is maintained only the ostial CBs, which form inflow openings (ostia) in the heart proper. LG, lymph gland (hematopoietic organ). (B) Live fluorescence of GFP and RFP reporter genes to label the DV and AMs. RFP is driven by a somatic muscle enhancer (Hollfelder, Frasch, & Reim, 2014; Schaub, Nagaso, Jin, & Frasch, 2012) of the *Tbx1* ortholog *org-1*, active in AMs, the midgut-connecting thoracic alary-related muscles (TARMs) and additional somatic muscles (^), and GFP by an enhancer from the *tup/Islet-1* locus (Tao, Wang, Tokusumi, Gajewski, & Schulz, 2007), which is active in all heart cells, the lymph gland, AMs, and some dorsal somatic muscles (*). (C) Antibody staining showing expression of H15 proteins in all CBs and Doc in the ostial (oCB) subset of CBs. Antibodies from Reim et al. (2003) and Leal, Qian, Lacin, Bodmer, and Skeath (2009).

muscles. Strikingly, relatives of all of these genes are involved in mammalian cardiogenesis and have been linked to congenital heart defects in humans (Ghosh, Brook, & Wilsdon, 2017; Greulich, Rudat, & Kispert, 2011).

3.1 Early Cardiac Specification Involves the *Dorsocross* T-Box Genes as Part of a Conserved Cardiogenic Network

The first step toward cardiac specification is the subdivision of the mesoderm and the emergence of the pCMdSM domain in a dorsal/posterior quadrant of each parasegment. Cells from this domain will give rise to all cardial, pericardial, lymph gland, as well as dorsal somatic muscle progenitors. As described in Section 2.2, this allocation requires Dpp/BMP and canonical Wg/Wnt signaling; thus, both pathways are also essential for *Drosophila* cardiogenesis and the decisions of cardiac vs visceral mesoderm development are intimately connected (Frasch, 1995; Lockwood & Bodmer, 2002; Park, Wu, Golden, Axelrod, & Bodmer, 1996; Wu, Golden, & Bodmer, 1995; Yin & Frasch, 1998). Like the neighboring presumptive TVM, the pCMdSM expresses the *Nkx2-5* ortholog *tin*. However, due to the reception of Wnt signals, the *Fox*G gene *slp* becomes specifically activated in the posterior of each parasegment where it inhibits TVM development, thereby enabling the acquisition of cardiac and somatic fates. The procardiogenic function of *slp* is permissive rather than instructive, as embryos that ectopically express *slp* in the whole mesoderm do not display a significant increase of cardiac cells even though visceral-specific gene expression and TVM fates are abolished in this situation (Lee & Frasch, 2000). This is because Wg/Wnt signals are additionally required, in combination with Dpp/BMP signals, as direct inputs for the induction of cardiogenic genes within these domains. Indeed, the *Doc* T-box genes are among the key examples of cardiogenic genes that respond to combined Dpp and Wg signaling in this manner and actively promote cardiac fates. *Doc* expression coincides with the regions where dorsal Dpp and segmental Wg inputs intersect in both dorsal mesoderm and dorsal ectoderm (Reim & Frasch, 2005; Reim et al., 2003) (Fig. 1B and F). Since *Doc* expression initiates independently of *tin* in the dorsal mesoderm (Reim & Frasch, 2005), and the *Doc* locus contains regions of predicted and ChIP-verified binding clusters for pMad (a *Drosophila* R-Smad) and dTCF (Junion et al., 2012), this early mesodermal *Doc* expression is likely to be directly activated by the synergistic inputs from these pathways (eg, Fig. 1H and I). Genetic removal of all three *Doc* genes leads to a loss of most cardiac cells, including nearly all CBs, PCs expressing the *odd-skipped* (*odd*) marker, and the lymph gland (Reim & Frasch, 2005). Similar phenotypes are also found in mutants of two other major cardiogenic genes, the prototypic heart gene *tin* and the *GATA4/5/6*-related gene *pannier* (*pnr*) (Alvarez, Shi, Wilson, & Skeath, 2003; Azpiazu &

Frasch, 1993; Bodmer, 1993; Gajewski, Fossett, Molkentin, & Schulz, 1999; Klinedinst & Bodmer, 2003; Mandal, Banerjee, & Hartenstein, 2004). After *Doc*- and *tin*-dependent activation of *pnr* and parallel refinement of the *tin* and *Doc* expression patterns during stage 11, all three cardiogenic genes are coexpressed in dorsal-most areas of the mesoderm. This sustained expression of all three genes, which now become interdependent on one another, specifies the definitive CM and subsequently promotes CB fates (Gajewski et al., 2001; Reim & Frasch, 2005) (Fig. 1C and D). Accordingly, ectopic expression of Doc together with Tin and Pnr leads to excessive CB formation in the dorsal mesoderm. Taken together, coexpression and cooperative action of the three factors Doc, Pnr, and Tin are a key element in early cardiogenesis that finds its vertebrate counterpart in the combination of Tbx5, Gata4, and Nkx2-5 first seen in the first heart field. Notably, factor combinations involving relatives of Pnr and Doc are also capable of promoting cardiac fates in vertebrates. For example, the combination of Gata4 and Tbx5 together with a cardiac-specific subunit of the BAF chromatin remodeling complex (Baf60c/Smarcd3) directs ectopic differentiation of mouse mesoderm into cardiomyocytes (Takeuchi & Bruneau, 2009). Likewise, the combination of Gata4, Tbx5, and myocyte-specific enhancer factor 2c (Mef2c) activity can reprogram mouse fibroblasts into cardiomyocytes (Ieda et al., 2010). The synergism of T-box, GATA, and Nkx factors is achieved at two levels, protein–protein interactions and clustering of the respective binding sites in common CRMs. Direct protein–protein interactions between the different factors were demonstrated for various combinations of the vertebrate homologs and for *Drosophila* Pnr and Tin, although direct interactions of either of these two *Drosophila* factors with Doc or the later expressed *Drosophila* Tbx20 proteins have not been reported so far (Durocher, Charron, Warren, Schwartz, & Nemer, 1997; Gajewski et al., 2001; Garg et al., 2003; Hiroi et al., 2001; Luna-Zurita et al., 2016). The expected cooperative binding to shared enhancers is evident from genome-wide chromatin immunoprecipitation (ChIP) data, obtained, eg, from in vitro-differentiated murine cell lines (Luna-Zurita et al., 2016) as well as from whole *Drosophila* embryos (Junion et al., 2012; Zinzen, Girardot, Gagneur, Braun, & Furlong, 2009). The *Drosophila* studies found that Tin, Pnr, and Doc preferentially bind as a collective, often together with the nuclear Dpp effector pMad and the Wg effector dTCF, to numerous regulatory regions, although only a fraction of those tested were associated with cardiac transcription (Junion et al., 2012). Another study focusing on in vivo Tin targets also found frequent cooccurrence of Tin, GATA, and T-box-binding motifs in Tin-bound CRMs active in cardiac tissues.[a] Mutational analysis of the Tin, GATA, and T-box domain-binding motifs, respectively, performed by Jin et al. (2013) implied that different cardiac enhancers vary in their functional requirement for binding of the different classes of cardiogenic factors and may use them in different combinations.

What are the functionally important target genes of the cardiogenic factors, in particular of the Doc T-box factors? Unfortunately, there are currently very few functionally verified targets of Doc. It was shown by overexpression experiments that Doc regulates the CB-specific enhancer of the *Toll* gene in conjunction with Tin, and this target was further validated by elimination of reporter expression in the ostial subset of CBs upon T-box-binding motif removal as well as by in vitro DNase I footprinting studies (Wang et al., 2005). Another example is the gene encoding the small G-protein RhoL, which is expressed in the CM and is regulated by an enhancer that depends on its T-box domain motifs (Jin et al., 2013). The *mid* gene has been a prime candidate for combinatorial activation by Doc, Pnr, and Tin since the onset of its expression at late stage 11 coincides with the time window at which the cardiac progenitors are committed to the CB fate (Fig. 1D). Furthermore, *mid* expression was shown to depend on mesodermal expression of Doc, Pnr, and Tin (Reim & Frasch, 2005; Reim et al., 2005). However, the mutual dependence of the cardiogenic factors at late stage 11 precludes a designation of its primary regulator by analyzing the respective mutants. Since a cardiac enhancer of *mid* was isolated independently by in silico searches for clusters of Tin-binding motifs and via Tin–ChIP (Jin et al., 2013; Ryu, Najand, & Brook, 2011), the question of which factor is essential for cardiac *mid* expression could be addressed at the molecular level. Analysis of respective reporter genes with mutations in the detected Tin, GATA, or T-box domain-binding motifs demonstrated that the activity of this enhancer is dependent on Tin but does not require Pnr (Jin et al., 2013; Ryu et al., 2011). The removal of T-box sites, which may be bound by Doc and/or by autoregulatory Mid, decreased *mid* enhancer activity. This effect points to a direct influence of either of these T-box domain proteins toward the expression of *mid* in CBs (Jin et al., 2013).

[a] It should be noted that the Doc and Mid/Tbx20 motifs obtained by SELEX are very similar to published T-box core motifs (Sebé-Pedrós & Ruiz-Trillo, 2017), but deviate from the de novo motif calculated from the Doc ChIP data set from Junion et al. (2012).

3.2 The *Dorsocross* and Tbx20-Related Genes Continue to Function in the Heart After Cardiac Specification

Once the cardiac progenitors have been committed to either CB or PC fate, the formerly coexpressed transcription factors Doc, Tin, and Pnr drastically change their expression patterns. During stages 12–13, Pnr expression fades entirely and the expression of Tin and Doc becomes mutually exclusive. Activity of the *tin* gene is maintained in the majority of CBs (herein referred to as generic CBs, gCBs) and a subset of PCs, whereas *Doc* activity is restricted to the CB subpopulation that expresses the COUP-TFII-related orphan nuclear receptor transcription factor Seven-up (Svp) and lack Tin (Lo & Frasch, 2001; Zaffran et al., 2006; see also Fig. 2A and C). Here, we refer to these $Svp^{+}Doc^{+}Tin^{-}$ CBs simply as ostial cardioblasts (oCBs) since they form the ostia in the posterior chamber-like portion of the embryonic/larval dorsal vessel and in the adult heart chambers that arise via remodeling during metamorphosis (Molina & Cripps, 2001). This distinction is crucial for establishing the patterned expression of certain differentiation genes, such as ion channels, enzymes, and structural components as well as for the acquisition of distinct cell shapes (see, eg, Iklé, Elwell, Bryantsev, & Cripps, 2008; Kremser, Gajewski, Schulz, & Renkawitz-Pohl, 1999; Monier, Tevy, Perrin, Capovilla, & Semeriva, 2007; Nasonkin et al., 1999).

Unlike *Doc* and *tin*, *mid* and *H15* are expressed in all CBs and mature cardiomyocytes (Fig. 2C) but are absent from the CM at least until mid-stage 11 (Miskolczi-McCallum, Scavetta, Svendsen, Soanes, & Brook, 2005; Qian, Liu, & Bodmer, 2005; Qian et al., 2008; Reim et al., 2005). There is a notable difference in the timing at which expression of these two *Tbx20* genes starts. The *mid* gene is activated during stage 11–12, starting in CB progenitors that are about to divide symmetrically into two gCBs, and subsequently in CBs derived from asymmetric divisions (gCBs of the anterior aorta and oCBs). Thus *mid* expression may be the first indicator of full CB commitment. *H15* expression starts after germ band retraction at the transition to stage 13. Accordingly, *mid* is expected to play a more important role during heart development than *H15*, which is confirmed by the observation of strong cardiac patterning defects in *mid* mutants as compared to only minor defects found in the viable *H15* single mutants (Miskolczi-McCallum et al., 2005; Qian et al., 2005; Reim et al., 2005). The three reports describing the embryonic heart defects in *mid* loss-of-function embryos address similar questions but disagree in some of their results, which could either be due to different interpretations or to the usage of different types of loss-of-function situations (eg, usage of RNAi may have included possible off-target effects; Qian et al., 2005). The data from the two studies that used genomic *mid* and *H15* mutations, both singly and doubly, and more recent analysis of additional *mid* mutants obtained from an EMS screen support the notion that the *Drosophila Tbx20* orthologs are largely dispensable for PC and CB specification (Miskolczi-McCallum et al., 2005; Reim et al., 2005; B. Schwartz & I. Reim, unpublished data). Even though overall cardiac cell numbers are essentially normal in *mid* and *mid+ H15* mutants, ectopic *mid* expression leads to an excess of Tin^+ cells and a subset of those extra Tin^+ cells appear to be committed to the CB fate, as based on CB-specific H15 and Mef2 coexpression (Miskolczi-McCallum et al., 2005; Qian et al., 2005; Reim et al., 2005). The expansion of *tin* expression may at least in part result from stimulated progenitor proliferation as suggested by some ectopic phospho-histone staining (Qian et al., 2005), even though in the normal situation *mid* is not required for the division of the CB progenitors. In addition or alternatively, ectopic *mid* may induce ectopic *tin* activation, which in turn may cause complete and incomplete fate switches. This possibility is supported by the strongly reduced *tin* expression in CBs of *mid* mutants.

Importantly, *mid* mutants feature dramatic patterning defects in the dorsal vessel, namely a decrease in the number of Tin-expressing gCBs and a concomitant increase of Doc-expressing oCBs. This effect ultimately leads to an increase in cardiomyocytes with ostia-like features in the heart proper as well as changes in the expression patterns of gCB- or oCB-specific differentiation genes (Reim et al., 2005). Thus, a major cardiac function of *mid* is to maintain the regular Tin pattern within four CBs per segment on either side of the myocardial tube. The consequences of the loss of Tin in CBs can be seen even more clearly in CB-specific *tin* mutants, in which the function of the endogenous *tin* gene was replaced by a *tin* transgene that fulfills the early cardiogenic function of *tin* but lacks the enhancer required for *tin* expression in CBs (Zaffran et al., 2006). The myocardial tube that lacks Tin in this situation consists entirely of Doc^+ cardiomyocytes and features similar patterning defects for certain differentiation markers as *mid* mutants, in which *tin* expression is not properly activated. The larval hearts of these mutants display an abnormal myofibril arrangement that apparently persists through metamorphosis. As a further consequence, the hearts of adult survivors are severely hypotrophic and feature abnormal functionality (Zaffran et al., 2006). These defects underscore the importance of specifying the correct subtype fates of cardiomyocytes.

How the distinct CB subtypes are established initially is still under investigation. Models that can be drawn from currently published data involve Svp as a repressor of *tin* in oCBs, Tin as a repressor of *Doc* in gCBs, and Doc as a (somewhat less potent) repressor of *tin*. Hence, Svp indirectly enables *Doc* expression within the two presumptive ostial CBs in each hemisegment (Lo & Frasch, 2001; Zaffran et al., 2006). Mid is then thought to contribute to *Doc* repression by stabilizing *tin* expression in the gCBs. Interestingly, mouse *Tbx20* mutants likewise feature massive expansion of another region-specific T-box gene, *Tbx2*, which is normally restricted to the atrioventricular canal and outflow tract (Cai et al., 2005; Christoffels et al., 2004; Singh et al., 2005; Stennard et al., 2005). The relationship between *Tbx20* and the nonchamber myocard-restricted expression of *Tbx2* is reminiscent of that of *mid* and ostia-specific *Doc* expression in *Drosophila*.

Notable parallels between *Drosophila* and mammalian *Tbx20* genes have also been reported for the later functions of these genes in maintaining cardiac functionality (Qian et al., 2008). Qian et al. detected an increase in the rate of pacing-induced heart arrest/fibrillation and other functional abnormalities in adult hearts of flies after CB-specific *mid* (*nmr2*) RNAi-mediated knockdown and, to a lesser extent, in viable *H15* (*nmr1*) null mutant flies (Qian et al., 2008). Impaired heart performance and an abnormal arrangement of the spiral myofibers were also detected when *mid* RNAi was used in combination with the TARGET system that limited the *mid* knockdown to the adult stage, and thus excluded defects resulting from abnormal cardiogenesis. Therefore, cardiac *mid* function continues to be required during the adult stage. It was not stated whether cardiac *tin* or *Doc* expression was affected in this situation. However, synergistic genetic interactions of the *Tbx20*-related genes with *tin* as well as with *Doc* were detected by scoring pacing-induced heart arrest/fibrillation rates in adult flies. Transheterozygous $Tbx20^{+/-}$ $tin^{+/-}$ combinations were also abnormal in their cardiomyocyte structure (Qian et al., 2008). Of note, synergistic interactions were also observed in compound heterozygous $Tbx20^{+/-}$ $Nkx2\text{-}5^{+/-}$ mice (Stennard et al., 2005).

4. THE ROLES OF THE *DROSOPHILA* ORTHOLOGS OF *Tbx1* (*ORG-1*) AND *Tbx20* (*MID* AND *H15*) DURING EMBRYONIC SOMATIC MUSCLE ASSIGNMENTS

Whereas the functions of T-box genes in early mesoderm development and cardiogenesis have long been known, more recently members of the T-box gene family have also been implicated in the regulation of skeletal (somatic) myogenesis (Kelly, Jerome-Majewska, & Papaioannou, 2004; Kong et al., 2014; Lee et al., 2015; Sambasivan et al., 2009; Windner et al., 2015). In *Drosophila*, two out of the eight *Drosophila* T-box genes are known to play important roles during the specification of myogenic lineages during larval somatic myogenesis. As described later, *Drosophila Tbx1* (*org-1*) and *Tbx20* (*mid*, with minor contributions of *H15*) mediate the commitment of muscle progenitor cells and FCs to distinct identities that are required for the myoblast-specific transcriptional activation of additional muscle identity factors and govern the formation of distinct muscle fibers.

In *Drosophila*, the larval somatic musculature is arranged in a highly stereotyped pattern of about 30 muscle fibers per abdominal hemisegment (Fig. 3A). The formation of each of these syncytial myofibers is dictated by an individual cell, termed muscle founder cell (Bate, 1990; Dohrmann, Azpiazu, & Frasch, 1990). Two different FCs (or, in some cases, one FC and one adult muscle progenitor, AMP) originate via asymmetric division from a muscle progenitor cell. These progenitor cells, in turn, segregate from competence clusters as a result of the antagonistic actions of receptor tyrosine kinase (RTK) and Delta/Notch signals. The remaining cells of the competence cluster become fusion-competent myoblasts (FCMs). During later stages of myogenesis, these will fuse with the FCs to built syncytial muscle fibers (Carmena & Baylies, 2006). Each FC is primed to an individual fate by the expression of a characteristic combination of transcription factors, the muscle identity factors, and thereby retains all the information required for the building of a particular muscle with its characteristic size, shape, pattern of innervation, and epidermal attachment. Upon myoblast fusion with a defined number of surrounding FCMs, the newly added nuclei from the FCMs are entrained into the genetic program of the FC, which results in the differentiation of a multinucleated, morphologically distinct muscle fiber. In some specific lineages, the sibling cell of a muscle founder becomes an AMP, ie, an imaginal cell that remains undifferentiated and is set aside for adult myogenesis (reviewed in Carmena & Baylies, 2006; de Joussineau, Bataille, Jagla, & Jagla, 2012; Frasch, 1999).

4.1 Org-1 is a Muscle Identity Gene in Specific Body Wall and Heart-Attached Muscles

The single *Drosophila* ortholog of vertebrate *Tbx1*, *org-1*, is expressed in a small subset of muscle progenitors, muscle FCs, muscle fibers, and adult

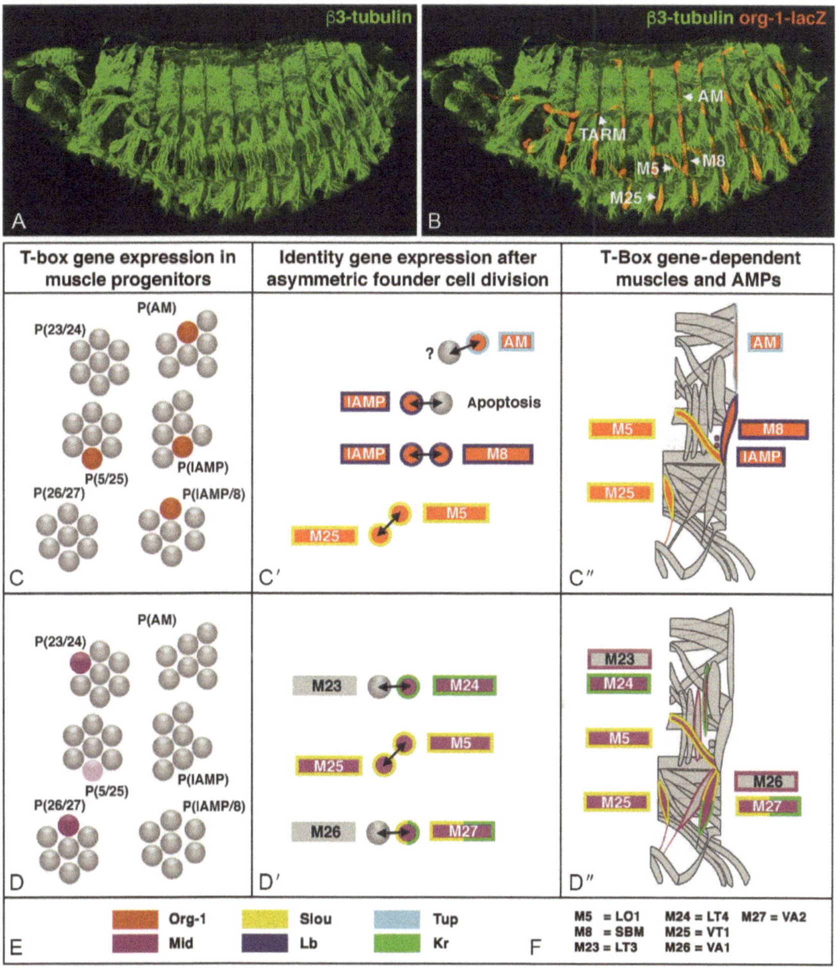

Fig. 3 Expression and functions of T-box domain proteins during larval somatic myogenesis. (A) The *Drosophila* embryonic somatic muscle pattern visualized by staining against the β3-Tubulin protein. (B) *org-1* (visualized through a specific *org-1-lacZ* reporter) is expressed in a subset of abdominal segmental muscles, namely in the muscles 5 (M5), 25 (M25), 8 (M8), and the alary muscles (AMs). (C, D) During muscle progenitor specification *org-1* (C) as well as *midline* (*mid*) (D) are expressed in a specific subset of muscle progenitors. Mid protein in P5/25 is not yet detectable. (C′, D′) Asymmetric founder cell division results in the expression of Org-1 in the founder cells of M5, M25, M8, and the AMs as well as in the lateral adult muscle precursors (lAMPs) (C′) and of Mid in the founder cells of M5, M25, M24, and M28. (D′) Differential coexpression with other muscle identity factors (Ladybird, *Lb*; Slouch, *Slou*; Tailup, *Tup*; and Krüppel, *Kr*) in the respective cells is indicated by *color coded* frames. (C″, D″) Schematic illustration of the muscles and AMPs that are dependent on T-box gene function for their formation or proper morphogenesis. Due to the functions of

(Continued)

muscle precursors, in addition to its expression in the visceral musculature that was discussed earlier (Schaub & Frasch, 2013). In specific sets of these myoblasts and muscles, *org-1* expression overlaps with the expression of the homeodomain factor-encoding muscle identity genes *ladybird* (ie, the paralogs *lbe* and *lbl*) (Jagla et al., 1998), *slouch* (*slou*) (Knirr, Azpiazu, & Frasch, 1999), *Ultrabithorax* (*Ubx*, a Hox gene) (Michelson, 1994), and the LIM homeodomain transcription factor-encoding *tailup* (*tup*, ortholog of *Islet1/Isl1*) (Boukhatmi et al., 2012) (Fig. 1C and D, Fig.2A and B, Fig. 3B and C–C″, D–D″). Unlike the other Org-1^+ FCs, the Org-1^+Tup$^+$ FCs form specialized muscle fibers that are not responsible for locomotion of the animal but connect the heart (dorsal vessel) and visceral organs to the larval exoskeleton. These include the so-called alary muscles (AMs) that affix the heart within the larval abdominal body cavity, clamp the main tracheal branch to the body wall, and provide guidance during morphogenesis of the Malpighian tubules (Boukhatmi et al., 2014; Weavers & Skaer, 2013). In addition, the Org-1^+ thoracic alary-related muscles (TARMs) connect the epidermis of thoracic segments to precise positions of the midgut (Fig. 2A and B).

As evident from the muscle phenotype of *org-1* mutant embryos, *org-1* is required for normal development of all muscles that derive from *org-1*-expressing muscle progenitors and FCs. In these cells, as well as in the lateral adult muscle precursors (lAMPs) that normally also express Org-1, loss of *org-1* function causes a failure to activate Lb, Slou, Ubx, and Tup, respectively (Boukhatmi et al., 2014; Busser et al., 2012; Schaub et al., 2012). As a consequence, the respective muscle fibers that normally arise from these FCs are either missing or malformed in *org-1*-deficient embryos, and the lAMPs are strongly reduced in number. In particular, this concerns the proper formation of muscle 5 (LO1), muscle 25 (VT1) (normally expressing Slou and Ubx), and muscle 8 (segment boundary muscle) (normally

Fig. 3—Cont'd *mid* in the ventral epidermis and concomitant broad disruptions of ventral muscle morphologies, it is unclear whether the effects of *mid* in M27 and M28 are cell autonomous. Likewise, the frequent absence of M8 in *mid* mutants and *mid H15* double mutants (Kumar, Dobi, Baylies, & Abmayr, 2015) could be due to disrupted influences from the ectoderm because Mid expression was not detectable in the lineage giving rise to this muscle. Known cell fate changes in the epidermis of *mid* mutants may, for example, disrupt the expression of critical Wingless signals (Buescher et al., 2004). (E) Color code for the expression pattern of Org-1, Mid, Lb, Slou, Tup, and Kr during (C–C″) and (D–D″). (F) Table of alternative terms for specific muscles as described in Bate (1993).

expressing Lb). Besides the abdominal body wall muscles, *org-1* is also required for proper formation of the abdominal AMs as well as the TARMs (Fig. 3B and C″). Both the AMs and the TARMs (normally also expressing Tup/Isl1) are missing or malformed in *org-1* mutant embryos, demonstrating a role of *org-1* in building and maintenance of the internal anatomy of the larva (Boukhatmi et al., 2014; Schaub et al., 2012).

The early expression of *org-1* in muscle progenitor cells and the dependence of the expression of *slou*, *lb*, and *tup*, respectively, on *org-1* gene activity indicated that Org-1 might be a direct upstream regulator of these muscle identity genes. Genetic analysis of FC and muscle-specific enhancer elements of *slou*, *lb*, and *tup* confirmed that these enhancer elements require *org-1* for their initial activation in Org-1$^+$ FCs and their maintained activity in the respective muscles. Ectopic expression of Org-1 induces ectopic activation of these enhancer elements and the mutation of T-Box domain-binding motifs (as defined by Org-1 SELEX; Schaub et al., 2012) leads to the abolishment of enhancer activity in the Org-1$^+$ lineages. Org-1 ChIP analysis further confirmed Org-1 in vivo binding to these T-Box-binding sites. Thus, the muscle identity genes *slou*, *lb*, and *tup* appear to be direct transcriptional targets of Org-1 in the FCs of the respective myogenic lineages, including the Org-1/Tup-expressing lineage of heart-attached AMs (Boukhatmi et al., 2014; Busser et al., 2012; Schaub et al., 2012). In the case of Slou$^+$ and Lb$^+$ muscle progenitors, the mutually exclusive activation of *slou* and *lb* in neighboring progenitors involves additional regulatory inputs that include mutual repression between *slou* and *lb* (Junion et al., 2007; Knirr et al., 1999).

4.2 Midline Functions as a Muscle Identity Gene in Specific Lateral Body Wall Muscles

Besides its expression in the developing heart, the *Tbx20*-related *midline* (*mid*) gene is expressed in a distinct subset of muscle progenitors, FCs, and muscle fibers, in which it acts as a muscle identity gene (Kumar et al., 2015). Mid expression overlaps with Org-1 and Slouch in the sibling FCs of muscles 5 (LO1) and 25 (VT1) (Fig. 3C′ and D′). In these cells, *mid* has a partially redundant role with its paralog *H15* to specify muscle identities, as evidenced by the absence of over half of the muscle 5 (LO1) and morphological abnormalities of most muscle 25 (VT1) in embryos that lack both genes (Kumar et al., 2015). Unlike Org-1 and Slouch, Mid is not yet present in the common progenitor of these two muscle founders (Fig. 3D). Together with the weaker penetrance of its mutant phenotype, this indicates that *mid* may function downstream of *org-1* and *slouch* in these cells. An even more critical role was ascribed to *mid* in the development of two lateral muscles, namely muscle 23 (M23/LT3) and muscle 24 (M24/LT4), which arise from a common muscle progenitor and its derived sibling FCs (Fig. 3D and D′). These two muscles are part of a set of four lateral transverse muscles that lie in parallel and have similar shapes and sizes. However, M24 (LT4) is distinguished from its neighbor M23 (LT3) by its dorsally shifted position and smaller number of nuclei. Mid is expressed in the common progenitor of these two muscles but its expression is only maintained in the founder and syncytium of M24 (LT4) after asymmetric cell division (Fig. 3D′ and D″). In *mid* mutant embryos, ~70% of the segments lack both muscles, suggesting that *mid* acts already in the common progenitor cell to regulate the development of the two muscles derived from it. Additional deletion of *H15* did not increase muscle loss, so in this case H15 does not seem to cooperate with its paralog Mid. Of note, in the ~30% of the segments that retained all four lateral muscles in *mid* mutant embryos, the muscle fiber at the normal site of M24 (LT4) featured the same shape, dorsoventral position, and nuclear number as M23 (LT3), thus suggesting a transformation of M24 (LT4) into M23 (LT3) identities. Conversely, forced expression of *mid* in all somatic myoblasts, including the founder of M23 (LT3), caused a transformation in the opposite direction, namely of M23 (LT3) into M24 (LT4). Hence, in the normal situation, the continued expression of Mid in the FC of M24 (LT4) but not in its sibling allows Mid to act as a muscle identity factor to implement the M24 (LT4) identity in this cell and its derived muscle. Mid presumably acts together with additional muscle identity factors known to be active in these lateral muscle lineages, particularly with the zinc finger transcription factor Krüppel (Kr), which shares its dynamic expression pattern in the progenitors, founders, and syncytia of M23 (LT3) and M24 (LT4) with that of Mid (Kumar et al., 2015) (Fig. 3D–D″).

5. THE ROLE OF *DROSOPHILA Tbx1* (*ORG-1*) IN LINEAGE PLASTICITY, REPROGRAMMING, AND COMMITMENT DURING ADULT MYOGENESIS

The majority of the *Drosophila* adult musculature forms from the AMPs that are set aside during embryogenesis and start proliferating and differentiating during metamorphosis. At the same time, most of the preexisting larval muscles are eliminated by histolysis. However, a small minority of particular larval muscles escape histolysis and are remodeled into adult muscles, as exemplified by the remodeling of anterior larval oblique muscles into dorsal longitudinal indirect flight muscles (Dutta & VijayRaghavan, 2006). Recently, an example of even more radical larval muscle remodeling was discovered, in which specific larval muscle syncytia undergo fragmentation into mononucleate cells that subsequently redifferentiate into the heart-associated VLMs (Schaub, März, Reim, & Frasch, 2015). Therefore, this process represents an interesting and rather unique case of syncytial muscle dedifferentiation and transdifferentiation (Frasch, 2016).

org-1 was found to be a major player in this latter process that guides the formation of the VLMs of the adult heart (Schaub et al., 2015). As discussed in the previous section, Org-1 is expressed as a muscle identity factor in a subset of somatic mesoderm-derived muscles, including the specialized AMs that attach to the heart (Schaub et al., 2012) (Fig. 2B and Fig. 4A). Org-1 expression in AMs fades away after larval myogenesis has been completed. At the onset of metamorphosis it reinitiates, but only in the first three among the seven segmental pairs of larval abdominal AMs (Fig. 4C). Genetic analysis demonstrated that *org-1* is essential to induce the dedifferentiation and fragmentation of these syncytial muscles into *org-1* expressing mononucleated myoblasts (termed alary muscle-derived cells, AMDCs; Fig. 4D). Downregulation of Org-1 expression via cell type-specific induction of RNAi prevents dedifferentiation of the AMs, clearly demonstrating that *org-1* function is required for the initiation of this process. Furthermore, temporal control of dedifferentiation is mediated by the nuclear hormone receptor for the steroid hormone ecdysone, EcR, and spatial restriction to the three anterior-most pairs of AMs is provided by the function of the Hox gene *Ubx*. Diminishing the function of these activities in the AMs leads to suppression of *org-1* expression and dedifferentiation. After dedifferentiation is completed, the bilaterally distributed clusters of AMDCs migrate to the ventral side of the dorsal vessel, which also undergoes extensive remodeling during metamorphosis (Monier, Astier, Semeriva, & Perrin, 2005; Zaffran et al., 2006) and provides yet unknown inductive cues to the ADMCs for their proper redifferentiation. The Org-1^{+} myoblasts, which under the influence of FGF-receptor signals become VLM FCs, subsequently fuse with FCMs, derived from the proliferating AMPs, and redifferentiate into the syncytial, striated, and Org-1^{+} ventral longitudinal heart-associated musculature (VLM) that is tightly juxtaposed to the adult dorsal vessel (Fig. 4B–E) (Schaub et al., 2015).

As described earlier, embryonic AM development depends on a transcriptional cascade in which Org-1 directly activates *tup* expression in the

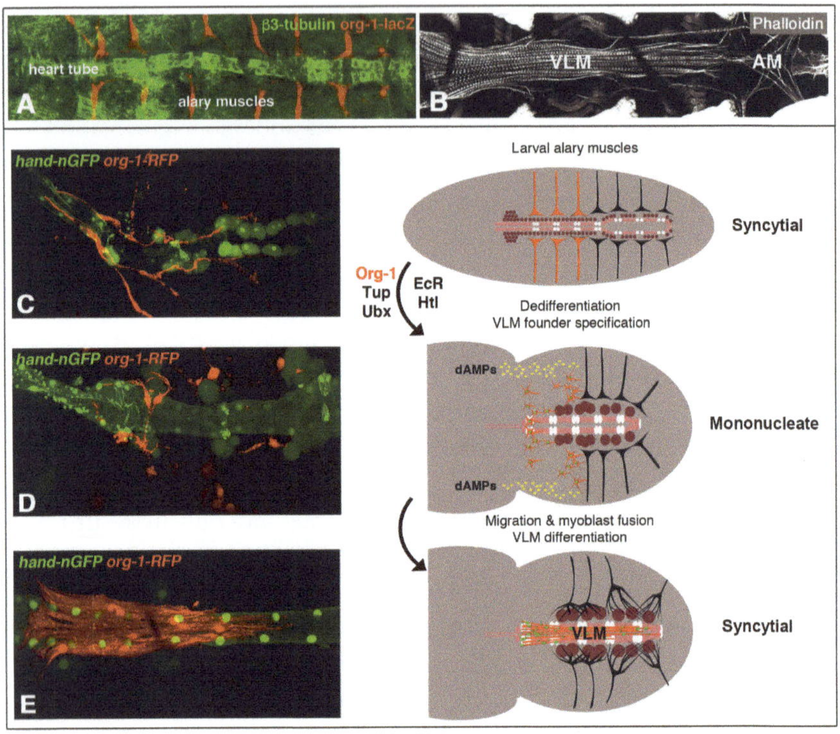

Fig. 4 The T-box transcription factor Org-1 is required for the transdifferentiation of larval alary muscles into the heart-associated ventral longitudinal musculature. (A) The *Drosophila* larval heart consists of two rows of cardiomyocytes and is anchored by seven pairs of alary muscles that express *org-1*. The somatic muscles and the heart are visualized with stainings against the β3-Tubulin protein and the *org-1* expression pattern is visualized by the activity of an *org-1-lacZ* reporter. (B) In the adult heart, the ventral longitudinal musculature (VLM) covers the adult heart ventrally and forms the dorsal diaphragm that separates the heart from the abdominal cavity. Only four of the seven pairs of alary muscles (AMs) from the larval heart remain in the adult heart after its remodeling. Actin is visualized with fluorescently labeled phalloidin. (C–E) Ex vivo life imaging of pupa carrying *org-1-RFP* and *hand-nGFP* (visualizing the heart, predominantly nuclei) reporters. (C) In early pupal stages, *org-1-RFP* reporter activity is found specifically in the first three pairs of syncytial alary muscles, but not in the posterior four pairs. (D) Subsequently, Org-1 is inducing dedifferentiation of these three anterior AMs and the syncytia fragment into mononucleate cells. The Hox gene *Ultrabithorax* (*Ubx*) provides spatial cues for this process whereas temporal coordination depends on ecdysone receptor (EcR)-mediated steroid signaling. In the mononucleate cells, Org-1 directly induces the expression of Tailup (Tup), whereas the activation of the FGFR Heartless (Htl) leads to the induction of founder cell fates in these. Dorsal adult muscle precursor cells (dAMPs) arise from Org-1$^-$ progenitors. (E) During the following pupal stages the AM-derived cells migrate to the heart tube and fuse with surrounding dAMPs to form syncytial muscle tubes, which differentiate into the syncytial VLM fibers.

FCs of the AMs (Boukhatmi et al., 2014). The VLMs also express both Org-1 and Tup, and knockdown of Tup function by cell type-specific RNAi disrupts transdifferentiation and perturbs differentiation of the few VLM fibers formed. The activation of *tup* expression during the transdifferentiation process is mediated by direct binding of Org-1 to the same *tup* enhancer element that is active during larval myogenesis since mutation of its T-Box-binding sites also abolishes its activity in the VLMs (Schaub et al., 2015). Hence, reinitiation of this transcriptional cascade during transdifferentiation of the AMs is required for the formation of the VLMs.

In sum, the *Drosophila* ortholog of *Tbx1*, *org-1* is required for the transdifferentiation of one differentiated syncytial muscle type into another through an intermediate of dedifferentiated, mononucleate myoblasts. *org-1* seems to play two opposite roles during this event. First, at the onset of metamorphosis, it is required in conjunction with Hox factors and nuclear hormone receptor signals for the initiation of dedifferentiation and fragmentation of the syncytial AMs into the mononucleate AMDCs, thereby conferring plasticity to the AM lineage. Second, after the induction of FC fates upon activation of FGFR signaling in the ADMCs, Org-1 together with its downstream target *tup* (*Drosophila Islet-1*) mediates the reprogramming and the commitment of these cells to the VLM lineage (Fig. 4C–E). The events described earlier have some parallel in the developing vertebrate heart, where cells of the second heart field that express and require the orthologs of *org-1* and *tup*, namely *Tbx1* and *Islet-1*, respectively, are added secondarily to the primitive heart tube derived from the *Tbx5/Nkx2-5* dependent first heart field. Although it would be too speculative to consider these homologous events, it is conceivable that kernels of regulatory events involving *org-1/Tbx1*, *tup/Islet-1*, and perhaps additional genes have been conserved and are being utilized during heart development in both vertebrates and invertebrates.

6. CONCLUDING REMARKS AND OUTLOOK

As we have illustrated in this review, T-box genes fulfill a large variety of important developmental roles in the *Drosophila* mesoderm and its derivatives, which begin at the earliest phases of embryonic mesoderm development and continue through metamorphosis. Whereas in one case, namely *byn*, this T-box gene is expressed only transiently and as far as known is never reused after its initial function in the early caudal mesoderm, more commonly individual T-box genes are reemployed during multiple stages and in different mesodermal tissue contexts. This complex temporal and spatial regulation appears largely to be driven by modular arrays of CRMs driving individual genes or, in the case of the *Doc* gene cluster, in some instances being shared by all three genes (as exemplified by the CVM enhancer of *Doc*). These CRMs are able to respond to different inputs from signaling pathways such as those triggered by localized Dpp, Wg, or RTK ligands, and from cell type-specific transcription factors that bind in combination with the corresponding signaling effectors. The work of defining these regulatory mechanisms in detail is currently still very much at the beginning. Another important issue concerns the question of how the same T-box domain protein is able to exert completely different functions in different tissue contexts. In part, these differential functions could be due to epigenetic effects that depend on the particular history of a specific cell and make the chromatin of target genes of individual T-box domain proteins differentially responsive to these factors. In addition, different combinations of cofactors, both signaling effectors and tissue-specific transcription factors, are likely to modulate the functions of T-box domain proteins. In the *Drosophila* mesoderm, this may include cell type-specific combinations of T-box domain proteins from different classes. For example, in oCBs of the heart, the Tbx20 factor Mid is coexpressed with the Tbx6-related factors Doc, whereas in the FCs of somatic muscles 5 and 25 Mid is coexpressed with the Tbx1-related factor Org-1. Likewise, the modulation of T-box domain protein outputs by differential combinations with transcription factors from other families has to be invoked. As an example, Org-1 functions in the visceral mesoderm in combination with the NK3 homeodomain protein Bagpipe and the FoxF factor Biniou, in the FCs of muscles 5 and 25 with the NK1 homeodomain protein Slouch, in those of muscle 8 with the Lbx-related homeodomain protein Ladybird, and in the AM progenitors with the Islet-1 related homeodomain protein Tup. Although in the case of Doc and its cardiogenic cofactors, namely the NK4 homeodomain factor Tin and the GATA factor Pannier, work to identify the global genomic targets has been initiated, it will take a lot of additional efforts to dissect the mechanistic interactions of these factors at the enhancer level and to single out the most crucial target genes that confer the developmental functions of these factors. If successful, in the future, it will be highly interesting to see whether the orthologous *Drosophila* and vertebrate T-box domain proteins (or in the CM perhaps the Tbx6-related Doc and vertebrate Tbx5) are activating some homologous target genes in the different phyla that mediate their developmental functions.

ACKNOWLEDGMENTS

We thank Johannes März for the drawings in Figure 4 and the Deutsche Forschungsgemeinschaft (DFG) for funding.

REFERENCES

Alvarez, A. D., Shi, W., Wilson, B. A., & Skeath, J. B. (2003). *Pannier* and *pointedP2* act sequentially to regulate *Drosophila* heart development. *Development, 130*, 3015–3026.

Azpiazu, N., & Frasch, M. (1993). *Tinman* and *bagpipe*: Two homeo box genes that determine cell fates in the dorsal mesoderm of *Drosophila*. *Genes & Development*, 7, 1325–1340.

Azpiazu, N., Lawrence, P. A., Vincent, J. P., & Frasch, M. (1996). Segmentation and specification of the *Drosophila* mesoderm. *Genes & Development, 10*, 3183–3194.

Bate, M. (1990). The embryonic development of larval muscles in *Drosophila*. *Development, 110*, 791–804.

Bate, M. (1993). The mesoderm and its derivatives. In M. Bate & A. Martinez-Arias (Eds.), *The development of* Drosophila melanogaster (pp. 1013–1090). Cold Spring Harbor: Cold Spring Harbor Laboratory Press.

Bodmer, R. (1993). The gene *tinman* is required for specification of the heart and visceral muscles in *Drosophila*. *Development, 118*, 719–729.

Bodmer, R., & Frasch, M. (2010). Development and aging of the *Drosophila* heart. In N. Rosenthal & R. P. Harvey (Eds.), *Heart development and regeneration: Vol. 1* (pp. 47–86). Oxford: Elsevier.

Boukhatmi, H., Frendo, J. L., Enriquez, J., Crozatier, M., Dubois, L., & Vincent, A. (2012). Tup/Islet1 integrates time and position to specify muscle identity in *Drosophila*. *Development, 139*, 3572–3582.

Boukhatmi, H., Schaub, C., Bataille, L., Reim, I., Frendo, J. L., Frasch, M., et al. (2014). An Org-1-Tup transcriptional cascade reveals different types of alary muscles connecting internal organs in *Drosophila*. *Development, 141*, 3761–3771.

Buescher, M., Svendsen, P. C., Tio, M., Miskolczi-McCallum, C., Tear, G., Brook, W. J., et al. (2004). *Drosophila* T box proteins break the symmetry of Hedgehog-dependent activation of Wingless. *Current Biology, 14*, 1694–1702.

Busser, B. W., Taher, L., Kim, Y., Tansey, T., Bloom, M. J., Ovcharenko, I., et al. (2012). A machine learning approach for identifying novel cell type-specific transcriptional regulators of myogenesis. *PLoS Genetics, 8*, e1002531.

Cai, C. L., Zhou, W., Yang, L., Bu, L., Qyang, Y., Zhang, X., et al. (2005). T-box genes coordinate regional rates of proliferation and regional specification during cardiogenesis. *Development, 132*, 2475–2487.

Campos-Ortega, J., & Hartenstein, V. (1997). *The embryonic development of* Drosophila melanogaster. Berlin: Springer-Verlag.

Capovilla, M., Brandt, M., & Botas, J. (1994). Direct regulation of *decapentaplegic* by Ultrabithorax and its role in *Drosophila* midgut morphogenesis. *Cell, 76*, 461–475.

Carmena, A., & Baylies, M. (2006). Development of the larval somatic musculature. In H. Sink (Ed.), Drosophila *muscle development* (pp. 79–91). Georgetown, TX: Landes Bioscience/Eurekah.com.

Carrasco-Rando, M., & Ruiz-Gomez, M. (2008). Mind bomb 2, a founder myoblast-specific protein, regulates myoblast fusion and muscle stability. *Development, 135*, 849–857.

Christoffels, V. M., Hoogaars, W. M., Tessari, A., Clout, D. E., Moorman, A. F., & Campione, M. (2004). T-box transcription factor Tbx2 represses differentiation and formation of the cardiac chambers. *Developmental Dynamics, 229*, 763–770.

de Joussineau, C., Bataille, L., Jagla, T., & Jagla, K. (2012). Diversification of muscle types in *Drosophila*: Upstream and downstream of identity genes. *Current Topics in Developmental Biology*, *98*, 277–301.

Dohrmann, C., Azpiazu, N., & Frasch, M. (1990). A new *Drosophila* homeo box gene is expressed in mesodermal precursor cells of distinct muscles during embryogenesis. *Genes & Development*, *4*, 2098–2111.

Duan, H., Skeath, J. B., & Nguyen, H. T. (2001). *Drosophila* Lame duck, a novel member of the Gli superfamily, acts as a key regulator of myogenesis by controlling fusion-competent myoblast development. *Development*, *128*, 4489–4500.

Durocher, D., Charron, F., Warren, R., Schwartz, R. J., & Nemer, M. (1997). The cardiac transcription factors Nkx2-5 and GATA-4 are mutual cofactors. *The EMBO Journal*, *16*, 5687–5696.

Dutta, D., & VijayRaghavan, K. (2006). Metamorphosis and the formation of the adult musculature. In H. Sink (Ed.), *Muscle development in* Drosophila (pp. 125–142). Georgetown, TX: Landes Bioscience/Eurekah.com.

Englund, C., Loren, C. E., Grabbe, C., Varshney, G. K., Deleuil, F., Hallberg, B., et al. (2003). Jeb signals through the Alk receptor tyrosine kinase to drive visceral muscle fusion. *Nature*, *425*, 512–516.

Frasch, M. (1995). Induction of visceral and cardiac mesoderm by ectodermal Dpp in the early *Drosophila* embryo. *Nature*, *374*, 464–467.

Frasch, M. (1999). Controls in patterning and diversification of somatic muscles during *Drosophila* embryogenesis. *Current Opinion in Genetics & Development*, *9*, 522–529.

Frasch, M. (2016). Dedifferentiation, redifferentiation, and transdifferentiation of striated muscles during regeneration and development. *Current Topics in Developmental Biology*, *116*, 331–355.

Fregoso Lomas, M., Hails, F., Lachance, J. F., & Nilson, L. A. (2013). Response to the dorsal anterior gradient of EGFR signaling in *Drosophila* oogenesis is prepatterned by earlier posterior EGFR activation. *Cell Reports*, *4*, 791–802.

Gajewski, K., Fossett, N., Molkentin, J. D., & Schulz, R. A. (1999). The zinc finger proteins Pannier and GATA4 function as cardiogenic factors in *Drosophila*. *Development*, *126*, 5679–5688.

Gajewski, K., Zhang, Q., Choi, C. Y., Fossett, N., Dang, A., Kim, Y. H., et al. (2001). Pannier is a transcriptional target and partner of Tinman during *Drosophila* cardiogenesis. *Developmental Biology*, *233*, 425–436.

Garg, V., Kathiriya, I. S., Barnes, R., Schluterman, M. K., King, I. N., Butler, C. A., et al. (2003). *GATA4* mutations cause human congenital heart defects and reveal an interaction with *TBX5*. *Nature*, *424*, 443–447.

Ghosh, T. K., Brook, J. D., & Wilsdon, A. (2017). T-Box genes in human development and disease. In M. Frasch (Ed.), *Current topics in developmental biology: Vol. 122* (pp. 383–415). Oxford: Elsevier.

Greulich, F., Rudat, C., & Kispert, A. (2011). Mechanisms of T-box gene function in the developing heart. *Cardiovascular Research*, *91*, 212–222.

Griffin, K. J., Stoller, J., Gibson, M., Chen, S., Yelon, D., Stainier, D. Y., et al. (2000). A conserved role for H15-related T-box transcription factors in zebrafish and *Drosophila* heart formation. *Developmental Biology*, *218*, 235–247.

Hamaguchi, T., Takashima, S., Okamoto, A., Imaoka, M., Okumura, T., & Murakami, R. (2012). Dorsoventral patterning of the *Drosophila* hindgut is determined by interaction of genes under the control of two independent gene regulatory systems, the dorsal and terminal systems. *Mechanisms of Development*, *129*, 236–243.

Hamaguchi, T., Yabe, S., Uchiyama, H., & Murakami, R. (2004). *Drosophila* Tbx6-related gene, *Dorsocross*, mediates high levels of Dpp and Scw signal required for the development of amnioserosa and wing disc primordium. *Developmental Biology*, *265*, 355–368.

Hiroi, Y., Kudoh, S., Monzen, K., Ikeda, Y., Yazaki, Y., Nagai, R., et al. (2001). Tbx5 associates with Nkx2-5 and synergistically promotes cardiomyocyte differentiation. *Nature Genetics*, *28*, 276–280.

Hollfelder, D., Frasch, M., & Reim, I. (2014). Distinct functions of the laminin beta LN domain and collagen IV during cardiac extracellular matrix formation and stabilization of alary muscle attachments revealed by EMS mutagenesis in *Drosophila*. *BMC Developmental Biology*, *14*, 26.

Hursh, D. A., Padgett, R. W., & Gelbart, W. M. (1993). Cross regulation of *decapentaplegic* and *Ultrabithorax* transcription in the embryonic visceral mesoderm of *Drosophila*. *Development*, *117*, 1211–1222.

Ieda, M., Fu, J. D., Delgado-Olguin, P., Vedantham, V., Hayashi, Y., Bruneau, B. G., et al. (2010). Direct reprogramming of fibroblasts into functional cardiomyocytes by defined factors. *Cell*, *142*, 375–386.

Iklé, J., Elwell, J. A., Bryantsev, A. L., & Cripps, R. M. (2008). Cardiac expression of the *Drosophila Transglutaminase* (*CG7356*) gene is directly controlled by myocyte enhancer factor-2. *Developmental Dynamics*, *237*, 2090–2099.

Ismat, A., Schaub, C., Reim, I., Kirchner, K., Schultheis, D., & Frasch, M. (2010). *HLH54F* is required for the specification and migration of longitudinal gut muscle founders from the caudal mesoderm of *Drosophila*. *Development*, *137*, 3107–3117.

Jagla, T., Bellard, F., Lutz, Y., Dretzen, G., Bellard, M., & Jagla, K. (1998). *ladybird* determines cell fate decisions during diversification of *Drosophila* somatic muscles. *Development*, *125*, 3699–3708.

Jin, H., Stojnic, R., Adryan, B., Ozdemir, A., Stathopoulos, A., & Frasch, M. (2013). Genome-wide screens for in vivo Tinman binding sites identify cardiac enhancers with diverse functional architectures. *PLoS Genetics*, *9*, e1003195.

Junion, G., Bataille, L., Jagla, T., Da Ponte, J. P., Tapin, R., & Jagla, K. (2007). Genome-wide view of cell fate specification: *Ladybird* acts at multiple levels during diversification of muscle and heart precursors. *Genes & Development*, *21*, 3163–3180.

Junion, G., Spivakov, M., Girardot, C., Braun, M., Gustafson, E. H., Birney, E., et al. (2012). A transcription factor collective defines cardiac cell fate and reflects lineage history. *Cell*, *148*, 473–486.

Kelly, R. G., Jerome-Majewska, L. A., & Papaioannou, V. E. (2004). The del22q11.2 candidate gene *Tbx1* regulates branchiomeric myogenesis. *Human Molecular Genetics*, *13*, 2829–2840.

Kispert, A., Herrmann, B. G., Leptin, M., & Reuter, R. (1994). Homologs of the mouse *Brachyury* gene are involved in the specification of posterior terminal structures in *Drosophila*, *Tribolium*, and *Locusta*. *Genes & Development*, *8*, 2137–2150.

Klapper, R., Stute, C., Schomaker, O., Strasser, T., Janning, W., Renkawitz-Pohl, R., et al. (2002). The formation of syncytia within the visceral musculature of the *Drosophila* midgut is dependent on *duf*, *sns* and *mbc*. *Mechanisms of Development*, *110*, 85–96.

Klinedinst, S. L., & Bodmer, R. (2003). Gata factor Pannier is required to establish competence for heart progenitor formation. *Development*, *130*, 3027–3038.

Knirr, S., Azpiazu, N., & Frasch, M. (1999). The role of the NK-homeobox gene *slouch* (*S59*) in somatic muscle patterning. *Development*, *126*, 4525–4535.

Kong, P., Racedo, S. E., Macchiarulo, S., Hu, Z., Carpenter, C., Guo, T., et al. (2014). *Tbx1* is required autonomously for cell survival and fate in the pharyngeal core mesoderm to form the muscles of mastication. *Human Molecular Genetics*, *23*, 4215–4231.

Kremser, T., Gajewski, K., Schulz, R. A., & Renkawitz-Pohl, R. (1999). Tinman regulates the transcription of the *beta3 tubulin gene* (*betaTub60D*) in the dorsal vessel of *Drosophila*. *Developmental Biology*, *216*, 327–339.

Kumar, R. P., Dobi, K. C., Baylies, M. K., & Abmayr, S. M. (2015). Muscle cell fate choice requires the T-box transcription factor Midline in *Drosophila*. *Genetics*, *199*, 777–791.

Kusch, T., & Reuter, R. (1999). Functions for *Drosophila brachyenteron* and *forkhead* in mesoderm specification and cell signalling. *Development*, *126*, 3991–4003.

Leal, S. M., Qian, L., Lacin, H., Bodmer, R., & Skeath, J. B. (2009). *Neuromancer1* and *Neuromancer2* regulate cell fate specification in the developing embryonic CNS of *Drosophila melanogaster*. *Developmental Biology*, *325*, 138–150.

Lee, H. H., & Frasch, M. (2000). Wingless effects mesoderm patterning and ectoderm segmentation events via induction of its downstream target *sloppy paired*. *Development*, *127*, 5497–5508.

Lee, H. H., & Frasch, M. (2005). Nuclear integration of positive Dpp signals, antagonistic Wg inputs and mesodermal competence factors during *Drosophila* visceral mesoderm induction. *Development*, *132*, 1429–1442.

Lee, H. H., Norris, A., Weiss, J. B., & Frasch, M. (2003). Jelly belly protein activates the receptor tyrosine kinase Alk to specify visceral muscle pioneers. *Nature*, *425*, 507–512.

Lee, K. Y., Singh, M. K., Ussar, S., Wetzel, P., Hirshman, M. F., Goodyear, L. J., et al. (2015). *Tbx15* controls skeletal muscle fibre-type determination and muscle metabolism. *Nature Communications*, *6*, 8054.

Lee, H.-H., Zaffran, S., & Frasch, M. (2006). The development of the larval visceral musculature. In H. Sink (Ed.), *Muscle development in* Drosophila (pp. 62–78). Georgetown, TX: Landes Bioscience/Eurekah.com.

Lehmacher, C., Abeln, B., & Paululat, A. (2012). The ultrastructure of *Drosophila* heart cells. *Arthropod Structure & Development*, *41*, 459–474.

Lo, P. C., & Frasch, M. (2001). A role for the COUP-TF-related gene *seven-up* in the diversification of cardioblast identities in the dorsal vessel of *Drosophila*. *Mechanisms of Development*, *104*, 49–60.

Lockwood, W. K., & Bodmer, R. (2002). The patterns of *wingless*, *decapentaplegic*, and *tinman* position the *Drosophila* heart. *Mechanisms of Development*, *114*, 13–26.

Luna-Zurita, L., Stirnimann, C. U., Glatt, S., Kaynak, B. L., Thomas, S., Baudin, F., et al. (2016). Complex interdependence regulates heterotypic transcription factor distribution and coordinates cardiogenesis. *Cell*, *164*, 999–1014.

Mandal, L., Banerjee, U., & Hartenstein, V. (2004). Evidence for a fruit fly hemangioblast and similarities between lymph-gland hematopoiesis in fruit fly and mammal aorta-gonadal-mesonephros mesoderm. *Nature Genetics*, *36*, 1019–1023.

Martin, B. L. (2016). Factors that coordinate mesoderm specification from neuromesodermal progenitors with segmentation during vertebrate axial extension. *Seminars in Cell & Developmental Biology*, *49*, 59–67.

Mathies, L. D., Kerridge, S., & Scott, M. P. (1994). Role of the *teashirt* gene in *Drosophila* midgut morphogenesis: Secreted proteins mediate the action of homeotic genes. *Development*, *120*, 2799–2809.

Michelson, A. M. (1994). Muscle pattern diversification in *Drosophila* is determined by the autonomous function of homeotic genes in the embryonic mesoderm. *Development*, *120*, 755–768.

Miskolczi-McCallum, C. M., Scavetta, R. J., Svendsen, P. C., Soanes, K. H., & Brook, W. J. (2005). The *Drosophila* melanogaster T-box genes *midline* and *H15* are conserved regulators of heart development. *Developmental Biology*, *278*, 459–472.

Molina, M. R., & Cripps, R. M. (2001). Ostia, the inflow tracts of the *Drosophila* heart, develop from a genetically distinct subset of cardial cells. *Mechanisms of Development*, *109*, 51–59.

Monier, B., Astier, M., Semeriva, M., & Perrin, L. (2005). Steroid-dependent modification of Hox function drives myocyte reprogramming in the *Drosophila* heart. *Development, 132*, 5283–5293.

Monier, B., Tevy, M. F., Perrin, L., Capovilla, M., & Semeriva, M. (2007). Downstream of homeotic genes: In the heart of Hox function. *Fly (Austin), 1*, 59–67.

Müller, J., Thuringer, F., Biggin, M., Zust, B., & Bienz, M. (1989). Coordinate action of a proximal homeoprotein binding site and a distal sequence confers the Ultrabithorax expression pattern in the visceral mesoderm. *The EMBO Journal, 8*, 4143–4151.

Murakami, R., Shigenaga, A., Kawakita, M., Takimoto, K., Yamaoka, I., Akasaka, K., et al. (1995). *Aproctous*, a locus that is necessary for the development of the proctodeum in *Drosophila* embryos, encodes a homolog of the vertebrate *Brachyury* gene. *Roux's Archives of Developmental Biology, 205*, 89–96.

Nasonkin, I., Alikasifoglu, A., Ambrose, C., Cahill, P., Cheng, M., Sarniak, A., et al. (1999). A novel sulfonylurea receptor family member expressed in the embryonic *Drosophila* dorsal vessel and tracheal system. *The Journal of Biological Chemistry, 274*, 29420–29425.

Nguyen, H. T., Voza, F., Ezzeddine, N., & Frasch, M. (2007). *Drosophila mind bomb2* is required for maintaining muscle integrity and survival. *The Journal of Cell Biology, 179*, 219–227.

Panganiban, G. E., Reuter, R., Scott, M. P., & Hoffmann, F. M. (1990). A *Drosophila* growth factor homolog, Decapentaplegic, regulates homeotic gene expression within and across germ layers during midgut morphogenesis. *Development, 110*, 1041–1050.

Papaioannou, V. E. (2014). The T-box gene family: Emerging roles in development, stem cells and cancer. *Development, 141*, 3819–3833.

Park, M., Wu, X., Golden, K., Axelrod, J. D., & Bodmer, R. (1996). The Wingless signaling pathway is directly involved in *Drosophila* heart development. *Developmental Biology, 177*, 104–116.

Pflugfelder, G. O., Schwarz, H., Roth, H., Poeck, B., Sigl, A., Kerscher, S., et al. (1990). Genetic and molecular characterization of the *optomotor-blind* gene locus in *Drosophila melanogaster*. *Genetics, 126*, 91–104.

Popichenko, D., Hugosson, F., Sjogren, C., Dogru, M., Yamazaki, Y., Wolfstetter, G., et al. (2013). Jeb/Alk signalling regulates the Lame duck GLI family transcription factor in the *Drosophila* visceral mesoderm. *Development, 140*, 3156–3166.

Porsch, M., Hofmeyer, K., Bausenwein, B. S., Grimm, S., Weber, B. H., Miassod, R., et al. (1998). Isolation of a *Drosophila* T-box gene closely related to human *TBX1*. *Gene, 212*, 237–248.

Qian, L., Liu, J., & Bodmer, R. (2005). *Neuromancer Tbx20*-related genes (*H15/midline*) promote cell fate specification and morphogenesis of the *Drosophila* heart. *Developmental Biology, 279*, 509–524.

Qian, L., Mohapatra, B., Akasaka, T., Liu, J., Ocorr, K., Towbin, J. A., et al. (2008). Transcription factor Neuromancer/TBX20 is required for cardiac function in *Drosophila* with implications for human heart disease. *Proceedings of the National Academy of Sciences of the United States of America, 105*, 19833–19838.

Reim, I., & Frasch, M. (2005). The *Dorsocross* T-box genes are key components of the regulatory network controlling early cardiogenesis in *Drosophila*. *Development, 132*, 4911–4925.

Reim, I., & Frasch, M. (2010). Genetic and genomic dissection of cardiogenesis in the *Drosophila* model. *Pediatric Cardiology, 31*, 325–334.

Reim, I., Lee, H. H., & Frasch, M. (2003). The T-box-encoding *Dorsocross* genes function in amnioserosa development and the patterning of the dorsolateral germ band downstream of Dpp. *Development, 130*, 3187–3204.

Reim, I., Mohler, J. P., & Frasch, M. (2005). *Tbx20*-related genes, *mid* and *H15*, are required for *tinman* expression, proper patterning, and normal differentiation of cardioblasts in *Drosophila*. *Mechanisms of Development, 122*, 1056–1069.

Reuter, R., Panganiban, G. E., Hoffmann, F. M., & Scott, M. P. (1990). Homeotic genes regulate the spatial expression of putative growth factors in the visceral mesoderm of *Drosophila* embryos. *Development, 110*, 1031–1040.

Reuter, R., & Scott, M. P. (1990). Expression and function of the homoeotic genes *Antennapedia* and *Sex combs reduced* in the embryonic midgut of *Drosophila*. *Development, 109*, 289–303.

Riechmann, V., Irion, U., Wilson, R., Grosskortenhaus, A., & Leptin, M. (1997). Control of cell fates and segmentation in the *Drosophila* mesoderm. *Development, 124*, 2915–2922.

Rotstein, B., & Paululat, A. (2016). On the morphology of the *Drosophila* heart. *Journal of Cardiovascular Development and Disease, 3*, 1–13.

Ryu, J. R., Najand, N., & Brook, W. J. (2011). Tinman is a direct activator of *midline* in the *Drosophila* dorsal vessel. *Developmental Dynamics, 240*, 86–95.

Sambasivan, R., Gayraud-Morel, B., Dumas, G., Cimper, C., Paisant, S., Kelly, R. G., et al. (2009). Distinct regulatory cascades govern extraocular and pharyngeal arch muscle progenitor cell fates. *Developmental Cell, 16*, 810–821.

San-Martin, B., Ruiz-Gomez, M., Landgraf, M., & Bate, M. (2001). A distinct set of founders and fusion-competent myoblasts make visceral muscles in the *Drosophila* embryo. *Development, 128*, 3331–3338.

Schaub, C., & Frasch, M. (2013). Org-1 is required for the diversification of circular visceral muscle founder cells and normal midgut morphogenesis. *Developmental Biology, 376*, 245–259.

Schaub, C., März, J., Reim, I., & Frasch, M. (2015). Org-1-dependent lineage reprogramming generates the ventral longitudinal musculature of the *Drosophila* heart. *Current Biology, 25*, 488–494.

Schaub, C., Nagaso, H., Jin, H., & Frasch, M. (2012). Org-1, the *Drosophila* ortholog of Tbx1, is a direct activator of known identity genes during muscle specification. *Development, 139*, 1001–1012.

Sebé-Pedrós, A., Ariza-Cosano, A., Weirauch, M. T., Leininger, S., Yang, A., Torruella, G., et al. (2013). Early evolution of the T-box transcription factor family. *Proceedings of the National Academy of Sciences of the United States of America, 110*, 16050–16055.

Sebé-Pedrós, A., & Ruiz-Trillo, I. (2017). Evolution and classification of the T-Box transcription factor family. In M. Frasch (Ed.), *Current topics in developmental biology: Vol. 122*. (pp. 1–26). Oxford: Elsevier.

Sellin, J., Albrecht, S., Kölsch, V., & Paululat, A. (2006). Dynamics of heart differentiation, visualized utilizing heart enhancer elements of the *Drosophila melanogaster* bHLH transcription factor Hand. *Gene Expression Patterns, 6*, 360–375.

Singer, J. B., Harbecke, R., Kusch, T., Reuter, R., & Lengyel, J. A. (1996). *Drosophila brachyenteron* regulates gene activity and morphogenesis in the gut. *Development, 122*, 3707–3718.

Singh, M. K., Christoffels, V. M., Dias, J. M., Trowe, M. O., Petry, M., Schuster-Gossler, K., et al. (2005). *Tbx20* is essential for cardiac chamber differentiation and repression of *Tbx2*. *Development, 132*, 2697–2707.

Stennard, F. A., Costa, M. W., Lai, D., Biben, C., Furtado, M. B., Solloway, M. J., et al. (2005). Murine T-box transcription factor *Tbx20* acts as a repressor during heart development, and is essential for adult heart integrity, function and adaptation. *Development, 132*, 2451–2462.

Stute, C., Schimmelpfeng, K., Renkawitz-Pohl, R., Palmer, R., & Holz, A. (2004). Myoblast determination in the somatic and visceral mesoderm depends on Notch signalling as well as on *milliways*(*mili*(*Alk*)) as receptor for Jeb signalling. *Development, 131*, 743–754.

Sun, B., Hursh, D. A., Jackson, D., & Beachy, P. A. (1995). Ultrabithorax protein is necessary but not sufficient for full activation of *decapentaplegic* expression in the visceral mesoderm. *The EMBO Journal, 14*, 520–535.

Takeuchi, J. K., & Bruneau, B. G. (2009). Directed transdifferentiation of mouse mesoderm to heart tissue by defined factors. *Nature, 459*, 708–711.

Tao, Y., Wang, J., Tokusumi, T., Gajewski, K., & Schulz, R. A. (2007). Requirement of the LIM homeodomain transcription factor Tailup for normal heart and hematopoietic organ formation in *Drosophila* melanogaster. *Molecular and Cellular Biology, 27*, 3962–3969.

Tepass, U., & Hartenstein, V. (1994). Epithelium formation in the *Drosophila* midgut depends on the interaction of endoderm and mesoderm. *Development, 120*, 579–590.

Tripathy, R., Kunwar, P. S., Sano, H., & Renault, A. D. (2014). Transcriptional regulation of *Drosophila* gonad formation. *Developmental Biology, 392*, 193–208.

van Doren, M. (2006). Development of the somatic gonad and fat bodies. In H. Sink (Ed.), *Muscle development in* Drosophila (pp. 51–61). Georgetown, TX: Landes Bioscience/ Eurekah.com.

Vogler, G., & Bodmer, R. (2015). Cellular mechanisms of *Drosophila* heart morphogenesis. *Journal of Cardiovascular Development and Disease, 2*, 2–16.

Wang, J., Tao, Y., Reim, I., Gajewski, K., Frasch, M., & Schulz, R. A. (2005). Expression, regulation, and requirement of the Toll transmembrane protein during dorsal vessel formation in *Drosophila melanogaster. Molecular and Cellular Biology, 25*, 4200–4210.

Weavers, H., & Skaer, H. (2013). Tip cells act as dynamic cellular anchors in the morphogenesis of looped renal tubules in *Drosophila. Developmental Cell, 27*, 331–344.

Windner, S. E., Doris, R. A., Ferguson, C. M., Nelson, A. C., Valentin, G., Tan, H., et al. (2015). *Tbx6, Mesp-b* and *Ripply1* regulate the onset of skeletal myogenesis in zebrafish. *Development, 142*, 1159–1168.

Wu, X., Golden, K., & Bodmer, R. (1995). Heart development in *Drosophila* requires the segment polarity gene *wingless. Developmental Biology, 169*, 619–628.

Xu, X., Yin, Z., Hudson, J. B., Ferguson, E. L., & Frasch, M. (1998). Smad proteins act in combination with synergistic and antagonistic regulators to target Dpp responses to the *Drosophila* mesoderm. *Genes & Development, 12*, 2354–2370.

Yin, Z., & Frasch, M. (1998). Regulation and function of *tinman* during dorsal mesoderm induction and heart specification in *Drosophila. Developmental Genetics, 22*, 187–200.

Zaffran, S., & Frasch, M. (2002). Early signals in cardiac development. *Circulation Research, 91*, 457–469.

Zaffran, S., Küchler, A., Lee, H. H., & Frasch, M. (2001). *biniou* (*FoxF*), a central component in a regulatory network controlling visceral mesoderm development and midgut morphogenesis in *Drosophila. Genes & Development, 15*, 2900–2915.

Zaffran, S., Reim, I., Qian, L., Lo, P. C., Bodmer, R., & Frasch, M. (2006). Cardioblast-intrinsic Tinman activity controls proper diversification and differentiation of myocardial cells in *Drosophila. Development, 133*, 4073–4083.

Zinzen, R. P., Girardot, C., Gagneur, J., Braun, M., & Furlong, E. E. (2009). Combinatorial binding predicts spatio-temporal cis-regulatory activity. *Nature, 462*, 65–70.

CHAPTER SEVEN

TBX5: A Key Regulator of Heart Development

J.D. Steimle, I.P. Moskowitz[1]
University of Chicago, Chicago, IL, United States
[1]Corresponding author: e-mail address: imoskowitz@peds.bsd.uchicago.edu

Contents

1. *TBX5*, a Member of the T-Box Family of Transcription Factors 196
2. *TBX5* Expression 198
   2.1 *TBX5* Expression Domains in the Embryonic and Adult Heart 198
   2.2 Extracardiac *TBX5* Expression 200
   2.3 Transcriptional Regulation of *TBX5* 200
3. *TBX5* Haploinsufficiency: Holt–Oram Syndrome 203
4. Animal Model of Holt–Oram Syndrome 205
5. *TBX5* in Cardiac Morphologic Development 205
   5.1 Ventricular Septum 206
   5.2 Atrial Septum 206
6. *TBX5* in Cardiac Conduction System Development 207
7. Homozygous *Tbx5* Mutations Reveal Novel Roles of TBX5 207
   7.1 Complete Loss of Mammalian *Tbx5* 208
   7.2 Zebrafish *tbx5a/tbx5b* 208
8. The TBX5 Gene Regulatory Network 209
   8.1 Positive Transcriptional Activation by TBX5 and Cofactors: Cardiomyocyte Specific Factors, Chromatin Modification, and Maturation 209
   8.2 TBX5-Mediated Repression: Inhibition of Noncardiomyocyte Fate Through Chromatin Remodeling 212
   8.3 Direct Targets of TBX5 Regulation 213
9. Concluding Remarks 214
References 214

Abstract

TBX5 is a member of the T-box transcription factor family and is primarily known for its role in cardiac and forelimb development. Human patients with dominant mutations in *TBX5* are characterized by Holt–Oram syndrome, and show defects of the cardiac septa, cardiac conduction system, and the anterior forelimb. The range of cardiac defects associated with *TBX5* mutations in humans suggests multiple roles for the transcription factor in cardiac development and function. Animal models demonstrate similar defects and have provided a useful platform for investigating the roles of *TBX5* during

Current Topics in Developmental Biology, Volume 122
ISSN 0070-2153
http://dx.doi.org/10.1016/bs.ctdb.2016.08.008

embryonic development. During early cardiac development, *TBX5* appears to act primarily as a transcriptional activator of genes associated with cardiomyocyte maturation and upstream of morphological signals for septation. During later cardiac development, *TBX5* is required for patterning of the cardiac conduction system and maintenance of mature cardiomyocyte function. A comprehensive understanding of the integral roles of *TBX5* throughout cardiac development and adult life will be critical for understanding human cardiac morphology and function.

1. *TBX5*, A MEMBER OF THE T-BOX FAMILY OF TRANSCRIPTION FACTORS

Ever since the association of *TBX5* mutations with Holt–Oram syndrome (Basson et al., 1997; Li et al., 1997), the *TBX5* gene has been a source of study, particularly with respect to cardiac and limb development. This review covers progress since those initial reports, published two decades ago.

The T-box family of transcription factors share a common T-box DNA-binding domain (Bollag et al., 1994) and are named after the founding member, *T*, which encodes the transcription factor Brachyury (Herrmann, Labeit, Poustka, King, & Lehrach, 1990; Pflugfelder, Roth, & Poeck, 1992). The T-box domain is approximately 170–200 amino acids in length (Agulnik, Bollag, & Silver, 1995; Agulnik et al., 1996; Bollag et al., 1994; Papaioannou, 2014), binds DNA directly (Pflugfelder et al., 1992), and is required for transcriptional activity (Kispert, Koschorz, & Herrmann, 1995). All members of the T-box gene family appear to bind the DNA consensus motif AGGTGHBA (Conlon, Fairclough, Price, Casey, & Smith, 2001; He, Kong, Ma, & Pu, 2011; Jolma et al., 2013; Kispert et al., 1995; Mathelier et al., 2016; Waldron et al., 2016; Wilson & Conlon, 2002). While some members, such as *T* and *EOMES*, are only able to bind palindromic sequences as dimers (Conlon et al., 2001; Ghosh et al., 2001; Kispert et al., 1995; Muller & Herrmann, 1997), others, including *TBX2*, *TBX3*, and *TBX5*, can bind individual motifs as monomers (Bruneau et al., 2001; Carreira, Dexter, Yavuzer, Easty, & Goding, 1998; Ghosh et al., 2001; He, Wen, Campbell, Wu, & Rao, 1999).

In the human genome, there are 17 coding genes that fall within five subfamilies of the T-box family (Agulnik et al., 1996; Papaioannou & Silver, 1998), most of which show sequence similarity between vertebrates and invertebrates (Agulnik et al., 1995, 1996; Papaioannou & Silver, 1998; Pflugfelder et al., 1992; Ruvinsky, Silver, & Gibson-Brown, 2000;

Wilson & Conlon, 2002). The human genes *TBX2*, *TBX3*, *TBX4*, and *TBX5* belong to one subfamily, and are homologous to the Drosophila gene *omb* (Agulnik et al., 1996). These four genes likely originated through tandem duplication of the ancestral gene through unequal crossover to form the ancestral *TBX2/3* and *TBX4/5* (Agulnik et al., 1996; Ruvinsky et al., 2000), followed by cluster duplication prior to the divergence of bony fish and tetrapods approximately 400 million years ago (Agulnik et al., 1996; Ruvinsky & Silver, 1997). This second duplication event generated one cluster containing *TBX2* and *TBX4* and a second containing *TBX3* and *TBX5*, which are located on human chromosomes 17 and 12, respectively (Agulnik et al., 1996; Ruvinsky & Silver, 1997).

The T-box family gene *TBX5* encodes a 518-amino acid protein with a 180-amino acid T-box domain located between amino acid residues 56 and 236 (Basson et al., 1997; Li et al., 1997). TBX5 contains two nuclear localization sequences (NLS): NLS1 located within the T-box domain (amino acids 78–90), and NLS2 located outside the T-box domain on the C-terminal end (325–340) (Collavoli et al., 2003; Zaragoza et al., 2004). While each NLS is sufficient to drive nuclear localization, they appear to work cooperatively (Collavoli et al., 2003). As a transcription factor, TBX5 also contains a transactivation domain located from amino acids 339–379 with functional requirement of amino acids 349–351 (Zaragoza et al., 2004). The sequence of amino acids 152–160 has been proposed to act as a nuclear export signal through the CRM1 export pathway by which TBX5 subcellular localization can be regulated through binding with the PDLIM7 protein (Camarata et al., 2006, 2010; Kulisz & Simon, 2008); however, this theory remains controversial as the crystal structure of TBX5 suggests this domain would be located on the inside of the protein and inaccessible without major protein rearrangements (Stirnimann, Ptchelkine, Grimm, & Muller, 2010). Aside from these domains, TBX5 also contains several other protein–protein interaction domains, discussed in detail later.

In addition to the best-described isoform, sometimes referred to as *TBX5a* (Georges, Nemer, Morin, Lefebvre, & Nemer, 2008), there have been four additional isoforms described in the literature (Georges et al., 2008; Yamak et al., 2015). These isoforms are derived from alternative splicing within thc *TBX5* locus and result in proteins of varying lengths, including either an N-terminal or a C-terminal truncated form as well as forms with varying C-terminal modifications (Georges et al., 2008; Yamak et al., 2015). Interestingly, all described isoforms retain the T-box domain

(Yamak et al., 2015). The alternative isoforms show differential expression and activity, including antagonism of TBX5a, and further investigation into these isoforms will be important for understanding the roles of *TBX5* in human development and health (Yamak et al., 2015).

2. *TBX5* EXPRESSION

The broad spatiotemporal expression domains of *TBX5* during development appear to be generally conserved throughout vertebrate evolution and consist of the heart, forelimb, and retina (Bruneau et al., 1999; Chapman et al., 1996; Gibson-Brown et al., 1996; Gibson-Brown, Agulnik, Silver, & Papaioannou, 1998; Horb & Thomsen, 1999; Showell, Christine, Mandel, & Conlon, 2006; Takabatake, Takabatake, & Takeshima, 2000); however, some tissue-specific expression differences occur between species. First, we will examine the expression domains of *TBX5* in the common tetrapod models as well as humans, with the greatest emphasis on subcardiac domains, and then we will discuss the regulation of *TBX5* gene expression based on evidence from mouse and human studies.

2.1 *TBX5* Expression Domains in the Embryonic and Adult Heart

The cardiac expression patterns of *TBX5* in human, mouse, chick, and frog are very similar. In human hearts, *TBX5* is expressed in the epicardium, myocardium, and endocardium of embryonic and adult hearts (Hatcher, Goldstein, Mah, Delia, & Basson, 2000). Human *TBX5* is expressed in the free walls and septa of all four chambers during development; however, atrial expression is much greater than ventricular, as seen in mouse and chicken (Hatcher et al., 2000). *TBX5* is expressed in the embryonic atrioventricular (AV) node, and is excluded from the AV valves (Hatcher et al., 2000). *TBX5* is expressed throughout the epicardium, but not in the endocardium of the left ventricle (Hatcher et al., 2000). Much like in animal models, *TBX5* expression is absent from the developing outflow tract of the heart (Hatcher et al., 2000). In human adults, *TBX5* expression is highest in the atrial appendages, followed by the lungs, left ventricle, and esophagus (GTEx Consortium, 2013; Mele et al., 2015).

In mice, *Tbx5* becomes abundantly expressed around E8.0 throughout the cardiac crescent (Bruneau et al., 1999), and this expression becomes restricted to the posterior portion of the forming heart tube, corresponding to the sinus venosa and future atria, between E8.25 and E8.5 (Bruneau et al., 1999; Chapman et al., 1996). At E9.0, *Tbx5* expression expands throughout the future left ventricle (Bruneau et al., 1999). Additionally, atrial expression of *Tbx5* is stronger than in the left ventricle, and in the ventricular free wall it is higher than in the trabeculae (Bruneau et al., 1999). *Tbx5* is also expressed in the right ventricular trabeculae, but not free wall (Bruneau et al., 1999). Expression of *Tbx5* in the left ventricle and atria is maintained throughout embryonic development (Chapman et al., 1996). During maturation of the mouse heart, like humans, *Tbx5* is expressed in and colocalizes with markers of the cardiac conduction system, including the AV bundle and bundle branch (Moskowitz et al., 2004). Genetic inducible fate mapping demonstrated that left ventricular *Tbx5* expression arises from the first heart field, specified prior to morphogenesis of the heart, whereas atrial and atrial septum *Tbx5* expression arises from *Mef2cAHF*$^{+}$ second heart field domain, suggesting potential independent roles of *Tbx5* in the first and second heart fields during development (Devine, Wythe, George, Koshiba-Takeuchi, & Bruneau, 2014).

In cardiac development of the chick, *Tbx5* expression is first detected throughout the entire bilateral cardiac primordia (Bruneau et al., 1999). This expression is maintained following fusion of the heart tube along the entire rostrocaudal length (Gibson-Brown et al., 1998), but adopts an anterior-to-posterior gradient shortly after (Bruneau et al., 1999). Although there appears to be a gradient to the expression, *Tbx5* is expressed throughout the whole heart during cardiac looping (Gibson-Brown et al., 1998). After looping is complete, *Tbx5* expression remains in the entire heart except the outflow tract, and this is the only major difference between mouse and chick heart expression. However, as cardiac development and maturation proceed, expression of *Tbx5* is restricted from the right ventricle, similar to the expression pattern in the mouse embryo (Bruneau et al., 1999).

In *Xenopus*, the earliest expression domains of *tbx5* are in two lateral stripes, corresponding to the cardiac primordia, on either side of the embryo and continue to be expressed in the migrating precardiac mesoderm (Horb & Thomsen, 1999; Showell et al., 2006). Similar to chick cardiogenesis, after fusion at the midline and formation of the early heart tube, *tbx5* is expressed throughout most of the cardiac tissue including the sinus venosus/inflow tract of the heart (Horb & Thomsen, 1999; Showell et al., 2006). As development continues, *tbx5* expression is lost from the most anterior structure, the bulbus cordis, and is strongly detected in the posterior regions of the heart, while it is also maintained in the ventricle (Horb & Thomsen, 1999; Showell et al., 2006). *tbx5* is expressed robustly in both the endocardium and myocardium and is detected in the epicardium (Horb & Thomsen, 1999).

2.2 Extracardiac *TBX5* Expression

In addition to the cardiac expression domains, *TBX5* is expressed in many noncardiac tissues. Perhaps the best-studied expression domain outside of the heart is that of the developing forelimb. *Tbx5* is expressed in the lateral plate mesoderm giving rise to the forelimb starting at E8.8 of mouse embryonic development (Gibson-Brown et al., 1996) and is robustly expressed in the forelimb bud at E9.5 (Chapman et al., 1996; Gibson-Brown et al., 1996). Expression throughout the developing limb is maintained until E11.5, when it then becomes restricted to the proximal portion of the forelimb (Gibson-Brown et al., 1996). *Tbx5* is also expressed in the periochondrium of the forelimb at E13.5 (Gibson-Brown et al., 1996).

Outside of the heart and forelimb, *TBX5* expression has been reported in several notable domains during development. The earliest reported expression domain for *Tbx5* during mouse development is in the allantois at E7.5 where it is transiently coexpressed with *Tbx4* (Chapman et al., 1996). Expression of *Tbx5* in the allantois has been suggested to be a mammalian-specific trait, as transcription of *Tbx5* is never observed in the allantois of chick embryos (Gibson-Brown et al., 1998). *Tbx5* is also expressed in the optic vesicle and the neural retina of the developing eye in mouse, chick, and *Xenopus* where it is coexpressed with the other members of the *omb* family of T-box genes (Chapman et al., 1996; Gibson-Brown et al., 1998; Horb & Thomsen, 1999; Showell et al., 2006). Additionally, *Tbx5* is expressed in the mesenchyme of the mandibular arch, the trachea, and the lung, as well as the body wall of the thorax (Chapman et al., 1996; Gibson-Brown et al., 1998). In both mouse and chicken, expression of *Tbx5* has been reported in the genital papilla (Chapman et al., 1996; Gibson-Brown et al., 1998). Specific to avian development, *Tbx5* expression is observed in the notochord during midembryonic development (Gibson-Brown et al., 1998).

2.3 Transcriptional Regulation of *TBX5*

The mechanisms governing spatiotemporal regulation of *TBX5* have begun to be addressed by defining *cis*-regulatory elements driving *TBX5* expression in the embryo and adult. Preliminary investigations have identified several elements that drive distinct spatial domains during development; however, this area is ripe for future investigations. Tiling experiments have identified three enhancers associated with in vivo expression of *Tbx5* in the mammalian heart (Minguillon et al., 2012; Smemo et al., 2012). The first enhancer, corresponding to hg19 chr12:114,463,712–114,464,080, drives expression of a reporter construct in both ventricular and atrial myocardium of E11.5 mouse hearts (Smemo et al., 2012). The second enhancer, hg19 chr12:114,701,207–114,704,691, drives expression in the posterior portion of the heart, including the ventricles, interventricular septum, and AV canal (Smemo et al., 2012). Additionally, this second enhancer contains a low-frequency SNP that abrogates the enhancer's ability to drive expression (Smemo et al., 2012). While this SNP was predicted to disrupt a TAL1 binding site, *Tal1* is not expressed in the myocardium, suggesting other members of the basic helix–loop–helix E-box-binding transcription factors may be driving expression of this enhancer (Smemo et al., 2012). The third *Tbx5* enhancer, hg19 chr12:114,853,271–114,858,238, is sufficient to drive expression in the ventricles, interventricular septum, and AV canal as well as the atria (Smemo et al., 2012).

In addition to the identified cardiac enhancers, there have been two additional enhancers identified that regulate limb expression. The first is located within intron 2 of *Tbx5* and drives expression within the lateral plate mesoderm of the forelimb, but not the heart (Minguillon et al., 2012). This forelimb enhancer is regulated in part through Hox4/5 genes, expressed in the region of the lateral plate mesoderm that gives rise to the forelimb, and has been proposed as the mechanism by which forelimb *Tbx5* expression is positionally defined along the anterior–posterior body axis (Minguillon et al., 2012). The second forelimb enhancer identified is known as CNS12 and is located approximately 120 kbp downstream of the *Tbx5* coding region (Adachi, Robinson, Goolsbee, & Shubin, 2016). The CNS12 enhancer drives expression in the lateral plate mesoderm and is sufficient to drive *Tbx5* expression for forelimb formation (Adachi et al., 2016).

Taken together, *Tbx5* expression in the developing heart and forelimb appears to be driven by distinct *cis*-regulatory elements, although the factors and transcriptional complexes that control the expression of these enhancers have yet to be uncovered.

Recent insight into the regulation of *Tbx5* expression has come from investigations of the local three-dimensional architecture of the *Tbx5* locus

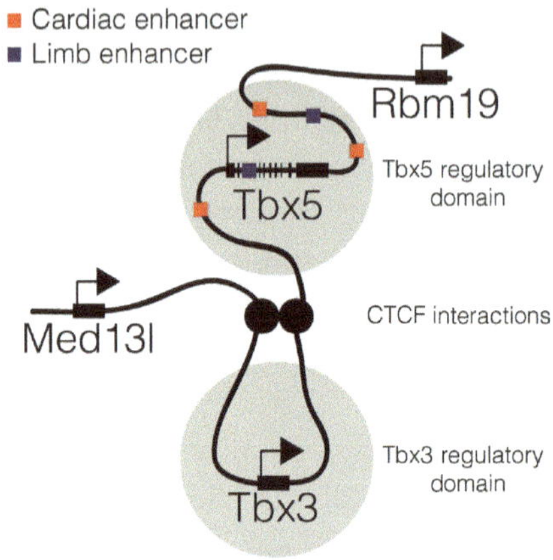

Fig. 1 Representation of *Tbx5* regulatory domain. Chromatin capture techniques suggest interactions between the promoter of *Tbx5* and *cis*-regulatory elements are limited to a 375-kbp region demarcated by functional CTCF sites located between *Rbm19* and *Tbx5* and *Tbx5* and *Tbx3* (van Weerd et al., 2014).

and its neighboring genes: *Rbm19*, *Tbx3*, and *Med13l* (Jin et al., 2013; van Weerd et al., 2014). Circular chromosome conformation capture with sequencing (4C-seq) data from the viewpoints of the *Tbx3* and *Tbx5* promoters, and the CTCF binding site between the two loci, suggests that the loci are in contact and that putative *cis*-regulatory elements for each gene are located almost exclusively within their own loci with CTCF sites acting as a regulatory barrier (Fig. 1) (van Weerd et al., 2014). Additionally, 4C-seq from the viewpoint of *Rbm19*, the nearest gene 3′ of *Tbx5*, suggests partially overlapping regulatory elements (van Weerd et al., 2014). Together, this suggests that most of the *cis*-regulatory information for *Tbx5* expression is located in the 375 kbp region between the CTCF sites demarcating the boundaries of the *Tbx3*/*Tbx5* and the *Tbx5*/*Rbm19* loci (Fig. 1) (van Weerd et al., 2014).

In addition to transcriptional regulation, *Tbx5* has also been shown to be regulated through microRNA-dependent mechanisms (Wang et al., 2014). In a screen of candidate human microRNAs, MiR-10a and MiR-10b were shown to bind to the 3′-UTR of the *Tbx5* and inhibit its translation (Wang et al., 2014). More recently, it has been shown that regulation through these two microRNAs may play a role in adult conduction defects and pathological remodeling in disease-state hearts (Torrado et al., 2015).

3. *TBX5* HAPLOINSUFFICIENCY: HOLT–ORAM SYNDROME

Holt–Oram syndrome is an autosomal dominant disorder caused primarily by dominant mutations in *TBX5*. Holt–Oram syndrome is a clinical diagnosis that includes completely penetrant, variably expressed upper-limb malformations including preaxial radial ray anomalies and congenital heart defects, typically septal and/or conduction defects (Holt & Oram, 1960). Although sometimes difficult to detect, upper-limb malformations are fully penetrant, while structural cardiac defects occur in 76% of patients with Holt–Oram syndrome (Basson et al., 1994, 1999; Holt & Oram, 1960; Newbury-Ecob, Leanage, Raeburn, & Young, 1996). Holt–Oram syndrome affects 1 in 100,000–135,000 live births in European populations (Barisic et al., 2014; Elek, Vitez, & Czeizel, 1991), although defects can occur in any population (Al-Qattan & Abou Al-Shaar, 2015; Ekure, Okoromah, Briggs, & Ajenifuja, 2004; Kimura, Kikuchi, Ichinoi, & Kure, 2015; Najjar, Mardini, Tabbaa, & Nyhan, 1988). Holt–Oram syndrome exhibits classic Mendelian inheritance for a dominant trait (Basson et al., 1997, 1994; Holt & Oram, 1960; McDermott et al., 2005), and the risk of nonaffected parents with an affected proband giving rise to a second child with a de novo pathogenic mutation is the same as the average population (McDermott, Fong, & Basson, 1993). Manifestations of upper-limb defects can include single or combinatorial abnormalities of the hand and digits, bones of the lower arm, humerus, or shoulder girdle (Basson et al., 1994; Newbury-Ecob et al., 1996). Defects of the hand and digits must include defects of the thumb for the Holt–Oram diagnosis, while defects of the lower arm are associated primarily with the radius (Basson et al., 1994; Holt & Oram, 1960; McDermott et al., 2005; Newbury-Ecob et al., 1996). Structural abnormalities of the heart can include secundum-type atrial septal defects, primum-type atrial septal defects, and/or ventricular septal defects (Basson et al., 1994; Holt & Oram, 1960; Newbury-Ecob et al., 1996). Conduction system defects manifest as long PR interval, AV block, bundle branch block, bradycardia, sick sinus syndrome, and atrial fibrillation (Basson et al., 1994; Holt & Oram, 1960; Newbury-Ecob et al., 1996), and these conduction defects can occur in the absence of overt structural defects (Basson et al., 1994; Newbury-Ecob et al., 1996). Holt–Oram syndrome is not associated with defects of the lower limb, postaxial upper limb, gastrointestinal system, genitourinary, or nervous system, which if present suggest an alternative diagnosis (Basson et al., 1994; Debeer, Race, Gewillig, Devriendt, & Frijns, 2007; Holt & Oram, 1960; McDermott et al., 2005; Newbury-Ecob et al., 1996).

Current evidence supports a model in which Holt–Oram syndrome is caused by *TBX5* haploinsufficiency (Basson et al., 1997; Li et al., 1997) with genetic abnormalities associated with *TBX5* coding or splice regulatory sequences underlying approximately 70% of patients meeting strict clinical diagnoses (Debeer et al., 2007; McDermott et al., 2005). In the latest collection of the Human Gene Mutation Database, there have been 103 reported mutations in coding, splicing, or regulatory sequences of *TBX5*, which result in Holt–Oram syndrome or other cardiac defects (Stenson et al., 2014), recently reviewed in Yamak et al. (2015). Additionally, there have been 44 pathologic point mutations reported in the coding region of *TBX5* (Stenson et al., 2014). Similar to the eight reported gross deletions (Stenson et al., 2014), some of these are nonsense mutations resulting in highly truncated proteins that are thought to act as null alleles (Basson et al., 1997; Fan et al., 2003; Gruenauer-Kloevekorn & Froster, 2003; McDermott et al., 2005). Missense mutations have been reported throughout much of the T-box domain, typically resulting in transcriptional decrements (Basson et al., 1999; Boogerd et al., 2010; McDermott et al., 2005; Postma et al., 2008). There are also several reported mutations that result in missplicing or alternative splicing (Basson et al., 1999; Borozdin et al., 2006; Cross et al., 2000; Heinritz et al., 2005; McDermott et al., 2005; Vianna, Miura, Pereira, & Jatene, 2011). Interestingly, duplications of *TBX5* are pathogenic, resulting in atypical Holt–Oram syndrome (Kimura et al., 2015; Patel, Silcock, McMullan, Brueton, & Cox, 2012). Additionally, a patient with a homozygous, single-base-pair mutation within a *cis*-regulatory element controlling *TBX5* displayed decreased *TBX5* expression and nonsyndromic congenital heart disease, raising the possibility that the etiology of some of the remaining 30% of Holt–Oram patients may result from *cis*-regulatory element mutations (Smemo et al., 2012).

While all cases of Holt–Oram syndrome result in both cardiac and forelimb defects, several cases have been reported in which defects in either the heart (Basson et al., 1997; Brassington et al., 2003; Li et al., 1997; Yang et al., 2000) or the limb (Brassington et al., 2003; Li et al., 1997; Yang et al., 2000) appear more severe in one tissue than the other, suggesting a potential tissue-specific role for distinct domains of the protein (Isphording, Leylek, Yeung, Mischel, & Simon, 2004). While the underlying mechanisms of these biased defects have not been identified, some proposed models include differences in binding partners and binding motif recognition (Basson et al., 1999; Camarata et al., 2006; Garg et al., 2003; Isphording et al., 2004; Krause et al., 2004).

4. ANIMAL MODEL OF HOLT–ORAM SYNDROME

Investigations into the role of *TBX5* in cardiac development have been undertaken in most major animal model systems: mouse, chick, frog, and zebrafish. Each system provides a unique set of tools for investigating the role of *TBX5* in the developmental etiology of Holt–Oram syndrome. The most well-characterized model of Holt–Oram syndrome is the mouse heterozygous for a *Tbx5* knockout allele. The $Tbx5^{tm1Jse}$ mouse allele contains loxP sites surrounding exon 3, which encodes a portion of the T-box DNA-binding domain, and upon Cre-mediated recombination, will generate truncated *Tbx5* transcripts (Bruneau et al., 2001). Germline deletion of exon 3 generates the $Tbx5^{tm1.1Jse}$ mouse (Bruneau et al., 2001). Heterozygous $Tbx5^{tm1.1Jse}$ mice exhibit the characteristic haploinsufficient phenotype of Holt–Oram syndrome, including anterior defects of the forelimb, septal defects of the heart, and defects of cardiac conduction (Bruneau et al., 2001; Moskowitz et al., 2004). The mouse animal model has provided a robust system in which to study Holt–Oram syndrome in vivo and will continue to provide a platform by which to study the disease and the role of *Tbx5* in cardiac and limb development.

5. *TBX5* IN CARDIAC MORPHOLOGIC DEVELOPMENT

The morphologic cardiac defects associated with Holt–Oram syndrome are most commonly malformations of the septa dividing the left and right sides of the heart (Basson et al., 1994; Holt & Oram, 1960; McDermott et al., 2005; Newbury-Ecob et al., 1996). From our current understanding, the ontogeny of the septa dividing the ventricular and atrial chambers is quite different (Anderson, Webb, Brown, Lamers, & Moorman, 2003), and yet defects in both arise from haploinsufficiency of *Tbx5* (Basson et al., 1997; Bruneau et al., 1999, 2001; Hoffmann et al., 2014; Koshiba-Takeuchi et al., 2009; Takeuchi et al., 2003; Xie et al., 2012).

5.1 Ventricular Septum

The morphology of the ventricular septum depends heavily on the localization of *Tbx5* expression during development. Both the left and the right ventricles contribute equally toward the formation of the interventricular septum, suggesting that a balance of left and right contributions may underlie development of the septum (Franco et al., 2006). During development of the ventricular chambers, *Tbx5* is unilaterally expressed on the left side, including the left side of the ventricular septum (Bruneau et al., 1999, 2001; Takeuchi et al., 2003). Overexpression of *Tbx5* bilaterally results in malformation of the ventricular chambers and absence of the ventricular septum (Koshiba-Takeuchi et al., 2009; Liberatore, Searcy-Schrick, & Yutzey, 2000; Takeuchi et al., 2003). However, it remains unclear whether *Tbx5*-dependent transcriptional regulation alone controls ventricular septum formation (Franco et al., 2006; Koshiba-Takeuchi et al., 2009; Takeuchi et al., 2003). Two additional T-box family genes, *Tbx18* and *Tbx20*, are unilaterally expressed in the left and right ventricles, respectively (Franco et al., 2006; Takeuchi et al., 2003). The boundary between *Tbx5*-positive, *Tbx20*-negative, and *Tbx5*-negative, *Tbx20*-positive myocardium appears to demarcate the location of ventricular septation and shifts in the expression levels can result in ventricular septum abnormalities; however, the exact mechanism by which the interface between *Tbx5* and the other T-box family genes instructs formation of the ventricular septum remains unclear (Koshiba-Takeuchi et al., 2009; Takeuchi et al., 2003).

5.2 Atrial Septum

Haploinsufficiency of *TBX5* in Holt–Oram patients results in atrial septal defects in approximately half of cases (Bruneau et al., 1999), and *Tbx5* haploinsufficient mice exhibit atrial septal defects approximately 40% of the time (Bruneau et al., 2001). The atrial septum is derived from the second heart field contributing to the inflow tract of the heart (Goddeeris et al., 2008; Hoffmann, Peterson, Friedland-Little, Anderson, & Moskowitz, 2009; Mommersteeg et al., 2006; Snarr, Wirrig, Phelps, Trusk, & Wessels, 2007; Wessels et al., 2000). *Sonic hedgehog* (*Shh*), secreted from the pulmonary endoderm, signals through GLI-dependent transcription factors and is essential for atrial septation (Goddeeris et al., 2008; Hoffmann et al., 2009, 2014; Xie et al., 2012). *Gli1* genetically interacts with *Tbx5* in the second heart field to coactivate downstream targets, including *Osr1* and *Foxf1* (Goddeeris et al., 2008; Hoffmann et al., 2009, 2014; Xie et al., 2012). Furthermore, deletion of one or both copies of *Tbx5* from

Shh-receiving cells results in primum-type atrial septal defects, which can be rescued through constitutive activation of hedgehog signaling (Xie et al., 2012). This supports a model in which *Tbx5* acts upstream of active *Shh* signaling in the second heart field and that both *Tbx5* and activating GLI factors coregulate transcription at the top of a hierarchy of atrial septation genes (Hoffmann et al., 2014; Xie et al., 2012).

6. *TBX5* IN CARDIAC CONDUCTION SYSTEM DEVELOPMENT

The cardiac conduction system is a highly specialized network of cardiomyocytes within the heart that generate and transmit electrical impulses throughout the heart to coordinate contraction. A majority of Holt–Oram syndrome patients present with conduction system abnormalities (Basson et al., 1994; Holt & Oram, 1960; Newbury-Ecob et al., 1996). Evidence suggests that *Tbx5* plays three key roles in the cardiac conduction system: specification of the conduction system during development, regulation of the conduction system transcriptome, and maintenance of conduction system identity in the adult (Arnolds et al., 2012; Moskowitz et al., 2007, 2004). *Tbx5* and *Nkx2-5* genetically interact to specify the ventricular cardiac conduction system in a ventricular subdomain with the highest expression of both factors (Moskowitz et al., 2004; Thomas et al., 2001). Furthermore, *Tbx5* and *Nkx2-5* are required to coregulate the transcriptional repressor *Id2*, which is required for proper formation and function of the conduction system (Moskowitz et al., 2007). Although the *Tbx5*-dependent transcriptome in the conduction system has not been well characterized to date, *Tbx5* is required for the regulation of critical conduction system ion channels, *Gja5* and *Scn5a*, and removal of *Tbx5* from the adult ventricular conduction system results in loss of these ion channels and altered ventricular conduction system function (Arnolds et al., 2012; Bruneau et al., 2001; Moskowitz et al., 2004; van den Boogaard et al., 2014).

7. HOMOZYGOUS *TBX5* MUTATIONS REVEAL NOVEL ROLES OF TBX5

While much attention has been focused on understanding *Tbx5* haploinsufficiency as it relates to human disease, homozygous deletion of *Tbx5* in animal models reveals novel requirements for *Tbx5* not uncovered by haploinsufficiency.

7.1 Complete Loss of Mammalian *Tbx5*

As *Tbx5* haploinsufficiency is associated with the phenotype of Holt–Oram syndrome in mice and *Tbx5* null embryos die in utero around E10.5, most studies examining the role of *Tbx5* in cardiac morphogenesis and transcription have focused on the *Tbx5* null heterozygote. In contrast, the null state provides the opportunity to understand critical roles of *Tbx5* not observed in heterozygotes (Agarwal et al., 2003; Bruneau et al., 2001; Hoffmann et al., 2014; Luna-Zurita et al., 2016; Mori et al., 2006; Moskowitz et al., 2004; Rallis et al., 2003; Xie et al., 2012). In the mouse, germline deletion of both copies of *Tbx5*, $Tbx5^{tm1.1Jse/tm1.1Jse}$, results in embryonic lethality by E10.5 (Bruneau et al., 2001). These animals exhibit a grossly abnormal, linear heart tube (Bruneau et al., 2001) and complete absence of the forelimb buds (Agarwal et al., 2003; Bruneau et al., 2001). The complete absence of the forelimb buds indicates a role for *Tbx5* in limb bud initiation, shown to be downstream of fibroblast growth factor signaling (Agarwal et al., 2003; Hasson, Del Buono, & Logan, 2007; Rallis et al., 2003). Similar to the germline homozygous null embryos, homozygous hypomorphs for *Tbx5* show embryonic lethality prior to E11.5, with hypoplastic left ventricles and sinoatrial structures (Mori et al., 2006). Distinctly, unlike the *Tbx5* homozygous null phenotype, homozygous hypomorphs still undergo heart looping and rudimentary formation of the left and right atrial chambers (Mori et al., 2006). These observations indicate important roles for *Tbx5* at sequential stages of cardiac development, although the distinctions between these roles have yet to be elucidated.

7.2 Zebrafish *tbx5a/tbx5b*

The *heartstrings* mutation is the first published mutation of *tbx5a* in zebrafish and was found as part of a screen for recessive lethal mutations affecting cardiac function (Garrity, Childs, & Fishman, 2002). The *tbx5a/heartstrings* mutants or morpholino knockdown of *tbx5a* recapitulate some aspects of Holt–Oram syndrome including forelimb defects and conduction defects (Ahn, Kourakis, Rohde, Silver, & Ho, 2002; Garrity et al., 2002). The hearts of these animals appear to develop normally during early cardiac development, only later displaying defects starting with the failure of heart looping and subsequent deterioration of chamber myocardium and heart failure (Ahn et al., 2002; Garrity et al., 2002).

More recently, a second copy of *tbx5* (*tbx5b*) was found in the genome of zebrafish (Albalat, Baquero, & Minguillon, 2010). While *tbx5a* is expressed in the eye, heart, and forelimb during early development, *tbx5b* is only expressed robustly in the eye and heart (Albalat et al., 2010), suggesting possible redundant functions for the two *tbx5* genes during early heart and eye development. Using morpholinos, *tbx5a*, *tbx5b*, or *tbx5a/tbx5b* double knockdowns all result in the heartstrings phenotype, i.e., normal heart tube formation, bradycardia, and progressive deterioration/heart failure (Garrity et al., 2002; Parrie, Renfrew, Wal, Mueller, & Garrity, 2013). Different downstream targets have been identified for *tbx5a* and *tbx5b* (Parrie et al., 2013). While *tbx5b* knockdown does not result in patterning defects of chamber formation or sinus venosus seen in either the *tbx5a* mutants or knockdown experiments, *tbx5b* knockdown results in abnormal expansion of two morphogenesis markers, *hand2* and *vcana*, similar to *tbx5a* (Garrity et al., 2002; Parrie et al., 2013). However, known direct targets of mammalian *TBX5* or zebrafish *tbx5a*, such as *bmp4*, *nppa*, *tbx2b*, and *hey2* (Bruneau et al., 2001; Camarata et al., 2010; Chi et al., 2008; Plageman & Yutzey, 2004; Puskaric et al., 2010), were not disrupted with *tbx5b* knockdown, and neither *tbx5a* overexpression nor *tbx5b* overexpression rescues the reciprocal knockdown, suggesting the role of *tbx5b* is different than that of *tbx5a* (Parrie et al., 2013). The apparent differences in *tbx5a* and *tbx5b* function suggest that following gene duplication there may have been evolutionarily beneficial subfunctionalization of the two copies.

8. THE TBX5 GENE REGULATORY NETWORK

As a T-box transcription factor, the primary role of TBX5 is thought to be the regulation of target gene transcription. Historically, TBX5 has thought to act a positive regulator of transcriptional activity; however, recent evidence suggests that TBX5 may have a role in both transcriptional activation and repression (Fig. 2). In this section, we will explore what is known about both roles.

8.1 Positive Transcriptional Activation by TBX5 and Cofactors: Cardiomyocyte Specific Factors, Chromatin Modification, and Maturation

TBX5 has long been known to act as a positive regulator of transcription in heart development and cardiomyocyte maturation (Bruneau et al., 2001; Goetz, Brown, & Conlon, 2006; Hiroi et al., 2001; Moskowitz et al., 2004). Some of the earliest identified direct targets of *TBX5* were *NPPA* (encoding ANF) and *GJA5* (encoding cx40), both of which are expressed

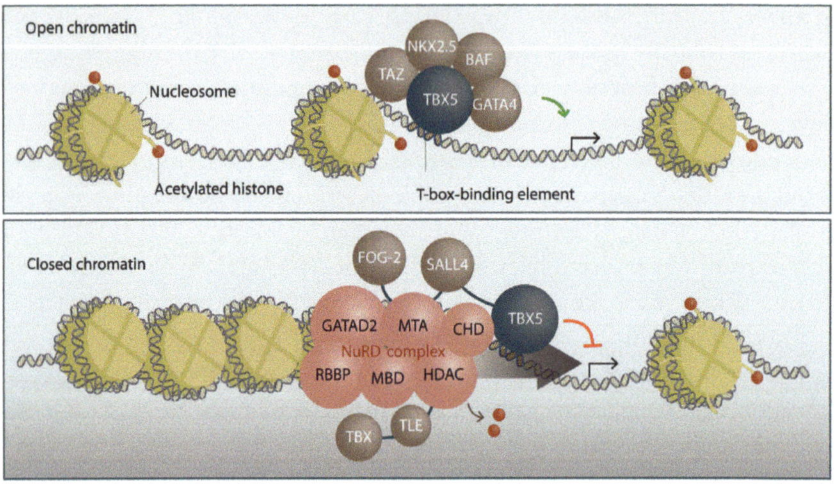

Fig. 2 Transcriptional regulation by TBX5. TBX5, through its interactions with other cardiac transcription factors, such as GATA4 and NKX2-5, and the BAF chromatin-remodeling complex drive active transcription of target cardiac genes in regions of open chromatin (*top panel*). TBX5, through its interactions with the NuRD complex and other transcriptional repressors, such as SALL4, remodel chromatin to a closed state, which represses gene expression of noncardiac genes (*bottom panel*). *Reprinted from Boogerd, C. J., & Evans, S. M. (2016). TBX5 and NuRD divide the heart.* Developmental Cell, *36(3), 242–244, with permission from Elsevier.*

in differentiating cardiomyocytes during development and are markers of cardiac chamber differentiation (Bruneau et al., 2001; Christoffels et al., 2000; Delorme et al., 1997; Hiroi et al., 2001). *Gja5* and *Nppa* are both highly sensitive markers of TBX5 activity with a nearly complete loss of *Gja5* in $Tbx5^{tm1.1Jse/+}$ embryos and a graded response of *Nppa* across a *Tbx5* allelic series (Bruneau et al., 2001; Mori et al., 2006).

The first identified interaction partner of TBX5 was the *tinman* transcription factor NKX2-5 (Hiroi et al., 2001). Identified by a classic yeast two-hybrid screen, TBX5 and NKX2-5 interact through the highly conserved C-terminus of NKX2-5, relying on four key amino acids in an α-helix, P139, D140, R150, and Q151 (Bruneau et al., 2001; Hiroi et al., 2001; Luna-Zurita et al., 2016). The interaction between TBX5 and NKX2-5 allows the proteins to synergistically activate targets such as *Nppa* through tandem transcription factor binding motifs (Bruneau et al., 2001; Hiroi et al., 2001). This interaction of TBX5 and NKX2-5 at tandem binding motifs induces bending in the DNA for transcriptional activity, a molecular mechanism by which synergistic activities of TBX5 and NKX2-5 interactions are determined at specific *cis*-regulatory elements (Luna-Zurita et al., 2016). Furthermore, the interaction between TBX5 and NKX2-5 is required to maintain fidelity in transcription factor binding throughout the genome, and the absence of either factor allows for inappropriate localization and activation of noncardiac genes by the other (Luna-Zurita et al., 2016). To this point, several sites within TBX5 are important for the synergistic activation of *Nppa*, which can be abrogated by HOS mutation in the α-helix mentioned earlier as well as G80R and R237W (Boogerd et al., 2010; Garg et al., 2003; Hiroi et al., 2001).

Besides NKX2-5, TBX5 also shows direct interaction with other major cardiac transcription factors, including GATA4 (Garg et al., 2003; Maitra et al., 2009), GATA6 (Maitra et al., 2009), TBX20 (Brown et al., 2005), MEF2C (Ghosh et al., 2009), and Myocardin (Wang, Cao, Wang, & Wang, 2011). The interaction between TBX5 and GATA4 was diminished by *GATA4* mutations causing nonsyndromic congenital heart defects (Garg et al., 2003), as well as by *TBX5* Holt–Oram mutations causing heart defects but not those causing only limb defects (Boogerd et al., 2010; Garg et al., 2003; Luna-Zurita et al., 2016). Both GATA4-TBX5 and MEF2C-TBX5 interactions are required for synergistic activation of α-cardiac myosin heavy chain encoded by *MYH6* (Ghosh et al., 2009; Maitra et al., 2009), though GATA6-TBX5 protein interactions are not, suggesting that TBX5 interaction partners generate tissue- and context-specific gene expression (Maitra et al., 2009).

TBX5 as a transcription factor appears to act as part of a multifactor transcriptional complex for the activation and maintenance of cardiac lineage genes; however, the ability of transcription factors to regulate gene targets requires the ability of the factors to bind open chromatin. Addition of *Gata4*, *Mef2c*, and *Tbx5* to fibroblasts is sufficient to drive reprogramming toward a cardiomyocyte fate (Ieda et al., 2010; Qian et al., 2012). This ability to reprogram cells suggests that this core set of transcription factors may drive chromatin accessibility. In support of this supposition, it has been shown that TBX5 interacts with Baf60c and Brg1, encoded by *Smarcd3* and *Smarca4*, respectively, members of the SWI/SNF family of proteins involved in chromatin remodeling to drive mesodermal cells to cardiomyocyte fate in vitro and in vivo (Lickert et al., 2004; Takeuchi & Bruneau, 2009; Takeuchi et al., 2011). Furthermore, in *Tbx5* haploinsufficient mice, there is a loss of chromatin remodeling complexes at the promoters of *Tbx5*-dependent cardiac genes (Takeuchi et al., 2011). For another T-box family member, T-bet, encoded by *Tbx21*, it has been previously shown that T-bet, the Brg1 chromatin-remodeling complex, and H3K27 demethylases physically interact (Miller, Mohn, & Weinmann, 2010), suggesting potential shared mechanisms with *Tbx5* in chromatin transitions seen in cardiac development (Wamstad et al., 2012).

8.2 TBX5-Mediated Repression: Inhibition of Noncardiomyocyte Fate Through Chromatin Remodeling

In addition to its positive role driving cardiac gene regulatory networks, recent genomic studies indicate that TBX5 acts as a direct transcriptional repressor during cardiac development where it is required for inhibition of inappropriate gene expression (Lewandowski et al., 2014; Waldron et al., 2016). While transcriptional repression by T-box factors in cardiac development and function has been well documented in the cases of TBX2 (Carreira et al., 1998; Christoffels et al., 2004) and TBX3 (He et al., 1999; Hoogaars et al., 2007, 2004; Lingbeek, Jacobs, & van Lohuizen, 2002), and it has been shown that TBX20 has roles in both transcriptional repression and activation (Kaltenbrun et al., 2013; Sakabe et al., 2012; Stennard et al., 2005), only recently has a repressive role for TBX5 been elucidated. Waldron et al. (2016) demonstrated that *Tbx5* inhibits noncardiac gene regulatory programs, including neuronal networks during early cardiac development. Through biochemical and genetic interaction studies, it was shown that TBX5 protein interacts with the nucleosome remodeling and deacetylase (NuRD) complex during embryonic development (Waldron et al., 2016). Similar to TBX20, the TBX5–NuRD interaction complex acts as an inhibitory mechanism by which TBX5 is able to repress noncardiogenic gene expression in the heart (Kaltenbrun et al., 2013; Waldron et al., 2016). TBX5 physically interacts with the NuRD complex through an evolutionarily conserved α-helix domain located from amino acids 255–264 and disruption of this domain can result in Holt–Oram syndrome (e.g., S261C; Brassington et al., 2003; Waldron et al., 2016). Using cardiomyocytes derived from murine embryonic stem cell differentiation, Luna-Zurita et al. (2016) demonstrated that TBX5 imparts specificity in cardiac transcription factor complexes by preventing off-target binding of other cardiac transcription factors. These findings suggest that the dual roles of T-box factors (i.e., TBX5 and TBX20 repression of noncardiomyocyte fate) may be a more common theme in cardiac development than previously thought.

8.3 Direct Targets of TBX5 Regulation

As a member of the T-box family of transcription factors, TBX5 regulates transcription through direct interaction with DNA. In recent years, multiple groups have turned to chromatin immunoprecipitation with sequencing (ChIP-sequencing) to identify direct targets of TBX5 binding and regulation (He et al., 2011; Luna-Zurita et al., 2016). The first ChIP-seq dataset was generated in the HL-1 atrial cardiomyocyte cell line using an overexpression construct of biotinylated TBX5, which identified over 56 k binding sites within the genome (He et al., 2011). The second set of data was generated using ChIP-exo technology in the context of mouse ES cell differentiation in both cardiac progenitors and cardiomyocytes, resulting in approximately 5 and 9 k bindings sites, respectively (Luna-Zurita et al., 2016). While these datasets share many of the same locations, each also identifies many unique sites, suggesting that binding site information will need to be generated in each *Tbx5* expression context in order to understand the direct *Tbx5* transcriptome. For example, whereas approximately 60% and 40% sites identified in the cardiac progenitor and cardiomyocyte mouse ES cell datasets are shared, only 4% of sites identified in the HL-1 dataset are shared with the mouse ES cell datasets. It is currently unclear to what degree the HL-1 dataset overestimates and the mouse embryonic stem cell datasets underestimate the total number of relevant binding sites, or whether both datasets overestimate functional binding events, HL-1 to a greater extent. Inclusion of additional markers, such as open chromatin, histone marks, and known TBX5-binding partners may allow the broad utilization of current datasets for identification of truly functional TBX5-binding sites.

To date, functionally confirmed direct targets of *TBX5* are almost exclusively in genes implicated in cardiac proliferation, maturation, and function, including *Nppa*, *Gja5*, and *Scn5a* (Arnolds et al., 2012; Bruneau et al., 2001; Goetz et al., 2006; Hatcher et al., 2001; Hiroi et al., 2001; Mori et al., 2006; Moskowitz et al., 2007, 2004; Puskaric et al., 2010; Xie et al., 2012). Interestingly, the direct targets mediating the morphogenesis requirement for *TBX5* remain unknown. While some candidate target genes may act prior to morphological changes (Hoffmann et al., 2014; Xie et al., 2012), no direct mediators of morphology have been uncovered. *TBX5* may therefore indirectly regulate these processes. With advances in genome-wide technology, understanding the basis by which *TBX5* regulates morphological change will be key to understanding the role of *TBX5* in cardiac development.

9. CONCLUDING REMARKS

The requirement of *TBX5* for normal human cardiac structure was identified over 20 years ago (Basson et al., 1997; Li et al., 1997); however, surprisingly, the mechanistic role of *TBX5* in cardiac development remains unclear. Little is known about the essential downstream targets of TBX5-mediated transcription in the context of cardiac development. Similarly, the complex temporal and spatial gene expression of *TBX5* has been mapped throughout development and into adult life; however, the *cis*-regulatory architecture governing this expression is just beginning to be described. From a biochemical perspective, significant strides have been made in recent years to understand how TBX5 activates gene expression. However, the mechanisms by which TBX5 and its cofactors are targeted to specific loci, the temporal recruitment of TBX5 and its cofactors, the interplay between TBX5 and it cotranscriptional partners, and the mechanisms distinguishing active and repressive TBX5 activity are just recently coming into focus, and provide opportunities for exciting mechanistic studies. These areas of investigation will contribute to a broader understanding of the mechanisms underlying the requirement for TBX5 in cardiac morphogenesis and more generally the transcriptional control of metazoan development.

REFERENCES

Adachi, N., Robinson, M., Goolsbee, A., & Shubin, N. H. (2016). Regulatory evolution of Tbx5 and the origin of paired appendages. *Proceedings of the National Academy of Sciences of the United States of America, 113*(36), 10115–10120.

Agarwal, P., Wylie, J. N., Galceran, J., Arkhitko, O., Li, C., Deng, C., et al. (2003). *Tbx5* is essential for forelimb bud initiation following patterning of the limb field in the mouse embryo. *Development, 130*(3), 623–633.

Agulnik, S. I., Bollag, R. J., & Silver, L. M. (1995). Conservation of the T-box gene family from Mus musculus to Caenorhabditis elegans. *Genomics, 25*(1), 214–219.

Agulnik, S. I., Garvey, N., Hancock, S., Ruvinsky, I., Chapman, D. L., Agulnik, I., et al. (1996). Evolution of mouse T-box genes by tandem duplication and cluster dispersion. *Genetics, 144*(1), 249–254.

Ahn, D. G., Kourakis, M. J., Rohde, L. A., Silver, L. M., & Ho, R. K. (2002). T-box gene *tbx5* is essential for formation of the pectoral limb bud. *Nature, 417*(6890), 754–758.

Albalat, R., Baquero, M., & Minguillon, C. (2010). Identification and characterisation of the developmental expression pattern of *tbx5b*, a novel *tbx5* gene in zebrafish. *Gene Expression Patterns, 10*(1), 24–30.

Al-Qattan, M. M., & Abou Al-Shaar, H. (2015). A novel missense mutation in the *TBX5* gene in a Saudi infant with Holt-Oram syndrome. *Saudi Medical Journal, 36*(8), 980–982.

Anderson, R. H., Webb, S., Brown, N. A., Lamers, W., & Moorman, A. (2003). Development of the heart: (2) Septation of the atriums and ventricles. *Heart, 89*(8), 949–958.

Arnolds, D. E., Liu, F., Fahrenbach, J. P., Kim, G. H., Schillinger, K. J., Smemo, S., et al. (2012). TBX5 drives *Scn5a* expression to regulate cardiac conduction system function. *Journal of Clinical Investigation*, *122*(7), 2509–2518.

Barisic, I., Boban, L., Greenlees, R., Garne, E., Wellesley, D., Calzolari, E., et al. (2014). Holt Oram syndrome: A registry-based study in Europe. *Orphanet Journal of Rare Diseases*, *9*, 156.

Basson, C. T., Bachinsky, D. R., Lin, R. C., Levi, T., Elkins, J. A., Soults, J., et al. (1997). Mutations in human *TBX5* [corrected] cause limb and cardiac malformation in Holt-Oram syndrome. *Nature Genetics*, *15*(1), 30–35.

Basson, C. T., Cowley, G. S., Solomon, S. D., Weissman, B., Poznanski, A. K., Traill, T. A., et al. (1994). The clinical and genetic spectrum of the Holt-Oram syndrome (heart-hand syndrome). *New England Journal of Medicine*, *330*(13), 885–891.

Basson, C. T., Huang, T., Lin, R. C., Bachinsky, D. R., Weremowicz, S., Vaglio, A., et al. (1999). Different *TBX5* interactions in heart and limb defined by Holt-Oram syndrome mutations. *Proceedings of the National Academy of Sciences of the United States of America*, *96*(6), 2919–2924.

Bollag, R. J., Siegfried, Z., Cebra-Thomas, J. A., Garvey, N., Davison, E. M., & Silver, L. M. (1994). An ancient family of embryonically expressed mouse genes sharing a conserved protein motif with the T locus. *Nature Genetics*, *7*(3), 383–389.

Boogerd, C. J., Dooijes, D., Ilgun, A., Mathijssen, I. B., Hordijk, R., van de Laar, I. M., et al. (2010). Functional analysis of novel *TBX5* T-box mutations associated with Holt-Oram syndrome. *Cardiovascular Research*, *88*(1), 130–139.

Borozdin, W., Bravo Ferrer Acosta, A. M., Bamshad, M. J., Botzenhart, E. M., Froster, U. G., Lemke, J., et al. (2006). Expanding the spectrum of *TBX5* mutations in Holt-Oram syndrome: Detection of two intragenic deletions by quantitative real time PCR, and report of eight novel point mutations. *Human Mutation*, *27*(9), 975–976.

Brassington, A. M., Sung, S. S., Toydemir, R. M., Le, T., Roeder, A. D., Rutherford, A. E., et al. (2003). Expressivity of Holt-Oram syndrome is not predicted by TBX5 genotype. *American Journal of Human Genetics*, *73*(1), 74–85.

Brown, D. D., Martz, S. N., Binder, O., Goetz, S. C., Price, B. M., Smith, J. C., et al. (2005). *Tbx5* and *Tbx20* act synergistically to control vertebrate heart morphogenesis. *Development*, *132*(3), 553–563.

Bruneau, B. G., Logan, M., Davis, N., Levi, T., Tabin, C. J., Seidman, J. G., et al. (1999). Chamber-specific cardiac expression of *Tbx5* and heart defects in Holt-Oram syndrome. *Developmental Biology*, *211*(1), 100–108.

Bruneau, B. G., Nemer, G., Schmitt, J. P., Charron, F., Robitaille, L., Caron, S., et al. (2001). A murine model of Holt-Oram syndrome defines roles of the T-box transcription factor *Tbx5* in cardiogenesis and disease. *Cell*, *106*(6), 709–721.

Camarata, T., Bimber, B., Kulisz, A., Chew, T. L., Yeung, J., & Simon, H. G. (2006). LMP4 regulates Tbx5 protein subcellular localization and activity. *The Journal of Cell Biology*, *174*(3), 339–348.

Camarata, T., Krcmery, J., Snyder, D., Park, S., Topczewski, J., & Simon, H. G. (2010). Pdlim7 (LMP4) regulation of Tbx5 specifies zebrafish heart atrio-ventricular boundary and valve formation. *Developmental Biology*, *337*(2), 233–245.

Carreira, S., Dexter, T. J., Yavuzer, U., Easty, D. J., & Goding, C. R. (1998). Brachyury-related transcription factor *Tbx2* and repression of the melanocyte-specific TRP-1 promoter. *Molecular and Cellular Biology*, *18*(9), 5099–5108.

Chapman, D. L., Garvey, N., Hancock, S., Alexiou, M., Agulnik, S. I., Gibson-Brown, J. J., et al. (1996). Expression of the T-box family genes, *Tbx1-Tbx5*, during early mouse development. *Developmental Dynamics*, *206*(4), 379–390.

Chi, N. C., Shaw, R. M., De Val, S., Kang, G., Jan, L. Y., Black, B. L., et al. (2008). Foxn4 directly regulates *tbx2b* expression and atrioventricular canal formation. *Genes & Development*, *22*(6), 734–739.

Christoffels, V. M., Habets, P. E., Franco, D., Campione, M., de Jong, F., Lamers, W. H., et al. (2000). Chamber formation and morphogenesis in the developing mammalian heart. *Developmental Biology*, *223*(2), 266–278.

Christoffels, V. M., Hoogaars, W. M., Tessari, A., Clout, D. E., Moorman, A. F., & Campione, M. (2004). T-box transcription factor Tbx2 represses differentiation and formation of the cardiac chambers. *Developmental Dynamics*, *229*(4), 763–770.

Collavoli, A., Hatcher, C. J., He, J., Okin, D., Deo, R., & Basson, C. T. (2003). TBX5 nuclear localization is mediated by dual cooperative intramolecular signals. *Journal of Molecular and Cellular Cardiology*, *35*(10), 1191–1195.

Conlon, F. L., Fairclough, L., Price, B. M., Casey, E. S., & Smith, J. C. (2001). Determinants of T box protein specificity. *Development*, *128*(19), 3749–3758.

Cross, S. J., Ching, Y. H., Li, Q. Y., Armstrong-Buisseret, L., Spranger, S., Lyonnet, S., et al. (2000). The mutation spectrum in Holt-Oram syndrome. *Journal of Medical Genetics*, *37*(10), 785–787.

Debeer, P., Race, V., Gewillig, M., Devriendt, K., & Frijns, J. P. (2007). Novel *TBX5* mutations in patients with Holt-Oram syndrome. *Clinical Orthopaedics and Related Research*, *462*, 20–26.

Delorme, B., Dahl, E., Jarry-Guichard, T., Briand, J. P., Willecke, K., Gros, D., et al. (1997). Expression pattern of connexin gene products at the early developmental stages of the mouse cardiovascular system. *Circulation Research*, *81*(3), 423–437.

Devine, W. P., Wythe, J. D., George, M., Koshiba-Takeuchi, K., & Bruneau, B. G. (2014). Early patterning and specification of cardiac progenitors in gastrulating mesoderm. *eLife*, *3*, 1–23.

Ekure, E. N., Okoromah, C. N., Briggs, E., & Ajenifuja, O. A. (2004). Holt-Oram syndrome with hypoplastic left heart syndrome in an African child. *The Nigerian Postgraduate Medical Journal*, *11*(3), 190–192.

Elek, C., Vitez, M., & Czeizel, E. (1991). Holt-Oram syndrome. *Orvosi Hetilap*, *132*(2), 73–74. 77–78.

Fan, C., Duhagon, M. A., Oberti, C., Chen, S., Hiroi, Y., Komuro, I., et al. (2003). Novel TBX5 mutations and molecular mechanism for Holt-Oram syndrome. *Journal of Medical Genetics*, *40*(3), e29.

Franco, D., Meilhac, S. M., Christoffels, V. M., Kispert, A., Buckingham, M., & Kelly, R. G. (2006). Left and right ventricular contributions to the formation of the interventricular septum in the mouse heart. *Developmental Biology*, *294*(2), 366–375.

Garg, V., Kathiriya, I. S., Barnes, R., Schluterman, M. K., King, I. N., Butler, C. A., et al. (2003). GATA4 mutations cause human congenital heart defects and reveal an interaction with TBX5. *Nature*, *424*(6947), 443–447.

Garrity, D. M., Childs, S., & Fishman, M. C. (2002). The heartstrings mutation in zebrafish causes heart/fin Tbx5 deficiency syndrome. *Development*, *129*(19), 4635–4645.

Georges, R., Nemer, G., Morin, M., Lefebvre, C., & Nemer, M. (2008). Distinct expression and function of alternatively spliced Tbx5 isoforms in cell growth and differentiation. *Molecular and Cellular Biology*, *28*(12), 4052–4067.

Ghosh, T. K., Packham, E. A., Bonser, A. J., Robinson, T. E., Cross, S. J., & Brook, J. D. (2001). Characterization of the TBX5 binding site and analysis of mutations that cause Holt-Oram syndrome. *Human Molecular Genetics*, *10*(18), 1983–1994.

Ghosh, T. K., Song, F. F., Packham, E. A., Buxton, S., Robinson, T. E., Ronksley, J., et al. (2009). Physical interaction between TBX5 and MEF2C is required for early heart development. *Molecular and Cellular Biology*, *29*(8), 2205–2218.

Gibson-Brown, J. J., Agulnik, S. I., Chapman, D. L., Alexiou, M., Garvey, N., Silver, L. M., et al. (1996). Evidence of a role for T-box genes in the evolution of limb morphogenesis and the specification of forelimb/hindlimb identity. *Mechanisms of Development*, *56*(1–2), 93–101.

Gibson-Brown, J. J., Agulnik, S. I., Silver, L. M., & Papaioannou, V. E. (1998). Expression of T-box genes Tbx2-Tbx5 during chick organogenesis. *Mechanisms of Development, 74*(1–2), 165–169.

Goddeeris, M. M., Rho, S., Petiet, A., Davenport, C. L., Johnson, G. A., Meyers, E. N., et al. (2008). Intracardiac septation requires hedgehog-dependent cellular contributions from outside the heart. *Development, 135*(10), 1887–1895.

Goetz, S. C., Brown, D. D., & Conlon, F. L. (2006). TBX5 is required for embryonic cardiac cell cycle progression. *Development, 133*(13), 2575–2584.

Gruenauer-Kloevekorn, C., & Froster, U. G. (2003). Holt-Oram syndrome: A new mutation in the TBX5 gene in two unrelated families. *Annales de Génétique, 46*(1), 19–23.

GTEx Consortium. (2013). The Genotype-Tissue Expression (GTEx) project. *Nature Genetics, 45*(6), 580–585.

Hasson, P., Del Buono, J., & Logan, M. P. (2007). Tbx5 is dispensable for forelimb outgrowth. *Development, 134*(1), 85–92.

Hatcher, C. J., Goldstein, M. M., Mah, C. S., Delia, C. S., & Basson, C. T. (2000). Identification and localization of TBX5 transcription factor during human cardiac morphogenesis. *Developmental Dynamics, 219*(1), 90–95.

Hatcher, C. J., Kim, M. S., Mah, C. S., Goldstein, M. M., Wong, B., Mikawa, T., et al. (2001). TBX5 transcription factor regulates cell proliferation during cardiogenesis. *Developmental Biology, 230*(2), 177–188.

He, A., Kong, S. W., Ma, Q., & Pu, W. T. (2011). Co-occupancy by multiple cardiac transcription factors identifies transcriptional enhancers active in heart. *Proceedings of the National Academy of Sciences of the United States of America, 108*(14), 5632–5637.

He, M., Wen, L., Campbell, C. E., Wu, J. Y., & Rao, Y. (1999). Transcription repression by Xenopus ET and its human ortholog TBX3, a gene involved in ulnar-mammary syndrome. *Proceedings of the National Academy of Sciences of the United States of America, 96*(18), 10212–10217.

Heinritz, W., Moschik, A., Kujat, A., Spranger, S., Heilbronner, H., Demuth, S., et al. (2005). Identification of new mutations in the TBX5 gene in patients with Holt-Oram syndrome. *Heart, 91*(3), 383–384.

Herrmann, B. G., Labeit, S., Poustka, A., King, T. R., & Lehrach, H. (1990). Cloning of the T gene required in mesoderm formation in the mouse. *Nature, 343*(6259), 617–622.

Hiroi, Y., Kudoh, S., Monzen, K., Ikeda, Y., Yazaki, Y., Nagai, R., et al. (2001). Tbx5 associates with Nkx2-5 and synergistically promotes cardiomyocyte differentiation. *Nature Genetics, 28*(3), 276–280.

Hoffmann, A. D., Peterson, M. A., Friedland-Little, J. M., Anderson, S. A., & Moskowitz, I. P. (2009). Sonic hedgehog is required in pulmonary endoderm for atrial septation. *Development, 136*(10), 1761–1770.

Hoffmann, A. D., Yang, X. H., Burnicka-Turek, O., Bosman, J. D., Ren, X., Steimle, J. D., et al. (2014). Foxf genes integrate tbx5 and hedgehog pathways in the second heart field for cardiac septation. *PLoS Genetics, 10*(10), e1004604.

Holt, M., & Oram, S. (1960). Familial heart disease with skeletal malformations. *British Heart Journal, 22*, 236–242.

Hoogaars, W. M., Engel, A., Brons, J. F., Verkerk, A. O., de Lange, F. J., Wong, L. Y., et al. (2007). Tbx3 controls the sinoatrial node gene program and imposes pacemaker function on the atria. *Genes & Development, 21*(9), 1098–1112.

Hoogaars, W. M., Tessari, A., Moorman, A. F., de Boer, P. A., Hagoort, J., Soufan, A. T., et al. (2004). The transcriptional repressor Tbx3 delineates the developing central conduction system of the heart. *Cardiovascular Research, 62*(3), 489–499.

Horb, M. E., & Thomsen, G. H. (1999). Tbx5 is essential for heart development. *Development, 126*(8), 1739–1751.

Ieda, M., Fu, J. D., Delgado-Olguin, P., Vedantham, V., Hayashi, Y., Bruneau, B. G., et al. (2010). Direct reprogramming of fibroblasts into functional cardiomyocytes by defined factors. *Cell, 142*(3), 375–386.

Isphording, D., Leylek, A. M., Yeung, J., Mischel, A., & Simon, H. G. (2004). T-box genes and congenital heart/limb malformations. *Clinical Genetics, 66*(4), 253–264.

Jin, F., Li, Y., Dixon, J. R., Selvaraj, S., Ye, Z., Lee, A. Y., et al. (2013). A high-resolution map of the three-dimensional chromatin interactome in human cells. *Nature, 503*(7475), 290–294.

Jolma, A., Yan, J., Whitington, T., Toivonen, J., Nitta, K. R., Rastas, P., et al. (2013). DNA-binding specificities of human transcription factors. *Cell, 152*(1–2), 327–339.

Kaltenbrun, E., Greco, T. M., Slagle, C. E., Kennedy, L. M., Li, T., Cristea, I. M., et al. (2013). A Gro/TLE-NuRD corepressor complex facilitates Tbx20-dependent transcriptional repression. *Journal of Proteome Research, 12*(12), 5395–5409.

Kimura, M., Kikuchi, A., Ichinoi, N., & Kure, S. (2015). Novel TBX5 duplication in a Japanese family with Holt-Oram syndrome. *Pediatric Cardiology, 36*(1), 244–247.

Kispert, A., Koschorz, B., & Herrmann, B. G. (1995). The T protein encoded by Brachyury is a tissue-specific transcription factor. *EMBO Journal, 14*(19), 4763–4772.

Koshiba-Takeuchi, K., Mori, A. D., Kaynak, B. L., Cebra-Thomas, J., Sukonnik, T., Georges, R. O., et al. (2009). Reptilian heart development and the molecular basis of cardiac chamber evolution. *Nature, 461*(7260), 95–98.

Krause, A., Zacharias, W., Camarata, T., Linkhart, B., Law, E., Lischke, A., et al. (2004). *Tbx5* and *Tbx4* transcription factors interact with a new chicken PDZ-LIM protein in limb and heart development. *Developmental Biology, 273*(1), 106–120.

Kulisz, A., & Simon, H. G. (2008). An evolutionarily conserved nuclear export signal facilitates cytoplasmic localization of the Tbx5 transcription factor. *Molecular and Cellular Biology, 28*(5), 1553–1564.

Lewandowski, S. L., Janardhan, H. P., Smee, K. M., Bachman, M., Sun, Z., Lazar, M. A., et al. (2014). Histone deacetylase 3 modulates Tbx5 activity to regulate early cardiogenesis. *Human Molecular Genetics, 23*(14), 3801–3809.

Li, Q. Y., Newbury-Ecob, R. A., Terrett, J. A., Wilson, D. I., Curtis, A. R., Yi, C. H., et al. (1997). Holt-Oram syndrome is caused by mutations in TBX5, a member of the Brachyury (T) gene family. *Nature Genetics, 15*(1), 21–29.

Liberatore, C. M., Searcy-Schrick, R. D., & Yutzey, K. E. (2000). Ventricular expression of tbx5 inhibits normal heart chamber development. *Developmental Biology, 223*(1), 169–180.

Lickert, H., Takeuchi, J. K., Von Both, I., Walls, J. R., McAuliffe, F., Adamson, S. L., et al. (2004). Baf60c is essential for function of BAF chromatin remodelling complexes in heart development. *Nature, 432*(7013), 107–112.

Lingbeek, M. E., Jacobs, J. J., & van Lohuizen, M. (2002). The T-box repressors *TBX2* and *TBX3* specifically regulate the tumor suppressor gene p14ARF via a variant T-site in the initiator. *Journal of Biological Chemistry, 277*(29), 26120–26127.

Luna-Zurita, L., Stirnimann, C. U., Glatt, S., Kaynak, B. L., Thomas, S., Baudin, F., et al. (2016). Complex interdependence regulates heterotypic transcription factor distribution and coordinates cardiogenesis. *Cell, 164*(5), 999–1014.

Maitra, M., Schluterman, M. K., Nichols, H. A., Richardson, J. A., Lo, C. W., Srivastava, D., et al. (2009). Interaction of *Gata4* and *Gata6* with *Tbx5* is critical for normal cardiac development. *Developmental Biology, 326*(2), 368–377.

Mathelier, A., Fornes, O., Arenillas, D. J., Chen, C. Y., Denay, G., Lee, J., et al. (2016). JASPAR 2016: A major expansion and update of the open-access database of transcription factor binding profiles. *Nucleic Acids Research, 44*(D1), D110–D115.

McDermott, D. A., Bressan, M. C., He, J., Lee, J. S., Aftimos, S., Brueckner, M., et al. (2005). TBX5 genetic testing validates strict clinical criteria for Holt-Oram syndrome. *Pediatric Research, 58*(5), 981–986.

McDermott, D. A., Fong, J. C., & Basson, C. T. (1993). Holt-Oram syndrome. In R. A. Pagon, M. P. Adam, H. H. Ardinger, S. E. Wallace, A. Amemiya, L. J. H. Bean, et al. (Eds.), *GeneReviews(R)*. Seattle, WA: University of Seattle.

Mele, M., Ferreira, P. G., Reverter, F., DeLuca, D. S., Monlong, J., Sammeth, M., et al. (2015). Human genomics. The human transcriptome across tissues and individuals. *Science*, *348*(6235), 660–665.

Miller, S. A., Mohn, S. E., & Weinmann, A. S. (2010). Jmjd3 and UTX play a demethylase-independent role in chromatin remodeling to regulate T-box family member-dependent gene expression. *Molecular Cell*, *40*(4), 594–605.

Minguillon, C., Nishimoto, S., Wood, S., Vendrell, E., Gibson-Brown, J. J., & Logan, M. P. (2012). Hox genes regulate the onset of *Tbx5* expression in the forelimb. *Development*, *139*(17), 3180–3188.

Mommersteeg, M. T., Soufan, A. T., de Lange, F. J., van den Hoff, M. J., Anderson, R. H., Christoffels, V. M., et al. (2006). Two distinct pools of mesenchyme contribute to the development of the atrial septum. *Circulation Research*, *99*(4), 351–353.

Mori, A. D., Zhu, Y., Vahora, I., Nieman, B., Koshiba-Takeuchi, K., Davidson, L., et al. (2006). *Tbx5*-dependent rheostatic control of cardiac gene expression and morphogenesis. *Developmental Biology*, *297*(2), 566–586.

Moskowitz, I. P., Kim, J. B., Moore, M. L., Wolf, C. M., Peterson, M. A., Shendure, J., et al. (2007). A molecular pathway including *Id2*, *Tbx5*, and *Nkx2-5* required for cardiac conduction system development. *Cell*, *129*(7), 1365–1376.

Moskowitz, I. P., Pizard, A., Patel, V. V., Bruneau, B. G., Kim, J. B., Kupershmidt, S., et al. (2004). The T-Box transcription factor *Tbx5* is required for the patterning and maturation of the murine cardiac conduction system. *Development*, *131*(16), 4107–4116.

Muller, C. W., & Herrmann, B. G. (1997). Crystallographic structure of the T domain-DNA complex of the Brachyury transcription factor. *Nature*, *389*(6653), 884–888.

Najjar, H., Mardini, M., Tabbaa, R., & Nyhan, W. L. (1988). Variability of the Holt-Oram syndrome in Saudi individuals. *American Journal of Medical Genetics*, *29*(4), 851–855.

Newbury-Ecob, R. A., Leanage, R., Raeburn, J. A., & Young, I. D. (1996). Holt-Oram syndrome: A clinical genetic study. *Journal of Medical Genetics*, *33*(4), 300–307.

Papaioannou, V. E. (2014). The T-box gene family: Emerging roles in development, stem cells and cancer. *Development*, *141*(20), 3819–3833.

Papaioannou, V. E., & Silver, L. M. (1998). The T-box gene family. *Bioessays*, *20*(1), 9–19.

Parrie, L. E., Renfrew, E. M., Wal, A. V., Mueller, R. L., & Garrity, D. M. (2013). Zebrafish *tbx5* paralogs demonstrate independent essential requirements in cardiac and pectoral fin development. *Developmental Dynamics*, *242*(5), 485–502.

Patel, C., Silcock, L., McMullan, D., Brueton, L., & Cox, H. (2012). TBX5 intragenic duplication: A family with an atypical Holt-Oram syndrome phenotype. *European Journal of Human Genetics*, *20*(8), 863–869.

Pflugfelder, G. O., Roth, H., & Poeck, B. (1992). A homology domain shared between Drosophila optomotor-blind and mouse Brachyury is involved in DNA binding. *Biochemical and Biophysical Research Communications*, *186*(2), 918–925.

Plageman, T. F., Jr., & Yutzey, K. E. (2004). Differential expression and function of *Tbx5* and *Tbx20* in cardiac development. *Journal of Biological Chemistry*, *279*(18), 19026–19034.

Postma, A. V., van de Meerakker, J. B., Mathijssen, I. B., Barnett, P., Christoffels, V. M., Ilgun, A., et al. (2008). A gain-of-function *TBX5* mutation is associated with atypical Holt-Oram syndrome and paroxysmal atrial fibrillation. *Circulation Research*, *102*(11), 1433–1442.

Puskaric, S., Schmitteckert, S., Mori, A. D., Glaser, A., Schneider, K. U., Bruneau, B. G., et al. (2010). Shox2 mediates Tbx5 activity by regulating Bmp4 in the pacemaker region of the developing heart. *Human Molecular Genetics*, *19*(23), 4625–4633.

Qian, L., Huang, Y., Spencer, C. I., Foley, A., Vedantham, V., Liu, L., et al. (2012). In vivo reprogramming of murine cardiac fibroblasts into induced cardiomyocytes. *Nature, 485*(7400), 593–598.

Rallis, C., Bruneau, B. G., Del Buono, J., Seidman, C. E., Seidman, J. G., Nissim, S., et al. (2003). *Tbx5* is required for forelimb bud formation and continued outgrowth. *Development, 130*(12), 2741–2751.

Ruvinsky, I., & Silver, L. M. (1997). Newly identified paralogous groups on mouse chromosomes 5 and 11 reveal the age of a T-box cluster duplication. *Genomics, 40*(2), 262–266.

Ruvinsky, I., Silver, L. M., & Gibson-Brown, J. J. (2000). Phylogenetic analysis of T-Box genes demonstrates the importance of amphioxus for understanding evolution of the vertebrate genome. *Genetics, 156*(3), 1249–1257.

Sakabe, N. J., Aneas, I., Shen, T., Shokri, L., Park, S. Y., Bulyk, M. L., et al. (2012). Dual transcriptional activator and repressor roles of TBX20 regulate adult cardiac structure and function. *Human Molecular Genetics, 21*(10), 2194–2204.

Showell, C., Christine, K. S., Mandel, E. M., & Conlon, F. L. (2006). Developmental expression patterns of *Tbx1*, *Tbx2*, *Tbx5*, and *Tbx20* in Xenopus tropicalis. *Developmental Dynamics, 235*(6), 1623–1630.

Smemo, S., Campos, L. C., Moskowitz, I. P., Krieger, J. E., Pereira, A. C., & Nobrega, M. A. (2012). Regulatory variation in a *TBX5* enhancer leads to isolated congenital heart disease. *Human Molecular Genetics, 21*(14), 3255–3263.

Snarr, B. S., Wirrig, E. E., Phelps, A. L., Trusk, T. C., & Wessels, A. (2007). A spatiotemporal evaluation of the contribution of the dorsal mesenchymal protrusion to cardiac development. *Developmental Dynamics, 236*(5), 1287–1294.

Stennard, F. A., Costa, M. W., Lai, D., Biben, C., Furtado, M. B., Solloway, M. J., et al. (2005). Murine T-box transcription factor Tbx20 acts as a repressor during heart development, and is essential for adult heart integrity, function and adaptation. *Development, 132*(10), 2451–2462.

Stenson, P. D., Mort, M., Ball, E. V., Shaw, K., Phillips, A., & Cooper, D. N. (2014). The Human Gene Mutation Database: Building a comprehensive mutation repository for clinical and molecular genetics, diagnostic testing and personalized genomic medicine. *Human Genetics, 133*(1), 1–9.

Stirnimann, C. U., Ptchelkine, D., Grimm, C., & Muller, C. W. (2010). Structural basis of TBX5-DNA recognition: The T-box domain in its DNA-bound and -unbound form. *Journal of Molecular Biology, 400*(1), 71–81.

Takabatake, Y., Takabatake, T., & Takeshima, K. (2000). Conserved and divergent expression of T-box genes Tbx2-Tbx5 in Xenopus. *Mechanisms of Development, 91*(1–2), 433–437.

Takeuchi, J. K., & Bruneau, B. G. (2009). Directed transdifferentiation of mouse mesoderm to heart tissue by defined factors. *Nature, 459*(7247), 708–711.

Takeuchi, J. K., Lou, X., Alexander, J. M., Sugizaki, H., Delgado-Olguin, P., Holloway, A. K., et al. (2011). Chromatin remodelling complex dosage modulates transcription factor function in heart development. *Nature Communications, 2*, 187.

Takeuchi, J. K., Ohgi, M., Koshiba-Takeuchi, K., Shiratori, H., Sakaki, I., Ogura, K., et al. (2003). *Tbx5* specifies the left/right ventricles and ventricular septum position during cardiogenesis. *Development, 130*(24), 5953–5964.

Thomas, P. S., Kasahara, H., Edmonson, A. M., Izumo, S., Yacoub, M. H., Barton, P. J., et al. (2001). Elevated expression of Nkx-2.5 in developing myocardial conduction cells. *Anatomical Record, 263*(3), 307–313.

Torrado, M., Franco, D., Lozano-Velasco, E., Hernandez-Torres, F., Calvino, R., Aldama, G., et al. (2015). A microRNA-transcription factor blueprint for early atrial arrhythmogenic remodeling. *BioMed Research International, 2015*, 263151.

van den Boogaard, M., Smemo, S., Burnicka-Turek, O., Arnolds, D. E., van de Werken, H. J., Klous, P., et al. (2014). A common genetic variant within *SCN10A* modulates cardiac *SCN5A* expression. *Journal of Clinical Investigation, 124*(4), 1844–1852.

van Weerd, J. H., Badi, I., van den Boogaard, M., Stefanovic, S., van de Werken, H. J., Gomez-Velazquez, M., et al. (2014). A large permissive regulatory domain exclusively controls Tbx3 expression in the cardiac conduction system. *Circulation Research, 115*(4), 432–441.

Vianna, C. B., Miura, N., Pereira, A. C., & Jatene, M. B. (2011). Holt-Oram syndrome: Novel *TBX5* mutation and associated anomalous right coronary artery. *Cardiology in the Young, 21*(3), 351–353.

Waldron, L., Steimle, J. D., Greco, T. M., Gomez, N. C., Dorr, K. M., Kweon, J., et al. (2016). The cardiac TBX5 interactome reveals a chromatin remodeling network essential for cardiac septation. *Developmental Cell, 36*(3), 262–275.

Wamstad, J. A., Alexander, J. M., Truty, R. M., Shrikumar, A., Li, F., Eilertson, K. E., et al. (2012). Dynamic and coordinated epigenetic regulation of developmental transitions in the cardiac lineage. *Cell, 151*(1), 206–220.

Wang, C., Cao, D., Wang, Q., & Wang, D. Z. (2011). Synergistic activation of cardiac genes by myocardin and Tbx5. *PloS One, 6*(8), e24242.

Wang, F., Yang, X. Y., Zhao, J. Y., Yu, L. W., Zhang, P., Duan, W. Y., et al. (2014). miR-10a and miR-10b target the 3'-untranslated region of TBX5 to repress its expression. *Pediatric Cardiology, 35*(6), 1072–1079.

Wessels, A., Anderson, R. H., Markwald, R. R., Webb, S., Brown, N. A., Viragh, S., et al. (2000). Atrial development in the human heart: An immunohistochemical study with emphasis on the role of mesenchymal tissues. *Anatomical Record, 259*(3), 288–300.

Wilson, V., & Conlon, F. L. (2002). The T-box family. *Genome Biology, 3*(6), 1–7. REVIEWS3008.

Xie, L., Hoffmann, A. D., Burnicka-Turek, O., Friedland-Little, J. M., Zhang, K., & Moskowitz, I. P. (2012). *Tbx5*-hedgehog molecular networks are essential in the second heart field for atrial septation. *Developmental Cell, 23*(2), 280–291.

Yamak, A., Georges, R. O., Sheikh-Hassani, M., Morin, M., Komati, H., & Nemer, M. (2015). Novel exons in the *tbx5* gene locus generate protein isoforms with distinct expression domains and function. *Journal of Biological Chemistry, 290*(11), 6844–6856.

Yang, J., Hu, D., Xia, J., Yang, Y., Ying, B., Hu, J., et al. (2000). Three novel *TBX5* mutations in Chinese patients with Holt-Oram syndrome. *American Journal of Medical Genetics, 92*(4), 237–240.

Zaragoza, M. V., Lewis, L. E., Sun, G., Wang, E., Li, L., Said-Salman, I., et al. (2004). Identification of the TBX5 transactivating domain and the nuclear localization signal. *Gene, 330*, 9–18.

CHAPTER EIGHT

Tbx1: Transcriptional and Developmental Functions

A. Baldini*,†,1, F.G. Fulcoli†, E. Illingworth‡,1
*University of Naples, Federico II, Naples, Italy
†Institute of Genetics and Biophysics of the CNR, Naples, Italy
‡University of Salerno, Fisciano, Italy
[1]Corresponding authors: e-mail address: antonio.baldini@igb.cnr.it; eillingworth@unisa.it

Contents

1. Tbx1: Structure and Function(s) 224
   1.1 DNA-Binding Properties 224
   1.2 Protein and Chromatin Interactions 226
   1.3 Transcription Regulation: Neither a Strong Activator nor a Strong Repressor 227
   1.4 From Transcription Regulation to Morphogenesis 228
2. Developmental Functions 228
   2.1 The PA and Cardiac Development 228
   2.2 *Tbx1* and Vascular Development 234
   2.3 *Tbx1* and Cerebral Cortex Development 236
3. Concluding Remarks and Future Perspectives 237
Acknowledgment 238
References 238

Abstract

Recent data have paved the way to mechanistic studies into the role of *Tbx1* during development. *Tbx1* is haploinsufficient and is involved in an important genetic disorder. The gene encodes a T-box transcription factor that is expressed from approximately E7.5 in mouse embryos and continues to be expressed in a highly dynamic manner. It is neither a strong transcriptional activator nor a strong repressor, but it regulates a large number of genes through epigenetic modifications. Here, we review recent literature concerning mechanisms of gene regulation by Tbx1 and its role in mammalian development, with a special focus on the cardiac, vascular, and central nervous systems.

The *Tbx1* gene is phylogenetically related to genes *Tbx10*, *Tbx15*, *Tbx18*, *Tbx20*, and *Tbx22* that together make up the "Tbx1" subfamily (Papaioannou, 2014). These genes, though related by nucleotide sequence, have different developmental roles. *Tbx1*, like several *Tbx* genes, is haploinsufficient in several species, including *Homo sapiens*, and *Tbx1*

Current Topics in Developmental Biology, Volume 122
ISSN 0070-2153
http://dx.doi.org/10.1016/bs.ctdb.2016.08.002

haploinsufficiency is the major genetic basis of a relatively frequent disease known as DiGeorge syndrome or 22q11.2 deletion syndrome (22q11.2DS). The "DiGeorge syndrome-like phenotype," from a mammalian developmental perspective, has become synonymous with severe developmental anomalies of the pharyngeal apparatus (PA) with disruption of the characteristic segmental-like sequence of pharyngeal arches and pouches. Loss of *Tbx1* causes one of the most severe disruptions of the PA ever described in mice. Progressively detailed analyses of *Tbx1* mutants have highlighted additional developmental anomalies, including in the central nervous system, the latter originating from loss of mesodermal *Tbx1* expression, as it will be detailed in this review. Overall, an intricate array of cell autonomous and nonautonomous functions suggests that *Tbx1* is a major regulator of tissue interactions and morphogenetic events that are not limited to the development of the PA.

1. TBX1: STRUCTURE AND FUNCTION(S)

1.1 DNA-Binding Properties

The DNA-binding motif of T-box proteins binds DNA in a sequence-specific manner. The T-box binding element is a palindromic DNA consensus sequence first defined as a sequence with high affinity for Brachyury (a T-box protein), that interacts with that sequence as a dimer, each monomer binding a half site called a T-half site (5′-AGGTGTGAAATT-3′) (Kispert & Herrmann, 1993). Crystallographic studies of several T-box proteins demonstrated that they engage the same amino acid residues to make comparable DNA contacts with few differences, indicating strong conservation of the underlying DNA-binding functions (Coll, Seidman, Müller, 2002; Müller & Herrmann, 1997). Moreover, this half site is recognized by many other T-box proteins with different optimal target sequences that in part confer target gene specificity (e.g., Ghosh et al., 2001; Lingbeek, Jacobs, & van Lohuizen, 2002; Sinha, Abraham, Gronostajski, & Campbell, 2000).

The consensus binding sequence of Tbx1 has been identified using two methods, Selex (Castellanos, Xie, Zheng, Cvekl, & Morrow, 2014) and ChIP-seq (Fulcoli et al., 2016). De novo motif discovery in ChIP-seq datasets uncovered a 8-bp consensus sequence, CC[A/C][G/C]CTCC that is similar to the complementary strand of those previously identified for Tbx5 using the same strategy (Table 1) (Luna-Zurita et al., 2016; Mori & Bruneau, 2004; Narlikar, 2013). Using Selex and the T-box domain of Tbx1 as a bait in vitro, an 8-bp optimal consensus motif (AGGTGT

Table 1 DNA Binding Motives Identified with the Techniques and for the T-Box Transcription Factors Indicated

T-box	References											Method
Tbx1	Fulcoli et al. (2016)	G	G	A	G	G/C	T	G	G			ChIP-Seq
Tbx1	Fulcoli et al. (2016)			A	G	G	T	G	T	G/T	A/T	SELEX
Org1	Schaub, Nagaso, Jin, and Frasch (2012)			A	G	G	T	G	T	G	A	SELEX
Tbx5	Narlikar (2013)	G	G/A	A	G	G/A	T	G	G/A			ChIP-Seq
Tbx5	Mori et al. (2006)			A/G/T	G/A	G	T	G	N	N	A	ChIP-Seq
Tbx5	Luna-Zurita et al. (2016)		G	A	G	G	T	G				ChIP-EXO

[G/T][A/T]), partially matching the ChIP-seq motif was identified (Castellanos et al., 2014), Table 1. The difference between the ChIP-seq- and Selex-generated motifs could have various explanations. First, the full-length protein may have different specificity than the T-box domain alone. Second, the two methods use fundamentally different environments, i.e., test-tube DNA–protein interaction and chromatin. It would be of interest to produce additional ChIP-seq data in different tissues and perhaps consider different binding sites for different chromatin contexts or different protein dosages.

T-box proteins may bind DNA as dimers, but whether or not this is the case for Tbx1 is unclear. The amino acid sequence of the dimerization domain is not highly conserved among T-box proteins, and Tbx1 crystallographic studies suggest that Tbx1 is more likely to bind DNA as monomer because the molecular surface for potential dimerization is probably too small to be of biological significance (El Omari et al., 2012).

DNA binding is not the only mechanism by which Tbx1 functions. Indeed, it has been shown that it can bind Smad1 independently from DNA binding (Fulcoli, Huynh, Scambler, & Baldini, 2009). This protein–protein interaction suppresses the ability of Smad1 to bind Smad4, thus reducing BMP signal transduction. Another example is the interaction with Srf that has a negative effect on Srf protein levels, possibly by directing it to the proteasome (Chen, Fulcoli, Tang, & Baldini, 2009). The relative impact of these noncanonical mechanisms on the developmental functions of Tbx1 is unknown.

1.2 Protein and Chromatin Interactions

A screen for Tbx1-interacting proteins using unbiased methods such as proteomics has not been reported yet, and most of the known interactors have been identified in hypothesis-driven studies. Nevertheless, a complex picture is starting to emerge.

Some protein interactions have been demonstrated outside of the chromatin context. Beside the already mentioned Smad1 and Srf, there is also an interaction with the methyltransferase Setd7, which occurs in the soluble nuclear fraction (Fulcoli et al., 2016). However, other known interactions are more clearly related to transcriptional functions, and in particular with histone modifiers and chromatin remodeling complexes. Specifically, Tbx1 interacts with Ash2l, a core component of a multimeric histone methyltransferase complex that is required for methylation of histone lysine residues (Stoller et al., 2010). Furthermore, a recent study demonstrated the ability of Tbx1 to interact at the chromatin level with the MLL3 complex and colocalization with the histone demethylase Lsd1. As a consequence of these interactions, Tbx1 has been shown to regulate H3K4 monomethylation of target genes, while Lsd1 enzymatic inhibition partially rescued the effects of reduced dosage of the *Tbx1* gene (Fulcoli et al., 2016).

Tbx1 also interacts with Baf60a (Smarcd1), a subunit of the SWI–SNF-like chromatin remodeling complex. Tbx1 coimmunoprecipitates with Baf60a and is capable of recruiting it to chromatin (Chen et al., 2012). In addition, knock-down of Baf60a affects the ability of Tbx1 to regulate some of its target genes in cultured cells. Baf60a is expressed in most tissues during mouse embryonic development, but, notably, its expression fades during late phases of cardiac differentiation while the expression of an alternative subunit, Baf60c (Smarcd3), increases. Tbx1 expression approximately parallels the tapering down of Baf60a expression (Chen et al., 2012). Together, these data indicate that Tbx1 regulates transcription through epigenetic modifications.

Intriguingly, Tbx1 coimmunoprecipitates with Nkx2-5, a critical cardiac transcription factor (Nowotschin et al., 2006). It would be of interest to validate this interaction using endogenous proteins and to explore its developmental significance in vivo. Interestingly, these two transcription factors have antagonistic effects in ascidian cardiac development (Wang,

Razy-Krajka, Siu, Ketcham, & Christiaen, 2013). Of particular interest is the genetic interaction observed between Chd7 and Tbx1 (Randall et al., 2009). *Chd7* encodes a chromodomain-containing chromatin remodeling protein that is mutated in CHARGE syndrome (Basson & van Ravenswaaij-Arts, 2015), which has some clinical overlap with DiGeorge syndrome. The molecular nature of the interaction is unknown, but it is intriguing that as for Tbx1, Chd7 tends to localize to H3K4me1-enriched regions (Schnetz et al., 2009).

1.3 Transcription Regulation: Neither a Strong Activator nor a Strong Repressor

Binding site maps and gene expression analyses after *Tbx1* knock-down have shown that Tbx1 elicits a modest transcription response of target genes within the time frame tested in the study, and this response can be activation or repression (Fulcoli et al., 2016). That Tbx1 is not a strong activator is also exemplified by the finding that the great majority of Tbx1 binding regions is characterized by absence or low levels of acetylation of H3K27, a marker of enhancer activation. In contrast, most binding sites overlap with monomethylation of H3K4 (H3K4me1), a marker of primed enhancers (Creyghton et al., 2010). In addition, reduced dosage of Tbx1 in cells or in mouse embryos is associated with reduced H3K4me1, likely because Tbx1 recruits histone methyltransferases, as detailed in the previous section. This activity is central to at least part of the mechanisms of action of Tbx1 because increasing H3K4me1 with an Lsd1 inhibitor is sufficient to compensate in part for reduced dosage of the transcription factor and to reduce significantly the severity of the mutant phenotype (Fulcoli et al., 2016).

Therefore, a possible model for Tbx1 function is to prime target enhancers and make them accessible to other regulators, which may be activators or repressors. Haploinsufficiency may be a direct consequence of reduced H3K4me1 and thus reduced accessibility of target enhancers.

Additional clues to possible mechanisms of action of Tbx1 derive from the finding that dosage reduction of p53 suppresses the *Tbx1* mutant phenotype (Caprio & Baldini, 2014). While phenotypic rescue may occur through different mechanisms, the findings of this study that p53 and Tbx1 colocalize on chromatin and that the two transcription factors have opposing effects on transcription of at least some target genes, suggest that there is a specific interaction that warrants further investigation.

1.4 From Transcription Regulation to Morphogenesis

Exhaustive knowledge of transcriptional targets may predict regulatory networks by which Tbx1 guides development (Ivins et al., 2005; Liao et al., 2008; Pane et al., 2012; Roberts, Ivins, Cook, Baldini, & Scambler, 2006). To date, the only reported systematic search for direct targets is in a cell culture model that has been partially validated in vivo (Fulcoli et al., 2016). Nevertheless, the results are intriguing because they show a clear connection between Tbx1 and basic mechanisms of cell morphology, cell dynamics, and cell–cell interactions and adhesion. These data suggest that Tbx1 is a direct regulator of morphogenetic processes and possibly tissue interactions, thus expanding the previously held view that Tbx1 primarily supports cell proliferation and inhibits cell differentiation. This "morphogenetic view" of the role of *Tbx1* in development is supported by recent phenotypic data that have revealed defects in cell polarity and tissue organization in *Tbx1* mutant mouse embryos (Francou, Saint-Michel, Mesbah, & Kelly, 2014).

2. DEVELOPMENTAL FUNCTIONS

2.1 The PA and Cardiac Development

Phenotypic analyses of mutants have shown broad consequences of *Tbx1* loss of function, most evident, but not limited to the development of the PA (Jerome & Papaioannou, 2001; Lindsay et al., 2001; Merscher et al., 2001; Vitelli, Morishima, Taddei, Lindsay, & Baldini, 2002). In *Tbx1* mouse mutants, the PA loses almost completely its characteristic segmentation in pharyngeal arches and pouches, and it is profoundly hypoplastic. The downstream consequences of this developmental anomaly are severe because (a) the pharyngeal region is a niche for cardiac progenitors of the second heart field (SHF) (Kelly, 2012), which contributes progenitors to a large portion of the mammalian heart; (b) The PA is the milieu within which cardiac neural crest cells migrate; (c) The pharyngeal arch arteries, which contribute to the definitive aortic arch and some of the great arteries, develop and remodel within the PA; (d) the mediastinic glands such as the thymus, parathyroids, and thyroid develop within the PA; (e) some nonsomitic muscles and lower face bones are also dependent on the PA for proper development. All of these organs and structures are affected by loss of *Tbx1*.

Within the PA, *Tbx1* is expressed in the surface ectoderm, lateral pharyngeal endoderm, and pharyngeal mesoderm. Extensive tissue-specific gene inactivation experiments have revealed distinct developmental roles in different tissues, as well as cell autonomous and nonautonomous roles (Arnold et al., 2006; Lania et al., 2009; Xu, Cerrato, & Baldini, 2005; Xu, Chen, & Baldini, 2007; Zhang et al., 2005). For example, the role in cardiac development (most likely mediated by expression in the SHF, as detailed in the next paragraph) is through mesodermal expression, as defined by the *Mesp1*^{Cre} expression domain (Zhang, Huynh, & Baldini, 2006). Expression in this domain is sufficient to support heart development in a mutant background, suggesting that ectodermal and endodermal *Tbx1* expression domains of *Tbx1* do not contribute significantly to heart development (Zhang et al., 2006). In contrast, the development of the 4th pharyngeal arch arteries, which is necessary for correct remodeling of the definitive aortic arch, requires *Tbx1* expression in the surface ectoderm, thus playing a cell nonautonomous function (Calmont et al., 2009; Zhang et al., 2005). Other examples of cell nonautonomous functions have been reported for the development of the thyroid (Lania et al., 2009), inner ear (Braunstein, Crenshaw Iii, Morrow, & Adams, 2008; Xu et al., 2007), and brain cortex (Flore, Cioffi, Bilio, & Illingworth, 2016). Thus, a substantial portion of the developmental roles of *Tbx1* occurs through regulation of signaling molecules, such as FGF (Vitelli, Taddei, et al., 2002; Watanabe et al., 2012), retinoic acid (Roberts et al., 2006), noncanonical Wnt (Chen et al., 2012), and perhaps other signaling pathways. However, in the light of the recent reports cited in the previous sections of this review, one should also consider additional mechanisms, such as defects in cell–cell adhesion and collective cell dynamics that may mimic cell nonautonomous functions.

2.1.1 Roles in the SHF

Some of the most intensely studied phenotypes associated with loss of *Tbx1* concern cardiac development. Examples of defects observed in preterm *Tbx1*$^{-/-}$ mouse embryos (E18.5) are shown in Fig. 1 and are mainly characterized by hypoplasia and lack of septation of the outflow tract (OFT), defective alignment of the OFT with respect to the heart chambers (the OFT remains aligned with the right ventricle rather than communicating with both ventricles), and defects of interventricular septation. *Tbx1* is not expressed, by in situ hybridization or immunostaining, in these structures at any time during development suggesting that the critical role of *Tbx1* is in precursor cells, before they reach their destination in the OFT or RV. This view is supported by time- and tissue-specific gene inactivation data, which have shown that *Tbx1* is not required in differentiated heart

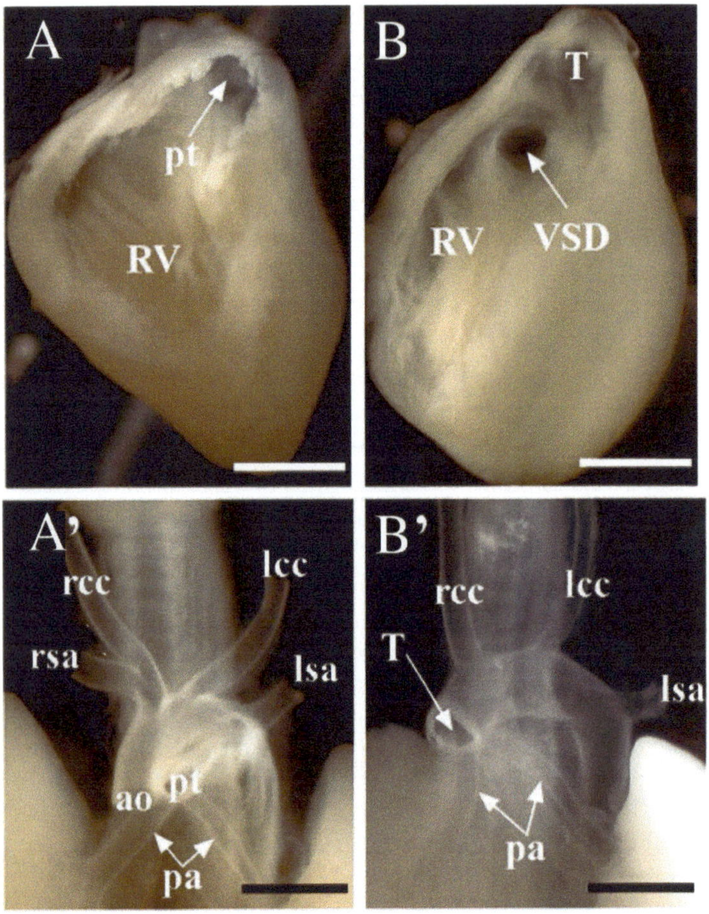

Fig. 1 Cardiac and great artery abnormalities in Tbx1 mutants. (A and B) Intracardiac phenotype in wild type (A) and *Tbx1*$^{-/-}$ (B) mutants. A′ and B′ aortic arch and great artery patterning in wild-type (A′) and *Tbx1*$^{-/-}$ (B′) mutants at E18.5. *RV*: right ventricle; *T*: truncus arteriosus; *ao*: aorta; *lcc*: left common carotid artery; *lsa*: left subclavian artery; *pt*: pulmonary trunk; *pa*: pulmonary arteries; *rcc*: right common carotid artery; *rsa*: right subclavian artery; *RV*: right ventricle; *VSD*: ventricular septum defect. Scale bars: 1 mm. *Reproduced with permission from reference Xu, H., Morishima, M., Wylie, J. N., Schwartz, R. J., Bruneau, B. G., Lindsay, E. A., et al. (2004). Tbx1 has a dual role in the morphogenesis of the cardiac outflow tract.* Development, 131, *3217–3227.*

tissue, and indicated a critical time window between E8.5 and E9.5, beyond which *Tbx1* is no longer required for heart development (Xu et al., 2004, 2005). The gene is expressed in the pharyngeal mesoderm, in part of the splanchnic mesoderm (peripharyngeal mesoderm), and in part of the posterior wall of the pericardiac cavity, including, but not limited to the SHF.

These domains are clearly delineated in a map of the descendants of *Tbx1*-expressing cells using a *Tbx1*^{Cre} knock-in driver and a Cre reporter (Huynh, Chen, Terrell, & Baldini, 2007) (Fig. 2). Later in development, descendants of *Tbx1*-expressing cells populate most of the OFT (including the endothelial layer), the subpulmonary myocardium, and part of the right ventricle (Huynh et al., 2007). Loss of *Tbx1* reduces, but does not eliminate, the contribution of progenitors to heart structures (Rana et al., 2014; Xu et al., 2004). Interestingly, reduced contribution of cardiac progenitors is also evident at the inflow tract/venous pole of the heart (Rana et al., 2014), where descendants of *Tbx1*-expressing cells are not present in a significant number. This suggests that the role of *Tbx1* in cardiac development is not limited to cell autonomous functions.

What does *Tbx1* do in the SHF? There are already excellent reviews concerning development and transcriptional programs in the SHF (Kelly, 2012; Vincent & Buckingham, 2010), thus we will not discuss general points about gene regulatory networks active in this population of progenitors. The prevailing view of the role of *Tbx1* in the SHF is to support cell proliferation and inhibit cell differentiation. The basic hypothesis is that *Tbx1* is required for self-renewal of cardiac progenitors and to promote the generation of a sufficient number of progenitors to support normal cardiac morphogenesis. The inhibitory effect would also be instrumental to this function by preventing premature differentiation, which does occur in *Tbx1* mutants (Chen et al., 2009).

The effects on cell proliferation may be mediated by upregulation of the FGF–MAPK signaling, BMP signaling downregulation, and/or p21 downregulation (Cao et al., 2010; Fulcoli et al., 2009; Guris, Duester, Papaioannou, & Imamoto, 2006; Moon et al., 2006; Papangeli & Scambler, 2013; Vitelli, Taddei, et al., 2002). Genetic and pharmacological approaches to phenotypic rescue approaches are associated with partial rescue of cell proliferation in the SHF (Caprio & Baldini, 2014; Fulcoli et al., 2016), and *Tbx1* overexpression is associated with increased cell proliferation (Chen et al., 2009). Thus, there is correlation between *Tbx1* dosage and cell proliferation in the SHF, but whether or not *Tbx1* has a direct effect on cell proliferation is still to be proven.

Inhibition of differentiation appears to be operated through a number of suppressive interactions with transcription factors important for cardiomyocyte differentiation. For example, *Gata4*, *Mef2c*, and Srf (Chen et al., 2009; Liao et al., 2008; Pane et al., 2012), at least some of these interactions are direct. A summary of interactions likely to mediate cell proliferation and differentiation effects of Tbx1 is shown in Fig. 3.

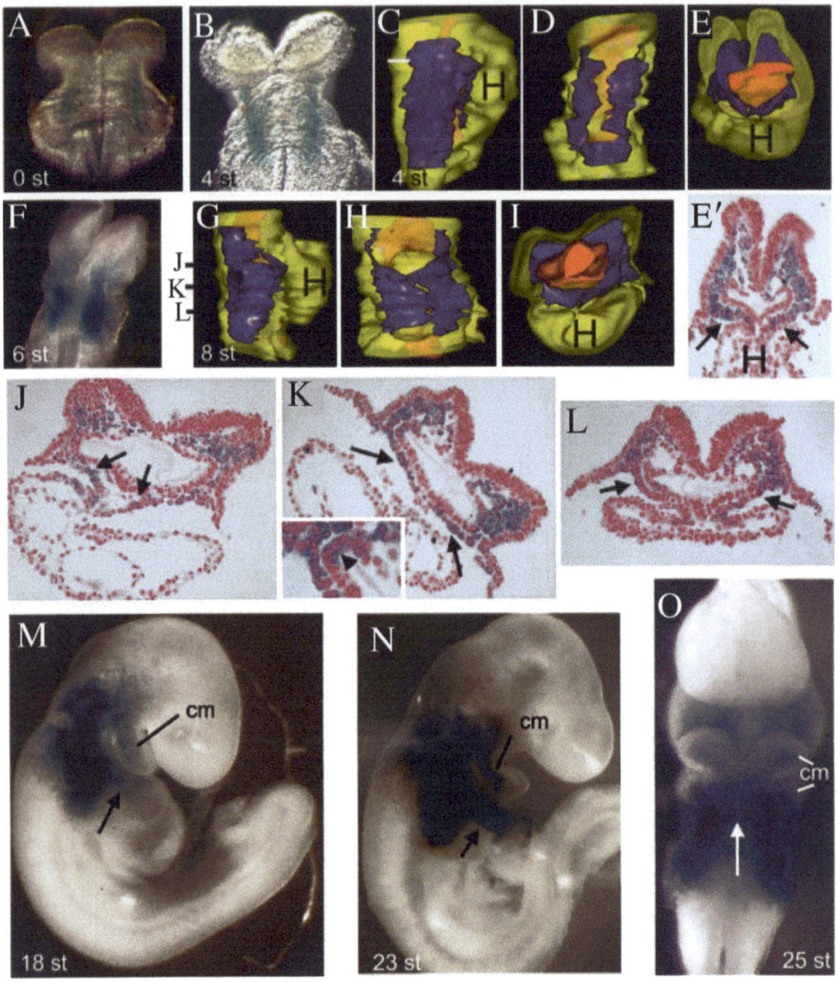

Fig. 2 A map of descendants of *Tbx1*-expressing cells in mouse embryos. (A and B) *Tbx1$^{Cre/+}$;R26R* embryos stained with X-Gal to reveal β-galactosidase activity derived from the activation of the R26R Cre reporter in the *Tbx1* expression domain. The developmental stage is indicated in somites (st). (C–E) Three dimensional reconstruction of the trunk of four-somite embryo viewed from the right (C), front (D), and from above (E). The surface of the embryo is rendered in *yellow*, β-gal staining is rendered in *blue*, and the pharyngeal cavity is shown in *red*. H is the heart (removed in panel D). (E′) Transverse section of the same embryo at the level of the *white line* shown in panel C. (F) Dorsal view of a *Tbx1$^{Cre/+}$;R26R* 6 st. (G–I) Three dimensional reconstruction of an 8-st embryo, same views and color coding as in panels (C–E). Note the virtual closing of the staining gap in the ventral aspect of the pharyngeal mesoderm/dorsal pericardial wall on panel H. (J–L) Transverse sections of the same embryo in correspondence of the OFT (J), intermediate (K), and inflow (L). *Arrows* indicate staining of the SHF/dorsal pericardial wall. The inset in panel K is a higher magnification detail showing pharyngeal endoderm staining (*arrowhead*). (M–O) *Tbx1$^{Cre/+}$;R26R* embryos of the stage indicated. The heart has been dissected out from the embryo shown in panel O (frontal view). *Arrows* indicate the OFT. *cm*: pharyngeal core mesoderm. *Reproduced with permission from Huynh, T., Chen, L., Terrell, P., & Baldini, A. (2007). A fate map of Tbx1-expressing cells reveals heterogeneity in the second cardiac field.* Genesis, 45, *470–475.*

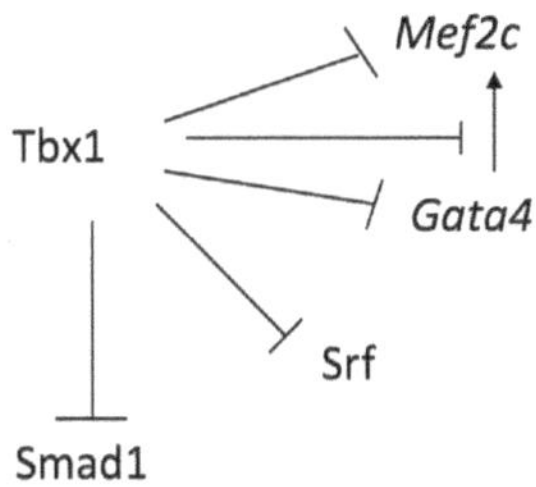

Fig. 3 Cartoon showing the interactions between Tbx1 and some of the genes or proteins involved in cardiac differentiation. Drawn on the basis of literature data (Chen et al., 2009; Fulcoli et al., 2009; Liao et al., 2008; Pane et al., 2012).

Recent data have added a new twist on the view of the developmental role of Tbx1. It has been shown that loss of Tbx1 is associated with a distinct cellular phenotype in the SHF that concerns loss of some epithelial characteristics of the SHF and loss of polarization (Francou et al., 2014). In addition, the finding that reduced dosage of *Tbx1* affects pathways involved in cell shape and adhesion (Fulcoli et al., 2016), clearly indicates that *Tbx1* can be directly involved in morphogenetic processes. Beside the already mentioned cell polarization and adhesion pathways, an important mediator of the morphogenetic functions of *Tbx1* is the planar cell polarity pathway (PCP), mainly through direct regulation of *Wnt5a* (Chen et al., 2012), *Ror2*, and possibly *Dvl2* (Fulcoli et al., 2016). *Wnt5a* is required for SHF development (Chen et al., 2012; Sinha et al., 2015). Loss of *Wnt5a* reduces the incorporation of cardiac progenitors into the epithelial-like layer of the SHF (Sinha et al., 2015). However, *Wnt5a* downregulation/suppression alone cannot explain the cardiovascular phenotype of *Tbx1* mutants because the latter is significantly more severe than in $Wnt5a^{-/-}$ mutants. In addition, $Tbx1^{-/-};Wnt5a^{-/-}$ embryos exhibit a catastrophic SHF defect, considerably more severe than the two individual mutants (Chen et al., 2012). This supports the view that Tbx1 affects SHF development through multiple pathways.

2.1.2 The Cardiopharyngeal Lineage: Emerging Concepts

The cardiopharyngeal lineage or field is defined as a population of cranial mesoderm cells that give rise to heart and branchiomeric muscles (Diogo et al., 2015). The SHF is, by definition, a subset of this population of cells. Tbx1 is clearly critical for the development of the cardiopharyngeal lineage because its mutation affects both the development of the heart and of the branchiomeric skeletal muscle (Grifone et al., 2008; Kelly,

Jerome-Majewska, & Papaioannou, 2004), and of the striated muscle of the esophagus, which is also derived from the cranial mesoderm (Gopalakrishnan et al., 2015). Loss of *Tbx1* impairs the normally robust onset of myogenic specification (Grifone et al., 2008). Tbx1 synergizes with the myogenic factor Myf5 (Sambasivan et al., 2009). The molecular mechanisms by which this synergy is enacted need to be clarified. Nevertheless, the finding that *Tbx1* is required for both branchiomeric and cardiac muscle development suggest that it functions upstream of the split of the two lineages from the cardiopharyngeal field. In addition, a subset of endothelial cells (ECs) of the heart, pharynx, and possibly brain, is derived from cranial mesoderm, and as *Tbx1* is required also in this cell population (details in the next section), it is likely that it has a developmental role in common progenitors, perhaps affecting their fate. This role would be compatible with the role in cell identity in *Drosophila* (Schaub, März, Reim, & Frasch, 2015; Schaub et al., 2012), in *Ciona intestinalis* (Wang et al., 2013), and it is supported by the epigenetic role discussed earlier, in the selection (or priming) of enhancers.

2.2 *Tbx1* and Vascular Development

Tbx1 is expressed in ECs of the dorsal aorta (Vitelli, Morishima, et al., 2002), brain (Paylor et al., 2006), and mesentery (Chen et al., 2010). In each case, expression was noted to be low level and in subpopulations of ECs or vessels. This is in stark contrast with results from fate mapping experiments that reveal an extensive contribution of *Tbx1*-expressing cells to the vascular networks of diverse organs, including the heart, intestine, and brain, where it has been shown that in preterm mouse embryos around 80% of blood vessels derives from *Tbx1*-expressing cells (Cioffi et al., 2014). The difference between active expression and descendant labeling indicates that Tbx1 is extensively expressed in mesodermal EC progenitors.

2.2.1 Does Tbx1 Have Essential Roles in ECs?

This question has been addressed in two in vivo contexts, namely, in mouse brain ECs and lymphatic ECs. In the mouse brain, *Tbx1* is expressed in ECs lining some brain vessels (Paylor et al., 2006). In order to establish the requirement for *Tbx1* in brain vessels, it has been inactivated in the mouse germ line and, separately, in ECs (Cioffi et al., 2014). In both cases this led to global brain vessel hyperplasia and disorganization, patchy hypoperfusion of brain vessels, and local tissue hypoxia. These phenotypes are not secondary to cardiac defects because mice with EC-specific deletion of *Tbx1* have normal heart (Chen et al., 2010). Thus, *Tbx1* expression in the EC is required for normal brain vascularization.

Outside the CNS, *Tbx1* has been shown to promote lymphangiogenesis. In fact, EC-specific inactivation of *Tbx1* leads to a severe loss of lymphatic vessels in many tissues and organs because they fail to maintain endothelial expression of *Vegfr3*, an essential lymphangiogenesis gene and transcriptional target of Tbx1 (Chen et al., 2010). Further studies are required to determine whether Vegfr3 is the sole transducer of *Tbx1* function in lymphatic ECs. The fact that some lymphatic vessels survive in *Tbx1* null embryos, and the severity of lymphatic vessel depletion in different tissues of *Tbx1* mutants is variable, may reflect the diverse origin of lymphatic ECs that has been reported recently (Klotz et al., 2015; Martinez-Corral et al., 2015).

Thus, inactivation of *Tbx1* in different EC populations has different, even opposite effects on vessel formation and growth. How may this come about? Recent data suggest that it is due, at least in part, to the contrasting roles of some Tbx1 targets in vascular development rather than to different Tbx1 functions in specific EC populations. In particular, in mice, Tbx1 positively regulates *Vegfr3* and negatively regulates *Vegfr2* in brain and lymphatic ECs (Chen et al., 2010; Cioffi et al., 2014; Lania, Ferrentino, & Baldini, 2015). However, while endothelial *Vegfr2* promotes both hemangiogenesis (Saharinen, Eklund, Pulkki, Bono, & Alitalo, 2011) and lymphangiogenesis in embryonic and adult tissues (Dellinger & Brekken, 2011; Dellinger, Meadows, Wynne, Cleaver, & Brekken, 2013; Saharinen et al., 2010), endothelial *Vegfr3* suppresses brain hemangiogenesis (Tammela et al., 2011) while promoting lymphangiogenesis (Dumont et al., 1998; Karkkainen, Jussila, Ferrell, Finegold, & Alitalo, 2001; Karkkainen et al., 2000). Thus, in *Tbx1* mutants, brain vessel hyperplasia likely results from the combined effects of increased expression of *Vegfr2* and decreased expression of *Vegfr3* in ECs, while lymphatic hypoplasia probably reflects the dominant effect of *Vegfr3* repression in lymphatic ECs, and the modest upregulation of *Vegfr2* is not sufficient to ameliorate the phenotype (Chen et al., 2010; Cioffi et al., 2014). These interactions and their phenotypic outcome in *Tbx1* mutants are shown in Fig. 4. Thus, while some of the molecular players are the same in CNS and non-CNS ECs, the way in which they are regulated (quantitatively) in different tissues may determine the vascular phenotype.

In addition to the regulation of the balance between Vegfr2 and Vegfr3, Tbx1 may have other roles in EC, consistent with what has been proposed for its role in the SHF. In particular, it may play a role in polarity and dynamic properties of leading ECs (tip cells) during angiogenesis. This might

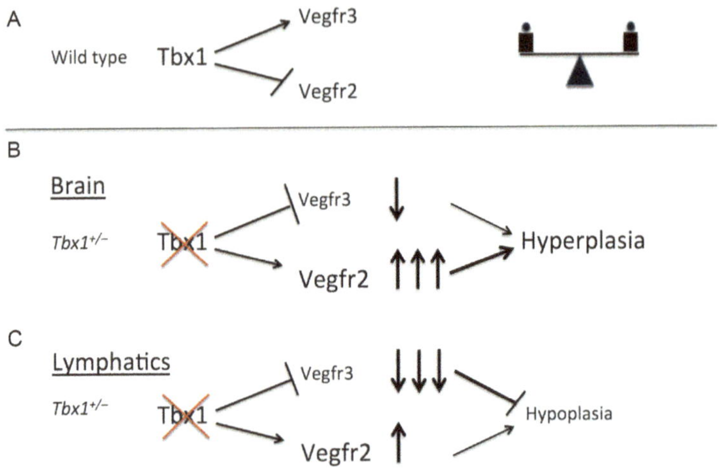

Fig. 4 Cartoon showing the interactions between Tbx1 and VEGF receptors involved in blood and lymphatic vessel development. The interpretation of the effect of these interactions on brain and lymphatic vessels is based on published data (Chen et al., 2010).

explain the disorganization of blood vessels in the brain of mutants. Furthermore, as discussed in early sections of this review, Tbx1 may have a role in cell fate determination. Determining whether brain hypervascularization in *Tbx1* mutants derives from increased differentiation of multipotent mesodermal progenitors toward an endothelial fate would be an interesting line of investigation.

2.3 *Tbx1* and Cerebral Cortex Development

Human genetic studies implicate a role for *TBX1* in brain (Ogata et al., 2014; Paylor et al., 2006) but there are no histopathological studies yet to investigate possible neurodevelopmental anomalies in patients with *TBX1* mutations. However, progress is being made in mouse models. Indeed, a requirement for *Tbx1* in brain was demonstrated 10 years ago when Paylor et al. (2006) showed that *Tbx1* heterozygous mice, and mice with a heterozygous multigene deletion that included *Tbx1*, had reduced prepulse inhibition (PPI) of the startle reflex, a finding that is synonymous with sensorimotor gating impairment. An independent study failed to identify PPI defects in *Tbx1* heterozygous animals, possibly because of genetic background-dependency of this phenotype (Long et al., 2006). Nevertheless, the finding was intriguing because reduced PPI has been reported in children and adults with 22q11.2DS (McCabe et al., 2014;

Sobin, Kiley-Brabeck, & Karayiorgou, 2005), in individuals with non-syndromic schizophrenia, reviewed in Braff, Geyer, and Swerdlow (2001) and Swerdlow, Weber, Qu, Light, and Braff (2008), and in individuals with Asperger syndrome (McAlonan et al., 2002). A more recent study showed that *Tbx1*$^{+/-}$ mice also have impairments in social interaction and memory-dependent behavior that were defined as autistic like (Hiramoto et al., 2011).

These studies, and the relevance for the human disease, encouraged further work into the characterization of neurodevelopmental anomalies in *Tbx1* mutants. Anomalies in the vascular compartment have already been discussed earlier. More recently, Flore et al. (2016) focused on the development of the primary somatosensory (S1) cortex (Flore et al., 2016). In a detailed morphological study conducted between embryonic day (E)13.5 and adulthood, the authors showed that *Tbx1* mutation leads to precocious differentiation of cortical progenitors in the medio-lateral cortex, which is the embryonic precursor of the S1 cortex, and in the ganglionic eminences. This was linked to abnormal maturation and distribution of excitatory (glutamatergic) neurons and inhibitory (GABAergic) interneurons. Intriguingly, both abnormalities were recapitulated in mesoderm-specific conditional *Tbx1* mutants, indicating that S1 development requires signals from, or interaction with mesoderm or mesoderm-derived tissues. The nature of this interaction and where and when it occurs is currently unknown. An obvious candidate mesodermal cell type is the EC lining brain capillaries, upon which nerves are highly dependent for the supply of nutrients and oxygen. This close relationship may be disrupted in *Tbx1* mutants due to brain vessel hyperplasia and disorganization or to functional defects such are altered permeability and hypoxia (Cioffi et al., 2014). Alternatively, expression of *Tbx1* in the cranial mesoderm adjacent to the developing brain may be critical for cortical development.

3. CONCLUDING REMARKS AND FUTURE PERSPECTIVES

Research of developmental roles, especially in the mouse, has dominated the scene of *Tbx1* research in the last 15 years. We expect an increased focus on molecular mechanisms by which Tbx1 regulates its target genes, particularly concerning epigenetic programming of enhancers involved in cell lineage control and cell differentiation. In addition, a related question is what is the mechanism of Tbx1 in haploinsufficiency, and are chromatin/histone modifications a sensitive readout of gene dosage? Why are certain developmental processes, even within the cardiovascular system, more sensitive than others to *Tbx1* dosage (Zhang & Baldini, 2008)?

The role of *Tbx1* in brain development should gain further attention, particularly as it appears that *Tbx1* may play a role in regulating (pharyngeal?) mesoderm–neural tissue interactions in cortical development.

A relatively unexplored field, but one that is mature for direct investigation is the cellular phenotype (and its underlying mechanisms) caused by loss or gain of function of *Tbx1*, especially concerning dynamic, polarity, and morphological response of cells to Tbx1 dosage.

Very little information is available as to the transcription factors and mechanisms by which the *Tbx1* gene is regulated. While there has been significant information derived from transgenic analyses (Brown et al., 2004; Garg et al., 2001; Yamagishi et al., 2003), the deletion in vivo of a Fox-binding enhancer of the *Tbx1* gene produced only very modest effects on *Tbx1* expression and no phenotypic changes (Zhang & Baldini, 2010). Thus, the critical transcription factors and enhancer elements driving the exquisitely regulated *Tbx1* gene expression are still to be identified.

In addition, the role of *Tbx1* in human disease, and the search for drug targets, remains a major field of interest. The 22q11.2 deletion syndrome, although defined a rare disease, is relatively frequent and has no specific treatment. The adult phenotype of this syndrome might benefit from targeted drugs.

Last but not least, some of the cited characteristics of Tbx1 function and interactions predict relevance in cancer biology. Indeed, it has already been shown that *Tbx1* is a target of Sox9 in basal cell carcinoma (BCC) tumorigenesis, and is overexpressed in this type of cancer in humans (Larsimont et al., 2015). While the role of Tbx1 in BCC is still to be defined, analyses of pathways perturbed by loss of *Tbx1* lists BCC as one of the most significantly affected (Fulcoli et al., 2016).

ACKNOWLEDGMENT

The authors' research is funded by the Fondation Leducq (TNE 15CVD01), Fondazione Telethon (GGP14211), and the Italian Association for Cancer Research (AIRC, IG2015-17529).

REFERENCES

Arnold, J. S., Braunstein, E. M., Ohyama, T., Groves, A. K., Adams, J. C., Brown, M. C., et al. (2006). Tissue-specific roles of Tbx1 in the development of the outer, middle and inner ear, defective in 22q11DS patients. *Human Molecular Genetics*, *15*, 1629–1639.

Basson, M. A., & van Ravenswaaij-Arts, C. (2015). Functional insights into chromatin remodelling from studies on CHARGE syndrome. *Trends in Genetics*, *31*, 600–611.

Braff, D. L., Geyer, M. A., & Swerdlow, N. R. (2001). Human studies of prepulse inhibition of startle: Normal subjects, patient groups, and pharmacological studies. *Psychopharmacology, 156*, 234–258.

Braunstein, E. M., Crenshaw Iii, E. B., Morrow, B. E., & Adams, J. C. (2008). Cooperative function of Tbx1 and Brn4 in the periotic mesenchyme is necessary for cochlea formation. *Journal of the Association for Research in Otolaryngology, 9*, 33–43.

Brown, C. B., Wenning, J. M., Lu, M. M., Epstein, D. J., Meyers, E. N., & Epstein, J. A. (2004). Cre-mediated excision of *Fgf8* in the *Tbx1* expression domain reveals a critical role for *Fgf8* in cardiovascular development in the mouse. *Developmental Biology, 267*, 190–202.

Calmont, A., Ivins, S., Van Bueren, K. L., Papangeli, I., Kyriakopoulou, V., Andrews, W. D., et al. (2009). Tbx1 controls cardiac neural crest cell migration during arch artery development by regulating *Gbx2* expression in the pharyngeal ectoderm. *Development, 136*, 3173–3183.

Cao, H., Florez, S., Amen, M., Huynh, T., Skobe, Z., Baldini, A., et al. (2010). Tbx1 regulates progenitor cell proliferation in the dental epithelium by modulating PITX2 activation of p21. *Developmental Biology, 347*, 289–300.

Caprio, C., & Baldini, A. (2014). p53 suppression partially rescues the mutant phenotype in mouse models of DiGeorge syndrome. *Proceedings of the National Academy of Sciences of the United States of America, 111*, 13385–13390.

Castellanos, R., Xie, Q., Zheng, D., Cvekl, A., & Morrow, B. E. (2014). Mammalian TBX1 preferentially binds and regulates downstream targets via a tandem T-site repeat. *PLoS One, 9*(5), e95151.

Chen, L., Fulcoli, F. G., Ferrentino, R., Martucciello, S., Illingworth, E. A., & Baldini, A. (2012). Transcriptional control in cardiac progenitors: Tbx1 interacts with the BAF chromatin remodeling complex and regulates *Wnt5a*. *PLoS Genetics, 8*(3), e1002571.

Chen, L., Fulcoli, F. G., Tang, S., & Baldini, A. (2009). Tbx1 regulates proliferation and differentiation of multipotent heart progenitors. *Circulation Research, 105*, 842–851.

Chen, L., Mupo, A., Huynh, T., Cioffi, S., Woods, M., Jin, C., et al. (2010). Tbx1 regulates *Vegfr3* and is required for lymphatic vessel development. *Journal of Cell Biology, 189*, 417–424.

Cioffi, S., Martucciello, S., Fulcoli, F. G., Bilio, M., Ferrentino, R., Nusco, E., et al. (2014). Tbx1 regulates brain vascularization. *Human Molecular Genetics, 23*, 78–89.

Coll, M., Seidman, J. G., & Müller, C. W. (2002). Structure of the DNA-bound T-box domain of human TBX3, a transcription factor responsible for ulnar-mammary syndrome. *Structure, 10*, 343–356.

Creyghton, M. P., Cheng, A. W., Welstead, G. G., Kooistra, T., Carey, B. W., Steine, E. J., et al. (2010). Histone H3K27ac separates active from poised enhancers and predicts developmental state. *Proceedings of the National Academy of Sciences of the United States of America, 107*, 21931–21936.

Dellinger, M. T., & Brekken, R. A. (2011). Phosphorylation of Akt and ERK1/2 is required for VEGF-A/VEGFR2-induced proliferation and migration of lymphatic endothelium. *PLoS One, 6*(12), e28947.

Dellinger, M. T., Meadows, S. M., Wynne, K., Cleaver, O., & Brekken, R. A. (2013). Vascular endothelial growth factor receptor-2 promotes the development of the lymphatic vasculature. *PLoS One, 8*(9), e74686.

Diogo, R., Kelly, R. G., Christiaen, L., Levine, M., Ziermann, J. M., Molnar, J. L., et al. (2015). A new heart for a new head in vertebrate cardiopharyngeal evolution. *Nature, 520*, 466–473.

Dumont, D. J., Jussila, L., Taipale, J., Lymboussaki, A., Mustonen, T., Pajusola, K., et al. (1998). Cardiovascular failure in mouse embryos deficient in VEGF receptor-3. *Science, 282*, 946–949.

El Omari, K., De Mesmaeker, J., Karia, D., Ginn, H., Bhattacharya, S., & Mancini, E. J. (2012). Structure of the DNA-bound T-box domain of human TBX1, a transcription factor associated with the DiGeorge syndrome. *Proteins*, *80*, 655–660.

Flore, G., Cioffi, S., Bilio, M., & Illingworth, E. (2016). Cortical development requires mesodermal expression of Tbx1, a gene haploinsufficient in 22q11.2 deletion syndrome. *Cerebral Cortex*. http://dx.doi.org/10.1093/cercor/bhw076, Epub ahead of print.

Francou, A., Saint-Michel, E., Mesbah, K., & Kelly, R. G. (2014). TBX1 regulates epithelial polarity and dynamic basal filopodia in the second heart field. *Development*, *141*, 4320–4331.

Fulcoli, F. G., Franzese, M., Liu, X., Zhang, Z., Angelini, C., & Baldini, A. (2016). Rebalancing gene haploinsufficiency in vivo by targeting chromatin. *Nature Communications*, 7, 11688.

Fulcoli, F. G., Huynh, T., Scambler, P. J., & Baldini, A. (2009). Tbx1 regulates the BMP-Smad1 pathway in a transcription independent manner. *PLoS One*, *4*, e6049.

Garg, V., Yamagishi, C., Hu, T., Kathiriya, I. S., Yamagishi, H., & Srivastava, D. (2001). *Tbx1*, a DiGeorge syndrome candidate gene, is regulated by sonic hedgehog during pharyngeal arch development. *Developmental Biology*, *235*, 62–73.

Ghosh, T. K., Packham, E. A., Bonser, A. J., Robinson, T. E., Cross, S. J., & Brook, J. D. (2001). Characterization of the TBX5 binding site and analysis of mutations that cause Holt-Oram syndrome. *Human Molecular Genetics*, *10*, 1983–1994.

Gopalakrishnan, S., Comai, G., Sambasivan, R., Francou, A., Kelly, R. G., & Tajbakhsh, S. (2015). A cranial mesoderm origin for esophagus striated muscles. *Developmental Cell*, *34*(6), 694–704.

Grifone, R., Jarry, T., Dandonneau, M., Grenier, J., Duprez, D., & Kelly, R. G. (2008). Properties of branchiomeric and somite-derived muscle development in *Tbx1* mutant embryos. *Developmental Dynamics*, *237*, 3071–3078.

Guris, D. L., Duester, G., Papaioannou, V. E., & Imamoto, A. (2006). Dose-dependent interaction of Tbx1 and Crkl and locally aberrant RA signaling in a model of del22q11 syndrome. *Developmental Cell*, *10*, 81–92.

Hiramoto, T., Kang, G., Suzuki, G., Satoh, Y., Kucherlapati, R., Watanabe, Y., et al. (2011). Tbx1: Identification of a 22q11.2 gene as a risk factor for autism spectrum disorder in a mouse model. *Human Molecular Genetics*, *20*, 4775–4785.

Huynh, T., Chen, L., Terrell, P., & Baldini, A. (2007). A fate map of Tbx1 expressing cells reveals heterogeneity in the second cardiac field. *Genesis*, *45*, 470–475.

Ivins, S., Lammerts van Beuren, K., Roberts, C., James, C., Lindsay, E., Baldini, A., et al. (2005). Microarray analysis detects differentially expressed genes in the pharyngeal region of mice lacking Tbx1. *Developmental Biology*, *285*(2), 554–569.

Jerome, L. A., & Papaioannou, V. E. (2001). DiGeorge syndrome phenotype in mice mutant for the T-box gene, Tbx1. *Nature Genetics*, *27*, 286–291.

Karkkainen, M. J., Ferrell, R. E., Lawrence, E. C., Kimak, M. A., Levinson, K. L., McTigue, M. A., et al. (2000). Missense mutations interfere with VEGFR-3 signalling in primary lymphoedema. *Nature Genetics*, *25*(2), 153–159.

Karkkainen, M. J., Jussila, L., Ferrell, R. E., Finegold, D. N., & Alitalo, K. (2001). Molecular regulation of lymphangiogenesis and targets for tissue oedema. *Trends in Molecular Medicine*, 7(1), 18–22.

Kelly, R. G. (2012). The second heart field. *Current Topics in Developmental Biology*, *100*, 33–65.

Kelly, R. G., Jerome-Majewska, L. A., & Papaioannou, V. E. (2004). The del22q11.2 candidate gene *Tbx1* regulates branchiomeric myogenesis. *Human Molecular Genetics*, *13*, 2829–2840.

Kispert, A., & Herrmann, B. G. (1993). The Brachyury gene encodes a novel DNA binding protein. *EMBO Journal*, *12*, 3211–3220.

Klotz, L., Norman, S., Vieira, J. M., Masters, M., Rohling, M., Dubé, K. N., et al. (2015). Cardiac lymphatics are heterogeneous in origin and respond to injury. *Nature*, *522*, 62–67.

Lania, G., Ferrentino, R., & Baldini, A. (2015). TBX1 Represses *Vegfr2* Gene Expression and Enhances the Cardiac Fate of VEGFR2+ Cells. *PLoS One*, *10*, e0138525.

Lania, G., Zhang, Z., Huynh, T., Caprio, C., Moon, A. M., Vitelli, F., et al. (2009). Early thyroid development requires a Tbx1-Fgf8 pathway. *Developmental Biology*, *328*, 109–117.

Larsimont, J.-C., Youssef, K. K., Sánchez-Danés, A., Sukumaran, V., Defrance, M., Delatte, B., et al. (2015). Sox9 controls self-renewal of oncogene targeted cells and links tumor initiation and invasion. *Cell Stem Cell*, *17*, 60–73.

Liao, J., Aggarwal, V. S., Nowotschin, S., Bondarev, A., Lipner, S., & Morrow, B. E. (2008). Identification of downstream genetic pathways of Tbx1 in the second heart field. *Developmental Biology*, *316*, 524–537.

Lindsay, E. A., Vitelli, F., Su, H., Morishima, M., Huynh, T., Pramparo, T., et al. (2001). Tbx1 haploinsufficieny in the DiGeorge syndrome region causes aortic arch defects in mice. *Nature*, *410*, 97–101.

Lingbeek, M. E., Jacobs, J. J., & van Lohuizen, M. (2002). The T-box repressors TBX2 and TBX3 specifically regulate the tumor suppressor gene p14ARF via a variant T-site in the initiator. *Journal of Biological Chemistry*, *277*, 26120–26127.

Long, J. M., LaPorte, P., Merscher, S., Funke, B., Saint-Jore, B., Puech, A., et al. (2006). Behavior of mice with mutations in the conserved region deleted in velocardiofacial/DiGeorge syndrome. *Neurogenetics*, *7*, 247–257.

Luna-Zurita, L., Stirnimann, C. U., Glatt, S., Kaynak, B. L., Thomas, S., Baudin, F., et al. (2016). Complex interdependence regulates heterotypic transcription factor distribution and coordinates cardiogenesis. *Cell*, *164*, 999–1014.

Martinez-Corral, I., Ulvmar, M. H., Stanczuk, L., Tatin, F., Kizhatil, K., John, S. W. M., et al. (2015). Nonvenous origin of dermal lymphatic vasculature. *Circulation Research*, *116*, 1649–1654.

McAlonan, G. M., Daly, E., Kumari, V., Critchley, H. D., van Amelsvoort, T., Suckling, J., et al. (2002). Brain anatomy and sensorimotor gating in Asperger's syndrome. *Brain: A Journal of Neurology*, *125*, 1594–1606.

McCabe, K. L., Atkinson, R. J., Cooper, G., Melville, J. L., Harris, J., Schall, U., et al. (2014). Pre-pulse inhibition and antisaccade performance indicate impaired attention modulation of cognitive inhibition in 22q11.2 deletion syndrome (22q11DS). *Journal of Neurodevelopmental Disorders*, *6*, 38.

Merscher, S., Funke, B., Epstein, J. A., Heyer, J., Puech, A., Min Lu, M. M., et al. (2001). TBX1 is responsible for cardiovascular defects in velo-cardio-facial/DiGeorge syndrome. *Cell*, *104*, 619–629.

Moon, A. M., Guris, D. L., Seo, J. H., Li, L., Hammond, J., Talbot, A., et al. (2006). Crkl deficiency disrupts fgf8 signaling in a mouse model of 22q11 deletion syndromes. *Developmental Cell*, *10*, 71–80.

Mori, A. D., & Bruneau, B. G. (2004). TBX5 mutations and congenital heart disease: Holt-Oram syndrome revealed. *Current Opinion in Cardiology*, *19*, 211–215.

Mori, A. D., Zhu, Y., Vahora, I., Nieman, B., Koshiba-Takeuchi, K., Davidson, L., et al. (2006). Tbx5-dependent rheostatic control of cardiac gene expression and morphogenesis. *Developmental Biology*, *297*, 566–586.

Müller, C. W., & Herrmann, B. G. (1997). Crystallographic structure of the T domain-DNA complex of the Brachyury transcription factor. *Nature*, *389*, 884–888.

Narlikar, L. (2013). MuMoD: A Bayesian approach to detect multiple modes of protein-DNA binding from genome-wide ChIP data. *Nucleic Acids Research*, *41*, 21–32.

Nowotschin, S., Liao, J., Gage, P. J., Epstein, J. A., Campione, M., & Morrow, B. E. (2006). Tbx1 affects asymmetric cardiac morphogenesis by regulating Pitx2 in the secondary heart field. *Development*, *133*, 1565–1573.

Ogata, T., Niihori, T., Tanaka, N., Kawai, M., Nagashima, T., Funayama, R., et al. (2014). TBX1 mutation identified by exome sequencing in a Japanese family with 22q11.2 deletion syndrome-like craniofacial features and hypocalcemia. *PLoS One*, *9*(3), e91598.

Pane, L. S., Zhang, Z., Ferrentino, R., Huynh, T., Cutillo, L., & Baldini, A. (2012). Tbx1 is a negative modulator of *Mef2c*. *Human Molecular Genetics*, *21*, 2485–2496.

Papaioannou, V. E. (2014). The T-box gene family: Emerging roles in development, stem cells and cancer. *Development*, *141*, 3819–3833.

Papangeli, I., & Scambler, P. J. (2013). Tbx1 genetically interacts with the transforming growth factor-β/bone morphogenetic protein inhibitor Smad7 during great vessel remodeling. *Circulation Research*, *112*, 90–102.

Paylor, R., Glaser, B., Mupo, A., Ataliotis, P., Spencer, C., Sobotka, A., et al. (2006). *Tbx1* haploinsufficiency is linked to behavioral disorders in mice and humans: Implications for 22q11 deletion syndrome. *Proceedings of the National Academy of Sciences of the United States of America*, *103*, 7729–7734.

Rana, M. S., Théveniau-Ruissy, M., De Bono, C., Mesbah, K., Francou, A., Rammah, M., et al. (2014). Tbx1 coordinates addition of posterior second heart field progenitor cells to the arterial and venous poles of the heart. *Circulation Research*, *115*, 790–799.

Randall, V., McCue, K., Roberts, C., Kyriakopoulou, V., Beddow, S., Barrett, A. N., et al. (2009). Great vessel development requires biallelic expression of Chd7 and Tbx1 in pharyngeal ectoderm in mice. *Journal of Clinical Investigation*, *119*, 3301–3310.

Roberts, C., Ivins, S., Cook, A. C., Baldini, A., & Scambler, P. J. (2006). Cyp26 genes a1, b1 and c1 are down-regulated in *Tbx1* null mice and inhibition of Cyp26 enzyme function produces a phenocopy of DiGeorge syndrome in the chick. *Human Molecular Genetics*, *15*, 3394–3410.

Saharinen, P., Eklund, L., Pulkki, K., Bono, P., & Alitalo, K. (2011). VEGF and angiopoietin signaling in tumor angiogenesis and metastasis. *Trends in Molecular Medicine*, *17*, 347–362.

Saharinen, P., Helotera, H., Miettinen, J., Norrmen, C., D'Amico, G., Jeltsch, M., et al. (2010). Claudin-like protein 24 interacts with the VEGFR-2 and VEGFR-3 pathways and regulates lymphatic vessel development. *Genes & Development*, *24*, 875–880.

Sambasivan, R., Gayraud-Morel, B., Dumas, G., Cimper, C., Paisant, S., Kelly, R. G., et al. (2009). Distinct regulatory cascades govern extraocular and pharyngeal arch muscle progenitor cell fates. *Developmental Cell*, *16*, 810–821.

Schaub, C., März, J., Reim, I., & Frasch, M. (2015). Org-1-dependent lineage reprogramming generates the ventral longitudinal musculature of the Drosophila heart. *Current Biology*, *25*, 488–494.

Schaub, C., Nagaso, H., Jin, H., & Frasch, M. (2012). Org-1, the Drosophila ortholog of Tbx1, is a direct activator of known identity genes during muscle specification. *Development*, *139*(5), 1001–1012.

Schnetz, M. P., Bartels, C. F., Shastri, K., Balasubramanian, D., Zentner, G. E., Balaji, R., et al. (2009). Genomic distribution of CHD7 on chromatin tracks H3K4 methylation patterns. *Genome Research*, *19*, 590–601.

Sinha, S., Abraham, S., Gronostajski, R. M., & Campbell, C. E. (2000). Differential DNA binding and transcription modulation by three T-box proteins, T, TBX1 and TBX2. *Gene*, *258*, 15–29.

Sinha, T., Li, D., Théveniau-Ruissy, M., Hutson, M. R., Kelly, R. G., & Wang, J. (2015). Loss of Wnt5a disrupts second heart field cell deployment and may contribute to OFT malformations in DiGeorge syndrome. *Human Molecular Genetics*, *24*, 1704–1716.

Sobin, C., Kiley-Brabeck, K., & Karayiorgou, M. (2005). Lower prepulse inhibition in children with the 22q11 deletion syndrome. *The American Journal of Psychiatry*, *162*, 1090–1099.

Stoller, J. Z., Huang, L., Tan, C. C., Huang, F., Zhou, D. D., Yang, J., et al. (2010). Ash2l interacts with Tbx1 and is required during early embryogenesis. *Experimental Biology and Medicine*, *235*, 569–576.

Swerdlow, N. R., Weber, M., Qu, Y., Light, G. A., & Braff, D. L. (2008). Realistic expectations of prepulse inhibition in translational models for schizophrenia research. *Psychopharmacology*, *199*, 331–388.

Tammela, T., Zarkada, G., Nurmi, H., Jakobsson, L., Heinolainen, K., Tvorogov, D., et al. (2011). VEGFR-3 controls tip to stalk conversion at vessel fusion sites by reinforcing Notch signalling. *Nature Cell Biology*, *13*, 1202–1213.

Vincent, S. D., & Buckingham, M. E. (2010). How to make a heart: The origin and regulation of cardiac progenitor cells. *Current Topics in Developmental Biology*, *90*, 1–41.

Vitelli, F., Morishima, M., Taddei, I., Lindsay, E. A., & Baldini, A. (2002). Tbx1 mutation causes multiple cardiovascular defects and disrupts neural crest and cranial nerve migratory pathways. *Human Molecular Genetics*, *11*, 915–922.

Vitelli, F., Taddei, I., Morishima, M., Meyers, E. N., Lindsay, E. A., & Baldini, A. (2002). A genetic link between Tbx1 and fibroblast growth factor signaling. *Development*, *129*, 4605–4611.

Wang, W., Razy-Krajka, F., Siu, E., Ketcham, A., & Christiaen, L. (2013). NK4 antagonizes Tbx1/10 to promote cardiac versus pharyngeal muscle fate in the ascidian second heart field. *PLoS Biology*, *11*, e1001725.

Watanabe, Y., Zaffran, S., Kuroiwa, A., Higuchi, H., Ogura, T., Harvey, R. P., et al. (2012). Fibroblast growth factor 10 gene regulation in the second heart field by Tbx1, Nkx2-5, and Islet1 reveals a genetic switch for down-regulation in the myocardium. *Proceedings of the National Academy of Sciences of the United States of America*, *109*, 18273–18280.

Xu, H., Cerrato, F., & Baldini, A. (2005). Timed mutation and cell-fate mapping reveal reiterated roles of *Tbx1* during embryogenesis, and a crucial function during segmentation of the pharyngeal system via regulation of endoderm expansion. *Development*, *132*, 4387–4395.

Xu, H., Chen, L., & Baldini, A. (2007). In vivo genetic ablation of the periotic mesoderm affects cell proliferation survival and differentiation in the cochlea. *Developmental Biology*, *310*, 329–340.

Xu, H., Morishima, M., Wylie, J. N., Schwartz, R. J., Bruneau, B. G., Lindsay, E. A., et al. (2004). Tbx1 has a dual role in the morphogenesis of the cardiac outflow tract. *Development*, *131*, 3217–3227.

Yamagishi, H., Maeda, J., Hu, T., McAnally, J., Conway, S. J., Kume, T., et al. (2003). Tbx1 is regulated by tissue-specific forkhead proteins through a common Sonic hedgehog-responsive enhancer. *Genes & Development*, *17*, 269–281.

Zhang, Z., & Baldini, A. (2008). In vivo response to high-resolution variation of Tbx1 mRNA dosage. *Human Molecular Genetics*, *17*(1), 150–157.

Zhang, Z., & Baldini, A. (2010). Manipulation of endogenous regulatory elements and transgenic analyses of the *Tbx1* gene. *Mammalian Genome*, *21*, 556–564.

Zhang, Z., Cerrato, F., Xu, H., Vitelli, F., Morishima, M., Vincentz, J., et al. (2005). *Tbx1* expression in pharyngeal epithelia is necessary for pharyngeal arch artery development. *Development*, *132*, 5307–5315.

Zhang, Z., Huynh, T., & Baldini, A. (2006). Mesodermal expression of Tbx1 is necessary and sufficient for pharyngeal arch and cardiac outflow tract development. *Development*, *133*, 3587–3595.

CHAPTER NINE

T-Box Genes in the Kidney and Urinary Tract

A. Kispert[1]
Institut für Molekularbiologie, Medizinische Hochschule Hannover, Hannover, Germany
[1]Corresponding author: e-mail address: kispert.andreas@mh-hannover.de

Contents

1. Introduction 246
2. T-Box Genes, an Evolutionary Conserved Family of Developmental Regulators 249
3. *Tbx18* in Ureter and Kidney Development 253
   3.1 *Tbx18*, a Member of a Vertebrate-Specific Subgroup of *Tbx* Genes, Acts as a Repressing Patterning Factor in Different Developmental Contexts 253
   3.2 *Tbx18* in Ureter Development 257
   3.3 *Tbx18* in Kidney Development 261
4. *Tbx1* in Kidney Development 262
   4.1 *Tbx1*, a Member of the *Tbx1*-Subfamily, Regulates Cardiofacial Development 262
   4.2 *Tbx1* in Kidney Development 263
5. *Tbx20* in Bladder/Cloacal Development 264
   5.1 *Tbx20*, a Member of the *Tbx1*-Subfamily with Crucial Function in Heart Development and Homeostasis 264
   5.2 *Tbx20* (*HrT*) in the Development of the Zebrafish Cloaca 265
6. *Tbx2* and *Tbx3*, a Closely Related Pair of T-Box Transcription Factor Genes in Kidney and Bladder Development 266
   6.1 *Tbx2* and *Tbx3* Regulate Proliferation and Differentiation Processes in Diverse Programs of Organogenesis 266
   6.2 *Tbx2* in *Xenopus* Pronephros Development 268
   6.3 *Tbx2* and *Tbx3* in Bladder Development and Homeostasis 269
7. Conclusion and Outlook 270
Acknowledgment 270
References 271

Abstract

T-box (*Tbx*) genes encode an ancient group of transcription factors that play important roles in patterning, specification, proliferation, and differentiation programs in vertebrate organogenesis. This is testified by severe organ malformation syndromes in mice homozygous for engineered null alleles of specific T-box genes and by the large number of human inherited organ-specific diseases that have been linked to mutations in these genes. One of the organ systems that has not been associated with loss of specific

Current Topics in Developmental Biology, Volume 122
ISSN 0070-2153
http://dx.doi.org/10.1016/bs.ctdb.2016.06.002

T-box gene function in human disease for long is the excretory system. However, this has changed with the finding that mutations in *TBX18*, a member of a vertebrate-specific subgroup within the *Tbx1*-subfamily of T-box transcription factor genes, cause congenital anomalies of the kidney and urinary tract, predominantly hydroureter and ureteropelvic junction obstruction. Gene expression analyses, loss-of-function studies, and lineage tracing in the mouse suggest a primary role for this transcription factor in specifying the ureteric mesenchyme in the common anlage of the kidney, the ureter, and the bladder. We review the function of *Tbx18* in ureterogenesis and discuss the body of evidence that *Tbx18* and other members of the T-box gene family, namely, *Tbx1*, *Tbx2*, *Tbx3*, and *Tbx20*, play additional roles in development and homeostasis of other components of the excretory system in vertebrates.

1. INTRODUCTION

The control of the water and ion balance of the blood and the excretion of excess water, solutes, and catabolites is of crucial importance to maintain body homeostasis in vertebrate life. The task is met by a two-component system consisting of kidneys where the blood is filtered and the filtrate modified, and a drainage unit that removes the urine from the kidney and expels it to the outside. In vertebrate evolution such an excretory (or urinary) system has evolved to different levels of complexity. The simplest realization is found in fish and amphibian embryos where a pair of single nephrons (the pronephros) is attached to a duct, which in turn is linked to the cloaca. The pronephros is a tubular epithelial system that obtains blood filtrate produced from vascular balls, the glomeruli, and deposited into the coelom, and modifies it by different solute transporters localized in its proximal, intermediate, and distal tubular aspect. The pronephric duct is a straight tube, and the cloaca a simple opening through the epidermis to the outside. Of intermediate complexity is the excretory system in adult fish and amphibia with a mesonephric kidney harboring a largely increased number of nephrons, while in amniotes a metanephric kidney and a drainage system has evolved that is adapted for water resorption and production of a highly concentrated urine in terrestrial life (Kardong, 2014).

In mouse and man, the excretory system is a multiorgan array consisting of two metanephric kidneys, two ureters, a single bladder, and the urethra. In the metanephric kidneys, thousands (in the mouse) to millions (in humans) of nephrons are interwoven in a complex spatial manner with the vascular system to filter the blood from glomeruli with specialized endothelial cells through a basal membrane into a double-layered epithelial sheet

(Bowman's capsule) with highly interdigitated podocytes in the inner leaflet. The filtrate is then modified and concentrated by selective secretion and resorption systems along the three main aspects of the tubular nephron, the proximal tubule, the loop of Henle, and the distal tubule. Through a connecting piece the urine is transferred into the highly arborized collecting duct system and funneled to the renal pelvis. The ureters are straight tubes that by means of their muscular, peristaltically active coat propel the urine from the renal pelvis to the bladder. The bladder is a muscular ball that can collect a large volume of urine before it is willingly released through another tube, the urethra. All components of the highly elaborated urinary drainage system, the ureters, the bladder, and the urethra are lined by a specialized urothelium with a three-layered organization of superficial cells, intermediate cells, and basal cells providing sealing toward the interstitial space.

Although the contiguous and highly integrated nature of the organs of the urinary system suggests a common developmental origin, they are, in fact, derived from different epithelial and mesenchymal progenitor pools of different germ layers. The lower urinary tract (bladder and urethra) derives from an endodermal infolding and its surrounding mesenchyme, whereas the upper urinary system (the kidneys and ureters) originates from the epithelial nephric duct and its mesenchymal surrounding in the intermediate mesoderm. In the mouse, development of the upper urinary system starts around embryonic day (E) 8.5. At this stage, several pronephric tubules form within the urogenital ridge at the level of the future forelimb buds. These tubules are very short-lived but give rise to a more permanent structure that is crucial for the development of the upper urinary tract, the nephric (or Wolffian) duct. The nephric duct is a straight tube that elongates posteriorly in the urogenital ridge while inducing mesonephric tubules in the adjacent mesenchyme. Nephric duct and mesonephric tubules constitute the mesonephric kidney that, however, does not become a functional organ in the mouse. At E9.5, the nephric duct reaches the endodermal infolding of the cloaca and the epithelia of the two tissues fuse. Slightly anterior to this point, at the level of the future hindlimb, a bud emerges from the nephric duct at E10.5 and extends into the surrounding mesenchyme. The proximal end of this ureter bud engages into repetitive rounds of branching and elongation steps and generates the collecting duct system of the kidney. From the distal ureter bud derives the epithelial component of the ureter by continuous elongation in the absence of bifurcation events and by differentiation into the three urothelial cell layers from E15.5 onward. The mesenchyme in which the ureter bud invades is subdivided into at least four lineages to drive and support the development of the different epithelial tissues. The mesenchyme engulfing the nephric duct will become the coat of the future vas deferens in male embryos, while in females it will degenerate together with the epithelial tube. The mesenchyme surrounding the ureter stalk is patterned into three compartments between E12.5 and E16.5 from which the inner *lamina propria* with its fibrocytes, the intermediate thick layer of smooth muscles cells, and an outer fibroblast layer of the *tunica adventitia* will subsequently differentiate. The mesenchyme surrounding the proximal ureter tip will generate nephrons by a repetitive process of aggregation, epithelial transition, morphogenesis, and differentiation along the proximal–distal tubular axis. An additional mesenchymal layer engulfing the cap mesenchyme will largely contribute to stromal and mesangial cells within the kidney (Bohnenpoll et al., 2013; Mugford, Sipila, McMahon, & McMahon, 2008; for reviews on ureter and kidney development, see Bohnenpoll & Kispert, 2014; Little & McMahon, 2012; McMahon, 2016).

The endodermal infolding of the cloaca does not only represent the exit point of the urinary system, it is also the terminus of the genital and alimentary tract in the early embryo (in the mouse at E9.5–E10.5). Starting from E10.5 to E13.5, the cloaca is partitioned into an anorectal canal dorsally (from which the rectum arises), and a urogenital sinus ventrally with the mesodermal urorectal septum lying between. This developmental program is mediated by ingrowth of three endodermally lined mesenchymal tissue folds, and their fusion at the midline (for a review on bladder development, see Georgas et al., 2015).

Connectivity between the upper and lower urinary system, ie, between the ureter and the bladder, is established through a complex morphogenetic program. As mentioned, the nephric duct contacts the cloacal epithelium at E9.5. The piece distal to the ureteric bud, which is the common nephric duct, is then degraded by apoptosis. The distal end of the remaining ureter subsequently lies down onto the bladder wall to be removed by programmed cell death as well, allowing the integration of the ureter into the bladder distant from the nephric duct/vas deferens, and establishing patency of the vesicoureteric junction (for a review on distal ureter development, see Uetani & Bouchard, 2009).

Considering the complexity of cell interactions between the different primordial tissues in the early metanephric field, it is obvious that numerous signaling systems operate in coordinating the behavior of the different cellular lineages (Combes, Davies, & Little, 2015). Transcription factors integrate the diverse signaling inputs and translate them into adjustments of gene expression profiles that precede and cause the changes of cell behavior including specification, differentiation, and proliferation. Genetic analyses in the mouse as well as insight from human disease models have identified a large set of transcription factors that pattern and specify the various mesenchymal and epithelial cell lineages and direct their proper temporal differentiation. These include members of the homeobox, basic helix–loop–helix, and zinc-finger classes of transcription factors (Little & McMahon, 2012; McMahon, 2016). For some time it seemed that T-box transcription factor genes that have been implicated in the development of most other organs and tissues are a notable exception. Work in recent years, however, has provided compelling evidence that members of this gene family are expressed in different tissues of the developing and mature urinary system and are of functional relevance at least for some of the numerous cellular programs therein.

2. T-BOX GENES, AN EVOLUTIONARY CONSERVED FAMILY OF DEVELOPMENTAL REGULATORS

The prototypical member of this family surfaced in 1990 when a positional cloning effort isolated the gene underlying the murine developmental mutation *Brachyury* (*T*) that causes a short-tail phenotype in heterozygous mice and a severe posterior truncation in homozygous embryos (Herrmann, Labeit, Poustka, King, & Lehrach, 1990). The *T* gene is expressed in the notochord and the primitive streak of mouse embryos, but the sequence of the protein did not offer any clues for a possible cellular function at that time (Wilkinson, Bhatt, & Herrmann, 1990). Subsequent molecular analysis, however, revealed that the T protein localizes to the nucleus and binds specifically to DNA fragments harboring the sequence AGGTGTGAAA as a palindrome. The region required for specific binding to this site was ascribed to the stretch of 180 N-terminal amino acid residues and coined T-box (Kispert & Herrmann, 1993). Crystallographic analysis revealed that the Brachyury T-box binds to this palindromic T-box-binding element (TBE) as a dimer contacting via α-helices the minor and major groove of the double helix (Muller & Herrmann, 1997). Later studies identified variations of TBEs in promotor/enhancer regions of target genes upon binding to which Brachyury activates transcriptional programs important for notochord maintenance and mesoderm formation (Kispert, Koschorz, & Herrmann, 1995; Morley et al., 2009). While it became quickly clear that this role of Brachyury is conserved throughout vertebrates (Herrmann & Kispert, 1994), it came as a surprise when orthologs of the gene were also found in invertebrates like insects, crustacea, and sea urchins which lack a notochord and have very different modes of mesoderm formation (Harada, Yasuo, & Satoh, 1995; Kispert, Herrmann, Leptin, & Reuter, 1994).

Shortly before the characterization of the T-box as a novel DNA-binding motif in the Brachyury protein (Kispert & Herrmann, 1993), it was realized that the *Drosophila* Optomotor-blind (Omb) protein harbors a region with general DNA-binding specificity and sequence similarity to the N-terminal region of Brachyury, suggesting that *Brachyury* and *omb* are members of a larger gene family with a conserved T-box (Pflugfelder, Roth, & Poeck, 1992). Exploiting the sequence similarity between the T-box of T and Omb, Bollag and coworkers used a PCR-based approach and cloned a small set of related T-box containing genes, *Tbx1*, *Tbx2* (which turned out to be the ortholog of *omb*), and *Tbx3* in the mouse (Bollag et al., 1994). Further physical and bioinformatical screening identified in the following years a large number of additional T-box genes in both vertebrates and invertebrates. To the present day, 17 family members have been identified in mammals, which can be subdivided in 5 subfamilies (*T*, *Tbx1*, *Tbx2*, *Tbx6*, *Tbr1*) based on sequence conservation within the T-box (Fig. 1) (Papaioannou, 2014). Four *Tbx* subfamilies (*T*, *Tbx1*, *Tbx2*, *Tbx7/8*) do not only predate vertebrate radiation but were present at the onset of metazoan evolution. *Brachyury* orthologs were also found in sponges, unicellular organisms, and filamentous fungi, arguing that this gene is the most ancient member of the family predating the evolutionary appearance of metazoa (Degnan, Vervoort, Larroux, & Richards, 2009; Papaioannou, 2014; Sebe-Pedros et al., 2013). It should be noted that evolution of the T-box gene family is highly dynamic, with secondary losses, rapid evolution, and expansion of some members in certain species or phyla (Sebe-Pedros et al., 2013). Nonetheless, the specific DNA-binding property of the T-box has been retained throughout evolution. Most if not all T-box transcription factors bind to the TBE initially identified for Brachyury in vitro albeit sequence variants exist, and the orientation, spacing, and number of TBEs in promotor/enhancer regions vary (Sebe-Pedros et al., 2013). Specificity of T-box gene function may, therefore, not arise from recognizing different binding sites but from specific expression patterns and variable interaction with transcriptional cofactors. In fact, T-box genes exhibit highly specific developmental expression patterns. Functional analysis of the T-box gene family particularly in the mouse has revealed important

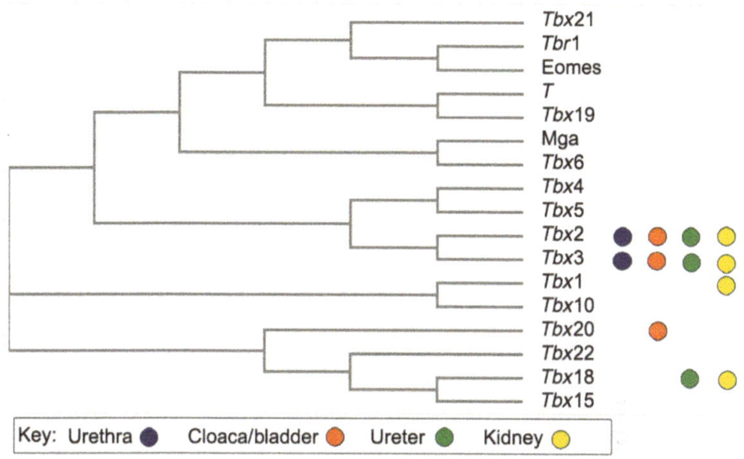

Fig. 1 Phylogenetic tree of the T-box gene family in the mouse. Cladogram of the phylogenetic relation of mammalian T-box proteins based on sequence comparison of the amino acid sequence of the T-box using the program Clustal Omega with the neighbor-joining clustering method. *Colored circles* on the *right* indicate expression and likely involvement of a particular gene in the specific organ of the excretory system as discussed in this review. Urethra, *blue*; bladder, *red*; ureter, *green*; kidney, *yellow*.

functions in regulating patterning, cell fate decisions, and proliferation in many embryonic tissues. One T-box gene can modulate transcription in different developmental contexts, thus regulating different cellular programs. Different T-box genes can act in a combinatorial or hierarchical fashion to regulate specific aspects of development of one organ. Prominent examples for the latter are the heart and the limb where six T-box genes act together to regulate the patterning and axial extension of the embryonic rudiment (Greulich, Rudat, & Kispert, 2011; Papaioannou, 2014). Expression and functional analyses now suggest that some members of the family are also essential in the development of the excretory system (Fig. 1; Table 1). We will describe in detail the findings on the role of *Tbx18* in ureter and kidney development, and discuss the body of evidence that *Tbx1*, *Tbx2*, *Tbx3*, and *Tbx20* regulate various aspects of kidney and bladder development and homeostasis in vertebrates. We precede the sections on excretory system-specific function of these T-box genes with short descriptions of their evolution, extrarenal expression and function, and biochemical activity to provide additional background for the reader.

Table 1 Summary of Expression and Associated Phenotypes of T-Box Genes in the Excretory System of Vertebrates

Tbx Gene	Species	Expression in Excretory System	Associated Phenotype in the Excretory System	References
Tbx1	Mouse	Adult kidney	Affects Tgfβ-signaling	Fu et al. (2014) and Jiang, Li, Yang, et al. (2014)
	Rat	Tubules in fetal renal cortex	–	Jiang, Li, Li-Ling, et al. (2014)
Tbx2	*Xenopus*	Mesoderm surrounding the pronephros, proctodeum	Increased size of pronephric nephrons	Hayata, Kuroda, Eisaki, and Asashima (1999) and Cho, Choi, Park, and Han (2011)
	Zebrafish	Proctodeum	–	Pyati, Cooper, Davidson, Nechiporuk, and Kimelman (2006)
	Mouse	Nephric duct, urethra	–	Chapman et al. (1996) and Douglas, Heng, Sauer, and Papaioannou (2012)
	Man	Fetal and adult kidney		Campbell, Goodrich, Casey, and Beatty (1995) and Law, Gebuhr, Garvey, Agulnik, and Silver (1995)
Tbx3	Mouse	Nephric duct, urethra	–	Chapman et al. (1996) and Douglas et al. (2012)
	Rat	Bladder urothelium		Ito, Asamoto, Hokaiwado, Takahashi, and Shirai (2005)
	Human	–	UMS, urinary tract defects, kidney agenesis	Gonzalez, Herrmann, and Opitz (1976) and Pallister, Herrmann, and Opitz (1976)

Table 1 Summary of Expression and Associated Phenotypes of T-Box Genes in the Excretory System of Vertebrates—cont'd

Tbx Gene	Species	Expression in Excretory System	Associated Phenotype in the Excretory System	References
	Human	Bladder urothelium	Silencing in bladder tumors, tumor progression	Kandimalla et al. (2012), Beukers et al. (2015), and Du et al. (2014)
Tbx18	Mouse	Urogenital ridge, ureteric mesenchyme	Short hydroureter, hydronephrosis	Kraus, Haenig, and Kispert (2001a), Airik, Bussen, Singh, Petry, and Kispert (2006), Bohnenpoll et al. (2013), and Xu, Nie, Cai, Cai, and Xu (2014)
	Mouse	Renal stromal cells	Reduced vascular density, dilation of glomeruli, reduced smooth muscle and mesangial cells	Xu et al. (2014)
	Human	–	UPJO, VUJO, hydroureter, megaureter, ureterocoel, renal hypoplasia, renal dysplasia, duplex kidneys	Vivante et al. (2015)
Tbx20	Zebrafish	Ventral mesoderm, proctodeum/cloaca	Dysmorphic cloaca	Pyati et al. (2006)

UMS, ulnar-mammary syndrome; *UPJO*, ureteropelvic junction obstruction; *VUJO*, vesicoureteric junction obstruction.

3. *Tbx18* IN URETER AND KIDNEY DEVELOPMENT

3.1 *Tbx18*, a Member of a Vertebrate-Specific Subgroup of *Tbx* Genes, Acts as a Repressing Patterning Factor in Different Developmental Contexts

The history of *Tbx18* can be traced back to 1999 when a database homology search identified four novel human T-box-expressed sequence tags (ESTs)

including one for *Tbx18* (Yi et al., 1999). Degenerate RT-PCR approaches using primers against conserved regions of the T-box shortly thereafter isolated highly related cDNAs in the mouse, zebrafish, and chick (Begemann, Gibert, Meyer, & Ingham, 2002; Haenig & Kispert, 2004; Kraus et al., 2001a). A *Xenopus Tbx18* cDNA was identified some time later by Blast search in the EST Database, complementing the set of orthologs in the main vertebrate model systems (Jahr, Schlueter, Brand, & Manner, 2008). Sequence comparisons with other T-box genes in the mouse showed that *Tbx18* is most closely related to *Tbx15. Tbx15* and *Tbx18* share the biggest similarity with *Tbx22*, together forming a subgroup within the *Tbx1*-subfamily of murine T-box genes (Fig. 1) (Papaioannou, 2014). In contrast to the other two subgroups of this subfamily, *Tbx1/Tbx10* and *Tbx20*, which are conserved in invertebrates, the *Tbx15/Tbx18/Tbx22*-subgroup is vertebrate specific. A single gene related to the *Tbx15/Tbx18/Tbx22*-subgroup was found in the genome of the tunicate *Ciona intestinalis* (*Ci-Tbx15/18/22*) and in the lancelet *Branchiostoma floridae* (*Bf-Tbx15/Tbx18/Tbx22*), suggesting that the three present day genes in vertebrates arose from a common chordate-specific precursor by two consecutive gene duplication events prior or at the onset of fish evolution (Paps, Holland, & Shimeld, 2012; Ruvinsky, Silver, & Gibson-Brown, 2000). *Tbx18* orthologs share a highly similar organization with eight coding exons distributed over some 30 kbps of genomic DNA. Transcription and splicing generate a single large transcript that is translated into a protein with a centrally located T-box flanked by large N- and C-terminal domains (Fig. 2A).

Insight in the basic activity of Tbx18 as a transcription factor has been obtained from detailed biochemical analysis of the mouse protein (Farin et al., 2007). Deletion analysis in cells uncovered that nuclear localization of Tbx18 depends on a short stretch of basic amino acid residues localized at the N-terminus of the protein, resembling a classical nuclear localization signal. A PCR-based in vitro cyclic-binding site selection protocol revealed that the Tbx18 protein prefers an imperfect palindrome of two antiparallel TBEs for optimal DNA binding. Direct and palindromic repeats of TBEs are also bound but binding depends on regions outside the T-box. Binding to the different DNA-sites occurs as a dimer; heterodimerization with the closely related Tbx15 protein was also observed. Tbx18 acts as a transcriptional repressor, with the repressive activity depending at least partly on a conserved eh1-motif in the N-terminus of the protein (Fig. 2A). This motif is a stretch of seven amino acid residues that mediates binding to the different members of the groucho family of corepressors. Interestingly, deletion of the eh1-motif abrogates the repressive activity of Tbx18 only partially,

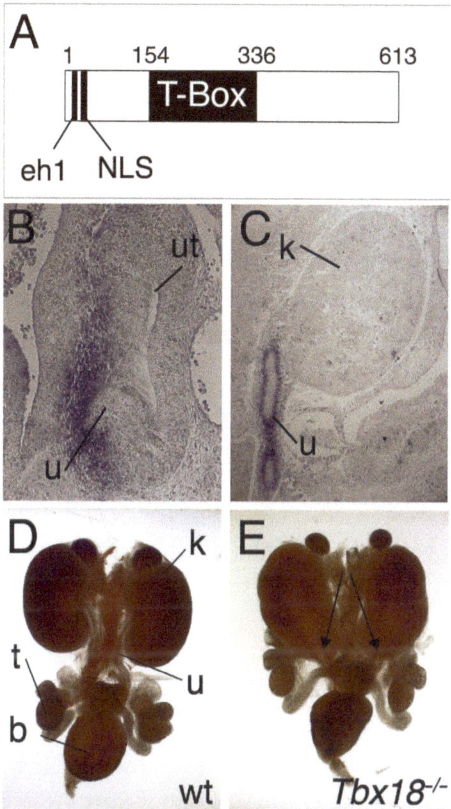

Fig. 2 *Tbx18* in the excretory system. (A) Domain structure of the Tbx18 protein in the mouse. Numbers refer to the position in the amino acid sequence. The eh1-motif (position 18–28) mediates binding to groucho corepressors. The NLS (position 32–45) is a short stretch of basic amino acid residues essential for nuclear localization. (B and C) In situ hybridization analysis on paraffin sections reveals expression of *Tbx18* in the developing upper urinary tract. (B) At E11.5, *Tbx18* expression is confined to a band of mesenchyme between the nephric duct and the mesenchyme surrounding the ureter tips. (C) At E14.5, *Tbx18* expression is restricted to the mesenchyme surrounding the ureteric epithelium. Expression in the kidney is not detected. (D and E) Morphological inspection of whole urogenital systems of wild-type (wt) (D) and $Tbx18^{-/-}$ (E) embryos. (E) *Arrows* point to the short dilated ureter in *Tbx18*-deficient mice. Also note that the kidney is widened due to hydronephrosis and that the testis is attached with tissue ligaments to the kidney in the mutant situation. *b*, bladder; *k*, kidney; *t*, testis, *u*, ureter; *ut*, ureteric tips. *Images in (B–E) are a courtesy of Anna-Carina Weiss.*

indicating that additional regions of the protein, most likely in the C-terminus are implicated in this activity (Farin et al., 2007). Binding assays performed for different T-box proteins showed that the T-box is not only a DNA-binding domain but also mediates interaction with transcription factors and cofactors. This could also be confirmed for Tbx18 for which specific binding to the homeodomain transcription factors Nkx2-5, Pax3, and Six1 was reported (Farin et al., 2007; Farin, Mansouri, Petry, & Kispert, 2008; Nie, Sun, Gordon, Cai, & Xu, 2010).

Detailed expression analysis using in situ hybridization on whole embryos and on sections described *Tbx18* expression at highly restricted sites mostly of mesodermal origin during mouse development. Expression starts in the paraxial mesoderm at E7.75 and is maintained therein at least until E14.5 when somitogenesis comes to a halt. Expression is found in two anterior stripes in the anterior presomitic mesoderm, in the anterior compartment of the forming and differentiating somites where it becomes confined to the lateral sclerotomal compartment, and in the unsegmented cranial paraxial mesoderm. There, expression is restricted to the inner ring of mesenchyme surrounding the otic vesicle from which the otic fibroblast compartment will arise (Kraus et al., 2001a). *Tbx18* is very highly expressed in the developing epicardium of the heart. Expression starts in the septum transversum at E8.0, continues in the proepicardium, and is found in the epicardium and pericardium until at least E14.5 (Christoffels et al., 2009; Greulich, Farin, Schuster-Gossler, & Kispert, 2012; Kraus et al., 2001a). Around E10.5–E14.5 expression starts at several other cardiac sites where it is maintained until E18.5. These include the myocardium of the septum transversum and the left ventricle, as well as the developing sinus horns and the sinoatrial node (Christoffels et al., 2009). Expression of *Tbx18* also marks dermal papilla precursor cells during embryonic hair follicle morphogenesis (Grisanti et al., 2013; Kraus et al., 2001a).

Tbx18 is also strongly expressed in limb development. At E9.25, expression starts in the anterior-proximal region within the forelimb bud mesenchyme. At E10.5 and E11.5, an additional weaker expression domain is present in the posterior-distal region. At E12.5, *Tbx18* expression is largely excluded from the precartilagenous condensations but presents in the surrounding mesenchyme (Kraus et al., 2001a). Expression of *Tbx18* is conserved at least at some of these sites in vertebrates including the somites, the appendages (fin/limb), and particularly in the developing epicardium (Begemann et al., 2002; Haenig & Kispert, 2004).

The functional significance of these expression domains was analyzed in the mouse using gene-targeting approaches. From these studies we know that *Tbx18*-deficient mice die shortly after birth exhibiting severe defects in a number of organ systems. The vertebral column is malformed due to expansion of the lateral elements of the vertebrae, the pedicles, and the proximal rib compartment. These defects were traced to a role of *Tbx18* in maintaining the anterior somite compartment (Bussen et al., 2004). *Tbx18*-deficient mice are also deaf due to a failure in otic fibrocyte differentiation (Trowe, Maier, Schweizer, & Kispert, 2008) and have a smaller sinoatrial node probably due to less precursor recruitment and differentiation (Christoffels et al., 2006). Together these findings suggest a role for *Tbx18* in patterning primordial tissues in different organ contexts.

3.2 *Tbx18* in Ureter Development

Phenotypic characterization of *Tbx18*-deficient embryos also uncovered severe anomalies of the urinary system that were correlated with expression of the gene in the ureteric mesenchyme. Initial analysis identified expression of *Tbx18* in the urogenital ridge in mouse embryos (Kraus et al., 2001a). Using comparative marker analysis, the domain of *Tbx18* within the urogenital ridge was further resolved (Bohnenpoll et al., 2013). At E9.5, *Tbx18* expression is excluded from the mesonephric mesenchyme and confined to the medial mesenchyme of the ridge close to the dorsal aorta. Transient expression occurs in the coelomic epithelium at E9.5. Interestingly, this domain gives rise to steroidogenic precursor cells of the gonads and the adrenal glands as detected by genetic lineage tracing using a mouse line with a *cre* knock-in into the *Tbx18* locus (Bohnenpoll et al., 2013; Hafner, Bohnenpoll, Rudat, Schultheiss, & Kispert, 2015). At E10.5, the mesenchymal expression of *Tbx18* is maintained in the urogenital ridge and expanded posteriorly into the early metanephric field. Here, expression is restricted to a small band of mesenchymal cells between the mesenchyme abutting the nephric duct and the nephrogenic mesenchyme that surrounds the ureteric tips (Fig. 2B). Comparative marker analysis at this stage and at E11.5 showed that *Tbx18* expression is clearly excluded from the nephrogenic mesenchymal lineage marked by *Uncx* and *Osr1* expression and the renal stromal lineage that is positive for *Foxd1*. At E12.5, *Tbx18* is found in the entire mesenchyme surrounding the epithelium of the distal ureter stalk before expression is restricted to the inner ring of mesenchymal cells adjacent to the ureteric epithelium. After E14.5, *Tbx18* is down-regulated in the ureteric mesenchyme but persists at low levels at least until birth (Fig. 2C) (Bohnenpoll et al., 2013). Together, these data show that *Tbx18* marks a molecularly distinct subpopulation of mesenchymal cells in the early metanephric field. Using again a *Tbx18*^{cre}-based genetic lineage tracing, it was found that all mesenchymal cells of the ureter derive from the

Tbx18-positive domain. However, *Tbx18*-descendants also contribute to a large degree to the stromal, mesangial, and smooth muscle cells of the medial region of the kidney and to the bladder mesenchyme, arguing that the *Tbx18*-positive cells of the early metanephric field are initially multipotent. Further DiI injection experiments in E11.5 kidney rudiments found out that only the cells directly adjacent to the ureteric epithelium contribute to the ureteric wall, whereas *Tbx18*-positive cells up to 200 μm distance from the epithelium contribute to the renal stroma. Cells beyond this distance die by apoptosis and may play a role in severing the kidney from the ureter and the nephric duct (Bohnenpoll et al., 2013).

Given the restriction of *Tbx18* expression to prospective and definitive ureteric mesenchyme, it is intriguing to ask how this restriction is achieved and what its relevance is. While the latter point has not yet been addressed experimentally, insight into the signals and factors that restrict *Tbx18* expression has been gained from embryological manipulation experiments. Removal of the ureteric epithelium resulted in loss of *Tbx18* expression, whereas transplantation of a ureteric epithelium into the *Tbx18*-positive lateral domain maintained *Tbx18* expression, arguing that *Tbx18* transcription is induced or at least maintained by epithelial signals (Bohnenpoll et al., 2013). Work by Trowe and coworkers showed that loss of canonical Wnt-signaling in the ureteric mesenchyme leads to loss of *Tbx18* expression, suggesting that epithelial Wnt signals (most likely Wnt9b and/or Wnt7b) act in trans to maintain *Tbx18* expression in the adjacent mesenchyme (Trowe et al., 2012). It must be noted that Wnt-signaling is not present in the *Tbx18*-positive domain at E10.5 and E11.5. Therefore, other factors are likely to induce the localized expression of *Tbx18* at this stage before epithelial Wnt-signaling kicks-in at E12.5 (Trowe et al., 2012).

The regulatory elements that account for the integration of Wnt- and other signaling inputs to precisely localize *Tbx18* to the ureteric mesenchyme have remained enigmatic. From Bac transgenesis we know that most but not all regulatory elements for *Tbx18* expression are located within a region 102 kbp upstream and 81 kbp downstream of the transcription unit (Wang, Tripathi, et al., 2009). Further, analysis of a novel allele of *Tbx18* generated by a reciprocal translocation breaking 78 kbp downstream of the gene, *12Gso*, identified an enhancer element at the breakage point that directs expression in the ureteric mesenchyme, opening up avenues to identify the binding sites for transcription factors involved in regulating *Tbx18* expression in this tissue (Bolt et al., 2014).

Mice homozygous for a null allele of *Tbx18* are born but die within the next day (Bussen et al., 2004). The urogenital system is severely affected, featuring a very short highly dilated ureter (hydroureter), and probably as a secondary consequence a dilated renal pelvis, commonly referred to as hydronephrosis (Fig. 2D and E). The ureteric mesenchyme is sparse, and smooth muscle cells are completely absent; urothelial differentiation is not apparent. Ectopic fibrous tissue is present around the kidney tethering the gonads to the kidney in both sexes (Airik et al., 2006; Wu, Dong, Regan, Su, & Majesky, 2013; Xu et al., 2014). Ink injection experiments showed that the vesicoureteric junction is obstructed in the majority of the cases (Airik et al., 2006). Reinspection of a larger number of urogenital systems of *Tbx18*-deficient newborn mice in a later study detected ureteropelvic junction obstruction in 74% of newborn embryos; the remaining 26% featured a short bilateral hydroureter. Only 7% of *Tbx18*-heterozygous mice had a mild proximal hydroureter, demonstrating a very minor role for haploinsufficiency (Vivante et al., 2015). Analysis at earlier embryonic stages traced the onset of phenotypic alterations back to the period between E11.5 and E12.5. (Prospective) ureteric mesenchymal cells are present at E11.5 but fail to condense around the ureteric epithelium. Instead these cells disperse and largely contribute to interstitial stroma and capsule tissue of the kidney (Airik et al., 2006). *Tbx18*-deficient mesenchymal cells no longer respond to epithelial signals and do not activate sonic hedgehog (Shh)- and Wnt-signaling, they also fail to maintain/express markers specific for the ureteric mesenchyme including *Bmp4* and *Sox9* (Airik et al., 2006; Bohnenpoll et al., 2013). Since both genes are required for smooth muscle differentiation in the ureter, their downregulation may contribute to the differentiation defects of the *Tbx18*-deficient ureteric mesenchyme (Airik et al., 2010; Wang, Brenner-Anantharam, Vaughan, & Herzlinger, 2009).

Together, these findings indicate that *Tbx18* is required to maintain the ureteric mesenchymal cell lineage. Given the finding that Tbx18 acts as a transcriptional repressor in vitro (Farin et al., 2007), it seems likely that Tbx18 does not actively induce a ureteric mesenchymal fate but represses genes crucial for the specification of adjacent mesenchymal cell lineages including the stromal and nephrogenic mesenchyme. Microarray analysis of *Tbx18*-deficient ureteric mesenchyme in combination with ChIP-seq analysis may help to unravel the targets of Tbx18 activity in the ureteric mesenchyme in the future.

Insight into a relevant protein interaction partner for Tbx18 in the ureteric mesenchyme has recently been obtained. Mice deficient for the homeodomain transcription factor gene *Six1* exhibit a delay in ureteric smooth muscle differentiation resulting in weak hydroureter formation at newborn stages. Mice with an additional loss of one allele of *Tbx18* have a strongly aggravated phenotype with bilateral megaureter. *Six1* is coexpressed with *Tbx18* in the undifferentiated ureteric mesenchyme, and Six1 physically interacts with Tbx18 in vitro and in vivo (Nie et al., 2010). It remains to be seen which specific molecular subprogram is regulated by Six1/Tbx18 complexes in the ureteric mesenchyme.

Congenital abnormalities of the ureter frequently occur in human newborns, affecting up to 1% of the population. They comprise a large set of disease manifestations including ureteropelvic and vesicoureteral junction obstruction, megaureter, duplex ureter, or vesicoureteral reflux, with the first being the most prevalent (Smith, Stablein, Munoz, Hebert, & McDonald, 2007). In combination or alone they frequently lead to hydronephrosis, dilatation of the calyces, and renal pelvis, which cause chronic kidney disease and may require kidney transplantation. Congenital ureter anomalies are a subgroup of congenital anomalies of the kidney and urinary tract (CAKUT) that all arise from perturbations of the genetic programs that guide the development of the tissue architecture of the organs of the excretory system or their connectivity (Chen, 2009; Woolf, 2000). To date, only 10% of cases can be explained by mutations in specific regulators of these programs, leaving the large bulk of cases without genetic explanation (Hwang et al., 2014; Saisawat et al., 2014). Considering the strong urinary tract defects associated with loss of *Tbx18* in the mouse, mutations in *TBX18* were obvious candidates for human CAKUT as well. The Hildebrandt lab recently identified a large Hispanic family with an autosomal dominant form of CAKUT with predominant ureteropelvic junction obstruction. Whole exome sequencing uncovered a heterozygous mutation leading to a TBX18 protein truncated after the T-box region in affected family members. Screening of additional CAKUT families identified two heterozygous *TBX18* mutations in another three families. All mutant proteins exhibited normal nuclear localization; homodimerization and heterodimerization with homeobox transcription factors including Six1 and Pax3 were unaffected. DNA binding was affected in one of the mutants; the protein half-life was largely increased and transcriptional repression activity reduced in all three mutant proteins. These data indicate that the mutant proteins act in a dominant-negative fashion probably by sequestering the wild-type protein in an inactive complex (Vivante et al., 2015). While the cellular changes that cause ureteropelvic junction obstruction and hydroureter in these patients can obviously not been analyzed, it is tempting to assume that defects in smooth muscle differentiation may contribute to the observed phenotypic changes.

3.3 *Tbx18* in Kidney Development

Xu and coworkers recently analyzed whether *Tbx18* has an independent function in kidney development (Xu et al., 2014). Using a reporter analysis from a $Tbx18^{lacZ}$ allele they detected expression of *Tbx18* in stromal cells of the kidney including the renal capsule, vascular smooth muscle cells, and mesangial cells. *In situ* hybridization on sections confirmed this finding albeit the signal was spotty and not in complete overlap with that provided by the reporter. Analysis of *Tbx18*-deficient embryos revealed a dilatation of the glomerular loop and reduced mesangial cells, defects that were subsequently traced to reduced proliferation of the mesenchymal glomerular core. Disorganization of the renal capsule and reduced smooth muscle investment of the vessels was described as another hallmark of the mutant kidney, with the latter being associated with enhanced apoptosis of perivascular mesenchyme. The authors concluded that *Tbx18* has an independent function in renal stromal cell development (Xu et al., 2014).

While the phenotypic changes observed in these *Tbx18*-deficient kidneys to a large degree support an earlier report with a different mutant allele (Airik et al., 2006), a claim of an independent function of *Tbx18* in stromal cell development seems precocious at this point for a couple of reasons. Previous work showed that in *Tbx18*-deficient kidneys, prospective ureteric mesenchymal cells adopt a stromal fate and largely contribute to all stromal cell compartments, particularly the renal capsule. Dislocation of cells resulted in severe changes of the tissue architecture of the mutant kidney with loss of the pelvic structure, mispositioning of the ureteric orifice and altered branching of the collecting duct tree from E12.5 onward. With onset of urine production at E15.5, dilation of the pelvis occurred leading to typical hydronephrotic lesions including cyst formation (Airik et al., 2006; Bohnenpoll et al., 2013). Since the loss of *Tbx18* in the ureteric mesenchyme has such dramatic consequences for renal stromal development from at least E12.5 onward, an independent analysis of a renal requirement of *Tbx18* seems difficult. Furthermore, *in situ* hybridization and reporter gene analysis did not detect *Tbx18* expression in the kidney (Airik et al., 2006; see also Fig. 2C). This discrepancy may reflect different sensitivity or specificity of the probes used. Moreover, reporter activity from the *Tbx18*^{lacZ} allele used in the study by Xu et al. (2014) does not recapitulate endogenous expression of *Tbx18* in other expression domains of the gene (Cai et al., 2008), unlike the allele used in the study by Airik et al. (2006). In fact, it seems plausible that the *Tbx18*^{lacZ} allele used by Xu et al. lacks a repressive element for *Tbx18* itself since the pattern of the reporter activity resembles that observed in *Tbx18*-deficient mice (Airik et al., 2006). A stromal cell-specific inactivation of *Tbx18* using a conditional approach with a *Foxd1*^{cre} line may solve the controversy on an independent role of *Tbx18* in the stromal cell lineage in the future.

4. *Tbx1* in Kidney Development

4.1 *Tbx1*, a Member of the *Tbx1*-Subfamily, Regulates Cardiofacial Development

Tbx1 is a member of the *Tbx1*-subfamily of vertebrate T-box transcription factor genes most closely related to *Tbx10*. Evolutionary emergence of *Tbx1* predates vertebrate evolution as orthologs of the gene are also found in invertebrates (Papaioannou, 2014; Paps et al., 2012). In the mouse, *Tbx1* is expressed in the pharyngeal endoderm and mesoderm, and in the cardiac outflow tract (Chapman et al., 1996), a complex field of progenitor cells that gives rise to skeletal muscles of the head and adds myocardium to the right ventricle and outflow tract at the arterial pole and atrial myocardium at the cardiac venous pole (Huynh, Chen, Terrell, & Baldini, 2007). Other areas of *Tbx1* expression include the lung epithelium, the sclerotome throughout the length of the spine, and the core of the tongue (Chapman et al., 1996). The human *TBX1* gene localizes to the region 22q11 that is deleted or duplicated to different extend in humans with DiGeorge or velocardiofacial syndrome (Chieffo et al., 1997). Heterozygous *Tbx1*-mutant mice have fourth pharyngeal arch artery hypo/aplasia, hypoplasia of the parathyroid and thyroid glands and behavorial defects, recapitulating malformations seen in 22q11 deletion syndrome patients. Mice homozygous for a *Tbx1* null allele die perinatally with largely aggravated cardiofacial defects including persistent truncus arteriosus, aplasia of the thymus and parathyroid glands, loss of the third, fourth, and sixth pharyngeal arch artery, small ears, and craniofacial muscle defects. This and further genetic evidence clearly showed that changed level of *TBX1* is a major contributor to the phenotypic alterations observed in DiGeorge-syndrome patients (Jerome & Papaioannou, 2001; Lindsay et al.,

2001; Merscher et al., 2001). *Tbx1* expression in the cardiopharyngeal region depends on Shh and forkhead transcription factors Foxc1 and Foxc2 (Yamagishi et al., 2003). Tbx1 activates transcription of the fibroblast growth factor genes *Fgf8* and *Fgf10* to maintain proliferative expansion and inhibit differentiation of cardiopharyngeal precursor cells (Chen, Fulcoli, Tang, & Baldini, 2009; Hu et al., 2004). It also activates transcription of *Pitx2* important for cardiac-left-right patterning and negatively affects retinoic acid signaling (Nowotschin et al., 2006; Roberts, Ivins, Cook, Baldini, & Scambler, 2006). Interestingly, the *Drosophila* ortholog of *Tbx1*, *org-1* is expressed in specific sets of visceral muscles and affects their specification, suggesting an evolutionary conserved role for this gene in cardiopharyngeal muscle development (Schaub, Nagaso, Jin, & Frasch, 2012).

4.2 *Tbx1* in Kidney Development

Renal malformations are commonly found among patients carrying a 22q11 deletion (Czarnecki, Van Dyke, Vats, & Feldman, 1998; Shprintzen, 2008), putting up the question whether reduced activity or inactivation of *TBX1* may contribute to this phenotypic defect. Initial in situ hybridization analysis did not describe expression of *Tbx1* in the developing kidney (Chapman et al., 1996), nor did the phenotypic description of homozygous *Tbx1*-mutant mice report on associated renal malformations (Jerome & Papaioannou, 2001; Lindsay et al., 2001; Merscher et al., 2001), arguing that other genes within the 22q11 deletion account for renal defects. However, recent studies gave a hint that *Tbx1* may play a direct role in kidney development. RT-PCR analysis detected transcripts of *Tbx1* in mouse and human renal tissue, and through a RNA interference experiment using a human embryonic kidney HEK293 cell line, it was demonstrated that TBX1 can alter TGFβ-signaling through interaction with HOXD10 (Fu et al., 2014).

A second study reported expression of Tbx1 in fetal rat kidneys at E16.5. As the expression was predominantly distributed in the cytoplasm of renal tubular epithelial cells in the cortex and antibody controls were not performed, the significance of this finding remains unclear. Further in vitro studies showed that Pax2, a crucial renal transcription factor, can bind to the *Tbx1* promotor and activate its transcription in vitro (Jiang, Li, Yang, et al., 2014).

In a third study, members of the same group reported on expression of *Tbx1* again in the cytoplasm of renal tubular epithelial and collecting duct cells in the cortex and medulla in adult mice that was enhanced by treatment with gentamycin. Further in vitro studies suggested that enhanced *Tbx1* expression correlates with Tgfβ-signaling in acute kidney injury (Jiang, Li, Li-Ling, et al., 2014a).

While these studies lack conclusiveness due to the poorly resolved expression of Tbx1 in renal tissue and the in vitro character of the functional assays, they at least should spur interest to reanalyze *Tbx1*-deficient mice for defects in kidney development and homeostasis.

5. *Tbx20* in Bladder/Cloacal Development

5.1 *Tbx20*, a Member of the *Tbx1*-Subfamily with Crucial Function in Heart Development and Homeostasis

Tbx20 is the single member of a third evolutionary conserved subgroup within the *Tbx1*-subfamily. In *Drosophila melanogaster*, the *Tbx20* homologues *midline* and *H15* are expressed in the cardioblasts of the dorsal vessel, the insect organ equivalent to the vertebrate heart where they participate in cardiac fate specification (Miskolczi-McCallum, Scavetta, Svendsen, Soanes, & Brook, 2005; Reim, Mohler, & Frasch, 2005). In vertebrates *Tbx20* (also termed *Tbx12* and *HrT*) is expressed in the cardiac mesoderm, the linear heart tube and the chambered heart during embryogenesis (Ahn, Ruvinsky, Oates, Silver, & Ho, 2000; Carson, Kinzler, & Parr, 2000; Kraus, Haenig, & Kispert, 2001b). In *Tbx20*-deficient mice, heart development arrests at the linear heart tube stage leading to hemodynamic failure shortly later. Derepression of the transcriptional repressor gene *Tbx2* is likely to contribute to this phenotype as does the lack of induction of chamber myocardial genes (Brown et al., 2005; Cai et al., 2005; Singh et al., 2005; Stennard et al., 2005; Takeuchi et al., 2005). Heterozygous *Tbx20*-knockout mice have mild atrial septal anomalies, including an increased prevalence of patent foramen ovale and primum atrial septal aneurysm (Stennard et al., 2005). Mutations in *TBX20* have been associated with atrial and ventricular septal defects, tetralogy of Fallot, valvular defects, cardiomyopathy, and arrhythmias in human patients (Kirk et al., 2007; Monroy-Munoz et al., 2015; Pan et al., 2015). To investigate whether cardiac disease in patients with these conditions results from an embryonic or ongoing requirement for *Tbx20* in myocardium, Shen and coworkers inactivated *Tbx20* specifically in adult cardiomyocytes in mice. This conditional gene targeting resulted in the onset of severe cardiomyopathy accompanied by arrhythmias, with death ensuing within 1–2 weeks of *Tbx20* ablation. Expression of a number of genes encoding critical transcription factors, ion channels, and myofibrillar proteins was downregulated (Shen et al., 2011). Analysis of genome-wide transcriptional changes combined with ChIP-Seq analysis confirmed earlier biochemical work that Tbx20 is unusual among T-box proteins in acting both as a transcriptional repressor and activator and that each of these functions regulates genes with very specialized and distinct molecular roles in the heart (Sakabe et al., 2012; Stennard et al., 2003, 2005). Transcriptional repression is probably mediated by a groucho/Nucleosome Remodeling and Deacetylase complex that binds to an eh1-motif in Tbx20 (Kaltenbrun et al., 2013).

5.2 *Tbx20* (*HrT*) in the Development of the Zebrafish Cloaca

In zebrafish and other teleosts, the larval kidney, the pronephros, is a pair of single nephrons that collect the glomerular filtrate of the blood and after further modification guide it to the outside. The end-segment of the nephron, the pronephric duct opens at a specialized structure, the cloaca that also harbors the end-piece of the alimentary tract, the rectum. The cloaca arises from the proctodeum, an inward fold on the surface of the embryonic ectoderm at the ventral posterior aspect of the embryo. At 24-h post fertilization (hpf), the proctodeum is visible as a mass of cells at the site of the pronephric opening. It surrounds the opening of the pronephric duct at all subsequent stages. By 48 hpf, the posterior gut has reached the cloaca and starts to fuse with it. As the posterior gut becomes canalized from anterior to posterior, the proctodeum invaginates to form the anus. The anal canal, the most distal part of the gastrointestinal tract, is formed by the fusion of the proctodeum with the posterior gut endoderm. By ~96 hpf the posterior gut lumen is in contact with the ventral edge of the embryo, but is not yet open. At 120 hpf the gastrointestinal tract opens externally, adjacent to the pronephric duct (Parkin, Allen, & Ingham, 2009).

Formation of the cloaca depends on Bmp-signaling at the posterior ventral aspect of the gastrulation stage zebrafish embryo. Bmp-signaling dose-dependently specifies ventral tissues including the ventral mesoderm. Probably by signals from ventral mesodermal cells the proctodeum can interact with the distal pronephric duct. This, in turn, allows the ventral migration of the duct and its fusion with and opening into the future cloaca. One of the essential targets of Bmp-signaling in the extreme ventral mesoderm is *Tbx20* (HrT), which acts to suppress early blood gene expression (*gata1*) in ventral mesodermal cells and is required for morphogenesis of the proctodeum to form the cloaca. In *HrT* morphants the cloaca is malformed but less severely than in *Bmp* mutants, indicating that HrT mediates one aspect of Bmp function in this region (Pyati et al., 2006).

To date, expression of *Tbx20* in the cloaca and/or the developing bladder has not been reported in mouse embryos, and anorectal defects have not been associated with mutations in *TBX20* in humans. It therefore, remains to be seen whether *Tbx20* plays a conserved role in mammals in the formation of the lower urinary tract.

6. *Tbx2* and *Tbx3*, a Closely Related Pair of T-Box Transcription Factor Genes in Kidney and Bladder Development

6.1 *Tbx2* and *Tbx3* Regulate Proliferation and Differentiation Processes in Diverse Programs of Organogenesis

Tbx2 and *Tbx3* encode a structurally highly related pair of T-box genes that phylogenetic analysis grouped with the structurally related *Tbx4* and *Tbx5* genes as the separate *Tbx2*-subfamily of vertebrate T-box genes. As far as chromosomal localization is concerned *Tbx2* is linked to *Tbx4* while *Tbx3* is linked to *Tbx5*. Based on these data it was suggested that an initial duplication of a single ancestral gene by unequal crossing over occurred to form a linear array of two *Tbx2/Tbx3* and *Tbx4/Tbx5* precursor genes that was later duplicated, and one copy dispersed to a different chromosomal location. Given the fact that *Tbx2/Tbx3* and *Tbx4/Tbx5* are represented by single genes in invertebrates like flies and nematodes the duplication of the ancestral *Tbx2/3/4/5* gene must have occurred before the divergence of the vertebrate and nematode lineages over 600 million years ago, whereas the duplication of the two-gene cluster occurred probably somewhere along the vertebrate lineage (Agulnik et al., 1996).

Consistent with this evolutionary relationship is the finding that the spatial and temporal expression of *Tbx2* is more similar to *Tbx3* than to *Tbx4* and that of *Tbx3* is more similar to *Tbx2* than to *Tbx5* (Chapman et al., 1996). Moreover, the biochemical properties of the two subgroups diverged dramatically with respect to their transcription modulation properties. While all four proteins recognize the conserved TBE, Tbx2 and Tbx3 repress transcription upon binding to this site (Carreira, Dexter, Yavuzer, Easty, & Goding, 1998; He, Wen, Campbell, Wu, & Rao, 1999), whereas Tbx4 and Tbx5 act as activators (Hiroi et al., 2001; Ouimette, Jolin,

L'Honore, Gifuni, & Drouin, 2010) leading to the interesting situation that at sites of coexpression (such as the atrioventricular canal of the heart) the proteins compete for target gene regulation (Habets et al., 2002; Hoogaars et al., 2008).

The transcriptional repression domain of the *Xenopus* ortholog of Tbx3 (ET) was mapped to a region C-terminal to the T-box (He et al., 1999). This domain is conserved in the Tbx2 protein but only represses transcription together with a domain N-terminal to the T-box in Tbx2 (Paxton, Zhao, Chin, Langner, & Reecy, 2002). The C-terminal domain binds histone deacetylases (HDACs), an interaction essential for repressive activity of the protein (Yarosh et al., 2008). Tbx3 (and therefore probably also Tbx2) does not only function as DNA-binding-dependent transcription factors, they also act as splice factors and even as ciliary proteins (Emechebe et al., 2016; Kumar et al., 2014).

Tbx2 and *Tbx3* share the same expression domains but also possess unique sites of expression both in development and adulthood. *Tbx3*-homozygous mice die at E14.5 presenting severe defects in limb and mammary gland development (Davenport, Jerome-Majewska, & Papaioannou, 2003). Additional requirements in the development of the palate, the conduction system and the liver have been reported and linked to molecular functions of *Tbx3* in early cell fate decisions (Hoogaars et al., 2007; Ludtke, Christoffels, Petry, & Kispert, 2009; Zirzow et al., 2009). Mice homozygous for null mutations of *Tbx2* survive until birth and display duplication of digit 4, cleft palate, and weak heart defects reflecting sites of unique expression of this gene (Farin et al., 2013; Harrelson et al., 2004; Zirzow et al., 2009). These requirements were traced to a role of the gene in patterning processes. Mice with combined loss of *Tbx2* and *Tbx3* die at E10.5 due to severe cardiovascular defects that relate to a redundant role of the two factors to locally inhibit chamber myocardial differentiation, and thus, allow the formation of the atrioventricular canal (Singh et al., 2012).

Both factors are overexpressed in numerous human cancers, including ovarian, cervical, pancreatic, breast, and melanoma indicating an additional role in growth control (Wansleben, Peres, Hare, Goding, & Prince, 2014). In fact, a number of in vitro studies suggested that TBX2 and TBX3 contribute to the enhanced proliferation and the oncogenic process by bypassing senescence through their ability to repress common targets including members of the family of cell-cycle-dependent kinase inhibitors (Jacobs et al., 2000; Lingbeek, Jacobs, & van Lohuizen, 2002; Prince, Carreira, Vance, Abrahams, & Goding, 2004). A role of Tbx2 in growth control in development has recently been reported in the lung mesenchyme of mice (Ludtke et al., 2013).

While mutations in *TBX2* have not yet been implicated in human congenital diseases, heterozygous loss-of-function mutations in *TBX3* underlie an autosomal dominant disorder known as ulnar-mammary syndrome (UMS) (Bamshad et al., 1997). UMS is fully penetrant but occurs with a highly variable clinical presentation, even among family members with the same mutation. As the name suggests, the syndrome is predominately characterized by posterior forelimb deficiencies or duplications involving the ulna and little finger, with rare involvement of the hindlimb, and apocrine/mammary gland hypoplasia or dysfunction with absent or abnormal nipples. Other common features are abnormal dentition, delayed puberty, genital abnormalities, and growth retardation (Bamshad et al., 1999).

6.2 *Tbx2* in *Xenopus* Pronephros Development

Nephrons arise from nephrogenic precursor cells within the intermediate mesoderm by aggregation and subsequent epithelialization. In the simplest case, namely, the pronephric nephron of fish and amphibian embryos, the size of the precursor field defines the size of the single nephron that forms. Work in *Xenopus* has shown that *Tbx2* plays an important role in delineating the territory of the pronephric nephron (Cho et al., 2011).

Tbx2 expression starts weakly in the presumptive pronephric anlage at stage 21 of *Xenopus* embryogenesis, which is when pronephric morphogenesis begins. Expression becomes gradually restricted to the nonnephric mesoderm surrounding the forming pronephric tubules and duct (Cho et al., 2011; Hayata et al., 1999). Injection of a hormone-inducible construct conferring mis/overexpression of *Tbx2* in the presumptive pronephric region abrogated expression of markers of the glomerulus, the pronephric duct, and tubule, indicating that exclusion of *Tbx2* from the pronephric anlage is important to allow pronephros formation. In contrast, injection of a construct encoding Tbx2 fused to the transcription activation domain of VP16 led to expanded expression of pronephric markers, providing evidence that Tbx2 acts as a repressor to control the formation of the pronephros. Depletion of *Tbx2* mRNA in the pronephric region by a morpholino approach as well as injection of a construct for a dominant-negative form of Tbx2 expanded the territory of the pronephric duct and glomerulus but not that of the pronephric tubule and led to markedly increased pronephric nephrons (Cho et al., 2011).

Expression of *Tbx2* in the nonnephric mesoderm is induced by Bmp-signaling. In turn, Tbx2 is required to repress expression of the Bmp antagonists *Gremlin1* and *Hey1* in this region, confining them to the nephric mesoderm. Further injection experiments showed that Hey1 induces expression of *Gremlin1*. In agreement with the fact that Gremlin and Hey1 inhibit Bmp-signaling, this resulted in downregulation of *Tbx2* expression in the nephric mesoderm. Together, this negative regulatory loop between Bmp2/Tbx2 and Gremlin and Hey1 can account for the mechanisms defining the territory of the pronephric nephron. The authors of this study also described that depletion of *Tbx2* affects the posterior migration of pronephric duct precursor cells. How Tbx2 is involved in cell movement and rearrangement is unclear, but similar to other systems, it could be that Tbx2 acts downstream of Wnt-signaling in this process (Cho et al., 2011).

Both *Tbx2* and *Tbx3* are expressed in the nephric duct and its derivative the vas deferens, and in the mesenchymal core of the developing urethra in mouse development (Chapman et al., 1996; Douglas et al., 2012). Northern blot analysis detected expression of *Tbx2* in human fetal and adult kidneys (Campbell et al., 1995; Law et al., 1995). Furthermore, in some UMS patients genital anomalies and urinary tract defects including unilateral absence of a kidney have been reported (Gonzalez et al., 1976; Pallister et al., 1976). Together these findings strongly argue for a functional and possibly redundant implication of *Tbx2* and *Tbx3* in kidney and/or urinary tract development and homeostasis.

6.3 *Tbx2* and *Tbx3* in Bladder Development and Homeostasis

Tbx2 and *Tbx3* transcripts are detected in the kidney, but expression analysis suggests that the two family members are also implicated in bladder development and homeostasis. Expression of *Tbx2* was found in the proctodeum, the ectodermal tissue fold that gives rise to the cloaca, both in zebrafish and *Xenopus* embryos (Cho et al., 2011; Hayata et al., 1999; Pyati et al., 2006). Expression of *Tbx3* occurs in basal and intermediate cells of the bladder urothelium in adult rats (Ito et al., 2005).

While overexpression of *Tbx2* and *Tbx3* has been correlated with formation of many tumors (Wansleben et al., 2014), the two proteins may act as tumor suppressors in the bladder urothelium as they are silenced in bladder cancer. Further, methylation of *TBX2* and *TBX3* in bladder cancer was associated with the progression of nonmuscle invasive tumors to muscle-invasive tumors; and patients, in which *TBX3* was not methylated, had a significantly better progression-free survival rate (Beukers et al., 2015; Kandimalla et al., 2012). On the molecular level, derepression of *E-cadherin* expression after silencing of *TBX2/TBX3* may contribute to migration and invasion of tumor cells (Du et al., 2014).

7. CONCLUSION AND OUTLOOK

T-box genes are crucial regulators of diverse cellular programs in the development of numerous organ systems. The analysis of the T-box gene function in the kidney and lower urinary tract has been relatively slow, probably due to the late discovery of the relevant family members and lack of conditional gene-targeting approaches in this region. Nonetheless, genetic experiments have now characterized a role for *Tbx18* in ureter development in mouse and man, and have suggested additional functions for *Tbx1*, *Tbx2*, *Tbx3*, *Tbx18*, and *Tbx20* in kidney and bladder development and homeostasis (Table 1). Further analysis should aim to characterize the molecular programs acting downstream of *Tbx18* in the ureteric mesenchyme to better understand the molecular control of lineage segregation in the early metanephric field. From an evolutionary point of view it would be highly interesting to learn about the phylogenetic emergence of *Tbx18* expression in the nephric field to correlate it with the emergence of a distinct ureter in amniotes. A possible requirement of *Tbx1* and *Tbx18* in kidney development should be addressed by conditional gene-targeting approaches. The role of *Tbx2*, *Tbx3*, and *Tbx20* in the development of the excretory system in the mouse is still completely unexplored. Again, tissue-specific gene inactivation approaches using available conditional alleles are likely to provide important insight into the regulation of mesenchymal and epithelial lineages in the development of the kidney and the bladder. Additional efforts should be directed toward understanding the significance of silencing of *TBX2* and *TBX3* expression in tumors of the bladder. In the end, we may learn that also in the excretory system a collection of T-box genes coordinately regulate important patterning and differentiation programs as they do in the heart and the limb.

ACKNOWLEDGMENT

The author likes to thank Anna-Carina Weiss in his laboratory for the images in Fig. 2B–E.

Funding: This work was supported by grants from the German Research Council [DFG KI728/7-2; DFG Ki728/9-1; DFG KI728/10-1] to A.K.

Conflict of Interest: The author declares that he has no conflict of interest.

REFERENCES

Agulnik, S. I., Garvey, N., Hancock, S., Ruvinsky, I., Chapman, D. L., Agulnik, I., et al. (1996). Evolution of mouse T-box genes by tandem duplication and cluster dispersion. *Genetics, 144*, 249–254.

Ahn, D. G., Ruvinsky, I., Oates, A. C., Silver, L. M., & Ho, R. K. (2000). tbx20, a new vertebrate T-box gene expressed in the cranial motor neurons and developing cardiovascular structures in zebrafish. *Mechanisms of Development, 95*, 253–258.

Airik, R., Bussen, M., Singh, M. K., Petry, M., & Kispert, A. (2006). Tbx18 regulates the development of the ureteral mesenchyme. *The Journal of Clinical Investigation, 116*, 663–674.

Airik, R., Trowe, M. O., Foik, A., Farin, H. F., Petry, M., Schuster-Gossler, K., et al. (2010). Hydroureternephrosis due to loss of Sox9-regulated smooth muscle cell differentiation of the ureteric mesenchyme. *Human Molecular Genetics, 19*, 4918–4929.

Bamshad, M., Le, T., Watkins, W. S., Dixon, M. E., Kramer, B. E., Roeder, A. D., et al. (1999). The spectrum of mutations in TBX3: Genotype/phenotype relationship in ulnar-mammary syndrome. *The American Journal of Human Genetics, 64*, 1550–1562.

Bamshad, M., Lin, R. C., Law, D. J., Watkins, W. C., Krakowiak, P. A., Moore, M. E., et al. (1997). Mutations in human TBX3 alter limb, apocrine and genital development in ulnar-mammary syndrome. *Nature Genetics, 16*, 311–315.

Begemann, G., Gibert, Y., Meyer, A., & Ingham, P. W. (2002). Cloning of zebrafish T-box genes tbx15 and tbx18 and their expression during embryonic development. *Mechanisms of Development, 114*, 137–141.

Beukers, W., Kandimalla, R., Masius, R. G., Vermeij, M., Kranse, R., van Leenders, G. J., et al. (2015). Stratification based on methylation of TBX2 and TBX3 into three molecular grades predicts progression in patients with pTa-bladder cancer. *Modern Pathology, 28*, 515–522.

Bohnenpoll, T., Bettenhausen, E., Weiss, A. C., Foik, A. B., Trowe, M. O., Blank, P., et al. (2013). Tbx18 expression demarcates multipotent precursor populations in the developing urogenital system but is exclusively required within the ureteric mesenchymal lineage to suppress a renal stromal fate. *Developmental Biology, 380*, 25–36.

Bohnenpoll, T., & Kispert, A. (2014). Ureter growth and differentiation. *Seminars in Cell & Developmental Biology, 36*, 21–30.

Bollag, R. J., Siegfried, Z., Cebra-Thomas, J. A., Garvey, N., Davison, E. M., & Silver, L. M. (1994). An ancient family of embryonically expressed mouse genes sharing a conserved protein motif with the T locus. *Nature Genetics*, 7, 383–389.

Bolt, C. C., Elso, C. M., Lu, X., Pan, F., Kispert, A., & Stubbs, L. (2014). A distant downstream enhancer directs essential expression of Tbx18 in urogenital tissues. *Developmental Biology, 392*, 483–493.

Brown, D. D., Martz, S. N., Binder, O., Goetz, S. C., Price, B. M., Smith, J. C., et al. (2005). Tbx5 and Tbx20 act synergistically to control vertebrate heart morphogenesis. *Development, 132*, 553–563.

Bussen, M., Petry, M., Schuster-Gossler, K., Leitges, M., Gossler, A., & Kispert, A. (2004). The T-box transcription factor Tbx18 maintains the separation of anterior and posterior somite compartments. *Genes & Development, 18*, 1209–1221.

Cai, C. L., Martin, J. C., Sun, Y., Cui, L., Wang, L., Ouyang, K., et al. (2008). A myocardial lineage derives from Tbx18 epicardial cells. *Nature, 454*, 104–108.

Cai, C. L., Zhou, W., Yang, L., Bu, L., Qyang, Y., Zhang, X., et al. (2005). T-box genes coordinate regional rates of proliferation and regional specification during cardiogenesis. *Development, 132*, 2475–2487.

Campbell, C., Goodrich, K., Casey, G., & Beatty, B. (1995). Cloning and mapping of a human gene (TBX2) sharing a highly conserved protein motif with the Drosophila omb gene. *Genomics, 28*, 255–260.

Carreira, S., Dexter, T. J., Yavuzer, U., Easty, D. J., & Goding, C. R. (1998). Brachyury-related transcription factor Tbx2 and repression of the melanocyte-specific TRP-1 promoter. *Molecular and Cellular Biology*, *18*, 5099–5108.

Carson, C. T., Kinzler, E. R., & Parr, B. A. (2000). Tbx12, a novel T-box gene, is expressed during early stages of heart and retinal development. *Mechanisms of Development*, *96*, 137–140.

Chapman, D. L., Garvey, N., Hancock, S., Alexiou, M., Agulnik, S. I., Gibson-Brown, J. J., et al. (1996). Expression of the T-box family genes, Tbx1-Tbx5, during early mouse development. *Developmental Dynamics*, *206*, 379–390.

Chen, F. (2009). Genetic and developmental basis for urinary tract obstruction. *Pediatric Nephrology*, *24*, 1621–1632.

Chen, L., Fulcoli, F. G., Tang, S., & Baldini, A. (2009). Tbx1 regulates proliferation and differentiation of multipotent heart progenitors. *Circulation Research*, *105*, 842–851.

Chieffo, C., Garvey, N., Gong, W., Roe, B., Zhang, G., Silver, L., et al. (1997). Isolation and characterization of a gene from the DiGeorge chromosomal region homologous to the mouse Tbx1 gene. *Genomics*, *43*, 267–277.

Cho, G. S., Choi, S. C., Park, E. C., & Han, J. K. (2011). Role of Tbx2 in defining the territory of the pronephric nephron. *Development*, *138*, 465–474.

Christoffels, V. M., Grieskamp, T., Norden, J., Mommersteeg, M. T., Rudat, C., & Kispert, A. (2009). Tbx18 and the fate of epicardial progenitors. *Nature*, *458*, E8–E9. discussion E9-10.

Christoffels, V. M., Mommersteeg, M. T., Trowe, M. O., Prall, O. W., de Gier-de Vries, C., Soufan, A. T., et al. (2006). Formation of the venous pole of the heart from an Nkx2-5-negative precursor population requires Tbx18. *Circulation Research*, *98*, 1555–1563.

Combes, A. N., Davies, J. A., & Little, M. H. (2015). Cell-cell interactions driving kidney morphogenesis. *Current Topics in Developmental Biology*, *112*, 467–508.

Czarnecki, P. M., Van Dyke, D. L., Vats, S., & Feldman, G. L. (1998). A mother with VCFS and unilateral dysplastic kidney and her fetus with multicystic dysplastic kidneys: additional evidence to support the association of renal malformations and VCFS. *Journal of Medical Genetics*, *35*, 348.

Davenport, T. G., Jerome-Majewska, L. A., & Papaioannou, V. E. (2003). Mammary gland, limb and yolk sac defects in mice lacking Tbx3, the gene mutated in human ulnar mammary syndrome. *Development*, *130*, 2263–2273.

Degnan, B. M., Vervoort, M., Larroux, C., & Richards, G. S. (2009). Early evolution of metazoan transcription factors. *Current Opinion in Genetics & Development*, *19*, 591–599.

Douglas, N. C., Heng, K., Sauer, M. V., & Papaioannou, V. E. (2012). Dynamic expression of Tbx2 subfamily genes in development of the mouse reproductive system. *Developmental Dynamics*, *241*, 365–375.

Du, H. F., Ou, L. P., Yang, X., Song, X. D., Fan, Y. R., Tan, B., et al. (2014). A new PKCalpha/beta/TBX3/E-cadherin pathway is involved in PLCepsilon-regulated invasion and migration in human bladder cancer cells. *Cellular Signalling*, *26*, 580–593.

Emechebe, U., Kumar, P. P., Rozenberg, J. M., Moore, B., Firment, A., Mirshahi, T., et al. (2016). T-box3 is a ciliary protein and regulates stability of the Gli3 transcription factor to control digit number. *eLife*, *5*, e07897.

Farin, H. F., Bussen, M., Schmidt, M. K., Singh, M. K., Schuster-Gossler, K., & Kispert, A. (2007). Transcriptional repression by the T-box proteins Tbx18 and Tbx15 depends on Groucho corepressors. *The Journal of Biological Chemistry*, *282*, 25748–25759.

Farin, H. F., Ludtke, T. H., Schmidt, M. K., Placzko, S., Schuster-Gossler, K., Petry, M., et al. (2013). Tbx2 terminates shh/fgf signaling in the developing mouse limb bud by direct repression of gremlin1. *PLoS Genetics*, *9*, e1003467.

Farin, H. F., Mansouri, A., Petry, M., & Kispert, A. (2008). T-box protein Tbx18 interacts with the paired box protein Pax3 in the development of the paraxial mesoderm. *The Journal of Biological Chemistry, 283*, 25372–25380.

Fu, Y., Li, F., Zhao, D. Y., Zhang, J. S., Lv, Y., & Li-Ling, J. (2014). Interaction between Tbx1 and Hoxd10 and connection with TGFbeta-BMP signal pathway during kidney development. *Gene, 536*, 197–202.

Georgas, K. M., Armstrong, J., Keast, J. R., Larkins, C. E., McHugh, K. M., Southard-Smith, E. M., et al. (2015). An illustrated anatomical ontology of the developing mouse lower urogenital tract. *Development, 142*, 1893–1908.

Gonzalez, C. H., Herrmann, J., & Opitz, J. M. (1976). Studies of malformation syndromes of man XXXXIIB: Mother and son affected with the ulnar-mammary syndrome type Pallister. *European Journal of Pediatrics, 123*, 225–235.

Greulich, F., Farin, H. F., Schuster-Gossler, K., & Kispert, A. (2012). Tbx18 function in epicardial development. *Cardiovascular Research, 96*, 476–483.

Greulich, F., Rudat, C., & Kispert, A. (2011). Mechanisms of T-box gene function in the developing heart. *Cardiovascular Research, 91*, 212–222.

Grisanti, L., Clavel, C., Cai, X., Rezza, A., Tsai, S. Y., Sennett, R., et al. (2013). Tbx18 targets dermal condensates for labeling, isolation, and gene ablation during embryonic hair follicle formation. *The Journal of Investigative Dermatology, 133*, 344–353.

Habets, P. E., Moorman, A. F., Clout, D. E., van Roon, M. A., Lingbeek, M., van Lohuizen, M., et al. (2002). Cooperative action of Tbx2 and Nkx2.5 inhibits ANF expression in the atrioventricular canal: implications for cardiac chamber formation. *Genes & Development, 16*, 1234–1246.

Haenig, B., & Kispert, A. (2004). Analysis of TBX18 expression in chick embryos. *Development Genes and Evolution, 214*, 407–411.

Hafner, R., Bohnenpoll, T., Rudat, C., Schultheiss, T. M., & Kispert, A. (2015). Fgfr2 is required for the expansion of the early adrenocortical primordium. *Molecular and Cellular Endocrinology, 413*, 168–177.

Harada, Y., Yasuo, H., & Satoh, N. (1995). A sea urchin homologue of the chordate Brachyury (T) gene is expressed in the secondary mesenchyme founder cells. *Development, 121*, 2747–2754.

Harrelson, Z., Kelly, R. G., Goldin, S. N., Gibson-Brown, J. J., Bollag, R. J., Silver, L. M., et al. (2004). Tbx2 is essential for patterning the atrioventricular canal and for morphogenesis of the outflow tract during heart development. *Development, 131*, 5041–5052.

Hayata, T., Kuroda, H., Eisaki, A., & Asashima, M. (1999). Expression of Xenopus T-box transcription factor, tbx2 in Xenopus embryo. *Development Genes and Evolution, 209*, 625–628.

He, M., Wen, L., Campbell, C. E., Wu, J. Y., & Rao, Y. (1999). Transcription repression by Xenopus ET and its human ortholog TBX3, a gene involved in ulnar-mammary syndrome. *Proceedings of the National Academy of Sciences of the United States of America, 96*, 10212–10217.

Herrmann, B. G., & Kispert, A. (1994). The T genes in embryogenesis. *Trends in Genetics, 10*, 280–286.

Herrmann, B. G., Labeit, S., Poustka, A., King, T. R., & Lehrach, H. (1990). Cloning of the T gene required in mesoderm formation in the mouse. *Nature, 343*, 617–622.

Hiroi, Y., Kudoh, S., Monzen, K., Ikeda, Y., Yazaki, Y., Nagai, R., et al. (2001). Tbx5 associates with Nkx2-5 and synergistically promotes cardiomyocyte differentiation. *Nature Genetics, 28*, 276–280.

Hoogaars, W. M., Barnett, P., Rodriguez, M., Clout, D. E., Moorman, A. F., Goding, C. R., et al. (2008). TBX3 and its splice variant TBX3 + exon 2a are functionally similar. *Pigment Cell & Melanoma Research, 21*, 379–387.

Hoogaars, W. M., Engel, A., Brons, J. F., Verkerk, A. O., de Lange, F. J., Wong, L. Y., et al. (2007). Tbx3 controls the sinoatrial node gene program and imposes pacemaker function on the atria. *Genes & Development, 21*, 1098–1112.

Hu, T., Yamagishi, H., Maeda, J., McAnally, J., Yamagishi, C., & Srivastava, D. (2004). Tbx1 regulates fibroblast growth factors in the anterior heart field through a reinforcing autoregulatory loop involving forkhead transcription factors. *Development, 131*, 5491–5502.

Huynh, T., Chen, L., Terrell, P., & Baldini, A. (2007). A fate map of Tbx1 expressing cells reveals heterogeneity in the second cardiac field. *Genesis, 45*, 470–475.

Hwang, D. Y., Dworschak, G. C., Kohl, S., Saisawat, P., Vivante, A., Hilger, A. C., et al. (2014). Mutations in 12 known dominant disease-causing genes clarify many congenital anomalies of the kidney and urinary tract. *Kidney International, 85*, 1429–1433.

Ito, A., Asamoto, M., Hokaiwado, N., Takahashi, S., & Shirai, T. (2005). Tbx3 expression is related to apoptosis and cell proliferation in rat bladder both hyperplastic epithelial cells and carcinoma cells. *Cancer Letters, 219*, 105–112.

Jacobs, J. J., Keblusek, P., Robanus-Maandag, E., Kristel, P., Lingbeek, M., Nederlof, P. M., et al. (2000). Senescence bypass screen identifies TBX2, which represses Cdkn2a (p19 (ARF)) and is amplified in a subset of human breast cancers. *Nature Genetics, 26*, 291–299.

Jahr, M., Schlueter, J., Brand, T., & Manner, J. (2008). Development of the proepicardium in Xenopus laevis. *Developmental Dynamics, 237*, 3088–3096.

Jerome, L. A., & Papaioannou, V. E. (2001). DiGeorge syndrome phenotype in mice mutant for the T-box gene, Tbx1. *Nature Genetics, 27*, 286–291.

Jiang, H., Li, L., Li-Ling, J., Qiu, G., Niu, Z., Jiang, H., et al. (2014a). Increased Tbx1 expression may play a role via TGFbeta-Smad2/3 signaling pathway in acute kidney injury induced by gentamicin. *International Journal of Clinical and Experimental Pathology, 7*, 1595–1605.

Jiang, H., Li, L., Yang, H., Bai, Y., Jiang, H., & Li, Y. (2014b). Pax2 may play a role in kidney development by regulating the expression of TBX1. *Molecular Biology Reports, 41*, 7491–7498.

Kaltenbrun, E., Greco, T. M., Slagle, C. E., Kennedy, L. M., Li, T., Cristea, I. M., et al. (2013). A Gro/TLE-NuRD corepressor complex facilitates Tbx20-dependent transcriptional repression. *Journal of Proteome Research, 12*, 5395–5409.

Kandimalla, R., van Tilborg, A. A., Kompier, L. C., Stumpel, D. J., Stam, R. W., Bangma, C. H., et al. (2012). Genome-wide analysis of CpG island methylation in bladder cancer identified TBX2, TBX3, GATA2, and ZIC4 as pTa-specific prognostic markers. *European Urology, 61*, 1245–1256.

Kardong, K. V. (2014). *Vertebrates: Comparative anatomy, function, evolution* (7th ed.). New York: McGraw-Hill Education.

Kirk, E. P., Sunde, M., Costa, M. W., Rankin, S. A., Wolstein, O., Castro, M. L., et al. (2007). Mutations in cardiac T-box factor gene TBX20 are associated with diverse cardiac pathologies, including defects of septation and valvulogenesis and cardiomyopathy. *The American Journal of Human Genetics, 81*, 280–291.

Kispert, A., & Herrmann, B. G. (1993). The Brachyury gene encodes a novel DNA binding protein. *The EMBO Journal, 12*, 3211–3220.

Kispert, A., Herrmann, B. G., Leptin, M., & Reuter, R. (1994). Homologs of the mouse Brachyury gene are involved in the specification of posterior terminal structures in Drosophila, Tribolium, and Locusta. *Genes & Development, 8*, 2137–2150.

Kispert, A., Koschorz, B., & Herrmann, B. G. (1995). The T protein encoded by Brachyury is a tissue-specific transcription factor. *The EMBO Journal, 14*, 4763–4772.

Kraus, F., Haenig, B., & Kispert, A. (2001a). Cloning and expression analysis of the mouse T-box gene Tbx18. *Mechanisms of Development, 100*, 83–86.

Kraus, F., Haenig, B., & Kispert, A. (2001b). Cloning and expression analysis of the mouse T-box gene tbx20. *Mechanisms of Development*, *100*, 87–91.

Kumar, P. P., Franklin, S., Emechebe, U., Hu, H., Moore, B., Lehman, C., et al. (2014). TBX3 regulates splicing in vivo: A novel molecular mechanism for ulnar-mammary syndrome. *PLoS Genetics*, *10*, e1004247.

Law, D. J., Gebuhr, T., Garvey, N., Agulnik, S. I., & Silver, L. M. (1995). Identification, characterization, and localization to chromosome 17q21-22 of the human TBX2 homolog, member of a conserved developmental gene family. *Mammalian Genome*, *6*, 793–797.

Lindsay, E. A., Vitelli, F., Su, H., Morishima, M., Huynh, T., Pramparo, T., et al. (2001). Tbx1 haploinsufficieny in the DiGeorge syndrome region causes aortic arch defects in mice. *Nature*, *410*, 97–101.

Lingbeek, M. E., Jacobs, J. J., & van Lohuizen, M. (2002). The T-box repressors TBX2 and TBX3 specifically regulate the tumor suppressor gene p14ARF via a variant T-site in the initiator. *The Journal of Biological Chemistry*, *277*, 26120–26127.

Little, M. H., & McMahon, A. P. (2012). Mammalian kidney development: Principles, progress, and projections. *Cold Spring Harbor Perspectives in Biology*, *4*, a008300.

Ludtke, T. H., Christoffels, V. M., Petry, M., & Kispert, A. (2009). Tbx3 promotes liver bud expansion during mouse development by suppression of cholangiocyte differentiation. *Hepatology*, *49*, 969–978.

Ludtke, T. H., Farin, H. F., Rudat, C., Schuster-Gossler, K., Petry, M., Barnett, P., et al. (2013). Tbx2 controls lung growth by direct repression of the cell cycle inhibitor genes Cdkn1a and Cdkn1b. *PLoS Genetics*, *9*, e1003189.

McMahon, A. P. (2016). Development of the mammalian kidney. *Current Topics in Developmental Biology*, *117*, 31–64.

Merscher, S., Funke, B., Epstein, J. A., Heyer, J., Puech, A., Lu, M. M., et al. (2001). TBX1 is responsible for cardiovascular defects in velo-cardio-facial/DiGeorge syndrome. *Cell*, *104*, 619–629.

Miskolczi-McCallum, C. M., Scavetta, R. J., Svendsen, P. C., Soanes, K. H., & Brook, W. J. (2005). The Drosophila melanogaster T-box genes midline and H15 are conserved regulators of heart development. *Developmental Biology*, *278*, 459–472.

Monroy-Munoz, I. E., Perez-Hernandez, N., Rodriguez-Perez, J. M., Munoz-Medina, J. E., Angeles-Martinez, J., Garcia-Trejo, J. J., et al. (2015). Novel mutations in the transcriptional activator domain of the human TBX20 in patients with atrial septal defect. *BioMed Research International*, *2015*, 718786.

Morley, R. H., Lachani, K., Keefe, D., Gilchrist, M. J., Flicek, P., Smith, J. C., et al. (2009). A gene regulatory network directed by zebrafish No tail accounts for its roles in mesoderm formation. *Proceedings of the National Academy of Sciences of the United States of America*, *106*, 3829–3834.

Mugford, J. W., Sipila, P., McMahon, J. A., & McMahon, A. P. (2008). Osr1 expression demarcates a multi-potent population of intermediate mesoderm that undergoes progressive restriction to an Osr1-dependent nephron progenitor compartment within the mammalian kidney. *Developmental Biology*, *324*, 88–98.

Muller, C. W., & Herrmann, B. G. (1997). Crystallographic structure of the T domain-DNA complex of the Brachyury transcription factor. *Nature*, *389*, 884–888.

Nie, X., Sun, J., Gordon, R. E., Cai, C. L., & Xu, P. X. (2010). SIX1 acts synergistically with TBX18 in mediating ureteral smooth muscle formation. *Development*, *137*, 755–765.

Nowotschin, S., Liao, J., Gage, P. J., Epstein, J. A., Campione, M., & Morrow, B. E. (2006). Tbx1 affects asymmetric cardiac morphogenesis by regulating Pitx2 in the secondary heart field. *Development*, *133*, 1565–1573.

Ouimette, J. F., Jolin, M. L., L'Honore, A., Gifuni, A., & Drouin, J. (2010). Divergent transcriptional activities determine limb identity. *Nature Communications*, *1*, 35.

Pallister, P. D., Herrmann, J., & Opitz, J. M. (1976). Studies of malformation syndromes in man XXXXII: A pleiotropic dominant mutation affecting skeletal, sexual and apocrine-mammary development. *Birth Defects Original Article Series*, *12*, 247–254.

Pan, Y., Geng, R., Zhou, N., Zheng, G. F., Zhao, H., Wang, J., et al. (2015). TBX20 loss-of-function mutation contributes to double outlet right ventricle. *International Journal of Molecular Medicine*, *35*, 1058–1066.

Papaioannou, V. E. (2014). The T-box gene family: Emerging roles in development, stem cells and cancer. *Development*, *141*, 3819–3833.

Paps, J., Holland, P. W., & Shimeld, S. M. (2012). A genome-wide view of transcription factor gene diversity in chordate evolution: Less gene loss in amphioxus? *Briefings in Functional Genomics*, *11*, 177–186.

Parkin, C. A., Allen, C. E., & Ingham, P. W. (2009). Hedgehog signalling is required for cloacal development in the zebrafish embryo. *The International Journal of Developmental Biology*, *53*, 45–57.

Paxton, C., Zhao, H., Chin, Y., Langner, K., & Reecy, J. (2002). Murine Tbx2 contains domains that activate and repress gene transcription. *Gene*, *283*, 117–124.

Pflugfelder, G. O., Roth, H., & Poeck, B. (1992). A homology domain shared between Drosophila optomotor-blind and mouse Brachyury is involved in DNA binding. *Biochemical and Biophysical Research Communications*, *186*, 918–925.

Prince, S., Carreira, S., Vance, K. W., Abrahams, A., & Goding, C. R. (2004). Tbx2 directly represses the expression of the p21(WAF1) cyclin-dependent kinase inhibitor. *Cancer Research*, *64*, 1669–1674.

Pyati, U. J., Cooper, M. S., Davidson, A. J., Nechiporuk, A., & Kimelman, D. (2006). Sustained Bmp signaling is essential for cloaca development in zebrafish. *Development*, *133*, 2275–2284.

Reim, I., Mohler, J. P., & Frasch, M. (2005). Tbx20-related genes, mid and H15, are required for tinman expression, proper patterning, and normal differentiation of cardioblasts in Drosophila. *Mechanisms of Development*, *122*, 1056–1069.

Roberts, C., Ivins, S., Cook, A. C., Baldini, A., & Scambler, P. J. (2006). Cyp26 genes a1, b1 and c1 are down-regulated in Tbx1 null mice and inhibition of Cyp26 enzyme function produces a phenocopy of DiGeorge syndrome in the chick. *Human Molecular Genetics*, *15*, 3394–3410.

Ruvinsky, I., Silver, L. M., & Gibson-Brown, J. J. (2000). Phylogenetic analysis of T-Box genes demonstrates the importance of amphioxus for understanding evolution of the vertebrate genome. *Genetics*, *156*, 1249–1257.

Saisawat, P., Kohl, S., Hilger, A. C., Hwang, D. Y., Yung Gee, H., Dworschak, G. C., et al. (2014). Whole-exome resequencing reveals recessive mutations in TRAP1 in individuals with CAKUT and VACTERL association. *Kidney International*, *85*, 1310–1317.

Sakabe, N. J., Aneas, I., Shen, T., Shokri, L., Park, S. Y., Bulyk, M. L., et al. (2012). Dual transcriptional activator and repressor roles of TBX20 regulate adult cardiac structure and function. *Human Molecular Genetics*, *21*, 2194–2204.

Schaub, C., Nagaso, H., Jin, H., & Frasch, M. (2012). Org-1, the Drosophila ortholog of Tbx1, is a direct activator of known identity genes during muscle specification. *Development*, *139*, 1001–1012.

Sebe-Pedros, A., Ariza-Cosano, A., Weirauch, M. T., Leininger, S., Yang, A., Torruella, G., et al. (2013). Early evolution of the T-box transcription factor family. *Proceedings of the National Academy of Sciences of the United States of America*, *110*, 16050–16055.

Shen, T., Aneas, I., Sakabe, N., Dirschinger, R. J., Wang, G., Smemo, S., et al. (2011). Tbx20 regulates a genetic program essential to adult mouse cardiomyocyte function. *The Journal of Clinical Investigation*, *121*, 4640–4654.

Shprintzen, R. J. (2008). Velo-cardio-facial syndrome: 30 Years of study. *Developmental Disabilities Research Reviews*, *14*, 3–10.

Singh, M. K., Christoffels, V. M., Dias, J. M., Trowe, M. O., Petry, M., Schuster-Gossler, K., et al. (2005). Tbx20 is essential for cardiac chamber differentiation and repression of Tbx2. *Development*, *132*, 2697–2707.

Singh, R., Hoogaars, W. M., Barnett, P., Grieskamp, T., Rana, M. S., Buermans, H., et al. (2012). Tbx2 and Tbx3 induce atrioventricular myocardial development and endocardial cushion formation. *Cellular and Molecular Life Sciences*, *69*, 1377–1389.

Smith, J. M., Stablein, D. M., Munoz, R., Hebert, D., & McDonald, R. A. (2007). Contributions of the transplant registry: The 2006 annual report of the North American pediatric renal trials and collaborative studies (NAPRTCS). *Pediatric Transplantation*, *11*, 366–373.

Stennard, F. A., Costa, M. W., Elliott, D. A., Rankin, S., Haast, S. J., Lai, D., et al. (2003). Cardiac T-box factor Tbx20 directly interacts with Nkx2-5, GATA4, and GATA5 in regulation of gene expression in the developing heart. *Developmental Biology*, *262*, 206–224.

Stennard, F. A., Costa, M. W., Lai, D., Biben, C., Furtado, M. B., Solloway, M. J., et al. (2005). Murine T-box transcription factor Tbx20 acts as a repressor during heart development, and is essential for adult heart integrity, function and adaptation. *Development*, *132*, 2451–2462.

Takeuchi, J. K., Mileikovskaia, M., Koshiba-Takeuchi, K., Heidt, A. B., Mori, A. D., Arruda, E. P., et al. (2005). Tbx20 dose-dependently regulates transcription factor networks required for mouse heart and motoneuron development. *Development*, *132*, 2463–2474.

Trowe, M. O., Airik, R., Weiss, A. C., Farin, H. F., Foik, A. B., Bettenhausen, E., et al. (2012). Canonical Wnt signaling regulates smooth muscle precursor development in the mouse ureter. *Development*, *139*, 3099–3108.

Trowe, M. O., Maier, H., Schweizer, M., & Kispert, A. (2008). Deafness in mice lacking the T-box transcription factor Tbx18 in otic fibrocytes. *Development*, *135*, 1725–1734.

Uetani, N., & Bouchard, M. (2009). Plumbing in the embryo: Developmental defects of the urinary tracts. *Clinical Genetics*, *75*, 307–317.

Vivante, A., Kleppa, M. J., Schulz, J., Kohl, S., Sharma, A., Chen, J., et al. (2015). Mutations in TBX18 cause dominant urinary tract malformations via transcriptional dysregulation of ureter development. *The American Journal of Human Genetics*, *97*, 291–301.

Wang, G. J., Brenner-Anantharam, A., Vaughan, E. D., & Herzlinger, D. (2009). Antagonism of BMP4 signaling disrupts smooth muscle investment of the ureter and ureteropelvic junction. *The Journal of Urology*, *181*, 401–407.

Wang, Y., Tripathi, P., Guo, Q., Coussens, M., Ma, L., & Chen, F. (2009). Cre/lox recombination in the lower urinary tract. *Genesis*, *47*, 409–413.

Wansleben, S., Peres, J., Hare, S., Goding, C. R., & Prince, S. (2014). T-box transcription factors in cancer biology. *Biochimica et Biophysica Acta*, *1846*, 380–391.

Wilkinson, D. G., Bhatt, S., & Herrmann, B. G. (1990). Expression pattern of the mouse T gene and its role in mesoderm formation. *Nature*, *343*, 657–659.

Woolf, A. S. (2000). A molecular and genetic view of human renal and urinary tract malformations. *Kidney International*, *58*, 500–512.

Wu, S. P., Dong, X. R., Regan, J. N., Su, C., & Majesky, M. W. (2013). Tbx18 regulates development of the epicardium and coronary vessels. *Developmental Biology*, *383*, 307–320.

Xu, J., Nie, X., Cai, X., Cai, C. L., & Xu, P. X. (2014). Tbx18 is essential for normal development of vasculature network and glomerular mesangium in the mammalian kidney. *Developmental Biology*, *391*, 17–31.

Yamagishi, H., Maeda, J., Hu, T., McAnally, J., Conway, S. J., Kume, T., et al. (2003). Tbx1 is regulated by tissue-specific forkhead proteins through a common Sonic hedgehog-responsive enhancer. *Genes & Development*, *17*, 269–281.

Yarosh, W., Barrientos, T., Esmailpour, T., Lin, L., Carpenter, P. M., Osann, K., et al. (2008). TBX3 is overexpressed in breast cancer and represses p14 ARF by interacting with histone deacetylases. *Cancer Research*, *68*, 693–699.

Yi, C. H., Terrett, J. A., Li, Q. Y., Ellington, K., Packham, E. A., Armstrong-Buisseret, L., et al. (1999). Identification, mapping, and phylogenomic analysis of four new human members of the T-box gene family: EOMES, TBX6, TBX18, and TBX19. *Genomics*, *55*, 10–20.

Zirzow, S., Ludtke, T. H., Brons, J. F., Petry, M., Christoffels, V. M., & Kispert, A. (2009). Expression and requirement of T-box transcription factors Tbx2 and Tbx3 during secondary palate development in the mouse. *Developmental Biology*, *336*, 145–155.

CHAPTER TEN

Control of Neuronal Development by T-Box Genes in the Brain

A.B. Mihalas*, R.F. Hevner*,†,1

*Center for Integrative Brain Research, Seattle Children's Research Institute, Seattle, WA, United States
†University of Washington School of Medicine, Seattle, WA, United States
[1]Corresponding author: e-mail addresses: rhevner@uw.edu; rhevner@u.washington.edu

Contents

1. Discovery of T-Box Gene Expression in the Developing Brain and Neural Retina 280
2. Expression of Tbr1 Subfamily T-Box Genes in Brain and Neural Retina 283
   2.1 Tbr2 and Tbr1 Are Part of a Transcription Factor Cascade in Glutamatergic Neurogenesis 284
   2.2 Tbr2 and Tbr1 Expression in Cortical Glutamatergic Neurogenesis 284
   2.3 Tbr2 and Tbr1 Expression in Developing and Adult DG 286
   2.4 Tbr2, Tbr1, and T-bet Expression in the OB and Adult SVZ 287
   2.5 Tbr2 and Tbr1 Expression in Developing Basal Forebrain, Amygdala, and Eminentia Thalami 288
   2.6 Tbr2 and Tbr1 Expression in the Cerebellum 289
   2.7 Tbr2 Expression in the Retina 289
   2.8 Tbr2 Expression in the Embryonic Midbrain 290
3. Functions of Tbr1 Subfamily T-Box Genes in Brain and Neural Retina 290
   3.1 Functions of Tbr1 Subfamily Members in Cerebral Cortex Development 290
   3.2 Functions of Tbr2 and Tbr1 in Developing and Adult DG 293
   3.3 Functions of Tbr1 Subfamily Members in OB Development and Adult SVZ Neurogenesis 294
   3.4 Functions of Tbr1 Subfamily Members in Basal Forebrain and Amygdala Development 295
   3.5 Functions of Tbr1 Subfamily Members in Cerebellum Development 295
   3.6 Functions of Tbr1 Subfamily Members in Neural Retina Development 295
   3.7 Functions of Tbr1 Subfamily Members in Mouse Behavior and Motor Control 296
4. Regulation of T-Box Genes 296
   4.1 Regulation of Tbr2 Gene Expression 297
   4.2 Regulation of Tbr1 Gene Expression 302
   4.3 Binding Partners of Tbr1 303
5. Downstream Target Genes Regulated by T-Box Genes in Cerebral Cortex 303
6. Human Disorders Caused by T-Box Gene Mutations in Brain 305
7. Summary 306

Acknowledgment 306

References 306

Current Topics in Developmental Biology, Volume 122
ISSN 0070-2153
http://dx.doi.org/10.1016/bs.ctdb.2016.08.001

Abstract

T-box transcription factors play key roles in the regulation of developmental processes such as cell differentiation and migration. Mammals have 17 T-box genes, of which several regulate brain development. The *Tbr1* subfamily of T-box genes is particularly important in development of the cerebral cortex, olfactory bulbs (OBs), and cerebellum. This subfamily is comprised of *Tbr1*, *Tbr2* (also known as *Eomes*), and *Tbx21*. In developing cerebral cortex, *Tbr2* and *Tbr1* are expressed during successive stages of differentiation in the pyramidal neuron lineage, from Tbr2+ intermediate progenitors to Tbr1+ postmitotic glutamatergic neurons. At each stage, Tbr2 and Tbr1 regulate laminar and regional identity of cortical projection neurons, cell migration, and axon guidance. In the OB, *Tbr1* subfamily genes regulate neurogenesis of mitral and tufted cells, and glutamatergic juxtaglomerular interneurons. Tbr2 is also prominent in the development of retinal ganglion cells in nonimage-forming pathways. Other regions that require Tbr2 or Tbr1 in development or adulthood include the cerebellum and adult dentate gyrus. In humans, de novo mutations in *TBR1* are important causes of sporadic autism and intellectual disability. Further studies of T-box transcription factors will enhance our understanding of neurodevelopmental disorders and inform approaches to new therapies.

1. DISCOVERY OF T-BOX GENE EXPRESSION IN THE DEVELOPING BRAIN AND NEURAL RETINA

T-box genes belong to a family of transcription factors that share a common DNA-binding domain, the T-box, which spans 180–200 amino acid residues and binds DNA in a sequence-specific manner to the half consensus sequence AGGTGTGAAA, called the T-box binding element. Unique among the T-box transcription factors is Mga that has both T-box and basic helix-loop-helix (bHLH)-zip DNA-binding domains. Based on homology in the T-box, these genes have been classified in five subfamilies: T (*T* and *Tbx19*), Tbx1 (*Tbx1*, *Tbx10*, *Tbx15*, *Tbx18*, *Tbx20*, *Tbx22*), Tbx2 (*Tbx2*, *Tbx3*, *Tbx4*, *Tbx5*), Tbx6 (*Tbx6*, *Mga*), and Tbr1 (*Tbr1*, *Eomes*, *Tbx21*) (reviewed by Papaioannou, 2014). In the developing brain, *Eomes* is known as *Tbr2*, and *Tbx21* as *T-bet*.

Tbr1 subfamily members exhibit a high degree of sequence similarity in the T-domain coding region, but less in N- and C-terminal coding regions. Tbr1 and Tbr2 have 86% similarity in the T-box domain (Fig. 1), 31% similarity in the N-terminal region, and 40% similarity in the C-terminal region. T-bet displays 72% similarity with Tbr1 in the T-box domain and 74% similarity with the Tbr2 T-box domain (Fig. 1). Interestingly, Tbr2 has a predicted

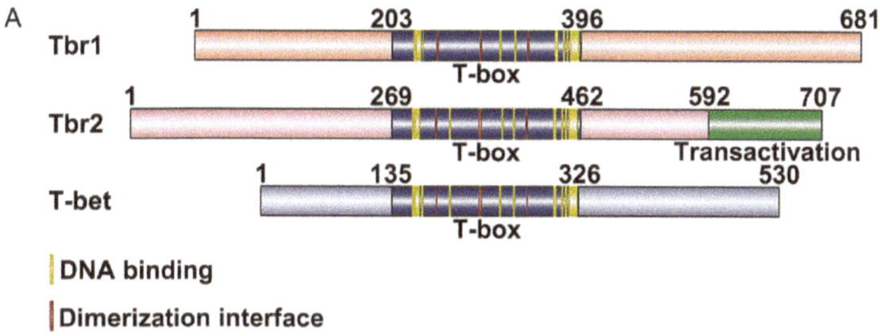

B

T-box multiple alignment

```
Tbr1  KAQVYLCNRPLWLKFHRHQTEMIITKQGRRMFPFLSFNISGLDPTAHYNIFVDVILADPN
Tbr2  -AHVYLCNRPLWLKFHRHQTEMIITKQGRRMFPFLSFNINGLNPTAHYNVFVEVVLADPN
T-bet KLRVALSNHLLWSKFNQHQTEMIITKQGRRMFPFLSFTVAGLEPTSHYRMFVDVVLVDQH
        :* *.*: ** **.:********************.: **:**:**.:**:*:*.* .

Tbr1  HWRFQGGKWVPCGKADTNVQGNRVYMHPDSPNTGAHWMRQEISFGKLKLTNNKGASNNNG
Tbr2  HWRFQGGKWVTCGKADNNMQGNKMYVHPESPNTGSHWMRQEISFGKLKLTNNKGANNNNT
T-bet HWRYQSGKWVQCGKAEGSMPGNRLYVHPDSPNTGAHWMRQEVSFGKLKLTNNKGASNNVT
      ***:*.**** ****:  .:**::*:**:*****:******:*************.**

Tbr1  QMVVLQSLHKYQPRLHVVEVNEDGTEDTSQPGRVQTFTFPETQFIAVTAYQNTDITQLKI
Tbr2  QMIVLQSLHKYQPRLHIVEVTEDGVEDLNEPSKTQTFTFSETQFIAVTAYQNTDITQLKI
T-bet QMIVLQSLHKYQPRLHIVEVNDGEPEAACSASNTHVFTFQETQFIAVTAYQNAEITQLKI
      **:*************:***.:   *   . ...:.*** ************::******

Tbr1  DHNPFAKGFRDNYD-
Tbr2  DHNPFAKGFRDNYDS
T-bet DNNPFAKGFREN---
      *.********:*
```

Fig. 1 Conservation among mouse Tbr1 subfamily members. (A) Domain organization of Tbr1 subfamily: Tbr1, T-box brain protein 1, NCBI accession number: NP_033348, Tbr2, eomesodermin homolog isoform 1, NCBI accession number: NP_034266, and T-bet, T-box transcription factor TBX21, NCBI accession number: NP_062380 (diagram constructed using Illustrator for Biological Sequences online software, http://ibs.biocuckoo.org/online.php, Liu et al., 2015). All share a conserved T-box domain in the middle of the protein. Within the T-box domain there are several DNA-binding sites (*yellow*) and dimerization interface sites (*red*).Tbr2 also has a predicted transcriptional activation domain in its C-terminal. (B) Multiple protein sequence alignment of the T-box domains of Tbr1 subfamily members using Clustal Omega (http://www.ebi.ac.uk/). There is 86% similarity between T-box domains of Tbr1 and Tbr2, 72% between Tbr1 and T-bet, and 74% between Tbr2 and T-bet. There are no significant similarities found within the N- or C-termini of either protein, except for 51% between the Tbr2 transactivation domain and Tbr1 C-terminus. Aminoacid residue color code is based on physicochemical properties: *red* (AVFPMILW), small + hydrophobic (including aromatic Y); *blue* (DE), acidic; *magenta* (RK), basic (including H); and *green* (STYHCNGQ), hydroxyl, sulfhydryl, and amine (including G). Consensus symbol meaning: (*), fully conserved residue; (:), conservation between groups of strongly similar properties; and (.), conservation between groups of weakly similar properties.

transactivation domain in its C-terminus. The Tbr2 transactivation domain has 51% similarity to the Tbr1 C-terminus, possibly revealing a conserved transcriptional activation function between Tbr1 and Tbr2.

Tbr1 was discovered by screening a cDNA library enriched for embryonic forebrain mRNA (Porteus et al., 1992). In situ hybridization (ISH) revealed specific expression of *Tbr1* in the cerebral cortex and a few other brain regions (Bulfone et al., 1995), starting at the onset of corticogenesis on embryonic day 10.5 (E10.5). In embryos, *Tbr1* was prominently expressed in the mantle zone and cortical plate (CP) of the telencephalic vesicle, as well as the olfactory bulb (OB) and cerebellum (Fig. 2) (Bulfone et al., 1995; Hevner et al., 2001). After birth, Tbr1 expression levels gradually decline (Bulfone et al., 1995).

Tbr2 (NCBI Gene: *Eomes*) was likewise identified by screening embryonic forebrain cDNA (Bulfone et al., 1999). ISH revealed *Tbr2* expression in mouse brain starting at E10.5, in most of the same areas as Tbr1, including cerebral cortex, OB, and cerebellum. In embryonic cortex, *Tbr2* exhibited complementary zonal expression to *Tbr1*, with *Tbr2* being restricted in the proliferative regions, indicating its expression in neuronal progenitors, and *Tbr1* in the mantle zone, where postmitotic neurons reside (Bulfone et al., 1999; Kimura, Nakashima, Ueno, Kiyama, & Taga, 1999) (Fig. 2). In retina, Tbr2 was recently found to be expressed by ganglion cells that contribute to nonimage-forming visual pathways (Mao et al., 2014; Sweeney, Tierney, & Feldheim, 2014).

T-bet (NCBI Gene: *Tbx21*) was identified by homology to *Tbr1* and *Tbr2*. Expression of *T-bet* in the brain is restricted to the OB (Faedo et al., 2002). T-bet protein expression first appears at E14.5 in the accessory olfactory bulb (AOB), and later extends to mitral and tufted cells of the main OB, where it persists through adulthood (Fig. 2).

Besides the Tbr1 subfamily, three other T-box genes have known brain-related expression or functions: *Tbx20*, *Tbx19*, and *Tbx18*. *Tbx20* plays a role in the migration of hindbrain motor neurons (Song et al., 2006). *Tbx19* is expressed in the rostral ventral diencephalon and adrenocorticotropin-producing, proopiomelanocortin (POMC) lineage cells of the anterior pituitary gland (Lamolet et al., 2001; Liu et al., 2001). *Tbx18* is required for otic mesenchyme to differentiate into fibrocytes and its deficiency is associated with deafness (Trowe, Maier, Schweizer, & Kispert, 2008). Another T-box family member, *Mga*, was reported to be important for brain morphogenesis in zebrafish (Rikin & Evans, 2010). Mouse *Mga* transcript is present in many regions of the developing brain, including cerebral cortex, according to

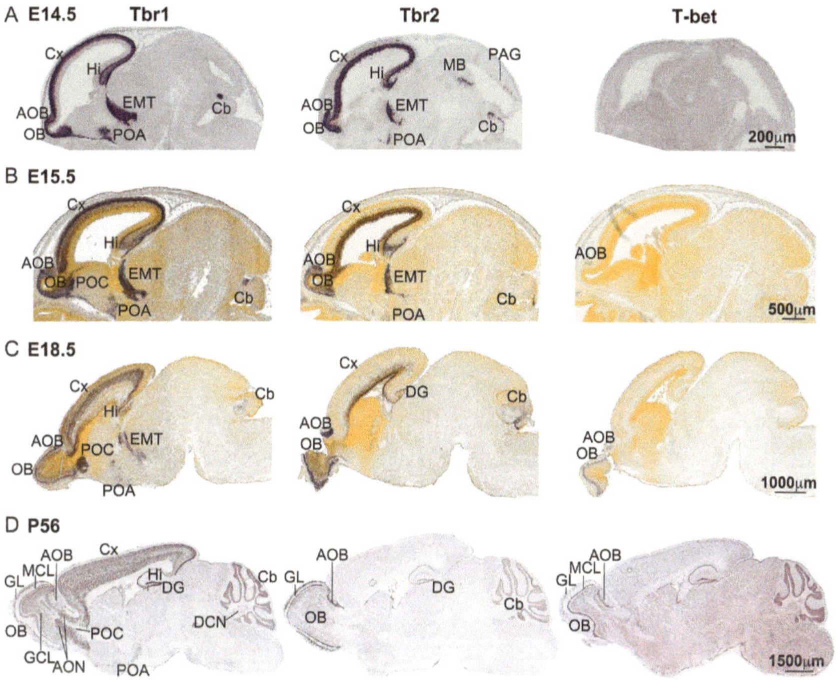

Fig. 2 Brain distributions of Tbr1 subfamily members at different developmental stages. In situ hybridization sagittal sections, adapted from (A) Gene Paint mouse atlas of gene expression (http://www.genepaint.org) and (B–D) Allen Developing Mouse Brain Atlas (http://developingmouse.brain-map.org), of E14.5, E15.5, E18.5 embryonic and P56 adult mouse *Tbr1*, *Tbr2*, and *T-bet* transcripts. The labeling represents the areas where the respective transcripts are present. Abbreviations: *AOB*, accessory olfactory bulb; *AON*, anterior olfactory nuclei; *Cb*, cerebellum; *Cx*, cortex; *DCN*, deep cerebellar nuclei; *DG*, dentate gyrus; *EMT*, eminentia thalami; *GCL*, granule cell layer; *GL*, glomerular layer; *Hi*, hippocampus; *MB*, midbrain; *MCL*, mitral cell layer; *OB*, olfactory bulb; *PAG*, periaqueductal gray; *POA*, preoptic area; *POC*, primary olfactory cortex.

ISH databases (Genepaint: genepaint.org; Allen Developing Mouse: developingmouse.brain-map.org). However, no detailed studies of *Mga* expression or functions in the brain have been reported.

2. EXPRESSION OF Tbr1 SUBFAMILY T-BOX GENES IN BRAIN AND NEURAL RETINA

Interestingly, detailed analysis has shown that Tbr1 subfamily genes appear to be expressed only in committed glutamatergic neuronal lineages, and not in other neuron types, glia, or neural stem cells (NSCs).

2.1 Tbr2 and Tbr1 Are Part of a Transcription Factor Cascade in Glutamatergic Neurogenesis

In the developing brain, neurons are produced through a series of fate choices and differentiation pathways, governed by the coordinated expression of a number of developmental transcription factors. In the cerebral cortex, Pax6, Neurogenin2, Tbr2, NeuroD, and Tbr1 form a transcription factor cascade that is related to stages of cellular differentiation, from NSCs to intermediate progenitors (IPs) to neurons (Fig. 3). Interestingly, the same transcription factors are expressed, in the same order, during glutamatergic neurogenesis in the adult subventricular zone (SVZ) and dentate gyrus (DG), and with variations in developing cerebellum and OB. Thus, *Tbr1* and *Tbr2* are part of a conserved genetic program for glutamatergic neurogenesis (reviewed by Hevner, Hodge, Daza, & Englund, 2006).

2.2 Tbr2 and Tbr1 Expression in Cortical Glutamatergic Neurogenesis

In developing cerebral cortex, excitatory pyramidal neurons are generated from NSC-like radial glial progenitors (RGPs) both directly and indirectly via committed neurogenic, transit-amplifying IPs (Fig. 3). RGPs are molecularly characterized by expression of Pax6 (Götz, Stoykova, Gruss, 1998), a homeobox and paired domain transcription factor. In turn, IPs are identified by expression of Tbr2, and declining or absent expression of Pax6 (Englund et al., 2005). In mice, RGPs are located almost exclusively in the ventricular zone (VZ), near the apical (ventricular) surface of the cortical neuroepithelium. Another type of RGP, known as basal or outer RGPs (bRGPs), detach from the ventricular surface and migrate to the SVZ, but retain molecular and cellular properties of RGPs, including neurogenesis. The bRGPs are a small minority of all RGPs in mice, but become abundant in some larger species, such as humans (Lui, Hansen, & Kriegstein, 2011).

Tbr2 was identified as a specific marker of IPs, which are the main cell type to form the embryonic SVZ (Englund et al., 2005). Nevertheless, IPs are distributed in both the VZ, where they are called apical IPs (IP-a), and in the SVZ, where they are called basal IPs (IP-b). The IP-a and IP-b types were discovered by unbiased hierarchical cluster analysis of precursor cells from embryonic mouse neocortex (Kawaguchi et al., 2008). Both IP-a and IP-b cells express Tbr2, but they also exhibit distinct cellular morphologies and molecular profiles. Unlike RGPs, which are multipotent NSCs capable of extensive proliferation, IPs undergo limited proliferation to

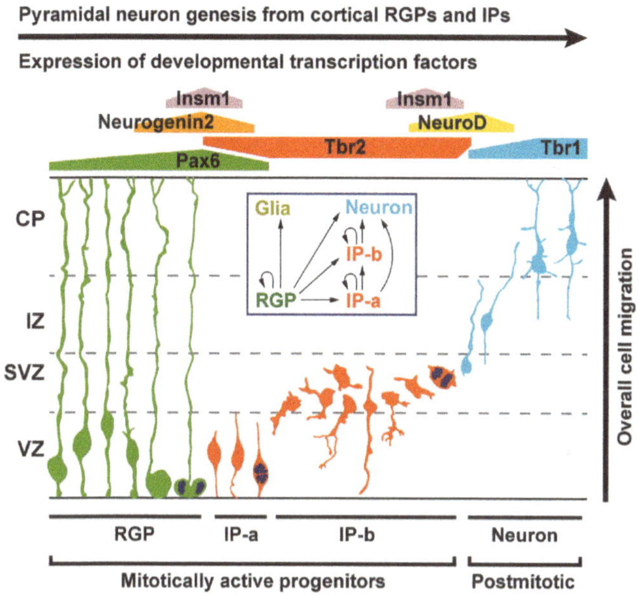

Fig. 3 A transcription factor cascade including Tbr1 and Tbr2 regulates glutamatergic neurogenesis in the cerebral cortex. Cortical excitatory neurons are derived from NSC-like RGPs both directly and indirectly via transit-amplifying IPs. Apical and basal types of IPs (IP-a and IP-b, respectively) were discovered by unbiased analysis of single-cell transcriptome profiles (Kawaguchi et al., 2008). IP-a cells exhibit "short radial" or "pin-like" morphology, contact the ventricular surface, and divide in the VZ, sometimes at the ventricular surface (Gal et al., 2006; Kowalczyk et al., 2009; Nelson, Hodge, Bedogni, & Hevner, 2013; Ochiai et al., 2009; Stancik, Navarro-Quiroga, Sellke, & Haydar, 2010). IP-b cells exhibit highly dynamic and diverse morphologies. They transiently extend processes, frequently from SVZ to VZ, which are important for long-distance Delta-Notch signaling from IPs to RGPs (Nelson et al., 2013). Expression of transcription factors correlates with transitions between cell types, as suggested by the bars indicating levels of transcription factor expression. The core sequential cascade (Pax6 → Neurogenin2 → Tbr2 → NeuroD → Tbr1) is punctuated by Insm1, which drives IP genesis and proliferation in cortex, and neurogenesis throughout the brain. Pax6 and Neurogenin2 directly bind and activate *Tbr2*, but other expression patterns represent order of expression, not necessarily direct regulation.

produce only glutamatergic projection neurons (Englund et al., 2005; Kawaguchi et al., 2008; Mihalas et al., 2016).

Following Tbr2, Tbr1 is expressed by new postmitotic cortical neurons as they travel through the intermediate zone (IZ) and enter the CP. The transition from IP to neuron appears to be rapid, as coexpression of Tbr2 and Tbr1 proteins in the same cells is rare, but detectable at the SVZ–IZ interface (Nelson et al., 2013). The transition from progenitor to neuron is also marked by expression of NeuroD, a bHLH transcription factor, which exhibits transient overlap with Tbr2 and Tbr1 (Hevner et al., 2006). Once in the CP, neurons mature and settle in layers in reverse order of their birth, starting from the bottom (layer 6) and progressively seeding more superficial layers. Tbr1 is expressed at least transiently in all cortical glutamatergic neurons, but at highest levels in layer 6, subplate, and Cajal–Retzius cells.

To summarize, differentiation in cortical lineages is marked by (1) stage-specific expression of developmental transcription factors, including Tbr2 and Tbr1; (2) changes in cell morphology; and (3) net migration from VZ to CP (Fig. 3). The minimal overlap of zonal expression patterns suggests that transcription factor expression is tightly regulated during transitions between cell types.

2.3 Tbr2 and Tbr1 Expression in Developing and Adult DG

DG development spans embryonic and postnatal stages and indeed continues seamlessly into adult neurogenesis. The DG originates from the embryonic dentate neuroepithelium (DNe) located along the medial boundary of the cortical vesicle. NSCs and IPs migrate from the DNe through the dentate migratory stream (DMS) to form several transient neurogenic zones or "niches," including the subgranular zone (SGZ), which is maintained as the adult neurogenic niche (Li, Kataoka, Coughlin, & Pleasure, 2009; Li & Pleasure, 2005; Pleasure, Collins, & Lowenstein, 2000).

In the developing DG, Tbr2 and Tbr1 are expressed in IPs and new neurons, much as in embryonic neocortex. However, Pax6 + NSCs and Tbr2 + IPs migrate extensively away from the ventricle and into the transient abventricular niches, where most of the Tbr1 + granule neurons are generated (Hodge et al., 2012, 2013). Tbr1 is expressed in postmitotic DG granule neurons during development and adulthood.

During postnatal development, NSCs in the SGZ and DMS continue to generate granule neurons that expand first the suprapyramidal, and then the infrapyramidal blades of the DG, conferring the characteristic V- or C-shaped morphology of this primordial gyrus. SGZ neurogenesis continues into adulthood in all mammals, except for cetaceans (whales, dolphins, and porpoises) in which the DG and adult SGZ neurogenesis have undergone evolutionary regression (Hevner, 2016).

In the adult DG, similar progenitor cell types and transcription factor cascades are observed as in developing cortex and DG (Hodge et al., 2008). Pax6 and Tbr2 are expressed in RGPs (also known as "type I" progenitors) and IPs, respectively. These progenitors are embedded in the SGZ. The Pax6+ NSCs produce Tbr2+ IPs by a tightly regulated process, such that the overall rate of adult SGZ neurogenesis is highly responsive to genetic and environmental factors, and correlates directly with the abundance of Tbr2+ IPs (Hodge et al., 2008). The transcription factor cascade Pax6 → Neurogenin2 → Tbr2 → NeuroD → Tbr1 is conserved in adult DG (Hodge et al., 2008; Roybon, Hjalt, et al., 2009). In adult mice, Tbr1 shows graded expression that is highest in granule neurons near the hilus, and lowest in cells near the molecular layer, suggesting that Tbr1 levels decline with granule neuron age and maturation.

2.4 Tbr2, Tbr1, and T-bet Expression in the OB and Adult SVZ

The OBs are rostral evaginations from the cerebral hemispheres and are composed of four key neuron types: projection neurons (the mitral and tufted cells), local inhibitory interneurons (periglomerular and granule cells), local excitatory interneurons (juxtaglomerular cells), and glial cells. Neurogenesis of GABAergic and glutamatergic interneurons continues in the postnatal and adult SVZ, source of neurons that migrate to the OB through the rostral migratory stream (RMS). While adult SVZ neurogenesis is prominent in rodents, some species, such as humans, lack adult SVZ neurogenesis (Sanai et al., 2011).

Tbr1 and Tbr2 are expressed in the mitral, tufted, and juxtaglomerular cell lineages of the OB (Bulfone et al., 1998; Faedo et al., 2002; Kimura et al., 1999). T-bet is uniquely expressed by postmitotic mitral cells and some tufted cells (Faedo et al., 2002; Kosaka & Kosaka, 2012). The expression of Tbr1 family genes is, however, highly dynamic during development and adulthood (Fig. 2).

On E12.5, Tbr1 and Tbr2 are expressed during initial stages in the generation of glutamatergic neurons from dividing progenitors (Kahoud, Elsen, Hevner, & Hodge, 2014). At E14.5, Tbr1 and Tbr2 are strongly expressed in many postmitotic mitral cells, but T-bet is initially detected only in a subset of AOB projection neurons. By E16.5, T-bet expression increases in the main OB, and mitral cells come to express different levels of Tbr1 subfamily members. By P7, tufted cells and juxtaglomerular neurons emerge in the external plexiform layer and glomerular layer, and the Tbr1 subfamily starts to show segregated expression patterns in these neurons. In the adult OB, an overlapping but distinct expression pattern of each Tbr1 subfamily member defines different subpopulations of OB neurons. The dynamic and unique combinatorial expression of Tbr1, Tbr2, and T-bet suggests their possible roles in the specification of different types of OB projection neurons and the formation of functional olfactory neural circuitry (Mizuguchi et al., 2012).

In adult mice, glutamatergic juxtaglomerular interneurons are produced in the SVZ and migrate to the OB via the RMS. Adult neurogenesis of these excitatory OB interneurons was unknown until our observation of Tbr2+ progenitors in the adult SVZ and RMS (Brill et al., 2009). These interneurons are also generated by the conserved Pax6 → Neurogenin2 → Tbr2 → NeuroD → Tbr1 cascade (Brill et al., 2009; Roybon, Deierborg, Brundin, & Li, 2009). In this lineage, NSC-like, Pax6+ progenitors express Ascl1 (another bHLH transcription factor, also known as Mash1), conferring the capacity to produce both neurons and oligodendrocytes. The choice between these fates is determined by expression of Neurogenin2 to generate neurons or Olig2 (oligodendrocyte transcription factor 2; a bHLH transcription factor) to produce oligodendrocytes. The neuronal lineages subsequently express Tbr2, NeuroD, and Tbr1 transiently (Brill et al., 2009; Roybon, Deierborg, et al., 2009; Takashima & Suzuki, 2013).

The excitatory and inhibitory interneurons produced in adult SVZ may be derived from distinct subsets of pallial- and subpallial-derived NSCs, which are already specified to produce glutamatergic or GABAergic neuron types, respectively (Tamamaki, 2005). The NSCs specified for glutamatergic fates, and their progeny Tbr2+ IPs, may serve as an endogenous source of progenitors that can be recruited to the cerebral cortex to produce pyramidal neurons after injury (Brill et al., 2009; Sequerra, Miyakoshi, Froes, Menezes, & Hedin-Pereira, 2010).

2.5 Tbr2 and Tbr1 Expression in Developing Basal Forebrain, Amygdala, and Eminentia Thalami

Tbr2 and *Tbr1* are both expressed in eminentia thalami, a transient precursor of bed nuclei of stria terminalis and other forebrain nuclei. Tbr2 and Tbr1 are also expressed in scattered cells of the amygdala, basal forebrain, preoptic area, and hypothalamus. The amygdala is a composite of nuclei derived from cortical and subcortical developmental origins. Some of these nuclei are derived from a caudal amygdaloid migration stream, in which Tbr1 is highly expressed (Remedios et al., 2007). Little is known about the cell types that express Tbr1 and Tbr2 in these regions.

2.6 Tbr2 and Tbr1 Expression in the Cerebellum

The cerebellum contains three main types of glutamatergic neurons, whose genesis involves Tbr1 subfamily genes. The first to be produced are projection neurons of the deep cerebellar nuclei (DCN), generated from E10.5 to E12.5 in mice (Miale & Sidman, 1961). The DCN lineages sequentially express Pax6, Tbr2, and Tbr1 in the rhombic lip, rostral rhombic lip stream, and nuclear transitory zone of the embryonic cerebellum; thus, the Pax6 → Tbr2 → Tbr1 cascade proceeds similarly in DCN as in pyramidal cell neurogenesis. In the medial DCN, Tbr1 expression is maintained to adulthood (Fink et al., 2006). One report described Tbr1 immunoreactivity in the cytoplasm of postnatal and adult Purkinje cells (Hong & Hsueh, 2007), but this was not confirmed by other studies (Englund et al., 2006; Fink et al., 2006).

The second wave of glutamatergic neurons produced from the rhombic lip are the unipolar brush cells (UBCs), a unique type of glutamatergic interneuron in the granule cell layer, generated from E13.5 to the early postnatal period in mice (Englund et al., 2006). UBCs express Tbr2 throughout life, from genesis in the rhombic lip through adulthood. However, neither UBCs nor their progenitors express Tbr1.

In the third wave of glutamatergic neurogenesis, granule neurons are produced during the first 3 postnatal weeks in mice (Miale & Sidman, 1961). Like UBCs, granule cells are derived from lineages that sequentially express Pax6 and Tbr2, but not Tbr1. However, Tbr2 is expressed only transiently, while Pax6 expression persists at moderately high levels in adult granule neurons. Thus, UBCs and granule neurons differ from other types of glutamatergic neurons discussed here, not only by apparent lack of Tbr1 expression at any stage in their maturation, but also by persistence of Tbr2 and/or Pax6 expression in mature postmitotic neurons. NeuroD is also expressed in UBC and granule cell lineages, and is essential for the development of granule neurons (Lee et al., 2000; Miyata, Maeda, & Lee, 1999).

2.7 Tbr2 Expression in the Retina

In the retina, Tbr2 is expressed by a minority (~10%) of retinal ganglion cells (RGCs), identified as components of nonimage-forming circuits (Mao et al., 2014; Sweeney et al., 2014). Among the ~20 different types of RGCs in mice, Tbr2 is expressed in ≥3 nonoverlapping types, representing a subset of the nonimage-forming pathways. Development of the retina also requires Pax6, but Tbr1 is apparently not expressed in the retina at any stage.

2.8 Tbr2 Expression in the Embryonic Midbrain

Transient *Tbr2* expression is observed in an unidentified group of cells in the midbrain during embryonic development (Fig. 2). These Tbr2+ cells are presumably glutamatergic neuronal precursors, but their properties and progeny are unknown.

In sum, the patterns of Tbr1 subfamily gene expression suggest these genes are involved in differentiation of glutamatergic neurons in several important brain regions.

3. FUNCTIONS OF Tbr1 SUBFAMILY T-BOX GENES IN BRAIN AND NEURAL RETINA

Tbr1 subfamily genes have critical functions in brain and retina development, as well as adult neurogenesis in the DG and SVZ. Many interesting phenotypes and novel mechanisms of development have been revealed by functional studies of Tbr2 and Tbr1, while T-bet has so far demonstrated no essential functions in brain development.

3.1 Functions of Tbr1 Subfamily Members in Cerebral Cortex Development

Functions of *Tbr2* in cortical development have been studied using conditional genetic approaches. Conditional knockout (cKO) of *Tbr2* is necessary to circumvent early embryonic lethality associated with homozygous constitutive *Tbr2* inactivation (Arnold, Hofmann, Bikoff, & Robertson, 2008; Russ et al., 2000).

Initial studies of mouse brain *Tbr2* cKO mutants found that basal mitoses (a marker of IPs in the SVZ) were reduced in the developing neocortex, suggesting that Tbr2 promotes IP genesis (Arnold, Huang, et al., 2008; Sessa, Mao, Hadjantonakis, Klein, & Broccoli, 2008). However, both of those previous studies used targeted Cre mice that interfered with cortical development by haploinsufficiency effects (see Discussion in Mihalas et al., 2016). More recent studies using a transgenic Cre found that basal mitoses actually increased in the absence of Tbr2, suggesting that Tbr2 is necessary for differentiation, not genesis of IPs (Mihalas et al., 2016). Also, cells expressing Pax6 and Insm1, transcription factors that drive *Tbr2* expression and IP genesis, were expressed ectopically in cortical differentiation zones, suggesting that Tbr2 normally feeds back to suppress upstream genes (Mihalas et al., 2016). Moreover, Tbr2 was found to modulate the balance of layer 6 and layer 5 neurons, such that PACAP-expressing layer 5 neurons differentiated precociously and exuberantly in Tbr2-deficient cortex (Mihalas et al., 2016).

In addition, *Tbr2* reporter alleles have been used to study the contributions of IP-mediated (indirect) neurogenesis to cortical layers. Experiments using various *Tbr2*-targeted alleles have shown that Tbr2+ IPs generate all subtypes of glutamatergic cortical neurons (including pyramidal projection neurons in all layers, as well as Cajal–Retzius and subplate neurons), and that overall, the majority of glutamatergic cortical neurons are produced from IPs (Kowalczyk et al., 2009; Mihalas et al., 2016; Vasistha et al., 2015). Furthermore, cohorts of IP progeny labeled with an inducible Cre reporter system ($Eomes^{CreERT2}$;*Ai14*; Pimeisl et al., 2013) make overlapping contributions to cortical neurogenesis (Mihalas et al., 2016). For example, some early generated IPs, produced from RGPs at the onset of cortical neurogenesis, persist in the cortex for at least 2 days before undergoing final mitosis and differentiating as upper layer neurons (Mihalas et al., 2016). Later in corticogenesis, IP laminar fates become progressively limited to upper layers (Mihalas et al., 2016).

Besides regulating cortical laminar fate, *Tbr2* plays an important role regulating rostro-caudal patterning of cortical areas. In the absence of Tbr2, rostral (motor) cortical areas, defined by gene expression domains, are reduced, and caudal (sensory) areas are expanded (Elsen et al., 2013).

While *Tbr2* is expressed only in glutamatergic lineages, inactivation of *Tbr2* also perturbs the development of other cell types that interact with IPs in developing cortex. Cortical IPs are one source of Cxcl12, a chemokine that acts as an attractive cue for tangential migration of interneurons from subpallial origins into the cortex, and for microglial accumulation in progenitor zones. As a result of reduced signaling from IPs in Tbr2-deficient cortex, tangential migration of GABAergic interneurons, and recruitment of microglia into the VZ and SVZ, are deficient (Arno et al., 2014; Sessa et al., 2010). Thus Tbr2, in the course of regulating glutamatergic neurogenesis, also modulates GABAergic interneuron migration and microglia recruitment. By coordinating these developmental processes, Tbr2 is important for maintaining the balance of excitatory and inhibitory neurons and may function as a rheostat to control the ratio of various cell lineages that comprise the cerebral cortex. In sum, Tbr2 regulates both laminar and rostro-caudal identity of cortical neurons, and their interactions with other cell types.

Tbr1 functions in brain development have been studied by constitutive gene inactivation, as no conditional or reporter alleles have been produced.

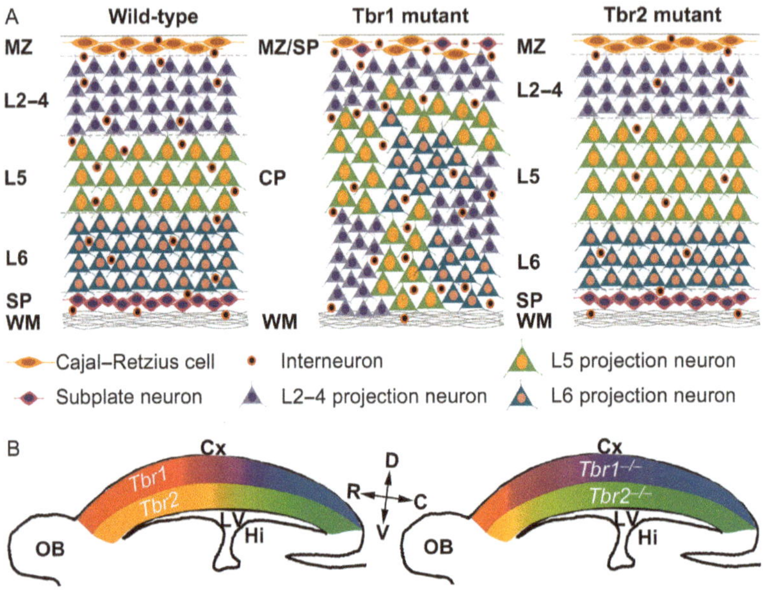

Fig. 4 Cortical phenotypes in Tbr1 and Tbr2 mutant mouse cortex. Cerebral cortex projection neurons are organized in six layers within the cortical plate (CP), which spans from the subplate (SP) to the marginal zone (MZ) at the pial surface. Cortical projection neurons are deposited in the cortical plate sequentially according to their birth. The cortex is formed in an inside-out fashion, with early born neurons populating the lower layers adjacent to the SP (L6 → L5) and later born neurons progressing more superficially toward the MZ. In the Tbr1 mutant cortex the preplate, which normally splits into the SP and MZ to allow the formation of the CP, fails to split, resulting in accumulation of neurons underneath the SP/MZ. The MZ contains Cajal–Retzius cells devoid of Reelin and the cortical layers are inverted similar to the Reeler phenotype in the Tbr1-deficient rostral cortex (not depicted here; Sheppard & Pearlman, 1997). In Tbr1 mutant caudal cortex (depicted here), however, the cortical layers are replaced by clusters of cells with similar molecular and birthdate identity. In the Tbr2 mutant cortex delayed neuronal differentiation results in laminar dysregulation that consists in L5 expansion at the expense of L6 and L2–4 neurons, and a reduction in interneuron tangential migration into the cortex from subpallial origins. Abbreviations: *C*, caudal; *Cx*, cortex; *D*, dorsal; *Hi*, hippocampus; *L*, layer; *LV*, lateral ventricle; *OB*, olfactory bulb; *R*, rostral; *V*, ventral; *WM*, white matter.

Loss of *Tbr1* function in mice causes defects of glutamatergic neuron migration and differentiation, with severe abnormalities of preplate and layer 6 (Fig. 4, Hevner et al., 2001). The Tbr1-deficient cortex is structurally disorganized, with abnormal neuron clustering. Major efferent (corticothalamic, corticobulbar, corticospinal, callosal) and afferent (thalamocortical) axon projections develop abnormally in *Tbr1* mutant brain. Corticothalamic axons end in the internal capsule, without crossing the boundary into diencephalon. Callosal fibers mostly terminate in Probst bundles without crossing the midline, and fewer cortical efferent axons reach the cerebral peduncles. Thalamocortical axons become misrouted in the internal capsule and remain in the basal telencephalon rather than entering the cortex.

Tbr1 also regulates cortical patterning and regional identity. In the absence of Tbr1, rostral genes are downregulated, while caudal genes are upregulated (Bedogni, Hodge, et al., 2010). Indeed, Tbr2 and Tbr1 act sequentially to promote rostral and suppress caudal identity in IPs and postmitotic neurons, respectively.

3.2 Functions of Tbr2 and Tbr1 in Developing and Adult DG

Tbr2 cKO mice have a severely atrophic DG, due to defects of hippocampal fissure morphogenesis, granule neuron differentiation, and NSC niche formation (Hodge et al., 2013). Agenesis of the hippocampal fissure was traced to defective migration of Cajal–Retzius cells, which help guide the migration of progenitors and granule neurons to the DG. In the absence of Tbr2, Cajal–Retzius cells did not migrate properly into the hippocampal fissure. Thus, the transhilar radial glial scaffold (which connects the DMS to the granule neuron layer) was compressed, and migration of progenitors and neuroblasts to the developing DG was impaired (Hodge et al., 2013). Also in *Tbr2* mutants, premature and ectopic granule cell neurogenesis was followed by increased apoptosis, and NSCs in the DG were depleted before proper establishment of the SGZ (Hodge et al., 2013).

In the adult SGZ, *Tbr2* ablation results in the cessation of neurogenesis, due to failure of differentiation (Hodge et al., 2012; Tsai, Tsai, Arnold, & Huang, 2015). Upon inactivation of *Tbr2*, SGZ production of new neurons is reduced despite enhanced proliferation and abundance of Sox2+ NSCs. Indeed, one function of Tbr2 is to directly repress *Sox2* expression in the granule neuron lineage. Molecular experiments indicated that Tbr2 directly binds and represses expression of *Sox2*, and thereby functions to repress NSC identity (Hodge et al., 2012). Thus, in adult DG neurogenesis as in developing cerebral neocortex, Tbr2 appears to play a dual role suppressing NSC identity and promoting neuronal differentiation. Together, the data indicate that *Tbr2* expression in IPs is critical for neuronal differentiation in the adult DG, and those IPs are an essential stage in the lineage from NSCs to new granule neurons.

The roles of Tbr1 in DG development and adult SGZ neurogenesis are virtually unknown, as *Tbr1* mutant mice rarely survive the neonatal period. Studies using conditional *Tbr1* knockout would be useful to unravel potential roles of Tbr1 in DG development (and adult neurogenesis), but no conditional *Tbr1* alleles have been reported.

3.3 Functions of Tbr1 Subfamily Members in OB Development and Adult SVZ Neurogenesis

Tbr1 subfamily genes are extremely important in OB development.

Mice with brain *Tbr2* cKO have severe OB hypoplasia and excessive accumulation of OB precursor cells in an expanded RMS (Arnold, Huang, et al., 2008; Kahoud et al., 2014). Moreover, in *Tbr2* (and *Tbr1*) mutant OB, T-bet expression is abolished (Faedo et al., 2002; Kahoud et al., 2014), consistent with upstream roles of Tbr2 and Tbr1 in mitral and tufted neuron development (Arnold, Huang, et al., 2008; Bulfone et al., 1998; Mizuguchi et al., 2012).

Conditional inactivation of *Tbr2* in late embryonic mitral and tufted cells caused changes in molecular expression, including a compensatory increase of Tbr1, and a concomitant shift of vesicular glutamate transporter (VGluT) subtypes from VGluT1 to VGluT2 (Mizuguchi et al., 2012). Tbr2-deficient mitral and tufted cells also exhibited structural abnormalities in their dendritic morphology and projection patterns. Furthermore, the number of dendrodendritic reciprocal synapses between mitral/tufted cells and GABAergic interneurons was significantly reduced. Upon stimulation with odorants, larger numbers of mitral and tufted cells were activated in *Tbr2* cKO mice. These results suggested that Tbr2 is required for not only the proper differentiation of mitral and tufted cells, but also for the establishment of functional neuronal circuitry in the OB and maintenance of excitatory–inhibitory balance crucial for odor information processing (Mizuguchi et al., 2012).

Conditional ablation of *Tbr2* in adult mice results in reduced genesis of Tbr1+ glutamatergic neuron precursors in the SVZ and RMS (Kahoud et al., 2014).

Tbr1 is necessary for the survival and differentiation of glutamatergic cells in the OB. Neonatal *Tbr1* mutants have small OBs, and lack OB projection neurons as well as the lateral olfactory tract (Bulfone et al., 1998). From experiments using retroviral-mediated overexpression of Tbr1 in cultured OB stem cells, Mendez-Gomez et al. (2011) reported that Tbr1 promotes the production of neurons and oligodendrocytes and suppresses astrocytic differentiation. However, Tbr1 expression has only been observed in neurons in vivo, not in oligodendrocytes or astrocytes (Hevner et al., 2001). Interestingly, Tbr1 is also necessary for development of a cell migration stream that produces part of the AOB (Huilgol et al., 2013).

Although T-bet is a specific marker of mitral and tufted projection neurons (Mitsui, Igarashi, Mori, & Yoshihara, 2011), no study has yet addressed the consequences of T-bet deficiency in the brain. Homozygous mouse *Tbx21* mutants are viable, but show asthma susceptibility (Finotto et al., 2002).

3.4 Functions of Tbr1 Subfamily Members in Basal Forebrain and Amygdala Development

The basal forebrain and amygdala appear histologically abnormal in *Tbr1* null mice (Hevner et al., 2001). However, these defects have not been studied in much detail, except in relation to behavioral anomalies in *Tbr1* heterozygous null mice (see later). Functions of *Tbr2* in amygdala and basal forebrain development have likewise not been determined.

3.5 Functions of Tbr1 Subfamily Members in Cerebellum Development

Although cerebellar phenotypes have not received much attention yet, evidence suggests that both *Tbr1* and *Tbr2* are functionally important in cerebellar development.

Deficiency of Tbr1 causes abnormal DCN morphogenesis, but no apparent defects of Reelin expression, glutamatergic differentiation, or axon pathfinding (Fink et al., 2006). Because *Tbr1* null mice die soon after birth, analysis of the cerebellar phenotype is limited to embryonic and perinatal periods. The mechanisms that control DCN cell migration and axon guidance during postnatal life are important subjects for future studies.

The *Tbr2* mutant cerebellum appears grossly normal, but detailed analysis of *Tbr2* null cerebellar phenotypes is still in progress. Preliminary studies suggest that UBCs are deficient in *Tbr2* cKO cerebellum (our unpublished observations).

3.6 Functions of Tbr1 Subfamily Members in Neural Retina Development

Tbr2 is necessary for the differentiation and survival of intrinsically photosensitive RGCs (Mao et al., 2014). Accordingly, *Tbr2* mutant mice have reduced retinal projections to nonimage-forming nuclei, and an attenuated pupillary light reflex (Mao et al., 2014; Sweeney et al., 2014).

3.7 Functions of Tbr1 Subfamily Members in Mouse Behavior and Motor Control

Tbr1 and *Tbr2* regulate the development of major brain regions involved in critical aspects of mammalian behavior and motor control. So, it is no surprise that mutations in these genes have been linked to behavioral and neurological abnormalities in mice. (Human disorders caused by *TBR1* and *TBR2* mutations are discussed separately later.)

In mice, *Tbr1* haploinsufficiency has been associated with impaired social interactions, abnormal ultrasonic vocalizations, and defects of associative memory and cognitive flexibility, together resembling autism-like phenotypes (Huang et al., 2014). These behavioral deficits were shown to be, at least in part, due to defective axonal projections of amygdalar neurons. Since stimulation of amygdalar neuronal activity by local infusion of a partial NMDA receptor agonist, D-cycloserine, ameliorated the behavioral defects of *Tbr1* haploinsufficient mice, reduced activity in the amygdala may be a key consequence of *Tbr1* deficiency. These results suggested that *Tbr1* is important in the regulation of amygdalar axonal connections and cognition. Tbr1 may also regulate neural plasticity by modulating *Grin2b* expression (Chuang, Huang, & Hsueh, 2014).

Mice lacking brain Tbr2 display increased aggressiveness, hyperactivity, and reduced grip strength (Arnold, Huang, et al., 2008), but no defects (and even improved performance) of repetitive motor tasks (Mihalas et al., 2016). The reduced grip strength may correlate with hypotonia in a human *TBR2* deficiency condition (Baala et al., 2007). Inactivation of *Tbr2* in adult hippocampus altered the response to stress (Tsai et al., 2015), and deficiency of Tbr2 in the retina caused defective pupillary reflex (Sweeney et al., 2014).

No behavioral abnormalities have been described in *Tbx21* mutants.

4. REGULATION OF T-BOX GENES

Much has been learned about how Tbr1 and Tbr2 fit into gene expression networks that regulate cerebral cortex neurogenesis (Hevner et al., 2006; Mihalas et al., 2016; Srinivasan et al., 2012). Upstream transcriptional regulators, binding partners, and direct downstream target genes of both *Tbr1* and *Tbr2* have been identified (Tables 1 and 2). In particular, these interactions illuminate the overall process of neurogenesis from RGP to IP to neuron in developing cerebral cortex.

4.1 Regulation of Tbr2 Gene Expression

Several studies have shown that cortical *Tbr2* expression is regulated by Pax6, which promotes IP genesis by regulating numerous genes, including *Tbr2* by direct transcriptional activation. In *Pax6* null (*Sey/Sey*) mice, *Tbr2* is among the most downregulated transcripts (*Tbr1* is also downregulated), and Tbr2+ IPs are greatly reduced (Quinn et al., 2007). Interestingly, abventricular mitoses (basal progenitors) are still numerous, but are converted to GABAergic cell fate in the absence of Pax6, as indicated by expression of Mash1/Ascl1 and Dlx2 (Quinn et al., 2007). Pax6 binds and regulates *Tbr2* in a dose-dependent manner in cortex (Sansom et al., 2009).

Pax6 also regulates *Tbr2* in the cerebellum (Fink et al., 2006; Ha et al., 2015). Microarrays and ISH studies of embryonic *Pax6* null (*Sey/Sey*) cerebellum revealed downregulation of *Tbr2* and *Tbr1* (Ha et al., 2015).

Under some conditions, Pax6 may negatively regulate expression of *Tbr2* and *Tbr1*, or prevent genesis of neurons that express these transcription factors. Exogenous expression of *Pax6* in embryonic OB postmitotic precursors decreased the number of cells that progressed to a mitral cell fate. Consequently, mitral cell precursors changed their identity and expressed molecular phenotypes characteristic of OB interneurons, including dopaminergic and GABAergic periglomerular cells (Imamura & Greer, 2013).

Another gene that directly regulates *Tbr2* in cerebral cortex is Neurogenin2. This bHLH transcription factor promotes IP genesis by activating the transcription of numerous genes, to initiate the transition from RGP to IP (Kovach et al., 2013). In *Neurogenin2* null cortex, *Tbr2* expression is decreased (Schuurmans et al., 2004). Furthermore, Neurogenin2 binds and directly activates the *Tbr2* gene (Ochiai et al., 2009).

A third key regulator of *Tbr2* in developing cerebral cortex is Insm1 (insulinoma-associated 1), a zinc finger transcription factor that is expressed in neurogenic progenitors and nascent neurons throughout the developing brain and spinal cord (Duggan et al., 2008). In developing cerebral cortex, Insm1 is expressed in IP-genic, as well as neurogenic progenitors (Farkas et al., 2008). Genetic loss of *Insm1* function in developing cortex decreases the number of Tbr2-positive IPs, while forced expression of Insm1

Table 1 Regulators of Tbr1 and Tbr2 in the Brain

Regulators	Gene Symbol	Brain Region	Effects	References
Pou4f2 (Pou domain, class 4, transcription factor 2)	*Pou4f2*[a]	Neural retina	Promotes Tbr2 expression and retinal ganglion cell differentiation	Mao et al. (2008)
Pax6 (paired box 6)	*Pax6*[a]	Cortex	Promotes expression of Tbr2. Promotes RGP to IP transition	Sansom et al. (2009)
		OB	Negative regulation of Tbr1 and Tbr2 expression in postmitotic mitral cell precursors	Imamura and Greer (2013)
		Cerebellum	Promotes expression of Tbr1 and Tbr2	Fink et al. (2006) and Ha et al. (2015)
Neurog2 (neurogenin 2)	*Neurog2*[a]	Cortex	Promotes expression of Tbr2. Promotes RGP to IP transition	Ochiai et al. (2009) and Kovach et al. (2013)
Insm1 (insulinoma-associated 1)	*Insm1*	Cortex	Promotes expression of Tbr2. Promotes RGP to IP transition	Farkas et al. (2008)
Cux2 (cut-like homeobox 2)	*Cux2*	Cortex	Promotes suppression of Tbr2. Promotes restriction of upper layer expansion	Cubelos et al. (2008)
AP2γ (activating enhancer-binding protein 2 gamma)	*Tfap2c*[a]	Cortex	Promotes expression of Tbr2	Pinto et al. (2009)
GSK-3 (glycogen synthase kinase-3)	*Gsk3b*	Cortex	Promotes IP differentiation and neurogenesis	Kim et al. (2009)
Axin	*Axin1*	Cortex	Promotes IP switch from proliferative to differentiative status	Fang, Chen, Fu, and Ip (2013)
Tox (thymocyte selection-associated high mobility group box)	*Tox*	Cortex	Inhibits IP switch from proliferative to differentiative status. Inhibits neurogenesis	Artegiani et al. (2015)

Ntf3 (neurotrophin 3)	*Ntf3*	Cortex	Promotes production of Tbr2-positive basal progenitors at the expense of apical progenitors. Promotes lower to upper neuron cell fate switch	Parthasarathy, Srivatsa, Nityanandam, and Tarabykin (2014)
MiR-92a and MiR-92b (microRNA 92a and b)	*Mir-92a* *Mir-92b*	Cortex	Promotes suppression of Tbr2	Bian et al. (2013) and Nowakowski et al. (2013)
Af9 (ALL1-fused gene from chromosome 9 protein)	*Mllt3*	Cortex	Promotes maintenance of Tbr2-positive progenitors. Suppresses Tbr1-positive cell fate mainly in upper layer neurons	Buttner, Johnsen, Kugler, and Vogel (2010)
CB1(cannabinoid receptor 1)	*Cnr1*	Cortex	Promotes RGP to IP transition	Diaz-Alonso et al. (2015)
Cocaine		Cortex	Transient decrease in Tbr2 and Tbr1 expression in the dorsal forebrain	McCarthy et al. (2011)
Gpsm2 (G-protein signaling modulator 2)	*Gpsm2*	Cortex	Limits IP genesis by regulating mitotic spindle orientation	Konno et al. (2008)
Insc (inscuteable homolog)	*Insc*	Cortex	Promotes IP genesis by regulating mitotic spindle orientation	Postiglione et al. (2011)
Satb2 (special AT-rich sequence-binding protein 2)	*Satb2*[a]	Cortex	Promotes expression of Tbr1	Srinivasan et al. (2012)
Ctip1 (COUP-TF-interacting protein 1)	*Bcl11a*[a]	Cortex	Promotes repression of Tbr1	Canovas et al. (2015)
Ctip2 (COUP-TF-interacting protein 2)	*Bcl11b*	Cortex	Promotes repression of Tbr1	Srinivasan et al. (2012)

[a]Direct transcriptional regulation.
Abbreviations: *IP*, intermediate progenitor; *OB*, olfactory bulb; *RGP*, radial glia progenitor.

Table 2 Targets of Tbr1 and Tbr2 in the Developing Cerebral Cortex

	Targets	Gene Symbol	Effect	Direct Target	Validation	References
Tbr1	Fezf2 (Fez family zinc finger 2)	*Fezf2*	Repression	Yes	ChIP	Han et al. (2011)
	Bhlhb5 (class B basic helix-loop-helix protein 5)	*Bhlhe22*	Repression	Yes	ChIP	Han et al. (2011)
	Reelin	*Reln*	Activation	Yes	ChIP	Hsueh, Wang, Yang, and Sheng (2000)
	Auts2 (autism susceptibility candidate 2)	*Auts2*	Activation	Yes	ChIP	Bedogni, Hodge, et al. (2010)
	NR2B (ionotropic glutamate receptor NMDA 2B)	*Grin2b*	Activation	Yes	ChIP	Wang et al. (2004)
Tbr2	Sox2 (SRY/sex determining region Y/-box 2)	*Sox2*	Repression	Yes	ChIP	Hodge et al. (2012)
	Pax6 (paired box 6)	*Pax6*	Repression	Yes	Luciferase assay	Kovach et al. (2013)
	Ebf1 (early B cell factor 1)	*Ebf1*	Repression	Yes	ChIP	Kovach et al. (2013)
	Insm1 (insulinoma-associated 1)	*Insm1*	Repression	Not tested	Protein (IHC), mRNA (microarrays)	Mihalas et al. (2016)
	Tbr1 (T-box brain 1)	*Tbr1*	Activation	Not tested	Protein (IHC), mRNA (microarrays)	Mihalas et al. (2016)
	Ctip2 (COUP-TF-interacting protein 2)	*Bcl11b*	Activation	Not tested	Protein (IHC), mRNA (microarrays)	Mihalas et al. (2016)
	Satb2 (special AT-rich sequence-binding protein 2)	*Satb2*	Activation	Not tested	Protein (IHC), mRNA (microarrays)	Mihalas et al. (2016)

Abbreviations: *ChIP*, chromatin immunoprecipitation; *IHC*, immunohistochemistry; *mRNA*, messenger RNA.

promotes genesis and proliferative expansion of Tbr2-positive IPs (Farkas et al., 2008). Microarray data comparing the transcriptome of *Insm1* null and control cerebral cortex (Farkas et al., 2008) indicated that *Tbr2* expression was significantly reduced by ~17% (our unpublished analysis of the data). However, it is currently unknown if changes in *Tbr2* expression are the result of direct binding and transcriptional activation by Insm1, or merely reflect overall changes in IP abundance.

Together, Pax6, Neurogenin2, and Insm1 appear to be the primary drivers of IP genesis and Tbr2 expression in developing cortex. Interestingly, IP genesis is not dependent on *Tbr2* function (Mihalas et al., 2016), but Tbr2 does feedback to suppress upstream IP-genic genes, including Pax6 and Insm1, although the exact mechanisms of this feedback are still unclear (Mihalas et al., 2016; Sessa et al., 2008).

The core genetic program of IP genesis is also modulated by transcription factors to achieve regional and temporal adjustments in IP genesis or proliferation. For example, AP2γ (an AP2 family transcription factor) binds and activates *Tbr2* in occipital cortex RGPs, to promote IP identity during upper layer neurogenesis (Pinto et al., 2009). In the absence of AP2γ, upper layer neurons in the occipital cortex are reduced, altering the function and plasticity of adult visual cortex. Cux2, a homeodomain transcription factor expressed in IPs (and some migrating interneurons), functions to suppress the proliferation of Tbr2-positive IPs during upper layer neurogenesis (Cubelos et al., 2008). In Cux2-deficient mice, selective expansion of SVZ neuronal precursors is accompanied by excess production of upper layer neurons. In contrast, the number of deep cortical layer neurons is not altered in Cux2-deficient mice (Cubelos et al., 2008).

Tbr2 is also a target of posttranscriptional regulation by microRNAs (miRs) during cortical neurogenesis. Interestingly, the 3′-UTR of *Tbr2* mRNA is targeted in RGPs by miR-92b (Nowakowski et al., 2013), and by miR-92a in the miR-17-92 cluster (Bian et al., 2013). The effect of these miRs is to suppress expression of *Tbr2* mRNA in RGPs, and thereby prevent premature differentiation of IPs, and concomitantly expand the RGP population.

In the retina, Tbr2 is regulated by Pou4f2 (Mao et al., 2008).

Upstream of transcriptional and posttranscriptional mechanisms that regulate *Tbr2*, developmentally important signaling pathways, such as Wnt, neurotrophin, and Shh, also influence IP genesis and proliferation, and *Tbr2* expression. For example, neurotrophin-3 (Gene: *Ntf3*) exerts positive control over IP expansion in developing cortex, presumably through TrkC signaling (Parthasarathy et al., 2014). In turn, *Ntf3* is regulated by Sip1 (Gene: *Zeb2*), a homeobox transcription factor, to modulate the balance of upper and lower layer neurogenesis (Parthasarathy et al., 2014).

Signaling through glycogen synthase kinase 3 (GSK-3), a protein kinase mediating Wnt, Shh, Notch, and PI3K pathways, is also essential in IP genesis and neurogenesis during brain development (Kim et al., 2009). In cortex lacking both α and β isozymes of GSK-3, RGPs massively hyperproliferate, IPs and postmitotic neurons are reduced, and Tbr2 expression is decreased (Kim et al., 2009). The individual importance of each pathway is further complicated by interactions among pathways, as well as context-dependent factors and differences among cell types (RGPs, IPs, neurons). For example, the canonical Wnt–β-catenin signaling pathway promotes neuronal differentiation of IPs (Kuwahara et al., 2010; Munji, Choe, Li, Siegenthaler, & Pleasure, 2011), but in RGPs can either promote or suppress IP genesis, depending on context (Chenn & Walsh, 2002; Fang et al., 2013; Kuwahara et al., 2010; Wrobel, Mutch, Swaminathan, Taketo, & Chenn, 2007).

Tox is a high mobility group (HMG)-box transcription factor proposed to bind and regulate *Tbr2* (and *Tbr1*), and thus regulate cortical progenitor proliferation and differentiation, although details of these interactions and their significance are uncertain (Artegiani et al., 2015).

4.2 Regulation of Tbr1 Gene Expression

Compared to Tbr2, somewhat less is known about the regulation of *Tbr1* gene expression. Ctip1 (Gene: *Bcl11a*), a zinc finger transcription factor related to Ctip2, functions as a direct repressor of *Tbr1* in layer 5 of developing cerebral cortex (Canovas et al., 2015). By repressing Tbr1, Ctip1 suppresses corticothalamic (layer 6) identity in favor of subcerebral projection neuron (SCPN) (layer 5) identity. Other transcription factors proposed to regulate *Tbr1* in developing cortex include Ctip2, Fezf2, and Satb2, although these effects may be direct or indirect (Srinivasan et al., 2012).

An epigenetic mode of *Tbr1* regulation has been demonstrated for Af9 (Gene: *Mllt3*), a multifunctional nuclear protein. In postmitotic neurons of upper cortical layers, Af9 represses *Tbr1* by interacting with Dot1l, a methyltransferase that negatively regulates transcription by methylation of histone H3 lysine 79 (H3K79) (Buttner et al., 2010). By this mechanism, Af9 limits *Tbr1* expression within low to moderate levels characteristic of upper layers, not the high levels seen in layer 6 and other early born neurons

(Cajal–Retzius and subplate). Like other T-box transcription factors, Tbr1 may bind to target genes in a concentration-dependent manner, and high levels of Tbr1 might be unnecessary in upper layer neurons, or even detrimental.

4.3 Binding Partners of Tbr1

The effects of Tbr1 subfamily members on gene expression are likely to be controlled in part by their binding partners, about which little is known. Recently, TBR1 was found to interact with FOXP2, a forkhead box transcription factor that is, like Tbr1, expressed in layer 6 (Ferland, Cherry, Preware, Morrisey, & Walsh, 2003). The interactions between TBR1 and FOXP2 could be functionally important, as they are disrupted by mutations that are pathogenic in humans (Deriziotis et al., 2014). CASK, a membrane-associated guanylate kinase, has also been proposed to interact with Tbr1 as a coactivator, to enhance transcription of *Reln* and other genes (Hsueh et al., 2000; Wang et al., 2004). According to this model, CASK enters the nucleus from the cytoplasm (where CASK has other functions) and binds T-element DNA sequences in a complex with Tbr1 and other proteins (Wang et al., 2004).

5. DOWNSTREAM TARGET GENES REGULATED BY T-BOX GENES IN CEREBRAL CORTEX

By activating or repressing specific genes, Tbr1 and Tbr2 play important roles in neurogenesis, axon guidance, laminar and regional specification of the cortex. Most T-box transcription factors act as activators at some genes and repressors at others, and Tbr1 and Tbr2 appear to be no exceptions.

Dual modes of gene regulation would seem well adapted for the dual roles of Tbr2 in cortical differentiation, to (1) suppress progenitor cell identity and (2) promote projection neuron identity and differentiation. Many potential target genes of Tbr2 activation or repression in the developing cortex have been identified by comparing transcriptome expression in control and *Tbr2*-null neocortex using microarrays (Elsen et al., 2013). However, for most genes, it is unknown which genes bind Tbr2 protein and are likely to be regulated directly.

Repression of specific progenitor genes by Tbr2 has been demonstrated in developing cortex and adult SGZ. Tbr2 directly binds and represses *Sox2*, an HMG-box transcription factor, in NSCs (Hodge et al., 2012). In *Tbr2* cKO DG, Sox2 expression is increased, granule neuron differentiation is disrupted, and neurogenesis is ineffective.

In developing neocortex, likely Tbr2 directly represses *Pax6* and *Insm1*, as indicated by several lines of evidence. First, *Pax6* and *Insm1* are upregulated and expressed ectopically in differentiation zones of the *Tbr2* cKO embryonic cortex (Mihalas et al., 2016). Second, forced expression of Tbr2 in the VZ causes rapid downregulation of *Pax6* (Sessa et al., 2008). Third, Tbr2 binding has been detected near the *Pax6* and *Insm1* promoters (Teo et al., 2011). Fourth, Tbr2 expression overlaps with Pax6 and Insm1 in fractions of IPs, presumably cells in transition between stages of cellular differentiation (Fig. 3; Englund et al., 2005; Farkas et al., 2008; Hevner et al., 2006; Mihalas et al., 2016). In addition, Tbr2 directly represses *Ebf1*, a bHLH transcription factor expressed in ventral forebrain (Kovach et al., 2013). By this mechanism, Tbr2 prevents cortical IPs from acquiring a noncortical fate.

Conversely, Tbr2 is also necessary to promote neuronal differentiation of cortical IPs by activating projection neuron genes such as *Tbr1*, *Ctip2*, and *Satb2* (Elsen et al., 2013; Kovach et al., 2013; Mihalas et al., 2016). However, it is unknown if these and other neuronal genes are direct targets of Tbr2 binding and transcriptional activation.

Several direct targets of Tbr1 binding and regulation are known. Transcriptome profiling of *Tbr1* mutant cortex identified hundreds of up- or downregulated genes (Bedogni, Hodge, et al., 2010), and some have been identified as direct targets. In developing neocortex, *Tbr1* binds and activates *Auts2* (*autism susceptibility candidate gene 2*), which encodes a nuclear protein and potential transcriptional regulator that is highly enriched in frontal cortex (Bedogni, Hodge, Nelson, et al., 2010; Gao et al., 2014). In *Tbr1* null mice, *Auts2* expression in cortex is virtually abolished (Bedogni, Hodge, et al., 2010). The Tbr1 → Auts2 cascade may be a principal effector of frontal cortex differentiation, originating from the protomap in RGPs (Pax6 → Tbr2 → Tbr1 → Auts2).

Also in developing neocortex, Tbr1 has been implicated as a direct repressor of *Fezf2*, a zinc finger transcription factor gene and marker of layer 5 SCPNs, including corticospinal motor neurons. Tbr1 binds and represses the *Fezf2* locus, and *Fezf2* mRNA expression is increased in neonatal *Tbr1* null cortex (Bedogni, Hodge, et al., 2010; Han et al., 2011; McKenna et al., 2011). Thus, Tbr1 suppresses SCPN fates from arising in layer 6, and restricts the laminar origin of corticospinal axons to layer 5, where Tbr1 expression is lower (Han et al., 2011). This mechanism complements the cross repression of Tbr1 in layer 5, by Ctip1 (Canovas et al., 2015).

Transcriptional control by Tbr1 has also been shown to depend on its interactions with other proteins. De novo truncating *TBR1* mutations in human autism spectrum disorders (ASDs) disrupt TBR1 protein interactions with FOXP2 and CASK observed in HEK293 and SHSY5Y cells, leading to abnormal gene regulation (Deriziotis et al., 2014).

6. HUMAN DISORDERS CAUSED BY T-BOX GENE MUTATIONS IN BRAIN

In the past 5 years, disruptive de novo mutations in the human *TBR1* gene have been repeatedly identified in patients with ASDs and intellectual disability (Hamdan et al., 2014; O'Roak et al., 2012; Sanders et al., 2015). The mutations in *TBR1* appear to disrupt protein functions and affect brain development by haploinsufficiency (Deriziotis et al., 2014). Given the large numbers of genes (hundreds) that can cause ASDs and intellectual disability, *TBR1* mutations likely account for $<1\%$ of all cases, but *TBR1* is nevertheless one of the genes that is most frequently linked to sporadic forms of these disorders.

Disruption of TBR1 interactions with FOXP2, a forkhead box transcription factor, constitutes a plausible mechanism for language impairment in autism (Deriziotis et al., 2014). Mutations in human *FOXP2* cause severe speech impairment characterized by inability to correctly perform the coordinated movements required for speech, and difficulties in comprehension (Fisher & Scharff, 2009). In the mammalian cortex, Tbr1 and FoxP2 show striking similarities in expression patterns (especially in layer 6), raising the possibility that they cooperate to regulate gene networks in deep layer cortical neurons, and other neural sites of coexpression. Given that communication deficits are a core feature of ASD, it has been suggested that Tbr1 and FoxP2 belong to a shared molecular network, which goes awry in different neurodevelopmental disorders involving impaired speech and/or language skills (Deriziotis et al., 2014).

In humans, *TBR2* (NCBI Gene: *EOMES*) homozygous loss of function in brain due to a balanced translocation results in severe brain malformation consisting of microcephaly, polymicrogyria, corpus callosum agenesis, small cerebellum, and enlarged ventricles (Baala et al., 2007). However, this phenotype has not been studied histologically, so the cellular and molecular abnormalities are unknown.

Human *T-BET* (NCBI Gene: *TBX21*) mutations are associated with aspirin-induced asthma (Akahoshi et al., 2005). No brain-related pathology has been linked to T-bet deficiency in mice or humans so far.

7. SUMMARY

- Tbr1 subfamily members: Tbr1/*Tbr1*, Tbr2/*Eomes*, and T-bet/*Tbx21* are the main T-box genes expressed in the brain.
- T-bet is specifically expressed in OB mitral and tufted cells in the brain.
- Tbr1 and Tbr2 are important regulators of glutamatergic neuron differentiation in cerebral cortex, OB, hippocampal DG, cerebellum, and adult SVZ.
- *TBR1* mutations in humans are linked to sporadic autism and intellectual disability, while *TBR2* mutations are linked to a severe neurological disorder that includes microcephaly.

ACKNOWLEDGMENT

This work was supported by NIH Grants R01 NS085081 and R01 NS092339 (R.F.H.).

REFERENCES

Akahoshi, M., Obara, K., Hirota, T., Matsuda, A., Hasegawa, K., Takahashi, N., et al. (2005). Functional promoter polymorphism in the TBX21 gene associated with aspirin-induced asthma. *Human Genetics*, *117*(1), 16–26.

Arno, B., Grassivaro, F., Rossi, C., Bergamaschi, A., Castiglioni, V., Furlan, R., et al. (2014). Neural progenitor cells orchestrate microglia migration and positioning into the developing cortex. *Nature Communications*, *5*, 5611.

Arnold, S. J., Hofmann, U. K., Bikoff, E. K., & Robertson, E. J. (2008). Pivotal roles for eomesodermin during axis formation, epithelium-to-mesenchyme transition and endoderm specification in the mouse. *Development*, *135*(3), 501–511.

Arnold, S. J., Huang, G. J., Cheung, A. F., Era, T., Nishikawa, S., Bikoff, E. K., et al. (2008). The T-box transcription factor Eomes/Tbr2 regulates neurogenesis in the cortical subventricular zone. *Genes & Development*, *22*(18), 2479–2484.

Artegiani, B., de Jesus Domingues, A. M., Bragado Alonso, S., Brandl, E., Massalini, S., Dahl, A., et al. (2015). Tox: A multifunctional transcription factor and novel regulator of mammalian corticogenesis. *EMBO Journal*, *34*(7), 896–910.

Baala, L., Briault, S., Etchevers, H. C., Laumonnier, F., Natiq, A., Amiel, J., et al. (2007). Homozygous silencing of T-box transcription factor EOMES leads to microcephaly with polymicrogyria and corpus callosum agenesis. *Nature Genetics*, *39*(4), 454–456.

Bedogni, F., Hodge, R. D., Elsen, G. E., Nelson, B. R., Daza, R. A., Beyer, R. P., et al. (2010a). Tbr1 regulates regional and laminar identity of postmitotic neurons in developing neocortex. *Proceedings of the National Academy of Sciences of the United States of America*, *107*(29), 13129–13134.

Bedogni, F., Hodge, R. D., Nelson, B. R., Frederick, E. A., Shiba, N., Daza, R. A., et al. (2010b). Autism susceptibility candidate 2 (Auts2) encodes a nuclear protein expressed in developing brain regions implicated in autism neuropathology. *Gene Expression Patterns*, *10*(1), 9–15.

Bian, S., Hong, J., Li, Q., Schebelle, L., Pollock, A., Knauss, J. L., et al. (2013). MicroRNA cluster miR-17-92 regulates neural stem cell expansion and transition to intermediate progenitors in the developing mouse neocortex. *Cell Reports*, *3*(5), 1398–1406.

Brill, M. S., Ninkovic, J., Winpenny, E., Hodge, R. D., Ozen, I., Yang, R., et al. (2009). Adult generation of glutamatergic olfactory bulb interneurons. *Nature Neuroscience*, *12*(12), 1524–1533.

Bulfone, A., Martinez, S., Marigo, V., Campanella, M., Basile, A., Quaderi, N., et al. (1999). Expression pattern of the Tbr2 (Eomesodermin) gene during mouse and chick brain development. *Mechanisms of Development*, *84*(1–2), 133–138.

Bulfone, A., Smiga, S. M., Shimamura, K., Peterson, A., Puelles, L., & Rubenstein, J. L. (1995). T-brain-1: A homolog of Brachyury whose expression defines molecularly distinct domains within the cerebral cortex. *Neuron*, *15*(1), 63–78.

Bulfone, A., Wang, F., Hevner, R., Anderson, S., Cutforth, T., Chen, S., et al. (1998). An olfactory sensory map develops in the absence of normal projection neurons or GABAergic interneurons. *Neuron*, *21*(6), 1273–1282.

Buttner, N., Johnsen, S. A., Kugler, S., & Vogel, T. (2010). Af9/Mllt3 interferes with Tbr1 expression through epigenetic modification of histone H3K79 during development of the cerebral cortex. *Proceedings of the National Academy of Sciences of the United States of America*, *107*(15), 7042–7047.

Canovas, J., Berndt, F. A., Sepulveda, H., Aguilar, R., Veloso, F. A., Montecino, M., et al. (2015). The specification of cortical subcerebral projection neurons depends on the direct repression of TBR1 by CTIP1/BCL11a. *Journal of Neuroscience*, *35*(19), 7552–7564.

Chenn, A., & Walsh, C. A. (2002). Regulation of cerebral cortical size by control of cell cycle exit in neural precursors. *Science*, *297*(5580), 365–369.

Chuang, H. C., Huang, T. N., & Hsueh, Y. P. (2014). Neuronal excitation upregulates Tbr1, a high-confidence risk gene of autism, mediating Grin2b expression in the adult brain. *Frontiers in Cellular Neuroscience*, *8*, 280.

Cubelos, B., Sebastian-Serrano, A., Kim, S., Moreno-Ortiz, C., Redondo, J. M., Walsh, C. A., et al. (2008). Cux-2 controls the proliferation of neuronal intermediate precursors of the cortical subventricular zone. *Cerebral Cortex*, *18*(8), 1758–1770.

Deriziotis, P., O'Roak, B. J., Graham, S. A., Estruch, S. B., Dimitropoulou, D., Bernier, R. A., et al. (2014). De novo TBR1 mutations in sporadic autism disrupt protein functions. *Nature Communications*, *5*, 4954.

Diaz-Alonso, J., Aguado, T., de Salas-Quiroga, A., Ortega, Z., Guzman, M., & Galve-Roperh, I. (2015). CB1 cannabinoid receptor-dependent activation of mTORC1/Pax6 signaling drives Tbr2 expression and basal progenitor expansion in the developing mouse cortex. *Cerebral Cortex*, *25*(9), 2395–2408.

Duggan, A., Madathany, T., de Castro, S. C., Gerrelli, D., Guddati, K., & Garcia-Anoveros, J. (2008). Transient expression of the conserved zinc finger gene INSM1 in progenitors and nascent neurons throughout embryonic and adult neurogenesis. *Journal of Comparative Neurology*, *507*(4), 1497–1520.

Elsen, G. E., Hodge, R. D., Bedogni, F., Daza, R. A., Nelson, B. R., Shiba, N., et al. (2013). The protomap is propagated to cortical plate neurons through an Eomes-dependent intermediate map. *Proceedings of the National Academy of Sciences of the United States of America*, *110*(10), 4081–4086.

Englund, C., Fink, A., Lau, C., Pham, D., Daza, R. A., Bulfone, A., et al. (2005). Pax6, Tbr2, and Tbr1 are expressed sequentially by radial glia, intermediate progenitor cells, and postmitotic neurons in developing neocortex. *Journal of Neuroscience*, *25*(1), 247–251.

Englund, C., Kowalczyk, T., Daza, R. A., Dagan, A., Lau, C., Rose, M. F., et al. (2006). Unipolar brush cells of the cerebellum are produced in the rhombic lip and migrate through developing white matter. *Journal of Neuroscience*, *26*(36), 9184–9195.

Faedo, A., Ficara, F., Ghiani, M., Aiuti, A., Rubenstein, J. L., & Bulfone, A. (2002). Developmental expression of the T-box transcription factor T-bet/Tbx21 during mouse embryogenesis. *Mechanisms of Development*, *116*(1–2), 157–160.

Fang, W. Q., Chen, W. W., Fu, A. K., & Ip, N. Y. (2013). Axin directs the amplification and differentiation of intermediate progenitors in the developing cerebral cortex. *Neuron*, *79*(4), 665–679.

Farkas, L. M., Haffner, C., Giger, T., Khaitovich, P., Nowick, K., Birchmeier, C., et al. (2008). Insulinoma-associated 1 has a panneurogenic role and promotes the generation and expansion of basal progenitors in the developing mouse neocortex. *Neuron*, *60*(1), 40–55.

Ferland, R. J., Cherry, T. J., Preware, P. O., Morrisey, E. E., & Walsh, C. A. (2003). Characterization of Foxp2 and Foxp1 mRNA and protein in the developing and mature brain. *Journal of Comparative Neurology*, *460*(2), 266–279.

Fink, A. J., Englund, C., Daza, R. A., Pham, D., Lau, C., Nivison, M., et al. (2006). Development of the deep cerebellar nuclei: Transcription factors and cell migration from the rhombic lip. *Journal of Neuroscience*, *26*(11), 3066–3076.

Finotto, S., Neurath, M. F., Glickman, J. N., Qin, S., Lehr, H. A., Green, F. H., et al. (2002). Development of spontaneous airway changes consistent with human asthma in mice lacking T-bet. *Science*, *295*(5553), 336–338.

Fisher, S. E., & Scharff, C. (2009). FOXP2 as a molecular window into speech and language. *Trends in Genetics*, *25*(4), 166–177.

Gal, J. S., Morozov, Y. M., Ayoub, A. E., Chatterjee, M., Rakic, P., & Haydar, T. F. (2006). Molecular and morphological heterogeneity of neural precursors in the mouse neocortical proliferative zones. *Journal of Neuroscience*, *26*(3), 1045–1056.

Gao, Z., Lee, P., Stafford, J. M., von Schimmelmann, M., Schaefer, A., & Reinberg, D. (2014). An AUTS2-Polycomb complex activates gene expression in the CNS. *Nature*, *516*(7531), 349–354.

Götz, M., Stoykova, A., & Gruss, P. (1998). Pax6 controls radial glia differentiation in the cerebral cortex. *Neuron*, *21*(5), 1031–1044.

Ha, T., Swanson, D., Larouche, M., Glenn, R., Weeden, D., Zhang, P., et al. (2015). CbGRiTS: Cerebellar gene regulation in time and space. *Developmental Biology*, *397*(1), 18–30.

Hamdan, F. F., Srour, M., Capo-Chichi, J. M., Daoud, H., Nassif, C., Patry, L., et al. (2014). De novo mutations in moderate or severe intellectual disability. *PLoS Genetics*, *10*(10), e1004772.

Han, W., Kwan, K. Y., Shim, S., Lam, M. M., Shin, Y., Xu, X., et al. (2011). TBR1 directly represses Fezf2 to control the laminar origin and development of the corticospinal tract. *Proceedings of the National Academy of Sciences of the United States of America*, *108*(7), 3041–3046.

Hevner, R. F. (2016). Evolution of the mammalian dentate gyrus. *Journal of Comparative Neurology*, *524*(3), 578–594.

Hevner, R. F., Hodge, R. D., Daza, R. A., & Englund, C. (2006). Transcription factors in glutamatergic neurogenesis: Conserved programs in neocortex, cerebellum, and adult hippocampus. *Neuroscience Research*, *55*(3), 223–233.

Hevner, R. F., Shi, L., Justice, N., Hsueh, Y., Sheng, M., Smiga, S., et al. (2001). Tbr1 regulates differentiation of the preplate and layer 6. *Neuron*, *29*(2), 353–366.

Hodge, R. D., Garcia, A. J., 3rd, Elsen, G. E., Nelson, B. R., Mussar, K. E., Reiner, S. L., et al. (2013). Tbr2 expression in Cajal-Retzius cells and intermediate neuronal progenitors is required for morphogenesis of the dentate gyrus. *Journal of Neuroscience*, *33*(9), 4165–4180.

Hodge, R. D., Kowalczyk, T. D., Wolf, S. A., Encinas, J. M., Rippey, C., Enikolopov, G., et al. (2008). Intermediate progenitors in adult hippocampal neurogenesis: Tbr2 expression and coordinate regulation of neuronal output. *Journal of Neuroscience*, *28*(14), 3707–3717.

Hodge, R. D., Nelson, B. R., Kahoud, R. J., Yang, R., Mussar, K. E., Reiner, S. L., et al. (2012). Tbr2 is essential for hippocampal lineage progression from neural stem cells to intermediate progenitors and neurons. *Journal of Neuroscience*, *32*(18), 6275–6287.

Hong, C. J., & Hsueh, Y. P. (2007). Cytoplasmic distribution of T-box transcription factor Tbr-1 in adult rodent brain. *Journal of Chemical Neuroanatomy*, *33*(3), 124–130.

Hsueh, Y. P., Wang, T. F., Yang, F. C., & Sheng, M. (2000). Nuclear translocation and transcription regulation by the membrane-associated guanylate kinase CASK/LIN-2. *Nature*, *404*(6775), 298–302.

Huang, T. N., Chuang, H. C., Chou, W. H., Chen, C. Y., Wang, H. F., Chou, S. J., et al. (2014). Tbr1 haploinsufficiency impairs amygdalar axonal projections and results in cognitive abnormality. *Nature Neuroscience*, *17*(2), 240–247.

Huilgol, D., Udin, S., Shimogori, T., Saha, B., Roy, A., Aizawa, S., et al. (2013). Dual origins of the mammalian accessory olfactory bulb revealed by an evolutionarily conserved migratory stream. *Nature Neuroscience*, *16*(2), 157–165.

Imamura, F., & Greer, C. A. (2013). Pax6 regulates Tbr1 and Tbr2 expressions in olfactory bulb mitral cells. *Molecular and Cellular Neuroscience*, *54*, 58–70.

Kahoud, R. J., Elsen, G. E., Hevner, R. F., & Hodge, R. D. (2014). Conditional ablation of Tbr2 results in abnormal development of the olfactory bulbs and subventricular zone-rostral migratory stream. *Developmental Dynamics*, *243*(3), 440–450.

Kawaguchi, A., Ikawa, T., Kasukawa, T., Ueda, H. R., Kurimoto, K., Saitou, M., et al. (2008). Single-cell gene profiling defines differential progenitor subclasses in mammalian neurogenesis. *Development*, *135*(18), 3113–3124.

Kim, W. Y., Wang, X., Wu, Y., Doble, B. W., Patel, S., Woodgett, J. R., et al. (2009). GSK-3 is a master regulator of neural progenitor homeostasis. *Nature Neuroscience*, *12*(11), 1390–1397.

Kimura, N., Nakashima, K., Ueno, M., Kiyama, H., & Taga, T. (1999). A novel mammalian T-box-containing gene, Tbr2, expressed in mouse developing brain. *Brain Research. Developmental Brain Research*, *115*(2), 183–193.

Konno, D., Shioi, G., Shitamukai, A., Mori, A., Kiyonari, H., Miyata, T., et al. (2008). Neuroepithelial progenitors undergo LGN-dependent planar divisions to maintain self-renewability during mammalian neurogenesis. *Nature Cell Biology*, *10*(1), 93–101.

Kosaka, T., & Kosaka, K. (2012). Further characterization of the juxtaglomerular neurons in the mouse main olfactory bulb by transcription factors, Sp8 and Tbx21. *Neuroscience Research*, *73*(1), 24–31.

Kovach, C., Dixit, R., Li, S., Mattar, P., Wilkinson, G., Elsen, G. E., et al. (2013). Neurog2 simultaneously activates and represses alternative gene expression programs in the developing neocortex. *Cerebral Cortex*, *23*(8), 1884–1900.

Kowalczyk, T., Pontious, A., Englund, C., Daza, R. A., Bedogni, F., Hodge, R., et al. (2009). Intermediate neuronal progenitors (basal progenitors) produce pyramidal-projection neurons for all layers of cerebral cortex. *Cerebral Cortex*, *19*(10), 2439–2450.

Kuwahara, A., Hirabayashi, Y., Knoepfler, P. S., Taketo, M. M., Sakai, J., Kodama, T., et al. (2010). Wnt signaling and its downstream target N-myc regulate basal progenitors in the developing neocortex. *Development*, *137*(7), 1035–1044.

Lamolet, B., Pulichino, A. M., Lamonerie, T., Gauthier, Y., Brue, T., Enjalbert, A., et al. (2001). A pituitary cell-restricted T box factor, Tpit, activates POMC transcription in cooperation with Pitx homeoproteins. *Cell*, *104*(6), 849–859.

Lee, J. K., Cho, J. H., Hwang, W. S., Lee, Y. D., Reu, D. S., & Suh-Kim, H. (2000). Expression of neuroD/BETA2 in mitotic and postmitotic neuronal cells during the development of nervous system. *Developmental Dynamics*, *217*(4), 361–367.

Li, G., Kataoka, H., Coughlin, S. R., & Pleasure, S. J. (2009). Identification of a transient subpial neurogenic zone in the developing dentate gyrus and its regulation by Cxcl12 and reelin signaling. *Development*, *136*(2), 327–335.

Li, G., & Pleasure, S. J. (2005). Morphogenesis of the dentate gyrus: What we are learning from mouse mutants. *Developmental Neuroscience*, *27*(2–4), 93–99.

Liu, J., Lin, C., Gleiberman, A., Ohgi, K. A., Herman, T., Huang, H. P., et al. (2001). Tbx19, a tissue-selective regulator of POMC gene expression. *Proceedings of the National Academy of Sciences of the United States of America*, *98*(15), 8674–8679.

Liu, W., Xie, Y., Ma, J., Luo, X., Nie, P., Zuo, Z., et al. (2015). IBS: An illustrator for the presentation and visualization of biological sequences. *Bioinformatics*, *31*(20), 3359–3361.

Lui, J. H., Hansen, D. V., & Kriegstein, A. R. (2011). Development and evolution of the human neocortex. *Cell*, *146*(1), 18–36.

Mao, C. A., Kiyama, T., Pan, P., Furuta, Y., Hadjantonakis, A. K., & Klein, W. H. (2008). Eomesodermin, a target gene of Pou4f2, is required for retinal ganglion cell and optic nerve development in the mouse. *Development*, *135*(2), 271–280.

Mao, C. A., Li, H., Zhang, Z., Kiyama, T., Panda, S., Hattar, S., et al. (2014). T-box transcription regulator Tbr2 is essential for the formation and maintenance of Opn4/melanopsin-expressing intrinsically photosensitive retinal ganglion cells. *Journal of Neuroscience*, *34*(39), 13083–13095.

McCarthy, D. M., Zhang, X., Darnell, S. B., Sangrey, G. R., Yanagawa, Y., Sadri-Vakili, G., et al. (2011). Cocaine alters BDNF expression and neuronal migration in the embryonic mouse forebrain. *Journal of Neuroscience*, *31*(38), 13400–13411.

McKenna, W. L., Betancourt, J., Larkin, K. A., Abrams, B., Guo, C., Rubenstein, J. L., et al. (2011). Tbr1 and Fezf2 regulate alternate corticofugal neuronal identities during neocortical development. *Journal of Neuroscience*, *31*(2), 549–564.

Mendez-Gomez, H. R., Vergano-Vera, E., Abad, J. L., Bulfone, A., Moratalla, R., de Pablo, F., et al. (2011). The T-box brain 1 (Tbr1) transcription factor inhibits astrocyte formation in the olfactory bulb and regulates neural stem cell fate. *Molecular and Cellular Neuroscience*, *46*(1), 108–121.

Miale, I. L., & Sidman, R. L. (1961). An autoradiographic analysis of histogenesis in the mouse cerebellum. *Experimental Neurology*, *4*, 277–296.

Mihalas, A. B., Elsen, G. E., Bedogni, F., Daza, R. A., Ramos-Laguna, K. A., Arnold, S. J., et al. (2016). Intermediate progenitor cohorts differentially generate cortical layers and require Tbr2 for timely acquisition of neuronal subtype identity. *Cell Reports*, *16*(1), 92–105.

Mitsui, S., Igarashi, K. M., Mori, K., & Yoshihara, Y. (2011). Genetic visualization of the secondary olfactory pathway in Tbx21 transgenic mice. *Neural Systems & Circuits*, *1*(1), 5.

Miyata, T., Maeda, T., & Lee, J. E. (1999). NeuroD is required for differentiation of the granule cells in the cerebellum and hippocampus. *Genes & Development*, *13*(13), 1647–1652.

Mizuguchi, R., Naritsuka, H., Mori, K., Mao, C. A., Klein, W. H., & Yoshihara, Y. (2012). Tbr2 deficiency in mitral and tufted cells disrupts excitatory-inhibitory balance of neural circuitry in the mouse olfactory bulb. *Journal of Neuroscience*, *32*(26), 8831–8844.

Munji, R. N., Choe, Y., Li, G., Siegenthaler, J. A., & Pleasure, S. J. (2011). Wnt signaling regulates neuronal differentiation of cortical intermediate progenitors. *Journal of Neuroscience*, *31*(5), 1676–1687.

Nelson, B. R., Hodge, R. D., Bedogni, F., & Hevner, R. F. (2013). Dynamic interactions between intermediate neurogenic progenitors and radial glia in embryonic mouse neocortex: Potential role in Dll1-Notch signaling. *Journal of Neuroscience*, *33*(21), 9122–9139.

Nowakowski, T. J., Fotaki, V., Pollock, A., Sun, T., Pratt, T., & Price, D. J. (2013). MicroRNA-92b regulates the development of intermediate cortical progenitors in embryonic mouse brain. *Proceedings of the National Academy of Sciences of the United States of America*, *110*(17), 7056–7061.

Ochiai, W., Nakatani, S., Takahara, T., Kainuma, M., Masaoka, M., Minobe, S., et al. (2009). Periventricular notch activation and asymmetric Ngn2 and Tbr2 expression in pair-generated neocortical daughter cells. *Molecular and Cellular Neuroscience, 40*(2), 225–233.

O'Roak, B. J., Vives, L., Fu, W., Egertson, J. D., Stanaway, I. B., Phelps, I. G., et al. (2012). Multiplex targeted sequencing identifies recurrently mutated genes in autism spectrum disorders. *Science, 338*(6114), 1619–1622.

Papaioannou, V. E. (2014). The T-box gene family: Emerging roles in development, stem cells and cancer. *Development, 141*(20), 3819–3833.

Parthasarathy, S., Srivatsa, S., Nityanandam, A., & Tarabykin, V. (2014). Ntf3 acts downstream of Sip1 in cortical postmitotic neurons to control progenitor cell fate through feedback signaling. *Development, 141*(17), 3324–3330.

Pimeisl, I. M., Tanriver, Y., Daza, R. A., Vauti, F., Hevner, R. F., Arnold, H. H., et al. (2013). Generation and characterization of a tamoxifen-inducible Eomes(CreER) mouse line. *Genesis, 51*(10), 725–733.

Pinto, L., Drechsel, D., Schmid, M. T., Ninkovic, J., Irmler, M., Brill, M. S., et al. (2009). AP2gamma regulates basal progenitor fate in a region- and layer-specific manner in the developing cortex. *Nature Neuroscience, 12*(10), 1229–1237.

Pleasure, S. J., Collins, A. E., & Lowenstein, D. H. (2000). Unique expression patterns of cell fate molecules delineate sequential stages of dentate gyrus development. *Journal of Neuroscience, 20*(16), 6095–6105.

Porteus, M. H., Brice, A. E., Bulfone, A., Usdin, T. B., Ciaranello, R. D., & Rubenstein, J. L. (1992). Isolation and characterization of a library of cDNA clones that are preferentially expressed in the embryonic telencephalon. *Brain Research Molecular Brain Research, 12*(1-3), 7–22.

Postiglione, M. P., Juschke, C., Xie, Y., Haas, G. A., Charalambous, C., & Knoblich, J. A. (2011). Mouse inscuteable induces apical-basal spindle orientation to facilitate intermediate progenitor generation in the developing neocortex. *Neuron, 72*(2), 269–284.

Quinn, J. C., Molinek, M., Martynoga, B. S., Zaki, P. A., Faedo, A., Bulfone, A., et al. (2007). Pax6 controls cerebral cortical cell number by regulating exit from the cell cycle and specifies cortical cell identity by a cell autonomous mechanism. *Developmental Biology, 302*(1), 50–65.

Remedios, R., Huilgol, D., Saha, B., Hari, P., Bhatnagar, L., Kowalczyk, T., et al. (2007). A stream of cells migrating from the caudal telencephalon reveals a link between the amygdala and neocortex. *Nature Neuroscience, 10*(9), 1141–1150.

Rikin, A., & Evans, T. (2010). The tbx/bHLH transcription factor mga regulates gata4 and organogenesis. *Developmental Dynamics, 239*(2), 535–547.

Roybon, L., Deierborg, T., Brundin, P., & Li, J. Y. (2009). Involvement of Ngn2, Tbr and NeuroD proteins during postnatal olfactory bulb neurogenesis. *European Journal of Neuroscience, 29*(2), 232–243.

Roybon, L., Hjalt, T., Stott, S., Guillemot, F., Li, J. Y., & Brundin, P. (2009). Neurogenin2 directs granule neuroblast production and amplification while NeuroD1 specifies neuronal fate during hippocampal neurogenesis. *PLoS One, 4*(3), e4779.

Russ, A. P., Wattler, S., Colledge, W. H., Aparicio, S. A., Carlton, M. B., Pearce, J. J., et al. (2000). Eomesodermin is required for mouse trophoblast development and mesoderm formation. *Nature, 404*(6773), 95–99.

Sanai, N., Nguyen, T., Ihrie, R. A., Mirzadeh, Z., Tsai, H. H., Wong, M., et al. (2011). Corridors of migrating neurons in the human brain and their decline during infancy. *Nature, 478*(7369), 382–386.

Sanders, S. J., He, X., Willsey, A. J., Ercan-Sencicek, A. G., Samocha, K. E., Cicek, A. E., et al. (2015). Insights into autism spectrum disorder genomic architecture and biology from 71 risk loci. *Neuron, 87*(6), 1215–1233.

Sansom, S. N., Griffiths, D. S., Faedo, A., Kleinjan, D. J., Ruan, Y., Smith, J., et al. (2009). The level of the transcription factor Pax6 is essential for controlling the balance between neural stem cell self-renewal and neurogenesis. *PLoS Genetics*, *5*(6), e1000511.

Schuurmans, C., Armant, O., Nieto, M., Stenman, J. M., Britz, O., Klenin, N., et al. (2004). Sequential phases of cortical specification involve Neurogenin-dependent and -independent pathways. *EMBO Journal*, *23*(14), 2892–2902.

Sequerra, E. B., Miyakoshi, L. M., Froes, M. M., Menezes, J. R., & Hedin-Pereira, C. (2010). Generation of glutamatergic neurons from postnatal and adult subventricular zone with pyramidal-like morphology. *Cerebral Cortex*, *20*(11), 2583–2591.

Sessa, A., Mao, C. A., Colasante, G., Nini, A., Klein, W. H., & Broccoli, V. (2010). Tbr2-positive intermediate (basal) neuronal progenitors safeguard cerebral cortex expansion by controlling amplification of pallial glutamatergic neurons and attraction of subpallial GABAergic interneurons. *Genes & Development*, *24*(16), 1816–1826.

Sessa, A., Mao, C. A., Hadjantonakis, A. K., Klein, W. H., & Broccoli, V. (2008). Tbr2 directs conversion of radial glia into basal precursors and guides neuronal amplification by indirect neurogenesis in the developing neocortex. *Neuron*, *60*(1), 56–69.

Sheppard, A. M., & Pearlman, A. L. (1997). Abnormal reorganization of preplate neurons and their associated extracellular matrix: An early manifestation of altered neocortical development in the reeler mutant mouse. *Journal of Comparative Neurology*, *378*(2), 173–179.

Song, M. R., Shirasaki, R., Cai, C. L., Ruiz, E. C., Evans, S. M., Lee, S. K., et al. (2006). T-Box transcription factor Tbx20 regulates a genetic program for cranial motor neuron cell body migration. *Development*, *133*(24), 4945–4955.

Srinivasan, K., Leone, D. P., Bateson, R. K., Dobreva, G., Kohwi, Y., Kohwi-Shigematsu, T., et al. (2012). A network of genetic repression and derepression specifies projection fates in the developing neocortex. *Proceedings of the National Academy of Sciences of the United States of America*, *109*(47), 19071–19078.

Stancik, E. K., Navarro-Quiroga, I., Sellke, R., & Haydar, T. F. (2010). Heterogeneity in ventricular zone neural precursors contributes to neuronal fate diversity in the postnatal neocortex. *Journal of Neuroscience*, *30*(20), 7028–7036.

Sweeney, N. T., Tierney, H., & Feldheim, D. A. (2014). Tbr2 is required to generate a neural circuit mediating the pupillary light reflex. *Journal of Neuroscience*, *34*(16), 5447–5453.

Takashima, Y., & Suzuki, A. (2013). Regulation of organogenesis and stem cell properties by T-box transcription factors. *Cellular and Molecular Life Sciences*, *70*(20), 3929–3945.

Tamamaki, N. (2005). Origin of the neocortical subependymal cells speculated by Emx1 and GAD67 expression. *Chemical Senses*, *30*(Suppl. 1), i111–i112.

Teo, A. K., Arnold, S. J., Trotter, M. W., Brown, S., Ang, L. T., Chng, Z., et al. (2011). Pluripotency factors regulate definitive endoderm specification through eomesodermin. *Genes & Development*, *25*(3), 238–250.

Trowe, M. O., Maier, H., Schweizer, M., & Kispert, A. (2008). Deafness in mice lacking the T-box transcription factor Tbx18 in otic fibrocytes. *Development*, *135*(9), 1725–1734.

Tsai, C. Y., Tsai, C. Y., Arnold, S. J., & Huang, G. J. (2015). Ablation of hippocampal neurogenesis in mice impairs the response to stress during the dark cycle. *Nature Communications*, *6*, 8373.

Vasistha, N. A., Garcia-Moreno, F., Arora, S., Cheung, A. F., Arnold, S. J., Robertson, E. J., et al. (2015). Cortical and clonal contribution of Tbr2 expressing progenitors in the developing mouse brain. *Cerebral Cortex*, *25*(10), 3290–3302.

Wang, T. F., Ding, C. N., Wang, G. S., Luo, S. C., Lin, Y. L., Ruan, Y., et al. (2004). Identification of Tbr-1/CASK complex target genes in neurons. *Journal of Neurochemistry*, *91*(6), 1483–1492.

Wrobel, C. N., Mutch, C. A., Swaminathan, S., Taketo, M. M., & Chenn, A. (2007). Persistent expression of stabilized beta-catenin delays maturation of radial glial cells into intermediate progenitors. *Developmental Biology*, *309*(2), 285–297.

CHAPTER ELEVEN

T-Box Genes in *Drosophila* Limb Development

G.O. Pflugfelder*,[1], F. Eichinger*, J. Shen†
*Institute of Genetics, Johannes Gutenberg University, Mainz, Germany
†China Agricultural University, Beijing, China
[1]Corresponding author: e-mail address: pflugfel@uni-mainz.de

Contents

1. Orthology Between *Drosophila* and Mammalian Tbx Genes 314
2. Embryonic Origin of the *Drosophila* Limb Imaginal Discs: The Role of Doc 316
3. Pattern Formation in the Larval Leg Disc 319
4. Tbx Genes in *Drosophila* Leg Development 320
5. Pattern Formation and Proliferation Control in the Larval Wing Disc 323
6. Tbx Genes in *Drosophila* Wing Development 325
   6.1 Omb, Doc, and Mid Expression in *Drosophila* Wing Development 325
   6.2 Functions of Omb in Wing Development 328
   6.3 Doc in Wing Development 339
   6.4 Mid in Wing Development 339
7. Tbx Gene Function in Haltere Development 340
8. Conclusion 342
Acknowledgments 342
References 342

Abstract

T-box genes are essential for limb development in vertebrates and arthropods. The *Drosophila* genome encodes eight T-box genes, six of which are expressed in limb ontogenesis. The *Tbx20*-related gene pair *midline* and *H15* is essential for dorso-ventral patterning of the *Drosophila* legs. The three *Tbx6*-related *Dorsocross* genes are required for epithelial remodeling during wing development. The *Drosophila* gene *optomotor-blind* (*omb*) is the only member of the Tbx2 subfamily in the fly and is predominantly involved in wing development. *Omb* is essential for wing development and is sufficient to promote the development of a second wing pair. Targeted manipulations of *omb* expression have shown that the bulk *omb* requirement for wing development can be deconstructed into a number of individual functions. Even though *omb* expression in the wing disc is symmetrical with regard to the anterior/posterior (A/P) compartment boundary, anterior and posterior knockdowns have distinct consequences: Anterior Omb is required for the maintenance of a straight A/P lineage restriction boundary. Posterior Omb suppresses formation of an apical epithelial fold along the A/P boundary. *Drosophila* T-box gene expression is not confined to the ectoderm-derived epithelia

Current Topics in Developmental Biology, Volume 122
ISSN 0070-2153
http://dx.doi.org/10.1016/bs.ctdb.2016.08.003

of the imaginal discs. Both Doc and Omb are prominently expressed in leg disc muscle precursor cells. Omb is also strongly expressed in a tracheal branch that invades the extracellular matrix of the wing disc. The function of Doc and Omb in the latter tissues is not known, indicative of the many questions still open in the field.

The relevance of T-box genes for *Drosophila* limb development is already apparent from the phenotype of the first mutation described for a T-box gene. In November 1911, Thomas Hunt Morgan observed a spontaneous X-linked *Drosophila* mutant which he named *bifid* (Morgan & Bridges, 1916). *Bifid* causes a slight alteration of the vein pattern in the hinge region of the adult wing. Many decades later it was recognized that *bifid* is caused by insertion of a transposable element in the T-box gene *optomotor-blind* (*omb*) and, thus, is a regulatory mutation in this gene (omb^{bi}) (Pflugfelder & Heisenberg, 1995; Pflugfelder et al., 1990). The roles of *omb* in wing development are described later.

1. ORTHOLOGY BETWEEN *DROSOPHILA* AND MAMMALIAN TBX GENES

There are 16 Tbx genes in human or mouse (17, if one includes the unusual fusion gene *MGA* (*Max gene associated*, Hurlin, Steingrimsson, Copeland, Jenkins, & Eisenman, 1999) which have been grouped phylogenetically into a variable number of subfamilies. Commonly between five and eight subfamilies are assumed (e.g., Ahn, You, & Kim, 2012) but the Tbx family is becoming more complex with the identification of additional Tbx genes in lower metazoan and even nonmetazoan species (Holstien et al., 2010; Martinelli & Spring, 2005; Sebe-Pedros et al., 2013). In Table 1, we use the subdivision into five major subfamilies of Naiche, Harrelson, Kelly, and Papaioannou (2005) to compare the Tbx gene complement between human and *Drosophila*.

Drosophila has at least one representative in four of the five subfamilies. No *Drosophila* ortholog exists in the Tbr1 subfamily. In the Tbx6 subfamily, *Drosophila* presents with the three paralogous *Dorsocross* (*Doc*) genes. The *Doc* paralogs are expressed in highly similar patterns in the embryo and in the wing imaginal disc (Butler et al., 2003; Hamaguchi, Yabe, Uchiyama, & Murakami, 2004; Stathopoulos, Van Drenth, Erives, Markstein, & Levine, 2002; Tomancak et al., 2002). Later in development their expression diverges (Reim, Lee, & Frasch, 2003). *Doc* genes show extensive functional redundancy among themselves (see below) and

Table 1 Mammalian and *Drosophila* Member in Tbx Subfamilies

Tbx Subfamily	**Mammalian Genes**	***Drosophila* Genes**
T	*T* *Tbx19* (*Tpit*)	*Brachyenteron* (*byn*)
Tbx1	*Tbx1* *Tbx10*	*Optomotor-blind-related-gene-1* (*org-1*)
	Tbx15 *Tbx18* *Tbx22*	
	Tbx20	*H15* (*H15*) *Midline* (*mid*)
Tbx2	*Tbx2* *Tbx3*	*Optomotor-blind* (*omb*)
	Tbx4 *Tbx5*	
Tbx6	*Tbx6*	*Dorsocross1, 2, 3* (*Doc1, 2, 3*)
Tbr1	*Tbr1* *Tbr2* (*Eomes*) *Tbx21* (*Tbet*)	

probably originated by relatively recent gene duplications. The beetle *Tribolium* has only a single *Doc* gene (Berns, Kusch, Schroder, & Reuter, 2008). In the Tbx1 subfamily, *Drosophila org-1* is most closely related to *Tbx1*. In the same subfamily the paralogous *Drosophila* gene pair *H15* and *midline* (*mid*) is most closely related to *Tbx20*. The *mid/H15* paralog pair exhibits partial redundancy (Buescher et al., 2006). Homologous pairs are found in other arthropods such as spiders and millipedes (Janssen, Feitosa, Damen, & Prpic, 2008). Functional differences were observed between *mid* and *H15* in *Drosophila* (Reim, Mohler, & Frasch, 2005; Svendsen, Formaz-Preston, Leal, & Brook, 2009). *Tbx1* and *Tbx20* are both expressed in the developing vertebrate limbs (Dastjerdi et al., 2007; Meins, Henderson, Bhattacharya, & Sowden, 2000).

Omb is the only *Drosophila* gene in the Tbx2 subfamily. The four vertebrate members in this subfamily probably arose by two sequential duplication steps from one *Tbx2* precursor gene: a tandem duplication on one chromosome, leading to *Tbx2/3* and *Tbx4/5* precursor genes, followed by duplication of the *Tbx2/3–Tbx4/5* cluster to a new chromosome.

The spatio-temporal expression patterns of mammalian *Tbx2* and *Tbx3* are still related, as are those of *Tbx4* and *Tbx5* indicating conservation of regulatory sequences since the second duplication (Agulnik et al., 1996; Ruvinsky & Silver, 1997). In a primitive metazoan, the sponge *Amphimedon queenslandica*, a single Tbx2 subfamily gene is found, but already in the eumetazoan cnidarian *Nematostella vectensis* a pair of genes (*Tbx2/3* and *Tbx4/5*) exists, suggesting that the first duplication event occurred early in metazoan evolution (Larroux et al., 2008) and that the solitary Tbx2 subfamily gene in insects results from secondary gene loss. Evolution of the T-box gene family and the role of T-box genes in vertebrate limb development are reviewed in this volume (Sheeba & Logan, 2017; Sebe-Pedrós & Ruiz-Trillo, 2017).

Expression of human *TBX2* and *TBX3* in the *Drosophila* wing imaginal discs induces similar transcriptional and cell biological effects as overexpression of *omb*. The intraepithelial cell motility and penetration of the extracellular matrix (ECM) induced by Omb or TBX2/3 expression resembles the metastasis promoting effect of TBX2 and TBX3 in melanoma, mamma carcinoma, and other tumors (Fischer & Pflugfelder, 2015; Shen et al., 2014). Hamaguchi and colleagues compared the effects of *Drosophila* Doc1 and *Xenopus* Tbx6 in a *Xenopus* animal cap RNA injection assay. The specificity of Doc1 and Tbx6 was similar, with Doc1 being less effective (Hamaguchi et al., 2004). The experiments with Doc1/Tbx6 and Omb/TBX2 suggest that functional properties of Tbx proteins are conserved in their respective subfamily. Since DNA-binding specificity of Tbx proteins is quite similar at least in vitro (Sen, Grimm, Hofmeyer, & Pflugfelder, 2014), the hypothesis of evolutionary conservation needs to be tested with less related Tbx proteins.

2. EMBRYONIC ORIGIN OF THE *DROSOPHILA* LIMB IMAGINAL DISCS: THE ROLE OF DOC

We consider as limbs the paired appendages of the insect thorax: the dorsal wings and halteres, and the three ventral pairs of legs (Fig. 1D and F). The appendages of the *Drosophila* head, antenna, proboscis, and maxillary palp, will not be discussed here. Insect limbs derive from the ectoderm and are largely specializations of the exoskeleton. In *Drosophila* embryogenesis, the limb anlagen are specified by the anterior–posterior and dorsoventral patterning systems. Around 9–10 h after egg laying, small groups

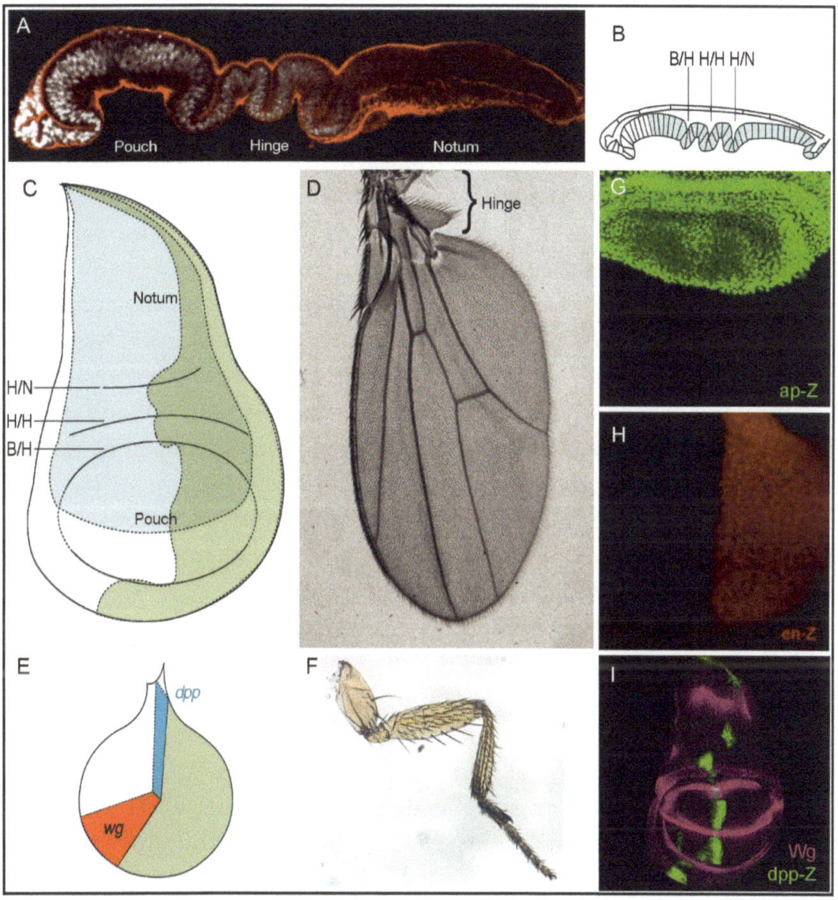

Fig. 1 Wing and leg development and structure. (A) Confocal micrograph of wing imaginal disc cryosectioned parallel to the A/P boundary and stained for the ECM marker laminin (*red*) and the nuclear marker DAPI (*white*). In the thin peripodial epithelium only few nuclei are visible. (B) Schematic drawing of the disc shown in (A). The dorsal compartment, which extends to the D/V boundary in the pouch, is in *light blue*. The three apical folds, blade/hinge (B/H), hinge/hinge (H/H), and hinge/notum (H/N) are indicated. The lumen between the peripodial and main epithelium is exaggerated. (C) Schematic drawing of the third instar wing imaginal disc. Dorsal compartment: *light blue*, posterior compartment: *light green*. Hinge folds are indicated in the dorsal hinge. In the wing pouch the compartment boundaries are roughly orthogonal. (D) Adult wing. The region proximal to the elliptical wing blade is marked as hinge. (E) Schematic drawing of the third instar leg imaginal disc. The main expression domains of *dpp* (*dark blue*) and *wg* (*dark red*) are indicated. The posterior compartment is in *light green*. Distal structures derive from the center of the disc. (F) Adult male foreleg with coxa, trochanter, femur, *(Continued)*

of cells invaginate from the lateral ectoderm by apical constriction while retaining contact with the surface epithelium of the embryo (Bate & Arias, 1991; Meise & Janning, 1994). The limb imaginal discs are formed by about 20–50 cells that span the parasegment border. Thus, disc cells with posterior identity derive from embryonic cells that express *engrailed* (*en*). This embryonically derived *en* expression is later maintained by epigenetic mechanisms (Randsholt, Maschat, & Santamaria, 2000). The limb disc primordia become distinct from the surrounding epithelium by the expression of several transcription factors. Lineage tracing indicated that, in the embryonic blastoderm, the anlagen of wing/haltere and leg primordia map very close or overlap. Use of molecular markers showed that cells which contribute to the wing and haltere discs separate from a ventral Distalless (Dll) domain by dorsal migration (Cohen, Simcox, & Cohen, 1993). Maintained *Dll* expression marks the future leg discs. Dorsal migration and appropriate specification of the wing and haltere precursor cells require a high level of the BMP2/4-related protein Decapentaplegic (Dpp) secreted from dorsal cells (Kubota, Goto, Eto, & Hayashi, 2000). The *Doc* T-box genes are expressed in the Dll domain before separation into dorsal and ventral disc anlagen. After their separation, Doc is only found in the wing and haltere precursors. Doc is required for the separation of the anlagen. Microinjection of dsRNAs directed against all three *Doc* family members curbs dorsal migration and the subsequent expression of wing/haltere-specific gene *snail*. *Doc* knockdown has little effect on the leg anlagen. Microinjection of dsRNAs directed against individual *Doc* transcripts was ineffective, indicating redundancy within the family (Hamaguchi et al., 2004).

The imaginal discs start out as bottle-shaped epithelia with mostly uniform cell morphology. During larval development, proliferation and morphological differentiation occur resulting in discs that are formed by two opposing epithelia connected by a transition zone of cuboidal cells: the main

Fig. 1—Cont'd tibia, and tarsae (five segments ending in claws). The proximalmost tarsal segment bears the male-specific sex combs. (G, H) Wing disc pouch stained for the dorsal determinant *apterous* (ap-lacZ) (G) and for expression of the posterior determinant *engrailed* (en-lacZ) (H). (I) Entire third larval instar wing disc stained for *dpp* (dpp-lacZ, *green*) and Wg expression (*magenta*). The wing pouch is encircled by two rings of Wg expression and bisected by Wg expressed on the D/V boundary. In C, E, and G–I dorsal is up and anterior left. *Panel (A): Reprinted from Sui, L., Pflugfelder, G. O., & Shen, J. (2012). The Dorsocross T-box transcription factors promote tissue morphogenesis in the Drosophila wing imaginal disc.* Development, 139*(15), 2773–2782, with permission from The Company of Biologists.*

epithelium (also known as disc proper) made up of columnar cells and the peripodial epithelium made up of thin squamous cells. The apical sides of both epithelia face a common lumen (Widmann & Dahmann, 2009) (Fig. 1A and B).

3. PATTERN FORMATION IN THE LARVAL LEG DISC

The leg imaginal disc, like the wing disc, is divided into anterior/posterior (A/P) and dorsal/ventral (D/V) compartments. In the leg disc, Hedgehog (Hh), secreted from posterior cells, induces *dpp* predominantly dorsally and *wingless* (*wg*) ventrally in the anterior compartment (Fig. 1E). This differs from the wing disc where *dpp* is induced in both the anterior dorsal and the anterior ventral compartment (Basler & Struhl, 1994; Diaz-Benjumea & Cohen, 1994). Loss of *wg* causes an intensification of *dpp* expression in the ventral sector (and vice versa) (Brook & Cohen, 1996). While *dpp* and *wg* are expressed along the entire proximo-distal axis, loss of *wg* and *dpp* predominantly affects the distal disc derivatives. The body wall, the most proximal leg disc derivative, is specified by the expression of *extradenticle* (*exd*) and *homothorax* (*hth*) whose expression is independent of Hh. Hth/Exd make proximal cells less responsive to Dpp and Wg (Gonzalez-Crespo & Morata, 1996; Wu & Cohen, 1999). Wg and Dpp jointly induce expression of the distal determinant *Dll* which in young discs directly borders on the Hth/Exd domain. Later, expression of several cross-repressive transcription factors defines additional domains along the proximo-distal (p-d) axis (Abu-Shaar & Mann, 1998; Azpiazu & Morata, 2002; Kojima, 2004; Lecuit & Cohen, 1997). This eventually leads to the articulated structure of the adult leg.

Adult leg D/V patterning is evinced by the arrangement of different types of bristles (Fig. 1F). Wg and Dpp, which synergize in the distalization of the leg disc, act antagonistically in dorso-ventral patterning (Brook & Cohen, 1996). In regulatory alles of *wg*, the ventral side of legs is replaced by a mirror copy of the dorsal side (correspondingly for *dpp* and the dorsal side) (Held, 1995). Unlike its role at the wing D/V boundary, Wg does not appear to act as a morphogen in leg D/V patterning (Diaz-Benjumea & Cohen, 1994; Wilder & Perrimon, 1995). In contrast, Hh is involved in this process, independent of its activation of *dpp* and *wg* (Hays, Buchanan, Neff, & Orenic, 1999), comparable to its role close to the A/P boundary in the wing disc (Vervoort, 2000). For a comparison of wing and leg disc development see Morata (2001).

4. TBX GENES IN *DROSOPHILA* LEG DEVELOPMENT

Tbx genes from three subfamilies are known to be expressed in the leg disc: *H15* and *mid*, *omb*, and *Doc*. The two closely related *H15* and *mid* genes are expressed in the ventral, and *omb* in the dorsal disc. *H15/mid* and *omb* are, thus, expressed in opposing patterns in the main epithelium (Brook & Cohen, 1996; Tsai et al., 1997). *Doc* is expressed in a patch of cells in the dorsal main epithelium and in basally located myoblasts (Reim et al., 2003) (Fig. 2C).

H15 expression is centered on the ventral wedge of *wg* expression (compare Fig. 1I and J). Wg signaling is necessary and sufficient for *H15* expression. The broader expression of *H15* compared to Wg is due to the diffusion of Wg (Zecca, Basler, & Struhl, 1996). Strong Dpp signaling antagonizes *H15* expression. Omb is expressed in a dorsal wedge, centered on the dorsal domain of strong dpp-lacZ expression (compare Figs. 1I and 2A). Published data on *omb* in the leg disc showed the expression of *omb* enhancer trap lines (lacZ and Gal4) (Abu-Shaar & Mann, 1998; Brook, 2010; Brook & Cohen, 1996; Maves & Schubiger, 1998; Tsai et al., 1997). While these lines correctly render the restriction of *omb* expression to the dorsal compartment, they suggest a narrower distribution along the A–P axis than what is observed with Omb antibody staining (compare Fig. 2A and B). We have previously noted the poor suitability of enhancer trap stainings to reveal graded expression patterns in the wing imaginal disc (Shen, Dahmann, & Pflugfelder, 2010). Antibody staining shows that Omb is also expressed in about 100 scattered cells in the ventral part of the disc. These cells lie basally to the main epithelium and hence probably are a subpopulation of imaginal myoblasts (Soler, Daczewska, Da Ponte, Dastugue, & Jagla, 2004) (Fig. 2A). Expression of *omb* in myoblasts has not been noted so far, partly because this part of the *omb* expression pattern is not rendered by all *omb* enhancer trap lines (enhancer trap infidelity) (Mayer, Diegelmann, Abassi, Eichinger, & Pflugfelder, 2013). The cluster of scattered Doc-positive cells also lies basally (Fig. 2C‴). Doc is known to be required for development of myocardial and pericardial cells of the embryonic heart (Reim & Frasch, 2005).

The mechanism by which *omb* is activated by Dpp differs between wing and leg discs; in the wing disc, Dpp induces *omb* by relieving the repression by Brinker (Brk), while in the leg disc Brk is not involved in controlling *omb* (Estella & Mann, 2008). In the ring of Hth/Exd expression that runs along

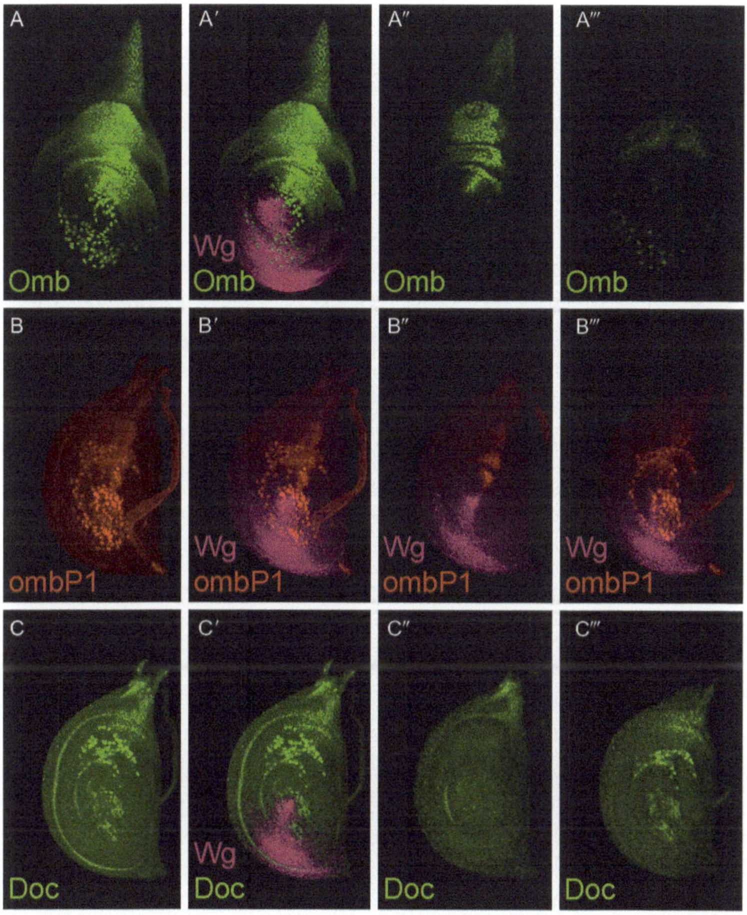

Fig. 2 Leg disc expression of Omb, omb-lacZ, and Doc. Confocal micrographs of third instar leg discs. (A) Omb expression (*green*). (A′) Combined with Wg (*magenta*). Apical (A″) and basal (A‴) sections. (B) Omb-lacZ expression (*red*), otherwise as in (A). (C) Doc expression (*green*), otherwise as in (A). In all panels dorsal is up and anterior left.

the periphery of the leg disc, response to Dpp and Wg is repressed such that the stripes of omb-lacZ and H15-lacZ expression do not reach the margin of the disc (Gonzalez-Crespo & Morata, 1996). In *hth* mutant clones, *H15* and *omb* expression can expand to the disc periphery (Abu-Shaar & Mann, 1998). Fig. 2A shows that peripheral repression of *omb* is not absolute.

Like in other developmental contexts, H15 and Mid in ventral leg development act redundantly to mediate part of Wg function. *H15 mid* double mutant clones cause dorsal transformation of ventral bristles without ectopic expression of *dpp* or *omb*. This differs from *wg* clones in which these genes are induced. The ventral-to-dorsal fate transformation is reflected in an altered expression pattern of, e.g., *Sex combs reduced* (*Scr*) which is required for development of a particular type of ventral bristles (Svendsen et al., 2009). The redundancy of H15 and Mid may be due to cross-repression by which loss of one gene leads to increased expression of the other (Svendsen, Ryu, & Brook, 2015).

Dorsal expression of genes orthologous to *omb* in leg development is conserved in arthropod evolution (based on the fly *Drosophila*, the spider *Cupiennius salei*, and the millipede *Glomeris marginata*) and even in pre-arthropod Onychophorans which posses unsegmented legs. For the ventral expression of *mid/H15* this holds to a lesser extent (Janssen, Jorgensen, Prpic, & Budd, 2015; Janssen et al., 2008; Prpic, Janssen, Damen, & Tautz, 2005; Prpic, Janssen, Wigand, Klingler, & Damen, 2003). Nonetheless, there is only a minor leg defect in the strongly hypmorphic *omb* allele combination $omb^{bi}/l(1)omb^{3198}$ in which the tibia exhibits a slight dorsal kink. Other imaginal structures, like eye or wing, where *omb* is also expressed, are more strongly affected (Grimm & Pflugfelder, 1996; Tsai et al., 2015). It is conceivable that *omb* and *Doc*, both of which are expressed in the dorsal leg disc downstream of Dpp, are partly redundant in leg development. Precedence for partial *omb*/*Doc* redundancy comes from studies on the development of abdominal tergites and the eye (Kopp & Duncan, 1997; Tomlinson, 2003). The *Doc* loss-of-function phenotype in the leg is not known.

The gain-of-function phenotype of ectopic dorsal *mid/H15* expression morphologically is nearly reciprocal to that of loss of function. Dorsal expression of *mid* can repress *omb* and induce development of ventral bristles in the dorsal leg. This transformation occurs predominantly in regions of the distal leg disc that are in the range of high Wg concentration. Suppression of Wg signaling in *mid* expressing clones suppresses the transformation (Brook, 2010; Svendsen et al., 2009).

Ectopic expression of *omb* in the ventral disc under dpp-Gal4 control induces *dpp*, represses *wg*, and causes dorsalization of the affected tissue in the adult integument to a similar extent as is observed upon ectopic expression of *dpp* (Maves & Schubiger, 1998). *H15* is repressed by ventral Omb (Svendsen et al., 2015). Omb overexpression, in addition, can cause truncation or bifurcation of the distal leg, depending on the Gal4 driver used for misexpression. In the antenna, too, the distalmost structure, the arista, is deleted or duplicated/triplicated ((Porsch et al., 2005) and our unpublished observations). Similar phenotypes are caused by clonal expression of *hh* (Basler & Struhl, 1994) and by ectopic activation of Dpp or Wg signaling (Diaz-Benjumea & Cohen, 1994; Morimura, Maves, Chen, & Hoffmann, 1996), suggesting that Omb acts downstream of Hh and Dpp in these processes. *Org-1*, whose role in limb development is not known, causes a drastic proximo-distal contraction and distal truncation of both adult legs and wings when ectopically expressed in the imaginal discs. In contrast, the arista undergoes transformation to leg fate when *org-1* is expressed in the anlage of the distal antenna (Porsch et al., 2005). Loss of function of several genes (*Dll, hth, spineless, cut, distal antenna*), some of which are controlled by Omb in the wing disc, can cause similar antenna-to-leg transformation (Dong, Dicks, & Panganiban, 2002; Ebacher, Todi, Eberl, & Boekhoff-Falk, 2007; Emerald, Curtiss, Mlodzik, & Cohen, 2003). Ectopic distal expression of *Doc2* causes proximo-distal contraction and distal truncation of wings, legs, and antennae (Reim et al., 2003). In domain swap experiments, we have shown that the functional specificity of Omb and Org-1 resides mainly outside of the T-domain (Porsch et al., 2005). Therefore, it is possible that the different or overlapping outcomes in such overexpression experiments result from differential or similar action of different Tbx proteins bound to the same target sites. Studies in vertebrates support this assumption (Goering et al., 2003; Hoogaars, Barnett, Moorman, & Christoffels, 2007; Wehn & Chapman, 2010).

5. PATTERN FORMATION AND PROLIFERATION CONTROL IN THE LARVAL WING DISC

During the three instars of *Drosophila* larval development (~4 days at 25°C) the cell number of the wing disc increases from about 50 to 50,000 (Garcia-Bellido & Merriam, 1971). Parallel to this massive proliferation, patterning occurs in the disc specifying position and fate of the future adult structures (Baena-Lopez, Nojima, & Vincent, 2012; Held, 2002). The wing disc is the primordium of part of the mesothoracic integument (scutum, scutellum, mesopleura) and of the wing proper, consisting of hinge and blade (Aldaz & Escudero, 2010; Garcia-Bellido, Ripoll, & Morata, 1976; Resino, Salama-Cohen, & Garcia-Bellido, 2002; Wang et al., 2004) (Fig. 1D).

The wing disc is made up of two cell populations with anterior (A) and posterior (P) fate, respectively (Fig. 1C). P fate is specified by the expression of the posterior selector gene *engrailed* (Fig. 1G). En in P cells induces the expression of *hh*. Hh spreads into the anterior compartment where Hh signaling induces the A fate. P cells cannot respond to Hh because En represses *Cubitus interruptus* (*Ci*), encoding a Hh signaling effector. The A/P compartment boundary, thus, is not simply defined by apposition of cells that differ in *en* expression but additionally requires signaling across the boundary. The A/P boundary prevents intermingling of cells from opposite sides of the boundary and serves as a major organizing center of disc development. Hh induces the expression of *dpp* in several rows of cells anterior to the A/P boundary (Dahmann & Basler, 1999, 2000) (Fig. 1I). Diffusible Dpp spreads into both compartments forming a nearly symmetrical gradient (Akiyama et al., 2008; Entchev, Schwabedissen, & Gonzalez-Gaitan, 2000; Teleman & Cohen, 2000). Dpp in the wing discs is a morphogen, specifying different cell fates depending on its concentration: e.g., low concentrations suffice to induce *omb*, while higher concentrations are required to induce *spalt* (*salm*) leading to a clustered expression pattern of these two target genes (Lecuit et al., 1996; Nellen, Burke, Struhl, & Basler, 1996). Hh, in addition to inducing *dpp* expression, directly patterns the anterior compartment close to the A/P boundary (Mullor, Calleja, Capdevila, & Guerrero, 1997; Strigini & Cohen, 1997; Vervoort, 2000).

Manipulations of Dpp signaling suggest that Dpp is necessary and sufficient for growth of the wing disc (Burke & Basler, 1996). However, growth is essentially uniform across the wing disc, in spite of the graded profile of Dpp pathway activity (Milan, Campuzano, & Garcia-Bellido, 1996; Rogulja & Irvine, 2005). In the third larval instar, Dpp may no longer be required for disc growth (Akiyama & Gibson, 2015). Numerous models have been proposed to reconcile these apparently contradictory findings (Hamaratoglu, Affolter, & Pyrowolakis, 2014).

Unlike the A/P boundary, which already exists at the time of larval hatching, the D/V compartment boundary arises during the second larval instar. Establishment of dorsal and ventral compartments in the wing discs occurs in the context of proximo-distal development (Klein, 2001). Apterous (Ap) starts to be expressed in the second larval instar by all cells of the dorsal compartment and acts as a dorsal selector (Blair, Brower, Thomas, & Zavortink, 1994; Diaz-Benjumea & Cohen, 1993; Milan & Cohen, 1999). The D/V compartment boundary is established by the apposition of cells that do and do not express Ap (Fig. 1H). Ap controls the differential expression of *fringe* (*fng*) and the Notch (N) ligands Delta (Dl) and Serrate (Ser) (Fleming, Gu, & Hukriede, 1997; Irvine & Wieschaus, 1994; Kim, Irvine, & Carroll, 1995; Klein & Arias, 1998b), leading to the localized activation of N at the D/V

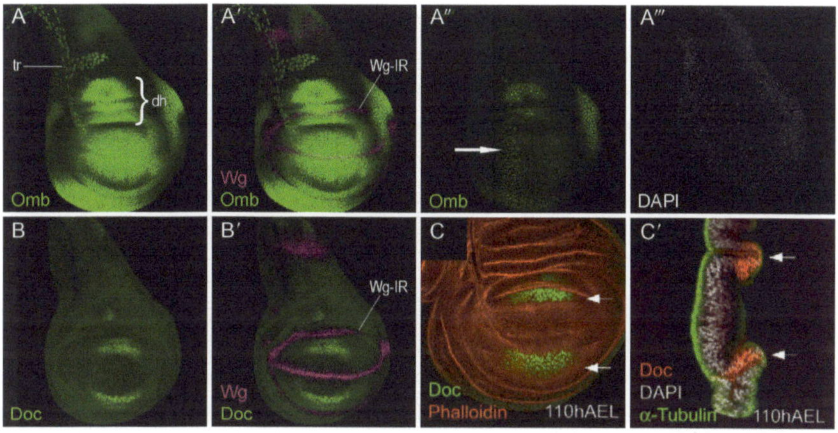

Fig. 3 Wing disc expression of Omb and Doc. Confocal micrographs of third instar wing discs. (A) Omb expression (*green*) (*tr*, trachea; *dh*, dorsal hinge). (A′) Inclusion of Wg (*magenta*) (Wg-IR, inner ring of Wg expression). (A″ and A‴) Apical focus. Omb (A″) and DAPI (A‴). Omb is expressed in an anterior subpopulation of peripodial cells. (B) Doc expression. (B′) Inclusion of Wg (*magenta*). (C) Doc expression (*green*) in relation to epithelial folds, visualized by increased phalloidin staining (*red*). *White arrows* point at Doc domains and adjacent folds. (C′) Confocal section parallel to A/P boundary through the pouch region. In (A) to (C) dorsal is up and anterior left; in (C′) the apical side of the main epithelium is left, dorsal is up. *Panels (C) and (C′): Reprinted from Sui, L., Pflugfelder, G. O., & Shen, J. (2012). The Dorsocross T-box transcription factors promote tissue morphogenesis in the Drosophila wing imaginal disc.* Development, 139*(15), 2773–2782, with permission from The Company of Biologists.*

boundary. Wg, which up to the second instar is expressed in a distal wedge of the wing disc, gains a new expression pattern by N activation along the D/V boundary of the wing pouch (Figs. 1I, 3A′, and B′). In the wing pouch, expression of diffusible Wg along the D/V boundary controls growth and patterning in the flanking dorsal and ventral compartments, comparable to the function of Dpp along the A–P axis (Couso, Bishop, & Martinez Arias, 1994; Strigini & Cohen, 2000; Zecca et al., 1996).

6. TBX GENES IN *DROSOPHILA* WING DEVELOPMENT

6.1 Omb, Doc, and Mid Expression in *Drosophila* Wing Development

In the wing imaginal disc, *omb* is prominently expressed (Fig. 3A) (Grimm & Pflugfelder, 1996). Several expression domains can be discerned. In the third larval instar, Omb expression nearly fills the elliptical wing pouch. Omb expression along the A–P axis is graded and laterally does not quite reach the inner ring of Wg expression (Wg-IR) that surrounds the pouch (Couso, Bate, & Martinez-Arias, 1993; Shen et al., 2010; Williams, Paddock, & Carroll, 1993) (Fig. 3A′). The wing pouch is surrounded by several concentric folds [wing/hinge (W/H), hinge/hinge (H/H), and hinge/notum (H/N)] which separate hinge from the future wing blade (W/H), and hinge from future notum (H/N), as well as proximal from distal hinge (H/H) (Fig. 1A and B). These folds can best be recognized in the central dorsal hinge (Johnstone, Wells, Strutt, & Zeidler, 2013) (Fig. 3C). Outside the pouch, Omb is strongly expressed in the central dorsal and ventral hinge. Ventrally, Omb is also expressed in the future mesopleura (cells at the ventral periphery of the disc which will form part of the body wall underneath the wing). These expression domains are symmetrical with regard to the A/P boundary where the patterning factor Dpp originates. Conditions that lead to the loss of *dpp* expression cause loss of the entire central *omb* domain (Grimm & Pflugfelder, 1996; Lecuit et al., 1996). Ectopic activation of Dpp signaling in the distal wing disc causes *omb* expression in all cells of the distal disc along with dramatic lateral overgrowth (Nellen et al., 1996). Clones mutant for hypomorphic *Mad*, a central component of the Dpp pathway, cause cell-autonomous reduction of *omb* expression (Lecuit et al., 1996). These experiments show necessity and sufficiency of Dpp signaling for central *omb* expression (cf. also Hadar et al., 2012). The symmetrical gradient of *omb* expression is not directly induced by the graded expression of Dpp. Rather, Dpp signaling represses transcription of *brk* which encodes a transcriptional repressor. Thus, an inverse gradient of Brk is formed along the A–P axis of the wing disc with its minimum in the center of the disc. *Omb* is directly repressed by Brk involving the corepressor Groucho leading to high *omb* expression at the A/P boundary and low expression laterally (Campbell & Tomlinson, 1999; Jazwinska, Kirov, Wieschaus, Roth, & Rushlow, 1999; Minami, Kinoshita, Kamoshida, Tanimoto, & Tabata, 1999; Moser & Campbell, 2005; Sivasankaran, Vigano, Muller, Affolter, & Basler, 2000; Song, Hasson, Paroush, & Courey, 2004; Winter & Campbell, 2004).

Omb expression in the pouch is also symmetrical with regard to the D/V boundary, the source of the Wg morphogen (Neumann & Cohen, 1996b; Zecca et al., 1996). Rearing larvae homozygous for a temperature-sensitive *wg* allele at the nonpermissive temperature causes loss of *omb* from the pouch, while the *omb* hinge expression is not affected (Grimm & Pflugfelder, 1996). Whether this reflects direct regulation of *omb* by Wg signaling or rather the role of Wg in specifying the pouch fate is not known (Ng, Diaz-Benjumea, Vincent, Wu, & Cohen, 1996).

Anterior to these central expression domains, there is weak Omb expression at the anterior margin of the disc separated from central Omb expression by Wg-IR and the later developing Wg outer ring (Wg-OR) (Fig. 3A′ and B′). This expression domain maps to the anlagen of proximal hinge elements (cf. Buratovich & Wilder, 2001; Karlsson, 1981; Neumann & Cohen, 1996a; Villa-Cuesta & Modolell, 2005). Omb is also expressed along the posterior margin of the wing disc. This expression is centered on a posterior *dpp* expression domain required for the development of alula and axillary cord (Foronda, Perez-Garijo, & Martin, 2009). In the notum, there is a weak stripe of Omb expression in the main epithelium partly overlapping with the notal Wg expression domain. This Wg expression domain is required for the development for a subset of notal bristles (Phillips & Whittle, 1993). Whether *omb* is required for the development of the structures in whose anlagen it is expressed is not clear, but at least costa, alula, and scutellar bristles are well developed in *omb* null mutant pharate adults (Grimm & Pflugfelder, 1996).

Omb is also expressed in cells of the trachea that enters the wing disc from the distal side to form the air sac primordium and two finer branches in the notum region of the disc (Guha, Lin, & Kornberg, 2009) (Fig. 3A). Strong *omb* expression in larval trachea is confirmed by FlyAtlas transcriptome data (http://flyatlas.org/atlas.cgi). The role of *omb* in tracheal development is unknown.

Lastly, *omb* is expressed in anterior peripodial cells (Fig. 3A″). Cells with squamous morphology contribute only about 1% to the total cell number of the wing disc (Pallavi & Shashidhara, 2003) (Fig. 3A‴). Accordingly, their contribution to the adult integument is small. Peripodial cells form part of body wall and ventral hinge (Milner, Bleasby, & Kelly, 1984). However, cells of the peripodial and main epithelium signal across the narrow lumen that separates the two epithelia such that ablation or manipulation of EGFR or Hh signaling in the peripodial epithelium will affect growth, survival, and patterning of the disc proper (Cho, Chern, Izaddoost, & Choi, 2000; Gibson & Schubiger, 2000; Pallavi & Shashidhara, 2003; Stultz, Lee, Ramon, & Hursh, 2006). Furthermore, morphological changes in peripodial cells are required in the biomechanical process of thorax closure and disc evagination during pupal development (Aldaz, Escudero, & Freeman, 2013; Tripura, Chandrika, Susmitha, Noselli, & Shashidhara, 2011). The latter two processes appear to occur normally in *omb* null mutant animals.

The three Doc genes are expressed in very similar or identical patterns in the wing disc, based on staining with an antibody directed against Doc2+3 and in situ hybridizations against *Doc1* and *Doc2* (Butler et al., 2003; Reim et al., 2003). After the embryonic Doc expression in the wing disc primordium has declined, it resumes in the third larval instar as two separate stripes at the dorsal and ventral edge of the wing pouch. Later a spot-like expression in the central dorsal hinge and a stripe in the posterior notum/pleura complete the pattern (Reim et al., 2003; Sui, Pflugfelder, & Shen, 2012) (Fig. 3B). The two Doc pouch expression stripes proximally border on the hinge marker Hth and distally on the expression domain of vg^{QE}. Doc and Hth as well as Doc and Vg exert mutual repression (Sui et al., 2012). In the center of the wing pouch, Doc is also repressed by epigenetic mechanisms (Gutierrez et al., 2012).

Mid is nearly uniformly expressed in the cells of the main and the peripodial epithelium (Fu et al., 2016).

6.2 Functions of Omb in Wing Development

Wing development in *omb* null mutants is grossly disturbed. Mutant animals that develop to the pharate adult stage only bear a tiny wing rudiment while the rest of the integument is comparatively well developed. In the residual winglet, anterior (costa) and posterior (alula) elements of the hinge are largely normal. Between these peripheral structures it is difficult to identify remnants of the regular hinge organization (Del Alamo Rodriguez, Terriente Felix, & Diaz-Benjumea, 2004; Grimm & Pflugfelder, 1996) (Fig. 4D). Large distal clones that lack *omb* are not viable (Grimm & Pflugfelder, 1996; Simon & Guerrero, 2015). Similarly, knockdown of *omb* can eliminate the distal wing blade (Shen, Dorner, Bahlo, & Pflugfelder, 2008). The larval wing imaginal disc of *omb* mutants is overgrown and expresses apoptotic markers such as the Jun N-terminal kinase (JNK) target *puckered* (*puc*) and activated Caspase-3. A mild reduction in *omb* expression, as in the allele omb^{bi}, is already sufficient to activate apoptotic JNK signaling (Adachi-Yamada, Fujimura-Kamada, Nishida, & Matsumoto, 1999; Del Alamo Rodriguez et al., 2004). Apparently a certain level of *omb* is required in cells of the *omb* expression domain to allow their survival in the epithelium during the pupal stage. These severe phenotypes demonstrate that *omb* is essential for wing development but provide little information in which developmental processes it is required. The use of hypomorphic mutations and targeted manipulations reveals that *omb* has many roles in wing development that differ in different parts of the wing disc. Thus, Omb contributes to uniform proliferation in the disc by promoting

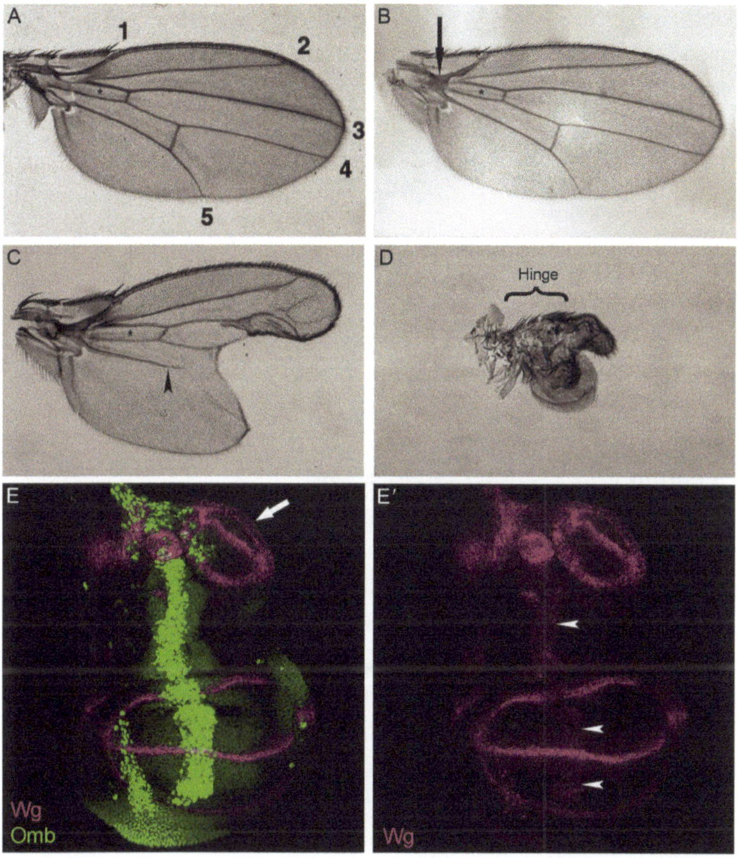

Fig. 4 Loss and gain of function of *omb*. (A) to (D) micrographs of adult wings. (A) *Wildtype*. The main longitudinal veins are marked 1–5. *Asterisk* indicates the basal radial cell. (B) Bifid-like phenotype induced by hinge-specific *omb* knockdown (CCKLR-17D3 > omb-RNAi). All longitudinal veins fuse in one common stalk in the hinge (*arrow*). (C) Enhanced bifid phenotype in $omb^{bi}/l(1)omb^{3198}$ wing. *Arrowhead* point to truncated vein L5. (D) Nearly complete loss of wing blade in wing of pharate adult $l(1)omb^{3198}$ fly. The *curly bracket* indicates the residual hinge domain. (E) and (F) Overexpression of *omb* (dpp > omb). (E) Wing disc. Omb (*green*), Wg (*magenta*). *Arrow* points to ectopic wing pouch (theta-like Wg expression) next to ectopic Omb. (E′) Separated Wg channel. *Arrowheads* point to ectopic Wg along the stripe of ectopic Omb.

proliferation in the lateral pouch, away from the A/P boundary, and by attenuating proliferation close to it (Zhang, Luo, Pflugfelder, & Shen, 2013).

Dworkin and Gibson determined how heterozygosity for mutations in components of developmental pathways that crucially govern wing development (EGFR, Notch, Hh, Tgf-β) affects wing shape. Of 50 mutations in 43 genes tested, the mutation omb^{P3} (originally omb^{md653}, Calleja, Moreno, Pelaz, & Morata, 1996; Mayer et al., 2013) had the strongest effect. The authors observed a slight venation defect even in the heterozygous state of this viable allele (Dworkin & Gibson, 2006). This shows that *omb* has an unusually strong and dose-sensitive effect on wing development. In *Drosophila*, only about 25 loci with a developmental haploinsufficient phenotype have been described (Cook et al., 2012). High dosage sensitivity is also seen with the *omb*-orthologous genes in human. In the Tbx2 subfamily, three of four genes are haploinsufficient for limb development (Bongers et al., 2004; Packham & Brook, 2003).

6.2.1 Omb in Wing Hinge Development

Bifid is a hypomorphic *omb* allele caused by insertion of a *412* retrotransposon and additional sequences into the first *omb* intron (Pflugfelder, 2009). In omb^{bi}, *omb* expression is reduced in all imaginal discs (Shen et al., 2008). However the wing, and in particular the wing hinge, is most strongly affected, yielding the "bifid" phenotype in which the five longitudinal veins, which normally merge in two distinct positions (L1–L3 in the anterior and L4, L5 in the posterior hinge, Fig. 4A), all fuse in the center of the hinge and then extend proximally in two stalks (Adachi-Yamada, Nakamura, et al., 1999; Grimm & Pflugfelder, 1996; Morgan & Bridges, 1916; Pflugfelder, 2009; Shen et al., 2008). In addition, omb^{bi} wings with variable expressivity exhibit a slight loss of tissue at the distal tip of the wing. Both phenotypes are aggravated when Omb dose is further reduced as in omb^{bi}/omb^{0} transheterozygotes (Grimm & Pflugfelder, 1996) (Fig. 4C). A bifid-like phenotype is also seen in the regulatory mutant $omb^{\Delta(P2\text{-}P3w)}$ in which about 30 kb upstream of the *omb* promoter is deleted (Shen et al., 2008). This deletion was engineered to remove a Dpp-controlled *omb* wing enhancer, previously analysed by Sivasankaran et al. (2000). The subtle phenotype of $omb^{\Delta(P2\text{-}P3w)}$ compared with that of amorphic *omb* alleles indicates the existence of additional enhancers for *omb* wing disc expression (several of which exist, our unpublished data). The strong hinge phenotype of omb^{bi}/omb^{0} is characterized by the apparent loss of the medial part (along the proximo-distal axis) of the basal radial cell that lies between the L1/2/3 and the L4/5 veins and vein stalks (compare panel A with B and C in Fig. 4). A strong hinge vein fusion phenotype can also be generated by hinge-specific *omb* knockdown (Fig. 4B). The apparent loss of intervein tissue in *omb*^{bi} can have several (mutual non-exclusive) causes: decreased proliferation, enhanced cell death, or fate change from intervein to vein tissue.

During larval development both proliferation and cell death are locally enhanced in *omb*^{bi}. This has previously been observed with attention mainly on pleura and ventral hinge (Umemori, Takemura, Maeda, Ohba, & Adachi-Yamada, 2007). The ventral compartment is strongly overgrown in *omb* amorphic wing discs. This overgrowth can be suppressed by expression of the JNK pathway antagonist *puc*, suggesting that ventral hyperproliferation is dependent on JNK-dependent apoptosis (Del Alamo Rodriguez et al., 2004). Apoptosis is known to elicit compensatory proliferation (Morata, Shlevkov, & Perez-Garijo, 2011). However, in the hypomorph *omb*^{bi}, attenuation of cell death by expression of *DIAP1* (*Death-associated inhibitor of apoptosis 1*) was reported not to reduce pleural overgrowth (Umemori et al., 2007).

It is conceivable that the vein fusion phenotype of *omb*^{bi} is caused by local vein broadening in the hinge. The adult vein phenotype is established during pupal development. The width of wing veins is specified during this stage by restricting Dpp signaling to the center of veins. In *omb* amorphic wing discs, the expression of *dpp* is deregulated and extends far into the anterior compartment (Del Alamo Rodriguez et al., 2004) implicating *dpp* deregulation as a potential cause of the *bifid* hinge phenotype. However, pupal overexpression of *dpp* or reduced expression of the Dpp antagonist *Smox* (*Smad on X*, aka *dSmad2*) in the wing hinge does not mimic the bifid phenotype (Sander, Eivers, Choi, & De Robertis, 2010; Sotillos & De Celis, 2005; Yu et al., 1996).

While manipulations of *omb* expression have confirmed the role of Omb in patterning the wing hinge along the A–P axis, mechanistic understanding of this process is still incomplete.

Omb expression in the dorsal hinge extends between the circumferential blade/hinge and hinge/notum folds (Figs. 1B and 3A). Omb hinge expression directly borders on the notal expression domain of *mirror* (*mirr*) and *araucan* (*ara*), members of the *Iroquois* gene complex (*Iro-C*). Omb prevents expansion of *Iro-C* expression into the hinge but *omb* expression does not expand into the notum region of the wing disc in clones deficient for *Iro-C*. Conditions that eliminate the sharp discontinuity of Iro-C and Omb expression at the H/N fold attenuate its formation (Wang, Li, Lu, Liu, & Shen, 2016) (see Section 6.3 for the role of Doc in hinge fold development).

6.2.2 Role of Omb in the Anterior Compartment of the Wing Disc

As described earlier, the expression of Omb in the wing pouch is symmetrical with regard to the A/P compartment boundary. Nonetheless, the function of Omb anterior and posterior to this boundary differs. Compartment boundaries are lineage boundaries: all progeny of a precursor cells will remain in the parental compartment. As outlined earlier, *engrailed* and its paralog *invected* (*inv*) are expressed in the posterior compartment of the wing and encode the posterior determinants. Posterior cells that lack En/Inv will cross into A; anterior cells ectopically expressing En will cross into P. Nonetheless, lack of En/Inv expression in A cells does not suffice to impart anterior fate. En induces the expression of *Hh* and represses *Ci*, a component of the Hh pathway, such that P cells cannot respond to Hh. If A cells are made unresponsive to Hh by elimination of the Hh signaling receptor gene *Smoothened* (*Smo*), they sort into the posterior compartment (Blair & Ralston, 1997; Rodriguez & Basler, 1997) without inducing *en* or *inv*. A cells must have all components of the Hh pathway, including the transcription factor Ci, in order to retain anterior fate. This indicates that transcription induced by Ci is of relevance for A specification (Dahmann & Basler, 2000). A prime transcription target of Hh signaling in A cells just anterior to the A/P boundary is *dpp*. Dpp and its pathway components Tkv (Thickveins) and Mad are required to maintain the A/P boundary in wing development (Hidalgo, 1994; Shen & Dahmann, 2005b). Shen and Dahmann showed that the Dpp target gene *omb* in the anterior compartment mediates the effect of Hh/Dpp signaling in maintaining the A/P boundary, even though Omb is symmetrically expressed in both compartments. *Omb* clones of anterior origin merge with the P compartment (Fig. 5B), while *omb* clones of posterior origin stay in their original compartment. In contrast, clones of the Dpp target *salm* do not perturb the A/P boundary (Shen & Dahmann, 2005b).

6.2.3 Role of Omb in the Posterior Compartment of the Wing Disc

Where cells of different fate or cells deriving from different lineage compartments border on one another in an epithelial field, they are often separated by a fold in the surface of the epithelium (Tepass, Godt, & Winklbauer, 2002). However, no fold is formed at the A/P boundary of the wing imaginal disc (Fig. 3C), in spite of the abrupt discontinuity of gene expression profiles at this junction and the strict separation of cells of anterior and posterior lineage (Fig. 1G). Similarly, the adult wing blade, excepting the corrugation of the wing veins, exhibits an even surface. The morphology

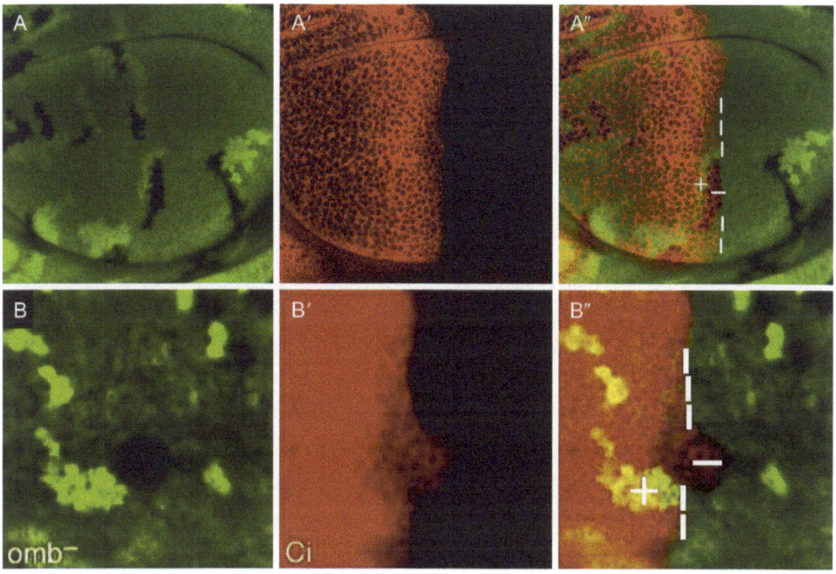

Fig. 5 Anterior loss of *omb* destabilizes the wing disc A/P boundary. (A–A″) Neutral clones respect the A/P boundary. The entire wing pouch is shown. (A) Clones are marked by loss of GFP (*green*) expression. *Twins spots* show increased green fluorescence. (A′) Ci (*red*) marks the anterior compartment and a straight A/P boundary. (A″) Combination of channels. The neutral clone (*minus*) sign has a straight outline where it meets the A/P boundary. The clone boundary that lies within the anterior compartment is wiggly. This also holds for the twin spot (*plus sign*). (B–B″) Anterior *omb* clone invades the posterior compartment. Only part of the wing pouch is shown. (B) *Omb* clones are marked by loss of GFP (*green*) expression. (B′) Ci (*red*) marks cells derived from the anterior compartment. (B″) Combination of channels. The *omb* clone (*minus sign*) violates the A/P boundary (*dashed line*), while the twin spot (*plus sign*) remains in the anterior compartment. *Panels (B–B″): Reprinted from Shen, J., & Dahmann, C. (2005a). Extrusion of cells with inappropriate Dpp signaling from Drosophila wing disc epithelia.* Science, 307*(5716), 1789–1790; Shen, J., & Dahmann, C. (2005b). The role of Dpp signaling in maintaining the Drosophila anteroposterior compartment boundary.* Developmental Biology, 279*(1), 31–43, with permission from Elsevier.*

of wing disc cells on both sides of the A/P boundary is not different except of those cells that are in contact with the A/P boundary (Blair, 2003; Brower, Smith, & Wilcox, 1982). These contact cells have a slightly larger apical cross section than cells within the interior of compartments. Cell bonds along the A/P boundary are under higher tension than bonds that lie within the compartments (Landsberg et al., 2009). Hh signaling across the A/P boundary is necessary for setting up tension at the border (Landsberg et al., 2009; Rudolf et al., 2015; Schilling et al., 2011).

It appears, that fold formation along the A/P boundary of the wing disc is suppressed during normal development and can become apparent under mutant conditions. Umemori and colleagues first noted that in *omb*^{bi} wing discs an epithelial fold develops along the A/P boundary of the pouch (Umemori et al., 2007). Based on the increased expression of the Hh target *dpp* in *omb*^{bi}, the authors proposed that Omb is part of a negative feedback loop from Dpp to Hh signaling such that reduced Omb expression will cause an increase in Hh signaling. Hyperactivation of the Hh pathway at the A/P boundary was proposed to cause A/P fold formation by local activation of JNK signaling, independent of JNK's role as an inducer of apoptotic cell death. Because Omb was postulated to affect Hh signaling, the Umemori model implicitly assumes that Omb acts in the anterior compartment to suppress fold formation.

We described the formation of an A/P fold in the slightly stronger hypomorph *omb*$^{\Delta(P2\text{-}P3w)}$. The fold forms directly on the A/P boundary, with both anterior and posterior cells contributing to it. Fold formation becomes dramatic, extending down to the basement membrane, when the Omb level is further reduced, such as in *omb*^{bi}/*omb*^{0}. Posterior knockdown or clonal loss of *omb* is sufficient to induce the A/P fold (Fig. 6B); anterior knockdown is ineffective in fold induction (Fig. 6A). During pupal development, a separation of the normally contiguous anterior and posterior compartments occurs in the pouch, leading to adult wings in which the a deep cleft runs through the blade from the most distal point down to the hinge (Shen et al., 2008). Reduced *omb* expression in the pouch results in increased cell death which, however, is not restricted to the A/P boundary. Expression of the cell death inhibitors DIAP1 or P35 does not prevent fold formation (Shen et al., 2008; Umemori et al., 2007). In cells that are drawn into the fold, the apical enrichment of α-tubulin (apical microtubule web, Eaton, Wepf, & Simons, 1996) is strongly decreased (Shen et al., 2008). The sufficiency of posterior Omb reduction for A/P fold formation rules out an involvement of Omb in the Hh signaling pathway downstream of the Hh receptors since Hh signaling is confined to the anterior compartment. It is conceivable, however, that Omb in the posterior compartment influences the complex process of Hh release and spread (Briscoe & Therond, 2013; Eaton, 2008; Jiang & Hui, 2008).

Bharathi and colleagues described the formation of an A/P fold in wing discs mutant for certain alleles of *coronin* (*coro*), which encodes an F-actin-binding protein. In *coro* discs, the gradient of Dpp is more shallow but

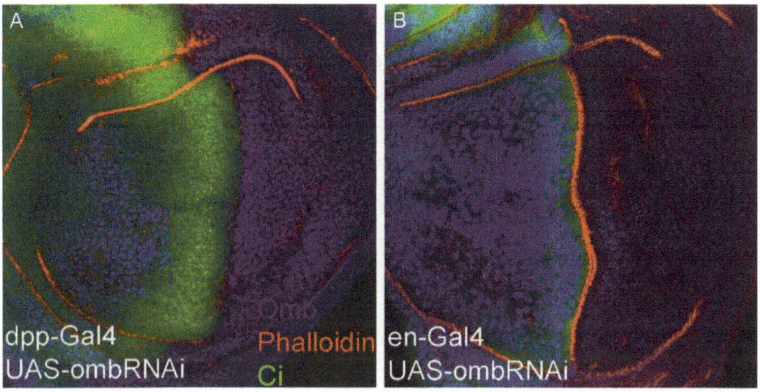

Fig. 6 Posterior reduction of *omb* induces a fold at the wing disc A/P boundary. (A) Omb knockdown in the anterior compartment does not elicit fold formation along the A/P boundary (dpp > omb-RNAi). The anterior compartment is marked by expression of Ci (*green*), folds are visualized by phalloidin staining (*red*), Omb (*blue*). (A′) A/P fold is induced by posterior *omb* knockdown (en > omb-RNAi). *Reprinted from Shen, J., Dorner, C., Bahlo, A., & Pflugfelder, G. O. (2008). Optomotor-blind suppresses instability at the A/P compartment boundary of the Drosophila wing.* Mechanisms of Development, 125*(3–4), 233–246, with permission from Elsevier.*

extends further than in *wildtype*. In the $coro^{ex11}$ background, *omb* is downregulated in the posterior compartment (Bharathi, Pallavi, Bajpai, Emerald, & Shashidhara, 2004). Since posterior reduction of *omb* expression is sufficient for fold formation, this may be the mechanism by which the A/P fold arises in *coro* discs. Bharathi et al. also described that overexpression of an activated form of the Dpp receptor Tkv along the anterior side of the A/P boundary induces a fold in the wing disc and a deep cleft in the adult wing. This phenotype is dominantly enhanced by *coro* mutations (Bharathi et al., 2004). Tkv overexpression leads to sequestration of Dpp (Lecuit & Cohen, 1998) and strongly reduces Omb in the posterior compartment (Shen et al., 2010); i.e., also this genetic manipulation is likely to act via *omb*.

Intriguingly, Gibson and colleagues observed similar phenotypes (fold in the wing disc and cleft in the adult wing) by downregulation of Dpp signaling in the peripodial epithelium. However, the authors state that folds can also arise parallel to but not on the A/P boundary, as well as parallel to the D/V boundary. Whether *omb* expression in the peripodial and main epithelium is affected by block of peripodial Dpp signaling was not reported (Gibson, Lehman, & Schubiger, 2002).

6.2.4 Graded Omb Expression Along the Wing Disc A–P Axis

The roughly symmetrical Dpp gradient with its central peak just anterior to the A/P boundary leads to nested expression of target genes such as *salm*, *omb*, and *vg* (Kim, Johnson, Chen, Carroll, & Laughon, 1997; Lecuit et al., 1996; Nellen et al., 1996). The Dpp concentration-dependent specification of different expression domains of Dpp target genes established the wing disc as a paradigm for threshold models of morphogen action (Affolter & Basler, 2007; Campbell & Tomlinson, 1999; Gurdon & Bourillot, 2001; Jazwinska et al., 1999; Podos & Ferguson, 1999). However, quantification of the Salm and Omb protein distribution shows that both proteins form smooth gradients along the A–P axis (Barrio & de Celis, 2004; Shen et al., 2010). This also holds for Brk, which forms an inverse gradient (Campbell & Tomlinson, 1999; Jazwinska et al., 1999; Minami et al., 1999). Another gradient downstream of Dpp is formed by the density of the apical microtubule web (Gibson & Perrimon, 2005; Shen & Dahmann, 2005a). Clones of cells with Dpp signaling levels that strongly differ from those in the flanking cells (i.e., loss of Dpp signaling in the center of the disc, hyperactivation laterally) are removed from the epithelial surface by basal retraction or extrusion (Adachi-Yamada & O'Connor, 2002; Gibson & Perrimon, 2005; Shen & Dahmann, 2005a). Much of these effects are mediated by Omb. Thus, the basal retraction of lateral clones in which Dpp signaling is ectopically activated by expression of the constitutively active Dpp receptor Tkv^{QD} is prevented by knockdown of *omb*. Inappropriate levels of Omb are sufficient to induce basal retraction or delamination of cells (Shen et al., 2010).

Clusters of epithelial cells that differ from surrounding cells in the expression of a cell adhesion molecule (quantitatively or qualitatively) will round up in order to gain a maximum of adhesive and/or a minimum of repulsive interactions (Garcia-Bellido, 1966; Steinberg, 1963). The rounding-up of cell clones can be quantified by measuring contour length and area of the clones (shape factor: Lawrence, Casal, & Struhl, 1999). *Omb* null mutant clones round up to a differing degree depending on their position along the A–P axis. Clone shape is nearly circular close to the A/P boundary and becomes increasingly wiggly toward the lateral periphery of the wing disc. The gradient of the shape factor does not differ between A and P compartments and runs parallel to the Omb gradient (Shen et al., 2010). In the framework of the cell affinity hypothesis, one could deduce that Omb controls the expression of cell surface adhesion molecules. Graded distribution of cell adhesion molecules may serve to stabilize the position of cells in an epithelial field. The shape factor of *salm* clones is also graded along the A–P axis. Spalt, whose expression requires Omb (Del Alamo Rodriguez et al., 2004), represses *capricious* (*caps*) and *tartan* (*trn*) in the third larval instar (Milan, Perez, & Cohen, 2002). *Caps* and *trn* encoded related leucin rich repeat transmembrane proteins, which, based on the sorting behavior of misexpression clones, were proposed to mediate cell adhesion. The spatial expression pattern of Caps and Trn, unlike that of Omb and Sal, is dynamic (Milan, Weihe, Perez, & Cohen, 2001). It is, therefore, not clear whether Caps/Trn mediates the sorting effect seen with *omb* or *salm* clones.

6.2.5 Omb in the Specification of the Posterior Wing Vein L5

$Omb^{bi}/l(1)omb^{3198}$ transheterozygous animals can still develop sizeable adult wings. These exhibit the hinge vein fusion phenotype and distal truncation of the wing blade. While veins L1–L4 generally reach the notched wing margin, L5 often does not, indicating a specific sensitivity of L5 development to *omb* dosage (Grimm & Pflugfelder, 1996) (Fig. 4C). A similar L5 phenotype is shown by a viable allele of *abrupt* (ab^1) (Cook, Biehs, & Bier, 2004).

Determination of epithelial cells as vein or intervein can first be recognized in the third larval instar as an alternating pattern of Dl and N expression, roughly orthogonal to the D/V boundary (the future wing margin). Activation of N signaling in the provein flanking domains represses *rhomboid* (*rho*, aka *veinlet*), leading to *rho* expression and EGFR activation in the center of the provein (Biehs, Francois, & Bier, 1996; de Celis, Bray, & Garcia-Bellido, 1997; Sturtevant, Roark, & Bier, 1993). Intervein territory is marked by the expression of *blistered* (*bs*, aka *Drosophila Serum Response Factor*) (Montagne et al., 1996). The positions of the longitudinal veins along the A–P axis are determined by the Hh/Dpp-patterning system whereby each provein is elaborated with the help of different downstream factors. L2 is positioned along the anterior expression boundary of the Dpp target Spalt (Lunde et al., 2003; Sturtevant, Biehs, Marin, & Bier, 1997). Hh is assumed to positions veins L3 and L4 (Biehs, Sturtevant, & Bier, 1998; Strigini & Cohen, 1997; Vervoort, 2000). L5 forms in the posterior compartment at a position where the Omb and Brk gradients intersect; i.e., L5 is flanked by high Omb anteriorly and high Brk posteriorly. In the position of the L5 provein, *ab* is specifically expressed. Omb^{bi} dominantly enhances the ab^1 L5 truncation phenotype. Increased expression of Brk, which will cause an anterior shift of the flank of the Omb gradient, also moves the position of L5 anteriorly such that it partly fuses with L4. Omb mutant clones that span

L5 cause loss of the L5 vein, whereas other veins are not affected (Cook, Biehs, & Bier, 2004). Cook and colleagues postulated that Brk induces the formation of a short-range vein-inducing factor that can act in regions of low Brk/high Omb to induce expression of the provein factor Abrupt.

6.2.6 Omb Can Promote Development of an Ectopic Wing Pair

When *omb* is ectopically expressed under dpp-Gal4 control, a second pair of wings develops, indicating sufficiency of Omb for notum to wing transformation (Grimm & Pflugfelder, 1996). Ectopic wings only develop from the wing disc. The phenotype differs from homeotic transformations in which (para)segmental identity is changed, as in *Ultrabithorax* (*Ubx*) flies in which the haltere is transformed to wing (Lewis, 1978). The *omb* gain-of-function phenotype differs also from that obtained by ectopic expression of master genes such as *eyeless* (*ey*) or *vestigial* in which ectopic eyes or ectopic wings, respectively, can develop from most imaginal discs (Baena-Lopez & Garcia-Bellido, 2003; Halder, Callaerts, & Gehring, 1995; Kim et al., 1996).

In the boundary model of limb development, limbs grow and undergo patterning using positional information that is generated at the intersection of compartment or determination boundaries (Gelbart, 1989; Meinhardt, 1983). In the pouch of the wing imaginal disc, Dpp and Wg are expressed in orthogonal stripes along or on compartmental boundaries and are required for growth and patterning (Campbell, Weaver, & Tomlinson, 1993). *Omb* which is downstream of both factors and required for distal development was assumed to organize ectopic wing development by a mechanism similar to its role in the formation of the endogenous wing (Grimm & Pflugfelder, 1996). The actual mechanism is likely more indirect. In the notum, like in the pouch, *Omb* is expressed at the intersection of stripes of Dpp and Wg expression (Fig. 3A′, cf. to Fig. 4E for dpp-Gal4 expression). Endogenous notal *omb* expression is weak and arises late compared to the pouch. It evidently does not suffice to promote pouch development.

Wing duplications can be produced when wg signaling is ectopically actived in the notum no later than early second instar (Klein & Arias, 1998a; Ng et al., 1996; Silver, Hagen, Okamura, Perrimon, & Lai, 2007). Ectopic expression of *omb* under dpp-Gal4 control induces a stripe of ectopic Wg expression that extends into the notum (Fig. 4E′, cf. Fig. 3B′ for endogenous Wg expression). Ectopic expression of *omb*, thus, appears just as a circuitous way to generate a dpp > wg phenotype. The notal domain of strong combined Dpp and Wg signaling then organizes a second pouch field (Ng et al., 1996) (arrow in Fig. 4E′).

6.3 Doc in Wing Development

The onset of Doc expression in the wing disc at the border between the future wing blade and hinge correlates with the formation of a deep apical fold between these territories (B/H fold) (Fig. 3C). Like Doc expression, the fold first forms in the center of the disc (close to the A/P boundary) and then spreads laterally. A double knockdown of *Doc1* and *Doc2* only allows formation of a shallow fold. Clones that eliminate all three *Doc* genes cause a shallow fold in the center of the disc and block the lateral extension of the fold. In the first 4 h after puparium formation, the wing disc undergoes a bending process that brings the dorsal and ventral parts of the disc in apposition, thus prefiguring the structure of the adult wing. The dorsal and ventral *Doc* expression domains in the pouch also become juxtaposed at the end of this process. In discs carrying large *Doc* clones across the hinge region, the folding is delayed. Adult flies with such clones present with a wing postural defect. Ectopic expression of *Doc* (e.g., along the A/P boundary) causes formation of an ectopic fold along the new Doc expression domain. Ectopic *Doc* expression in a ring around the pouch, proximal to the endogenous expression domain, initiates premature bending of the wing disc.

Cells that contribute to an apical fold have a reduced level of apical microtubules and a corresponding basal enrichment (Fig. 3C′). A similar redistribution is observed when folds are ectopically induced by *Doc* expression. In a time-resolved study we have followed cell morphological changes after *Doc* induction. Three sequential cell shape changes lead to the formation of a deep fold: first, a widening of the apical cell diameter, second, shortening of the cell height, third, widening of the basal cell diameter.

Endogenous or Doc-induced ectopic folds are associated with reduced levels of the integrins subunits αPS and βPS and the main ECM component laminin (Fig. 1A) suggesting that fold induction by Doc expression is mediated by local degradation of the ECM. This is confirmed by the observation of an increased level of the matrix metalloproteinase MMP2 in the basal region of endogenous and induced folds. Coexpression of the MMP inhibitor Timp with Doc suppresses the Doc-induced precocious bending of the wing disc, further supporting a role of Doc-induced MMP2 in fold formation (Sui et al., 2012).

6.4 Mid in Wing Development

Fu et al. (2016) describe expression and function of Mid in wing development. Mid is nearly uniformly expressed in main and peripodial epithelium of the wing disc. *Mid* knockdown during disc development results in wings whose margins are either up- or downturned, depending on the Gal4 driver used. *Mid* overexpression in the wing pouch causes stronger defects, in particular deletion of the wing margin, reminiscent of hypomorphic *wingless* alleles. The authors report a subtle increase in Wg expression in *mid* clones and a corresponding decrease in *mid* overexpressing clones. Also *hedgehog* (*hh*) appears repressed by *mid* overexpression, albeit in an odd non-autonomous fashion. In agreement with the repressive action of Mid on *hh*, the phenotype of the *hh* gain-of-function allele hh^{Mrt} is enhanced by *mid* heterozygosity. Thus Mid appears to attenuate both *wg* and *hh* (Fu et al., 2016).

7. TBX GENE FUNCTION IN HALTERE DEVELOPMENT

Wings and halteres are serially homologous structures. In evolution the halteres derived from the hindwings of a four-winged ancestral insect (Carroll, Weatherbee, & Langeland, 1995). During ontogenesis, wing and haltere are initially specified by common mechanisms (Sharma & Chopra, 1976; Williams, Bell, & Carroll, 1991). In adult *Drosophila*, the difference in size between wings and halteres is large, even though the difference in cell number is only fivefold (Martin, 1982). During metamorphosis, the shape of wing and haltere cells diverges such that at the end of pupal development the apical surface area of wing cells is eightfold larger than that of haltere cells (Roch & Akam, 2000). The divergent development of wing and haltere is solely due to the expression of the Hox gene *Ubx* which is expressed in most or all cells of the haltere, while in the wing disc *Ubx* is restricted to the peripodial epithelium (Simon & Guerrero, 2015; White & Wilcox, 1984). Halteres that lack *Ubx* develop as wings (Bender et al., 1983). Ubx directly controls components or targets of several signaling pathways involved in growth and patterning (Akam, 1998; de Navas et al., 2011; Galant, Walsh, & Carroll, 2002; Mohit, Bajpai, & Shashidhara, 2003; Pallavi, Kannan, & Shashidhara, 2006; Pavlopoulos & Akam, 2011; Shashidhara, Agrawal, Bajpai, Bharathi, & Sinha, 1999; Weatherbee, Halder, Kim, Hudson, & Carroll, 1998). In the haltere disc, Omb is expressed in a wing-like pattern (Fig. 7A), but Spalt expression is lacking from the capitellum region which corresponds to the wing disc pouch (Weatherbee et al., 1998). Ubx causes attenuation of Dpp signaling by reducing *dpp* expression and by hampering Dpp spread

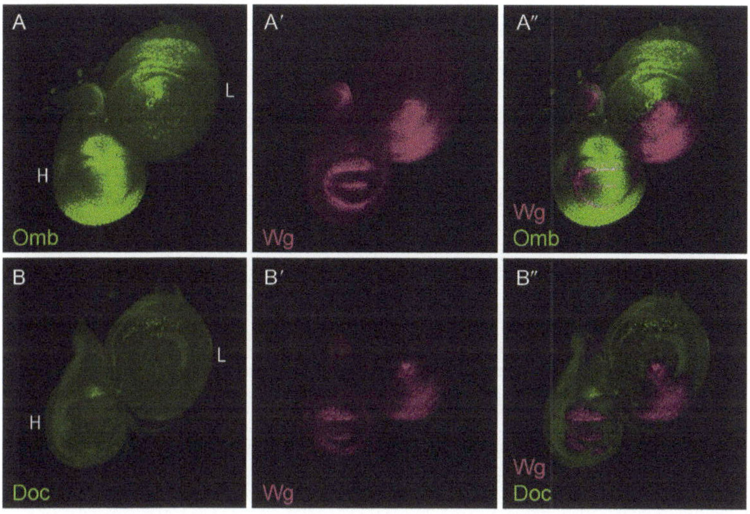

Fig. 7 Haltere disc expression of Omb and Doc. In (A) and (B), a pair of haltere (H) and leg (L) discs are shown. For the haltere disc anterior is left, dorsal up. (A) Omb expression (*green*) is similar to the wing disc (cf. Fig. 3A). (A′) Wg expression (*magenta*) differs from the wing disc in the posterior compartment (cf. Fig. 3A′). (A″) Omb and Wg channels combined. (B) Doc expression in the haltere is largely restricted to the hinge region. There is no expression in the pouch-equivalent capitellum (cf. Fig. 3B). (B′) Wg expression (*magenta*). (B″) Doc and Wg channels combined.

(via upregulation of *tkv*, downregulation of *dally* and *dlp*) (Crickmore & Mann, 2006, 2007; de Navas, Garaulet, & Sanchez-Herrero, 2006; Makhijani, Kalyani, Srividya, & Shashidhara, 2007; Mohit et al., 2006). Omb is involved in this regulation. In the haltere, Omb represses *dpp* as well as *dally* and *dlp* (Simon & Guerrero, 2015). *Salm* and vg^{QE} are Omb dependent in the wing pouch but are not expressed in the haltere capitellum primordium indicating insuffiency of Omb for *salm* and vg^{QE} expression in the capitellum. Rather, in the haltere, Omb represses *salm* in the future capitellum (Del Alamo Rodriguez et al., 2004; Simon & Guerrero, 2015). Doc, which is partly repressed by Omb in the wing disc, is also repressed by Omb in the haltere capitellum (Simon & Guerrero, 2015) (Fig. 7B, our unpublished data). These observations show that regulation of target genes by Omb is highly context dependent. Omb does not exert its effects on capitellar gene expression via regulation of Ubx, indicating that loss of *omb* does not cause a haltere to wing transformation (Simon & Guerrero, 2015).

8. CONCLUSION

This review summarizes current knowledge on expression and function of Tbx genes in *Drosophila* limb development. Sketching the solid ground highlights the gaps. For example, the limb expression of many genes is known to be affected by mutations in *mid/H15*, *Doc*, and *omb* (e.g., Del Alamo Rodriguez et al., 2004; Simon & Guerrero, 2015; Sui et al., 2012; Svendsen et al., 2009; Zhang et al., 2013). However, in no case has evidence for direct regulation been put forward, even though the DNA-binding specificity for these transcription factors is known (Jin et al., 2013; Liu et al., 2009; Najand, Ryu, & Brook, 2012; Sen et al., 2014). This knowledge will eventually help to untangle the regulatory networks. Furthermore, for parts of the Tbx expression patterns in the developing limbs, we do not yet know whether the Tbx proteins are actually functional there. Examples are Omb expression in leg disc myoblasts, and in wing disc peripodial and tracheal cells.

We refrained from pointing out parallels in the genetic machinery used in vertebrate and arthropod appendage development. For discussions of this aspect see Lemons, Fritzenwanker, Gerhart, Lowe, and McGinnis (2010), Mann and Casares (2002), Minelli (2003), Panganiban et al. (1997), Pueyo and Couso (2005), Shubin, Tabin, and Carroll (1977). Like their vertebrate counterparts, each *Drosophila* Tbx gene functions in several developmental contexts. This also had to be excluded to keep the review focused.

ACKNOWLEDGMENTS

We thank the members of our labs, past and present, for their contributions. Our work was supported by grants from the Deutsche Forschungsgemeinschaft, by intramural support from Johannes Gutenberg University, and by grants from the 973 Program, the National Natural Science Foundation of China, and the New Century Excellent Talent Award Program from the Ministry of Education of China.

REFERENCES

Abu-Shaar, M., & Mann, R. S. (1998). Generation of multiple antagonistic domains along the proximodistal axis during Drosophila leg development. *Development*, *125*, 3821–3830.

Adachi-Yamada, T., Fujimura-Kamada, K., Nishida, Y., & Matsumoto, K. (1999). Distortion of proximodistal information causes JNK-dependent apoptosis in Drosophila wing. *Nature*, *400*(6740), 166–169.

Adachi-Yamada, T., Nakamura, M., Irie, K., Tomoyasu, Y., Sano, Y., Mori, E., et al. (1999). p38 mitogen-activated protein kinase can be involved in transforming growth factor beta superfamily signal transduction in Drosophila wing morphogenesis. *Molecular and Cellular Biology*, *19*(3), 2322–2329.

Adachi-Yamada, T., & O'Connor, M. B. (2002). Morphogenetic apoptosis: A mechanism for correcting discontinuities in morphogen gradients. *Developmental Biology*, *251*(1), 74–90.

Affolter, M., & Basler, K. (2007). The Decapentaplegic morphogen gradient: From pattern formation to growth regulation. *Nature Reviews. Genetics*, *8*(9), 663–674.

Agulnik, S. I., Garvey, N., Hancock, S., Ruvinsky, I., Chapman, D. L., Agulnik, I., et al. (1996). Evolution of mouse T-box genes by tandem duplication and cluster dispersion. *Genetics*, *144*, 249–254.

Ahn, D., You, K. H., & Kim, C. H. (2012). Evolution of the tbx6/16 subfamily genes in vertebrates: Insights from zebrafish. *Molecular Biology and Evolution*, *29*(12), 3959–3983.

Akam, M. (1998). Hox genes: From master genes to micromanagers. *Current Biology*, *8*, R676–R678.

Akiyama, T., & Gibson, M. C. (2015). Decapentaplegic and growth control in the developing Drosophila wing. *Nature*, *527*(7578), 375–378.

Akiyama, T., Kamimura, K., Firkus, C., Takeo, S., Shimmi, O., & Nakato, H. (2008). Dally regulates Dpp morphogen gradient formation by stabilizing Dpp on the cell surface. *Developmental Biology*, *313*(1), 408–419.

Aldaz, S., & Escudero, L. M. (2010). Imaginal discs. *Current Biology*, *20*(10), R429–R431.

Aldaz, S., Escudero, L. M., & Freeman, M. (2013). Dual role of myosin II during Drosophila imaginal disc metamorphosis. *Nature Communications*, *4*, 1761.

Azpiazu, N., & Morata, G. (2002). Distinct functions of homothorax in leg development in Drosophila. *Mechanisms of Development*, *119*(1), 55.

Baena-Lopez, L. A., & Garcia-Bellido, A. (2003). Genetic requirements of vestigial in the regulation of Drosophila wing development. *Development*, *130*(1), 197–208.

Baena-Lopez, L. A., Nojima, H., & Vincent, J. P. (2012). Integration of morphogen signalling within the growth regulatory network. *Current Opinion in Cell Biology*, *24*(2), 166–172.

Barrio, R., & de Celis, J. F. (2004). Regulation of spalt expression in the Drosophila wing blade in response to the Decapentaplegic signaling pathway. *Proceedings of the National Academy of Sciences of the United States of America*, *101*(16), 6021–6026.

Basler, K., & Struhl, G. (1994). Compartment boundaries and the control of Drosophila limb pattern by hedgehog protein. *Nature*, *368*, 208–214.

Bate, M., & Arias, A. M. (1991). The embryonic origin of imaginal discs in Drosophila. *Development*, *112*(3), 755–761.

Bender, W., Akam, M., Karch, F., Beachy, P. A., Peifer, M., Spierer, P., et al. (1983). Molecular genetics of the bithorax complex in Drosophila melanogaster. *Science*, *221*(4605), 23–29.

Berns, N., Kusch, T., Schroder, R., & Reuter, R. (2008). Expression, function and regulation of Brachyenteron in the short germband insect Tribolium castaneum. *Development Genes and Evolution*, *218*(3–4), 169–179.

Bharathi, V., Pallavi, S. K., Bajpai, R., Emerald, B. S., & Shashidhara, L. S. (2004). Genetic characterization of the Drosophila homologue of coronin. *Journal of Cell Science*, *117*(Pt. 10), 1911–1922.

Biehs, B., Francois, V., & Bier, E. (1996). The Drosophila short gastrulation gene prevents Dpp from autoactivating and suppressing neurogenesis in the neuroectoderm. *Genes & Development*, *10*, 2922–2934.

Biehs, B., Sturtevant, M. A., & Bier, E. (1998). Boundaries in the Drosophila wing imaginal disc organize vein-specific genetic programs. *Development*, *125*(21), 4245–4257.

Blair, S. S. (2003). Lineage compartments in Drosophila. *Current Biology*, *13*(14), R548–R551.
Blair, S. S., Brower, D. L., Thomas, J. B., & Zavortink, M. (1994). The role of apterous in the control of dorsoventral compartmentalization and PS integrin gene expression in the developing wing of Drosophila. *Development*, *120*(7), 1805–1815.
Blair, S. S., & Ralston, A. (1997). Smoothened-mediated Hedgehog signalling is required for the maintenance of the anterior-posterior lineage restriction in the developing wing of Drosophila. *Development*, *124*(20), 4053–4063.
Bongers, E. M., Duijf, P. H., van Beersum, S. E., Schoots, J., Van Kampen, A., Burckhardt, A., et al. (2004). Mutations in the human TBX4 gene cause small patella syndrome. *The American Journal of Human Genetics*, *74*(6), 1239–1248.
Briscoe, J., & Therond, P. P. (2013). The mechanisms of Hedgehog signalling and its roles in development and disease. *Nature Reviews. Molecular Cell Biology*, *14*(7), 416–429.
Brook, W. J. (2010). T-box genes organize the dorsal ventral leg axis in Drosophila melanogaster. *Fly (Austin)*, *4*(2), 159–162.
Brook, W. J., & Cohen, S. M. (1996). Antagonistic interactions between wingless and decapentaplegic responsible for dorsal-ventral pattern in the Drosophila Leg. *Science*, *273*(5280), 1373–1377.
Brower, D. L., Smith, R. J., & Wilcox, M. (1982). Cell shapes on the surface of the Drosophila wing imaginal disc. *Journal of Embryology and Experimental Morphology*, *67*, 137–151.
Buescher, M., Tio, M., Tear, G., Overton, P. M., Brook, W. J., & Chia, W. (2006). Functions of the segment polarity genes midline and H15 in Drosophila melanogaster neurogenesis. *Developmental Biology*, *292*(2), 418–429.
Buratovich, M., & Wilder, E. (2001). The role of wingless in the development of the proximal wing hinge of Drosophila. *Drosophila Information Service*, *84*, 145–157.
Burke, R., & Basler, K. (1996). Dpp receptors are autonomously required for cell proliferation in the entire developing Drosophila wing. *Development*, *122*(7), 2261–2269.
Butler, M. J., Jacobsen, T. L., Cain, D. M., Jarman, M. G., Hubank, M., Whittle, J. R., et al. (2003). Discovery of genes with highly restricted expression patterns in the Drosophila wing disc using DNA oligonucleotide microarrays. *Development*, *130*(4), 659–670.
Calleja, M., Moreno, E., Pelaz, S., & Morata, G. (1996). Visualization of gene expression in living adult Drosophila. *Science*, *274*(5285), 252–255.
Campbell, G., & Tomlinson, A. (1999). Transducing the Dpp morphogen gradient in the wing of Drosophila: Regulation of Dpp targets by brinker. *Cell*, *96*(4), 553–562.
Campbell, G., Weaver, T., & Tomlinson, A. (1993). Axis specification in the developing Drosophila appendage: The role of wingless, decapentaplegic, and the homeobox gene aristaless. *Cell*, *74*, 1113–1123.
Carroll, S. B., Weatherbee, S. D., & Langeland, J. A. (1995). Homeotic genes and the regulation and evolution of insect wing number. *Nature*, *375*(6526), 58–61.
Cho, K. O., Chern, J., Izaddoost, S., & Choi, K. W. (2000). Novel signaling from the peripodial membrane is essential for eye disc patterning in Drosophila. *Cell*, *103*(2), 331–342.
Cohen, B., Simcox, A. A., & Cohen, S. M. (1993). Allocation of the thoracic imaginal primordia in the Drosophila embryo. *Development*, *117*, 597–608.
Cook, O., Biehs, B., & Bier, E. (2004). Brinker and optomotor-blind act coordinately to initiate development of the L5 wing vein primordium in Drosophila. *Development*, *131*(9), 2113–2124.
Cook, R. K., Christensen, S. J., Deal, J. A., Coburn, R. A., Deal, M. E., Gresens, J. M., et al. (2012). The generation of chromosomal deletions to provide extensive coverage and subdivision of the Drosophila melanogaster genome. *Genome Biology*, *13*(3), R21.

Couso, J. P., Bate, M., & Martinez-Arias, A. (1993). A wingless-dependent polar coordinate system in Drosophila imaginal discs. *Science, 259*(5094), 484–489.
Couso, J. P., Bishop, S. A., & Martinez Arias, A. (1994). The wingless signalling pathway and the patterning of the wing margin in Drosophila. *Development, 120*(3), 621–636.
Crickmore, M. A., & Mann, R. S. (2006). Hox control of organ size by regulation of morphogen production and mobility. *Science, 313*(5783), 63–68.
Crickmore, M. A., & Mann, R. S. (2007). Hox control of morphogen mobility and organ development through regulation of glypican expression. *Development, 134*(2), 327–334.
Dahmann, C., & Basler, K. (1999). Compartment boundaries at the edge of development. *Trends in Genetics, 15*, 320–326.
Dahmann, C., & Basler, K. (2000). Opposing transcriptional outputs of Hedgehog signaling and engrailed control compartmental cell sorting at the Drosophila A/P boundary. *Cell, 100*(4), 411–422.
Dastjerdi, A., Robson, L., Walker, R., Hadley, J., Zhang, Z., Rodriguez-Niedenfuhr, M., et al. (2007). Tbx1 regulation of myogenic differentiation in the limb and cranial mesoderm. *Developmental Dynamics, 236*(2), 353–363.
de Celis, J. F., Bray, S., & Garcia-Bellido, A. (1997). Notch signalling regulates veinlet expression and establishes boundaries between veins and interveins in the Drosophila wing. *Development, 124*(10), 1919–1928.
de Navas, L. F., Garaulet, D. L., & Sanchez-Herrero, E. (2006). The ultrabithorax Hox gene of Drosophila controls haltere size by regulating the Dpp pathway. *Development, 133*(22), 4495–4506.
de Navas, L. F., Reed, H., Akam, M., Barrio, R., Alonso, C. R., & Sanchez-Herrero, E. (2011). Integration of RNA processing and expression level control modulates the function of the Drosophila Hox gene Ultrabithorax during adult development. *Development, 138*(1), 107–116.
Del Alamo Rodriguez, D., Terriente Felix, J., & Diaz-Benjumea, F. J. (2004). The role of the T-box gene optomotor-blind in patterning the Drosophila wing. *Developmental Biology, 268*(2), 481–492.
Diaz-Benjumea, F. J., & Cohen, S. M. (1993). Interaction between dorsal and ventral cells in the imaginal disc directs wing development in Drosophila. *Cell, 75*(4), 741–752.
Diaz-Benjumea, F. J., & Cohen, S. M. (1994). Wingless acts through the shaggy/zeste-white 3 kinase to direct dorsal-ventral axis formation in the Drosophila leg. *Development, 120*, 1661–1670.
Dong, P. D., Dicks, J. S., & Panganiban, G. (2002). Distal-less and homothorax regulate multiple targets to pattern the Drosophila antenna. *Development, 129*(8), 1967–1974.
Dworkin, I., & Gibson, G. (2006). Epidermal growth factor receptor and transforming growth factor-beta signaling contributes to variation for wing shape in Drosophila melanogaster. *Genetics, 173*(3), 1417–1431.
Eaton, S. (2008). Multiple roles for lipids in the Hedgehog signalling pathway. *Nature Reviews. Molecular Cell Biology, 9*(6), 437–445.
Eaton, S., Wepf, R., & Simons, K. (1996). Roles for Rac1 and Cdc42 in planar polarization and hair outgrowth in the wing of Drosophila. *The Journal of Cell Biology, 135*(5), 1277–1289.
Ebacher, D. J., Todi, S. V., Eberl, D. F., & Boekhoff-Falk, G. E. (2007). Cut mutant Drosophila auditory organs differentiate abnormally and degenerate. *Fly (Austin), 1*(2), 86–94.
Emerald, B. S., Curtiss, J., Mlodzik, M., & Cohen, S. M. (2003). Distal antenna and distal antenna related encode nuclear proteins containing pipsqueak motifs involved in antenna development in Drosophila. *Development, 130*(6), 1171–1180.

Entchev, E. V., Schwabedissen, A., & Gonzalez-Gaitan, M. (2000). Gradient formation of the TGF-beta homolog Dpp. *Cell, 103*(6), 981–991.

Estella, C., & Mann, R. S. (2008). Logic of Wg and Dpp induction of distal and medial fates in the Drosophila leg. *Development, 134*(4), 627–636.

Fischer, K., & Pflugfelder, G. (2015). Putative breast cancer driver mutations in TBX3 cause impaired transcriptional repression. *Frontiers in Oncology, 5*(244), 244. 1–15.

Fleming, R. J., Gu, Y., & Hukriede, N. A. (1997). Serrate-mediated activation of Notch is specifically blocked by the product of the gene fringe in the dorsal compartment of the Drosophila wing imaginal disc. *Development, 124*(15), 2973–2981.

Foronda, D., Perez-Garijo, A., & Martin, F. A. (2009). Dpp of posterior origin patterns the proximal region of the wing. *Mechanisms of Development, 126*(3–4), 99–106.

Fu, C. L., Wang, X. F., Cheng, Q., Wang, D., Hirose, S., & Liu, Q. X. (2016). The T-box transcription factor Midline regulates wing development by repressing wingless and hedgehog in Drosophila. *Scientific Reports, 6*, 27981.

Galant, R., Walsh, C. M., & Carroll, S. B. (2002). Hox repression of a target gene: Extradenticle-independent, additive action through multiple monomer binding sites. *Development, 129*(13), 3115–3126.

Garcia-Bellido, A. (1966). Changes in selective affinity following transdetermination in imaginal disc cells of Drosophila melanogaster. *Experimental Cell Research, 44*(2), 382–392.

Garcia-Bellido, A., & Merriam, J. R. (1971). Parameters of the wing imaginal disc development of Drosophila melanogaster. *Developmental Biology, 24*, 61–87.

Garcia-Bellido, A., Ripoll, P., & Morata, G. (1976). Developmental compartimentalization in the dorsal mesothoracic disc of Drosophila. *Developmental Biology, 48*, 132–147.

Gelbart, W. M. (1989). The decapentaplegic gene: A TGF-beta homologue controlling pattern formation in Drosophila. *Development, 107*(Suppl.), 65–74.

Gibson, M. C., Lehman, D. A., & Schubiger, G. (2002). Lumenal transmission of decapentaplegic in Drosophila imaginal discs. *Developmental Cell, 3*(3), 451–460.

Gibson, M. C., & Perrimon, N. (2005). Extrusion and death of DPP/BMP-compromised epithelial cells in the developing Drosophila wing. *Science, 307*(5716), 1785–1789.

Gibson, M. C., & Schubiger, G. (2000). Peripodial cells regulate proliferation and patterning of Drosophila imaginal discs. *Cell, 103*(2), 343–350.

Goering, L. M., Hoshijima, K., Hug, B., Bisgrove, B., Kispert, A., & Grunwald, D. J. (2003). An interacting network of T-box genes directs gene expression and fate in the zebrafish mesoderm. *Proceedings of the National Academy of Sciences of the United States of America, 100*(16), 9410–9415.

Gonzalez-Crespo, S., & Morata, G. (1996). Genetic evidence for the subdivision of the arthropod limb into coxopodite and telopodite. *Development, 122*(12), 3921–3928.

Grimm, S., & Pflugfelder, G. O. (1996). Control of the gene optomotor-blind in Drosophila wing development by decapentaplegic and wingless. *Science, 271*(5255), 1601–1604.

Guha, A., Lin, L., & Kornberg, T. B. (2009). Regulation of Drosophila matrix metalloprotease Mmp2 is essential for wing imaginal disc:trachea association and air sac tubulogenesis. *Developmental Biology, 335*(2), 317–326.

Gurdon, J. B., & Bourillot, P. Y. (2001). Morphogen gradient interpretation. *Nature, 413*(6858), 797–803.

Gutierrez, L., Oktaba, K., Scheuermann, J. C., Gambetta, M. C., Ly-Hartig, N., & Muller, J. (2012). The role of the histone H2A ubiquitinase Sce in Polycomb repression. *Development, 139*(1), 117–127.

Hadar, N., Yaron, S., Oren, Z., Elly, O., Itamar, W., Johnathan, G., et al. (2012). A screen identifying genes responsive to Dpp and Wg signaling in the Drosophila developing wing. *Gene, 494*(1), 65–72.

Halder, G., Callaerts, P., & Gehring, W. J. (1995). Induction of ectopic eyes by targeted expression of the eyeless gene in Drosophila. *Science*, *267*(5205), 1788–1792.

Hamaguchi, T., Yabe, S., Uchiyama, H., & Murakami, R. (2004). Drosophila Tbx6-related gene, Dorsocross, mediates high levels of Dpp and Scw signal required for the development of amnioserosa and wing disc primordium. *Developmental Biology*, *265*(2), 355–368.

Hamaratoglu, F., Affolter, M., & Pyrowolakis, G. (2014). Dpp/BMP signaling in flies: From molecules to biology. *Seminars in Cell & Developmental Biology*, *32*, 128–136.

Hays, R., Buchanan, K. T., Neff, C., & Orenic, T. V. (1999). Patterning of Drosophila leg sensory organs through combinatorial signaling by hedgehog, decapentaplegic and wingless. *Development*, *126*(13), 2891–2899.

Held, L. I., Jr. (1995). Axes, boundaries and coordinates: The ABC of fly leg development. *BioEssays*, *17*, 721–732.

Held, L. I., Jr. (2002). *Imaginal discs. The genetic and cellular logic of pattern formation*. Cambridge, UK: Cambridge University Press.

Hidalgo, A. (1994). Three distinct roles for the engrailed gene in Drosophila wing development. *Current Biology*, *4*(12), 1087–1098.

Holstien, K., Rivera, A., Windsor, P., Ding, S., Leys, S. P., Hill, M., et al. (2010). Expansion, diversification, and expression of T-box family genes in Porifera. *Development Genes and Evolution*, *220*(9–10), 251–262.

Hoogaars, W. M., Barnett, P., Moorman, A. F., & Christoffels, V. M. (2007). T-box factors determine cardiac design. *Cellular and Molecular Life Sciences*, *64*(6), 646–660.

Hurlin, P. J., Steingrimsson, E., Copeland, N. G., Jenkins, N. A., & Eisenman, R. N. (1999). Mga, a dual-specificity transcription factor that interacts with Max and contains a T-domain DNA-binding motif. *The EMBO Journal*, *18*(24), 7019–7028.

Irvine, K. D., & Wieschaus, E. (1994). Fringe, a Boundary-specific signaling molecule, mediates interactions between dorsal and ventral cells during Drosophila wing development. *Cell*, *79*(4), 595–606.

Janssen, R., Feitosa, N. M., Damen, W. G., & Prpic, N. M. (2008). The T-box genes H15 and optomotor-blind in the spiders Cupiennius salei, Tegenaria atrica and Achaearanea tepidariorum and the dorsoventral axis of arthropod appendages. *Evolution & Development*, *10*(2), 143–154.

Janssen, R., Jorgensen, M., Prpic, N. M., & Budd, G. E. (2015). Aspects of dorso-ventral and proximo-distal limb patterning in onychophorans. *Evolution & Development*, *17*(1), 21–33.

Jazwinska, A., Kirov, N., Wieschaus, E., Roth, S., & Rushlow, C. (1999). The Drosophila gene brinker reveals a novel mechanism of Dpp target gene regulation. *Cell*, *96*(4), 563–573.

Jiang, J., & Hui, C. C. (2008). Hedgehog signaling in development and cancer. *Developmental Cell*, *15*(6), 801–812.

Jin, H., Stojnic, R., Adryan, B., Ozdemir, A., Stathopoulos, A., & Frasch, M. (2013). Genome-wide screens for in vivo Tinman binding sites identify cardiac enhancers with diverse functional architectures. *PLoS Genetics*, *9*(1), e1003195.

Johnstone, K., Wells, R. E., Strutt, D., & Zeidler, M. P. (2013). Localised JAK/STAT pathway activation is required for Drosophila wing hinge development. *PLoS One*, *8*(5), e65076.

Karlsson, J. (1981). The distribution of regenerative potential in the wing disc of Drosophila. *Journal of Embryology and Experimental Morphology*, *61*, 303–316.

Kim, J., Irvine, K. D., & Carroll, S. B. (1995). Cell recognition, signal induction, and symmetrical gene activation at the dorsal-ventral boundary of the developing Drosophila wing. *Cell*, *82*(5), 795–802.

Kim, J., Johnson, K., Chen, H. J., Carroll, S., & Laughon, A. (1997). Drosophila Mad binds to DNA and directly mediates activation of vestigial by Decapentaplegic. *Nature*, *388*(6639), 304–308.

Kim, J., Sebring, A., Esch, J. J., Kraus, M. E., Vorwerk, K., Magee, J., et al. (1996). Integration of positional signals and regulation of wing formation and identity by Drosophila vestigial gene. *Nature*, *382*(6587), 133–138.

Klein, T. (2001). Wing disc development in the fly: The early stages. *Current Opinion in Genetics & Development*, *11*(4), 470–475.

Klein, T., & Arias, A. M. (1998a). Different spatial and temporal interactions between Notch, wingless, and vestigial specify proximal and distal pattern elements of the wing in Drosophila. *Developmental Biology*, *194*(2), 196–212.

Klein, T., & Arias, A. M. (1998b). Interactions among Delta, Serrate and Fringe modulate Notch activity during Drosophila wing development. *Development*, *125*(15), 2951–2962.

Kojima, T. (2004). The mechanism of Drosophila leg development along the proximodistal axis. *Development, Growth & Differentiation*, *46*(2), 115–129.

Kopp, A., & Duncan, I. (1997). Control of cell fate and polarity in the adult abdominal segments of Drosophila by optomotor-blind. *Development*, *124*(19), 3715–3726.

Kubota, K., Goto, S., Eto, K., & Hayashi, S. (2000). EGF receptor attenuates Dpp signaling and helps to distinguish the wing and leg cell fates in Drosophila. *Development*, *127*(17), 3769–3776.

Landsberg, K. P., Farhadifar, R., Ranft, J., Umetsu, D., Widmann, T. J., Bittig, T., et al. (2009). Increased cell bond tension governs cell sorting at the Drosophila anteroposterior compartment boundary. *Current Biology*, *19*(22), 1950–1955.

Larroux, C., Luke, G. N., Koopman, P., Rokhsar, D. S., Shimeld, S. M., & Degnan, B. M. (2008). Genesis and expansion of metazoan transcription factor gene classes. *Molecular Biology and Evolution*, *25*(5), 980–996.

Lawrence, P. A., Casal, J., & Struhl, G. (1999). The hedgehog morphogen and gradients of cell affinity in the abdomen of Drosophila. *Development*, *126*(11), 2441–2449.

Lecuit, T., Brook, W. J., Ng, M., Calleja, M., Sun, H., & Cohen, S. M. (1996). Two distinct mechanisms for long-range patterning by Decapentaplegic in the Drosophila wing. *Nature*, *381*, 387–393.

Lecuit, T., & Cohen, S. M. (1997). Proximal-distal axis formation in the Drosophila leg. *Nature*, *388*(6638), 139–145.

Lecuit, T., & Cohen, S. M. (1998). Dpp receptor levels contribute to shaping the Dpp morphogen gradient in the Drosophila wing imaginal disc. *Development*, *125*(24), 4901–4907.

Lemons, D., Fritzenwanker, J. H., Gerhart, J., Lowe, C. J., & McGinnis, W. (2010). Co-option of an anteroposterior head axis patterning system for proximodistal patterning of appendages in early bilaterian evolution. *Developmental Biology*, *344*(1), 358–362.

Lewis, E. B. (1978). A gene complex controlling segmentation in Drosophila. *Nature*, *276*(5688), 565–570.

Liu, Q. X., Hiramoto, M., Ueda, H., Gojobori, T., Hiromi, Y., & Hirose, S. (2009). Midline governs axon pathfinding by coordinating expression of two major guidance systems. *Genes & Development*, *23*(10), 1165–1170.

Lunde, K., Trimble, J. L., Guichard, A., Guss, K. A., Nauber, U., & Bier, E. (2003). Activation of the knirps locus links patterning to morphogenesis of the second wing vein in Drosophila. *Development*, *130*(2), 235–248.

Makhijani, K., Kalyani, C., Srividya, T., & Shashidhara, L. S. (2007). Modulation of Decapentaplegic gradient during haltere specification in Drosophila. *Developmental Biology*, *302*(1), 243–255.

Mann, R. S., & Casares, F. (2002). Developmental biology: Signalling legacies. *Nature*, *418*(6899), 737–739.

Martin, P. (1982). Direct determination of the growth rate of Drosophila imaginal discs. *The Journal of Experimental Zoology*, *222*, 97–102.

Martinelli, C., & Spring, J. (2005). T-box and homeobox genes from the ctenophore Pleurobrachia pileus: Comparison of Brachyury, Tbx2/3 and Tlx in basal metazoans and bilaterians. *FEBS Letters*, *579*(22), 5024–5028.

Maves, L., & Schubiger, G. (1998). A molecular basis for transdetermination in Drosophila imaginal discs: Interactions between wingless and decapentaplegic signaling. *Development*, *125*(1), 115–124.

Mayer, L. R., Diegelmann, S., Abassi, Y., Eichinger, F., & Pflugfelder, G. O. (2013). Enhancer trap infidelity in Drosophila optomotor-blind. *Fly (Austin)*, 7(2), 1–11.

Meinhardt, H. (1983). A boundary model for pattern formation in vertebrate limbs. *Journal of Embryology and Experimental Morphology*, *76*, 115–137.

Meins, M., Henderson, D. J., Bhattacharya, S. S., & Sowden, J. C. (2000). Characterization of the human TBX20 gene, a new member of the T-Box gene family closely related to the Drosophila H15 gene. *Genomics*, *67*(3), 317–332.

Meise, M., & Janning, W. (1994). Localization of thoracic imaginal-disc precursor cells in the early embryo of Drosophila melanogaster. *Mechanisms of Development*, *48*(2), 109–117.

Milan, M., Campuzano, S., & Garcia-Bellido, A. (1996). Cell cycling and patterned cell proliferation in the wing primordium of Drosophila. *Proceedings of the National Academy of Sciences of the United States of America*, *93*(2), 640–645.

Milan, M., & Cohen, S. M. (1999). Notch signaling is not sufficient to define the affinity boundary between dorsal and ventral compartments. *Molecular Cell*, *4*(6), 1073–1078.

Milan, M., Perez, L., & Cohen, S. M. (2002). Short-range cell interactions and cell survival in the Drosophila wing. *Developmental Cell*, *2*(6), 797–805.

Milan, M., Weihe, U., Perez, L., & Cohen, S. M. (2001). The LRR proteins capricious and Tartan mediate cell interactions during DV boundary formation in the Drosophila wing. *Cell*, *106*(6), 785–794.

Milner, M. J., Bleasby, A., & Kelly, S. (1984). The role of the peripodial membrane of leg and wing imaginal discs of Drosophila melanogaster during evagination and differentiation in vitro. *Roux's Archives of Developmental Biology*, *193*, 180–186.

Minami, M., Kinoshita, N., Kamoshida, Y., Tanimoto, H., & Tabata, T. (1999). Brinker is a target of Dpp in Drosophila that negatively regulates Dpp-dependent genes. *Nature*, *398*(6724), 242–246.

Minelli, A. (2003). The origin and evolution of appendages. *The International Journal of Developmental Biology*, *47*(7–8), 573–581.

Mohit, P., Bajpai, R., & Shashidhara, L. S. (2003). Regulation of Wingless and Vestigial expression in wing and haltere discs of Drosophila. *Development*, *130*(8), 1537–1547.

Mohit, P., Makhijani, K., Madhavi, M. B., Bharathi, V., Lal, A., Sirdesai, G., et al. (2006). Modulation of AP and DV signaling pathways by the homeotic gene Ultrabithorax during haltere development in Drosophila. *Developmental Biology*, *291*(2), 356–367.

Montagne, J., Groppe, J., Guillemin, K., Krasnow, M. A., Gehring, W. J., & Affolter, M. (1996). The Drosophila Serum Response Factor gene is required for the formation of intervein tissue of the wing and is allelic to blistered. *Development*, *122*(9), 2589–2597.

Morata, G. (2001). How Drosophila appendages develop. *Nature Reviews. Molecular Cell Biology*, *2*(2), 89–97.

Morata, G., Shlevkov, E., & Perez-Garijo, A. (2011). Mitogenic signaling from apoptotic cells in Drosophila. *Development, Growth & Differentiation*, *53*(2), 168–176.

Morgan, T. H., & Bridges, C. B. (1916). *Sex-linked inheritance in Drosophila*. Carnegie Institution of Washington. Publication no. 237, 1–88.

Morimura, S., Maves, L., Chen, Y., & Hoffmann, F. M. (1996). Decapentaplegic overexpression affects Drosophila wing and leg imaginal disc development and wingless expression. *Developmental Biology*, *177*(1), 136–151.

Moser, M., & Campbell, G. (2005). Generating and interpreting the Brinker gradient in the Drosophila wing. *Developmental Biology, 286*(2), 647–658.

Mullor, J. L., Calleja, M., Capdevila, J., & Guerrero, I. (1997). Hedgehog activity, independent of Decapentaplegic, participates in wing disc patterning. *Development, 124*, 1227–1237.

Naiche, L. A., Harrelson, Z., Kelly, R. G., & Papaioannou, V. E. (2005). T-box genes in vertebrate development. *Annual Review of Genetics, 39*, 219–239.

Najand, N., Ryu, J. R., & Brook, W. J. (2012). In vitro site selection of a consensus binding site for the Drosophila melanogaster Tbx20 homolog midline. *PLoS One*, 7(10), e48176.

Nellen, D., Burke, R., Struhl, G., & Basler, K. (1996). Direct and long-range action of a DPP morphogen gradient. *Cell, 85*(3), 357–368.

Neumann, C. J., & Cohen, S. M. (1996a). Distinct mitogenic and cell fate specification functions of wingless in different regions of the wing. *Development, 122*(6), 1781–1789.

Neumann, C. J., & Cohen, S. M. (1996b). A hierarchy of cross-regulation involving Notch, wingless, vestigial and cut organizes the dorsal/ventral axis of the Drosophila wing. *Development, 122*, 3477–3485.

Ng, M., Diaz-Benjumea, F. J., Vincent, J. P., Wu, J., & Cohen, S. M. (1996). Specification of the wing by localized expression of wingless protein. *Nature, 381*(6580), 316–318.

Packham, E. A., & Brook, J. D. (2003). T-box genes in human disorders. T-box genes in human disorders. *Human Molecular Genetics, 12*(Spec No 1), R37–R44.

Pallavi, S. K., Kannan, R., & Shashidhara, L. S. (2006). Negative regulation of Egfr/Ras pathway by Ultrabithorax during haltere development in Drosophila. *Developmental Biology, 296*(2), 340–352.

Pallavi, S. K., & Shashidhara, L. S. (2003). Egfr/Ras pathway mediates interactions between peripodial and disc proper cells in Drosophila wing discs. *Development, 130*(20), 4931–4941.

Panganiban, G., Irvine, S. M., Lowe, C., Roehl, H., Corley, L. S., Sherbon, B., et al. (1997). The origin and evolution of animal appendages. *Proceedings of the National Academy of Sciences of the United States of America, 94*(10), 5162–5166.

Pavlopoulos, A., & Akam, M. (2011). Hox gene Ultrabithorax regulates distinct sets of target genes at successive stages of Drosophila haltere morphogenesis. *Proceedings of the National Academy of Sciences of the United States of America, 108*(7), 2855–2860.

Pflugfelder, G. O. (2009). omb and circumstance. *Journal of Neurogenetics, 23*(1), 15–33.

Pflugfelder, G. O., & Heisenberg, M. (1995). *Optomotor-blind* of *Drosophila melanogaster*: A neurogenetic approach to optic lobe development and optomotor behaviour. *Comparative Biochemistry and Physiology, 110A*, 185–202.

Pflugfelder, G. O., Schwarz, H., Roth, H., Poeck, B., Sigl, A., Kerscher, S., et al. (1990). Genetic and molecular characterization of the optomotor-blind gene locus in Drosophila melanogaster. *Genetics, 126*, 91–104.

Phillips, R. G., & Whittle, J. R. (1993). Wingless expression mediates determination of peripheral nervous system elements in late stages of Drosophila wing disc development. *Development, 118*(2), 427–438.

Podos, S. D., & Ferguson, E. L. (1999). Morphogen gradients. New insights from DPP. *Trends in Genetics, 15*, 396–402.

Porsch, M., Sauer, M., Schulze, S., Bahlo, A., Roth, M., & Pflugfelder, G. O. (2005). The relative role of the T-domain and flanking sequences for developmental control and transcriptional regulation in protein chimeras of Drosophila OMB and ORG-1. *Mechanisms of Development, 122*(1), 81–96.

Prpic, N. M., Janssen, R., Damen, W. G., & Tautz, D. (2005). Evolution of dorsal-ventral axis formation in arthropod appendages: H15 and optomotor-blind/bifid-type T-box genes in the millipede Glomeris marginata (Myriapoda: Diplopoda). *Evolution & Development*, 7(1), 51–57.

Prpic, N. M., Janssen, R., Wigand, B., Klingler, M., & Damen, W. G. (2003). Gene expression in spider appendages reveals reversal of exd/hth spatial specificity, altered leg gap gene dynamics, and suggests divergent distal morphogen signaling. *Developmental Biology, 264*(1), 119–140.

Pueyo, J. I., & Couso, J. P. (2005). Parallels between the proximal-distal development of vertebrate and arthropod appendages: Homology without an ancestor? *Current Opinion in Genetics & Development, 15*(4), 439–446.

Randsholt, N. B., Maschat, F., & Santamaria, P. (2000). Polyhomeotic controls engrailed expression and the hedgehog signaling pathway in imaginal discs. *Mechanisms of Development, 95*(1–2), 89–99.

Reim, I., & Frasch, M. (2005). The Dorsocross T-box genes are key components of the regulatory network controlling early cardiogenesis in Drosophila. *Development, 132*(22), 4911–4925.

Reim, I., Lee, H. H., & Frasch, M. (2003). The T-box-encoding Dorsocross genes function in amnioserosa development and the patterning of the dorsolateral germ band downstream of Dpp. *Development, 130*(14), 3187–3204.

Reim, I., Mohler, J. P., & Frasch, M. (2005). Tbx20-related genes, mid and H15, are required for tinman expression, proper patterning, and normal differentiation of cardioblasts in Drosophila. *Mechanisms of Development, 122*(9), 1056–1069.

Resino, J., Salama-Cohen, P., & Garcia-Bellido, A. (2002). Determining the role of patterned cell proliferation in the shape and size of the Drosophila wing. *Proceedings of the National Academy of Sciences of the United States of America, 99*(11), 7502–7507.

Roch, F., & Akam, M. (2000). Ultrabithorax and the control of cell morphology in Drosophila halteres. *Development, 127*(1), 97–107.

Rodriguez, I., & Basler, K. (1997). Control of compartmental affinity boundaries by Hedgehog. *Nature, 389*, 614–618.

Rogulja, D., & Irvine, K. D. (2005). Regulation of cell proliferation by a morphogen gradient. *Cell, 123*(3), 449–461.

Rudolf, K., Umetsu, D., Aliee, M., Sui, L., Julicher, F., & Dahmann, C. (2015). A local difference in Hedgehog signal transduction increases mechanical cell bond tension and biases cell intercalations along the Drosophila anteroposterior compartment boundary. *Development, 142*(22), 3845–3858.

Ruvinsky, I., & Silver, L. M. (1997). Newly identified paralogous groups on mouse chromosomes 5 and 11 reveal the age of a T-box cluster duplication. *Genomics, 40*, 262–266.

Sander, V., Eivers, E., Choi, R. H., & De Robertis, E. M. (2010). Drosophila Smad2 opposes Mad signaling during wing vein development. *PLoS One, 5*(4). e10383.

Schilling, S., Willecke, M., Aegerter-Wilmsen, T., Cirpka, O. A., Basler, K., & von Mering, C. (2011). Cell-sorting at the A/P boundary in the Drosophila wing primordium: A computational model to consolidate observed non-local effects of Hh signaling. *PLoS Computational Biology*, 7(4), e1002025.

Sebe-Pedros, A., Ariza-Cosano, A., Weirauch, M. T., Leininger, S., Yang, A., Torruella, G., et al. (2013). Early evolution of the T-box transcription factor family. *Proceedings of the National Academy of Sciences of the United States of America, 110*(40), 16050–16055.

Sebe-Pedrós, A., & Ruiz-Trillo, I. (2017). Evolution and classification of the T-Box transcription factor family. *Current Topics in Developmental Biology, 122*, pp. 1–26.

Sen, A., Grimm, S., Hofmeyer, K., & Pflugfelder, G. O. (2014). Optomotor-blind in the development of the Drosophila HS and VS lobula plate tangential cells. *Journal of Neurogenetics, 28*(3–4), 250–263.

Sharma, R. P., & Chopra, V. L. (1976). Effect of the Wingless (wg1) mutation on wing and haltere development in Drosophila melanogaster. *Developmental Biology, 48*(2), 461–465.

Shashidhara, L. S., Agrawal, N., Bajpai, R., Bharathi, V., & Sinha, P. (1999). Negative regulation of dorsoventral signaling by the homeotic gene Ultrabithorax during haltere development in Drosophila. *Developmental Biology, 212*(2), 491–502.

Sheeba, C. J., & Logan, M. P. O. (2017). The roles of T-Box genes in vertebrate limb development. *Current Topics in Developmental Biology, 122*, 355–381.
Shen, J., & Dahmann, C. (2005a). Extrusion of cells with inappropriate Dpp signaling from Drosophila wing disc epithelia. *Science, 307*(5716), 1789–1790.
Shen, J., & Dahmann, C. (2005b). The role of Dpp signaling in maintaining the Drosophila anteroposterior compartment boundary. *Developmental Biology, 279*(1), 31–43.
Shen, J., Dahmann, C., & Pflugfelder, G. O. (2010). Spatial discontinuity of optomotor-blind expression in the Drosophila wing imaginal disc disrupts epithelial architecture and promotes cell sorting. *BMC Developmental Biology, 10*, 23.
Shen, J., Dorner, C., Bahlo, A., & Pflugfelder, G. O. (2008). Optomotor-blind suppresses instability at the A/P compartment boundary of the Drosophila wing. *Mechanisms of Development, 125*(3–4), 233–246.
Shen, J., Lu, J., Sui, L., Wang, D., Yin, M., Hoffmann, I., et al. (2014). The orthologous Tbx transcription factors Omb and TBX2 induce epithelial cell migration and extrusion in vivo without involvement of matrix metalloproteinases. *Oncotarget, 5*(23), 11998–12015.
Shubin, N., Tabin, C., & Carroll, S. (1977). Fossils, genes and the evolution of animal limbs. *Nature, 388*, 639–649.
Silver, S. J., Hagen, J. W., Okamura, K., Perrimon, N., & Lai, E. C. (2007). Functional screening identifies miR-315 as a potent activator of Wingless signaling. *Proceedings of the National Academy of Sciences of the United States of America, 104*(46), 18151–18156.
Simon, E., & Guerrero, I. (2015). The transcription factor optomotor-blind antagonizes Drosophila haltere growth by repressing decapentaplegic and hedgehog targets. *PLoS One, 10*(3), e0121239.
Sivasankaran, R., Vigano, M. A., Muller, B., Affolter, M., & Basler, K. (2000). Direct transcriptional control of the Dpp target omb by the DNA binding protein Brinker. *The EMBO Journal, 19*(22), 6162–6172.
Soler, C., Daczewska, M., Da Ponte, J. P., Dastugue, B., & Jagla, K. (2004). Coordinated development of muscles and tendons of the Drosophila leg. *Development, 131*(24), 6041–6051.
Song, H., Hasson, P., Paroush, Z., & Courey, A. J. (2004). Groucho oligomerization is required for repression in vivo. *Molecular and Cellular Biology, 24*(10), 4341–4350.
Sotillos, S., & De Celis, J. F. (2005). Interactions between the Notch, EGFR, and decapentaplegic signaling pathways regulate vein differentiation during Drosophila pupal wing development. *Developmental Dynamics, 232*(3), 738–752.
Stathopoulos, A., Van Drenth, M., Erives, A., Markstein, M., & Levine, M. (2002). Whole-genome analysis of dorsal-ventral patterning in the Drosophila embryo. *Cell, 111*(5), 687–701.
Steinberg, M. S. (1963). Reconstruction of tissues by dissociated cells. Some morphogenetic tissue movements and the sorting out of embryonic cells may have a common explanation. *Science, 141*, 401–408.
Strigini, M., & Cohen, S. M. (1997). A Hedgehog activity gradient contributes to AP axial patterning of the Drosophila wing. *Development, 124*(22), 4697–4705.
Strigini, M., & Cohen, S. M. (2000). Wingless gradient formation in the Drosophila wing. *Current Biology, 10*(6), 293–300.
Stultz, B. G., Lee, H., Ramon, K., & Hursh, D. A. (2006). Decapentaplegic head capsule mutations disrupt novel peripodial expression controlling the morphogenesis of the Drosophila ventral head. *Developmental Biology, 296*(2), 329–339.
Sturtevant, M. A., Biehs, B., Marin, E., & Bier, E. (1997). The spalt gene links the A/P compartment boundary to a linear adult structure in the Drosophila wing. *Development, 124*(1), 21–32.

Sturtevant, M. A., Roark, M., & Bier, E. (1993). The Drosophila rhomboid gene mediates the localized formation of wing veins and interacts genetically with components of the EGF-R signaling pathway. *Genes & Development*, 7(6), 961–973.

Sui, L., Pflugfelder, G. O., & Shen, J. (2012). The Dorsocross T-box transcription factors promote tissue morphogenesis in the Drosophila wing imaginal disc. *Development*, *139*(15), 2773–2782.

Svendsen, P. C., Formaz-Preston, A., Leal, S. M., & Brook, W. J. (2009). The Tbx20 homologs midline and H15 specify ventral fate in the Drosophila melanogaster leg. *Development*, *136*(16), 2689–2693.

Svendsen, P. C., Ryu, J. R., & Brook, W. J. (2015). The expression of the T-box selector gene midline in the leg imaginal disc is controlled by both transcriptional regulation and cell lineage. *Biology Open*, *4*(12), 1707–1714.

Teleman, A. A., & Cohen, S. M. (2000). Dpp gradient formation in the Drosophila wing imaginal disc. *Cell*, *103*(6), 971–980.

Tepass, U., Godt, D., & Winklbauer, R. (2002). Cell sorting in animal development: Signalling and adhesive mechanisms in the formation of tissue boundaries. *Current Opinion in Genetics & Development*, *12*(5), 572–582.

Tomancak, P., Beaton, A., Weiszmann, R., Kwan, E., Shu, S., Lewis, S. E., et al. (2002). Systematic determination of patterns of gene expression during Drosophila embryogenesis. *Genome Biology*, *3*(12), 23.

Tomlinson, A. (2003). Patterning the peripheral retina of the fly: Decoding a gradient. *Developmental Cell*, *5*(5), 799–809.

Tripura, C., Chandrika, N. P., Susmitha, V. N., Noselli, S., & Shashidhara, L. S. (2011). Regulation and activity of JNK signaling in the wing disc peripodial membrane during adult morphogenesis in Drosophila. *The International Journal of Developmental Biology*, *55*(6), 583–590.

Tsai, Y. C., Grimm, S., Chao, J. L., Wang, S. C., Hofmeyer, K., Shen, J., et al. (2015). Optomotor-blind negatively regulates Drosophila eye development by blocking Jak/STAT signaling. *PLoS One*, *10*(3), e0120236.

Tsai, S. F., Jang, C. C., Prikhod'ko, G. G., Bessarab, D. A., Tang, C. Y., Pflugfelder, G. O., et al. (1997). Gypsy retrotransposon as a tool for the in vivo analysis of the regulatory region of the optomotor-blind gene in Drosophila. *Proceedings of the National Academy of Sciences of the United States of America*, *94*(8), 3837–3841.

Umemori, M., Takemura, M., Maeda, K., Ohba, K., & Adachi-Yamada, T. (2007). Drosophila T-box transcription factor Optomotor-blind prevents pathological folding and local overgrowth in wing epithelium through confining Hh signal. *Developmental Biology*, *308*(1), 68–81.

Vervoort, M. (2000). Hedgehog and wing development in Drosophila: A morphogen at work? *Bioessays*, *22*(5), 460–468.

Villa-Cuesta, E., & Modolell, J. (2005). Mutual repression between msh and Iro-C is an essential component of the boundary between body wall and wing in Drosophila. *Development*, *132*(18), 4087–4096.

Wang, T. F., Ding, C. N., Wang, G. S., Luo, S. C., Lin, Y. L., Ruan, Y., et al. (2004). Identification of Tbr-1/CASK complex target genes in neurons. *Journal of Neurochemistry*, *91*(6), 1483–1492.

Wang, D., Li, L., Lu, J., Liu, S., & Shen, J. (2016). Complementary expression of optomotor-blind and the Iroquois complex promotes fold formation to separate wing notum and hinge territories. *Developmental Biology*, *416*, 225–234.

Weatherbee, S. D., Halder, G., Kim, J., Hudson, A., & Carroll, S. (1998). Ultrabithorax regulates genes at several levels of the wing-patterning hierarchy to shape the development of the Drosophila haltere. *Genes & Development*, *12*, 1474–1482.

Wehn, A. K., & Chapman, D. L. (2010). Tbx18 and Tbx15 null-like phenotypes in mouse embryos expressing Tbx6 in somitic and lateral plate mesoderm. *Developmental Biology*, *347*(2), 404–413.
White, R. A., & Wilcox, M. (1984). Protein products of the bithorax complex in Drosophila. *Cell*, *39*(1), 163–171.
Widmann, T. J., & Dahmann, C. (2009). Wingless signaling and the control of cell shape in Drosophila wing imaginal discs. *Developmental Biology*, *334*(1), 161–173.
Wilder, E. L., & Perrimon, N. (1995). Dual functions of wingless in the Drosophila leg imaginal disc. *Development*, *121*(2), 477–488.
Williams, J. A., Bell, J. B., & Carroll, S. B. (1991). Control of Drosophila wing and haltere development by the nuclear vestigial gene product. *Genes & Development*, *5*(12B), 2481–2495.
Williams, J. A., Paddock, S. W., & Carroll, S. B. (1993). Pattern formation in a secondary field: A hierarchy of regulatory genes subdivides the developing Drosophila wing disc into discrete subregions. *Development*, *117*, 571–584.
Winter, S. E., & Campbell, G. (2004). Repression of Dpp targets in the Drosophila wing by Brinker. *Development*, *131*(24), 6071–6081.
Wu, J., & Cohen, S. M. (1999). Proximodistal axis formation in the Drosophila leg: Subdivision into proximal and distal domains by Homothorax and Distal-less. *Development*, *126*, 109–117.
Yu, K., Sturtevant, M. A., Biehs, B., Francois, V., Padgett, R. W., Blackman, R. K., et al. (1996). The Drosophila decapentaplegic and short gastrulation genes function antagonistically during adult wing vein development. *Development*, *122*, 4033–4044.
Zecca, M., Basler, K., & Struhl, G. (1996). Direct and long-range action of a wingless morphogen gradient. *Cell*, *87*(5), 833–844.
Zhang, X., Luo, D., Pflugfelder, G. O., & Shen, J. (2013). Dpp signaling inhibits proliferation in the Drosophila wing by Omb-dependent regional control of bantam. *Development*, *140*(14), 2917–2922.

CHAPTER TWELVE

The Roles of T-Box Genes in Vertebrate Limb Development

C.J. Sheeba, M.P.O. Logan[1]
King's College London, London, United Kingdom
[1]Corresponding author: e-mail address: malcolm.logan@kcl.ac.uk

Contents

1. Introduction 356
2. The *Tbx2/3/4/5* Subfamily Members Are Significant Players in Limb Development 357
3. The Paralogous T-Box Genes, *Tbx2* and *Tbx3* Have Distinct Functions During Limb Development 361
4. *Tbx4* and *Tbx5*, The Markers of the Hindlimb and Forelimb 367
   4.1 Regulation of *Tbx4* and *Tbx5* Expression and the Molecular Mechanism Underlying Limb Initiation 370
   4.2 *Tbx5* and *Tbx4* in Soft Tissue Patterning 371
5. Tbx1 Family Members Have Functions in Limb Chondrogenic Precursors 372
6. *Brachyury* (*T*) Has a Role in AER Maturation 373
7. *Eomes* (*Tbr2*), a Marker for Digit 4 374
8. Future Perspectives 374
References 375

Abstract

Members of the T-box gene family have diverse roles during embryogenesis and many play critical roles in the developing limb. This is exemplified by the fact that, in humans, mutations in T-box genes are associated with several congenital syndromes that include limb defects as part of their characteristic spectrum of abnormalities. T-box genes encode for evolutionary conserved transcription factors that include both transcriptional activators and repressors. The hallmark of T-box gene members is the presence of the eponymous DNA-binding T-box domain. There are 17 mammalian T-box genes, which based on the sequence homology of the T-box domain, are grouped into five subfamilies, namely, *T*, *Tbx1*, *Tbx2*, *Tbx6*, and *Tbr1*. At least nine T-box genes are expressed during limb development with distinct and dynamic expression patterns. All four members of Tbx2 subfamily (*Tbx2*, *Tbx*3, *Tbx4*, *Tbx5*) and three members of *Tbx1* (*Tbx1*, *Tbx15*, *Tbx18*), *Brachyury* (*T*) and *Eomes* (*Tbr2*) are expressed in the developing limb.

Current Topics in Developmental Biology, Volume 122
ISSN 0070-2153
http://dx.doi.org/10.1016/bs.ctdb.2016.08.009

ABBREVIATIONS

AER apical ectodermal ridge
AP anterior–posterior
BMP bone morphogenetic protein
DV dorsal ventral
E embryonic day
FGF fibroblast growth factor
LPM lateral plate mesoderm
PD proximal–distal
ZPA zone of polarizing activity

1. INTRODUCTION

In vertebrates overt limb development begins with the formation of a small bud of mesenchymal cells enveloped within an ectodermal sheet, at precise positions along the primary body axis. The relative positions of the limbs are fixed across a range of vertebrate species. The forelimb buds emerge at the level of the cervical–thoracic boundary and the hindlimbs at the lumbar–sacral boundary (Burke, Nelson, Morgan, & Tabin, 1995). This rigid conservation suggests that the mechanisms controlling limb position are tightly linked to patterning of the axial body plan. While the limb skeletal elements, tendon, connective tissues, and most vasculature are derived from the lateral plate mesoderm (LPM), the muscle precursors migrate into the limb from the hypaxial dermomyotome compartment of the somites positioned medial and adjacent to the limb bud. Cooperative feedforward and feedback loops are established as key regulators of the progressive steps of limb initiation, patterning, and outgrowth. Inductive signals from the axial mesoderm initiate *Fgf10* expression in the LPM, which induce *Fgf8* in the adjacent ectoderm. A positive feedback loop established between mesenchymal Fgf10 and the ectodermal Fgf8 is sufficient to initiate and maintain limb bud outgrowth. The limb develops along proximal–distal (PD), anterior–posterior (AP), and dorsal–ventral (DV) axes to attain its form and function. Patterning and growth along these axes are governed by three limb-signaling centers: the apical ectodermal ridge (AER), the zone of polarizing activity (ZPA), and the nonridge ectoderm. Thickening of the ectoderm at the distal tip of the limb creates the AER, which is the primary source of the distalizing fibroblast growth factor (FGF) signals that are crucial to maintain proliferation of limb progenitor cells and establish the limb PD axis that runs from shoulder (proximal) to digits tips (distal). As the limb develops, limb bone elements are laid down in a PD sequence, forming the proximal—stylopod (humerus/femur), middle—zeugopod (radius/ulna and tibia/fibula), and the distal—autopod (metacarpals and phalanges). The ZPA, located at the posterior distal margin of the limb produces the morphogen Sonic Hedgehog (Shh), which is essential for limb AP patterning that in the handplate runs from thumb (anterior) to little finger (posterior). Digit identity (denoted digits 1–5 from anterior to posterior) is a common readout of patterning along this axis. The DV axis of the limb, in the hand runs from the back of the hand to the palm, is specified in part by *Wnt7a* expressed in the dorsal ectoderm and *Lmx1b* in the dorsal mesenchyme. The bone morphogenetic protein (BMP) target *engrailed1*, expressed in the ventral ectoderm restricts *Wnt7a* to the dorsal ectoderm (Sheeba, Andrade, & Palmeirim, 2016). The timing of these key events establishing distinct axial morphologies begins with limb bud initiation. In mouse, limb induction occurs by embryonic day 9 (E9.0), which is followed by limb initiation (around E9–E9.5), outgrowth (E10.0–E11.5), and morphogenesis (from E11.5) events.

In this review, we discuss contemporary knowledge of the roles of T-box genes in limb development and those linked to human congenital diseases with associated limb abnormalities (Table 1). The review is structured around the T-box genes and the subfamilies to which they belong. In each case, we discuss the roles of individual T-box gene family members in the events of limb development.

2. THE *TBX2/3/4/5* SUBFAMILY MEMBERS ARE SIGNIFICANT PLAYERS IN LIMB DEVELOPMENT

The *Tbx2/3/4/5* subfamily (that is also referred to as the *Tbx2* subfamily) (Agulnik et al., 1996; Horton et al., 2008) consists of two linked paralogous pairs of genes, *Tbx2–Tbx4* and *Tbx3–Tbx5*. This cluster of four genes is thought to have derived from a single common ancestral gene through the process of tandem duplication followed by chromosomal duplication in *cis* (Agulnik et al., 1996). The *Tbx2–Tbx4* and *Tbx3–Tbx5* gene pairs have remained physically linked in the genome since before the divergence of protostomes and deuterostomes, suggesting this linkage has been important for the development of most metazoan animals. The precise reasons why this physical linkage has been retained remains unclear but could be the result of the necessity to retain shared regulatory elements in the loci of these genes which, as will be discussed, share common domains of expression.

Table 1 T-Box Genes Expressed in Embryonic Limb

T-Box Subfamily	Gene	Predominant Transcriptional Activity	References	Human Syndrome	References	Limb Mouse Mutants	References
T	T (Brachyury)	Activator	Lolas, Valenzuela, Tjian, and Liu (2014)			Midgestation lethal, disrupted posterior mesoderm and no hindlimb, forelimb-distorted AER	Liu et al. (2003)
Tbx1	Tbx1	Activator	Ataliotis, Ivins, Mohun, and Scambler (2005) and Hu et al. (2004)	DiGeorge[a]	OMIM: 188400; reviewed in Papangeli and Scambler (2013)		
	Tbx15	Repressor	Farin et al. (2007)	Cousin	OMIM: 260660; Cousin et al. (1982) and Lausch et al. (2008)	Tbx15$^{-/-}$-reduced scapula, in double Tbx15$^{-/-}$, Pax3$^{-/-}$—loss-of-posterior scapula	Kuijper et al. (2005), Singh et al. (2005), and Farin, Mansouri, Petry, and Kispert (2008)
	Tbx18	Repressor	Farin et al. (2007)			Tbx18$^{-/-}$no obvious phenotype, in double Tbx18$^{-/-}$, Pax3$^{-/-}$-reduced scapula	Farin et al. (2008)

Tbx2-	Tbx2	Repressor	Paxton, Zhao, Chin, Langner, and Reecy (2002) and Sinha, Abraham, Gronostajski, and Campbell (2000)			Midgestation lethal, bilateral hindlimb specific digit 4 duplication	Farin et al. (2013) and Harrelson et al. (2004)
	Tbx3	Repressor	He, Wen, Campbell, Wu, and Rao (1999), Carlson, Ota, Campbell, and Hurlin (2001), and Kumar et al. (2014)	Ulnar mammary	OMIM: 181450; Bamshad et al. (1997)	Midgestation lethal, both in $Tbx3^{-/-}$ and $Tbx3^{\Delta flox/\Delta flox}$—fore and hind limb defects, in $Tbx3^{\Delta 1-3/\Delta 1-3}$—mutant protein accumulates in cytoplasm, early limb conditional KO-disrupted limb initiation, Tbx3 deletion in posterior limb—digit loss, Tbx3 deletion in anterior limb—preaxial polydactyly	Davenport, Jerome-Majewska, and Papaioannou (2003), Frank, Emchebe, Thomas, and Moon (2013), and Emechebe et al. (2016)

Continued

Table 1 T-Box Genes Expressed in Embryonic Limb—cont'd

T-Box Subfamily	Gene	Predominant Transcriptional Activity	References	Human Syndrome	References	Limb Mouse Mutants	References
	Tbx4	Activator	Ouimette, Jolin, L'Honore, Gifuni, and Drouin (2010)	Small Patella	OMIM: 147891; Bongers et al. (2004)	Normal hindlimb induction and initial patterning but it fails to develop further	Naiche and Papaioannou (2003)
	Tbx5	Activator	Zaragoza et al. (2004)	Holt–Oram	OMIM: 142900; Basson et al. (1997) and Li et al. (1997)	Embryos die perinatally, forelimbs are completely absent	Agarwal et al. (2003), Rallis et al. (2003), and Minguillon, Gibson-Brown, and Logan (2009)
Tbr1	Eomes (Tbr2)	Activator	Bjornson et al. (2005)	Microcephaly[a]	OMIM: 604615; Baala et al. (2007)		

[a]Syndromes not associated with limb abnormalities.

Phylogenetic analyses of the *Tbx2/3/4/5* subfamily have helped illustrate and understand genome duplication events and gene retention that has occurred during evolution (Horton et al., 2008; Minguillon & Logan, 2003).

3. THE PARALOGOUS T-BOX GENES, *TBX2* AND *TBX3* HAVE DISTINCT FUNCTIONS DURING LIMB DEVELOPMENT

Tbx2 and Tbx3 share 90% amino acid similarity in the T-box domain (Agulnik et al., 1996) and they are believed to act predominantly as transcriptional repressors (Carreira, Dexter, Yavuzer, Easty, & Goding, 1998; He et al., 1999; Paxton et al., 2002; Rallis, Del Buono, & Logan, 2005; Sinha et al., 2000). Their expression patterns in the developing forelimb and hindlimbs of amniotes are similar (Fig. 1). As the limb develops, the broad domains of *Tbx2* and *Tbx3* in the LPM become restricted to striking, proximal-to-distal bands of expression in the anterior and posterior limb mesenchyme. Later expression becomes restricted to the interdigital domains (Chapman et al., 1996; Farin et al., 2013; Gibson-Brown et al., 1996; Gibson-Brown, Agulnik, Silver, & Papaioannou, 1998; Logan, Simon, & Tabin, 1998; Nissim, Allard, Bandyopadhyay, Harfe, & Tabin, 2007; Suzuki, Takeuchi, Koshiba-Takeuchi, & Ogura, 2004; Tümpel et al., 2002). *Tbx3* is additionally expressed in the AER (Farin et al., 2013; Gibson-Brown et al., 1996) and it is the only T-box gene described to be expressed in this important signaling center, although the functional significance of its expression in the AER has not been reported. The observation that *Tbx2* and *Tbx3* shares overlapping, or at least partially overlapping expression domains and that *Tbx2* and *Tbx3* are paralogous genes, together suggest they may share common and redundant functions in the limb. While this may be the case in some circumstances, both disease association in humans and analysis of mouse mutants suggest that *Tbx3* is the dominant player in the limb since mutations in this gene produce the clearest disruption of normal limb development.

The first direct evidence for the requirement of *Tbx3* in limb development came from studies that associated mutations in *TBX3* with the autosomal dominant disorder, ulnar-mammary syndrome (UMS; OMIM 181450) (Bamshad et al., 1997). The definitive defects associated with UMS include posterior forelimb defects (ulna and little finger), apocrine and mammary gland hypoplasia, dental and genital abnormalities. Limb defects are limited to the upper limb (arm) with no effects seen in lower limb

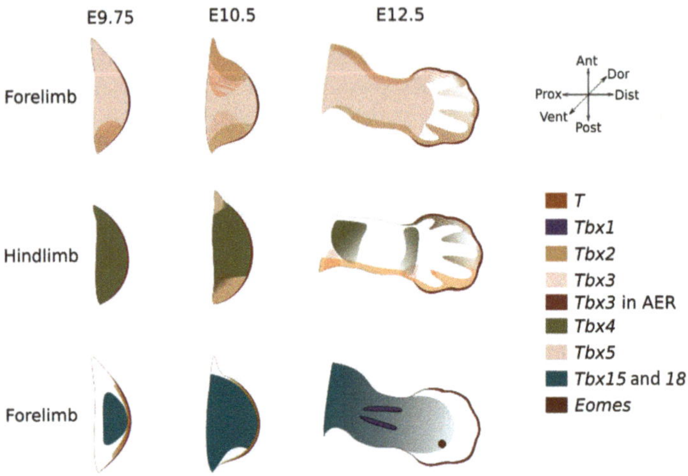

Fig. 1 Schematic summary of the expression domains of *Tbx* genes in the developing mouse limb. *Tbx5* and *Tbx4* are the earliest markers of the forelimb and hindlimb fields, respectively. They are homogenously expressed in the early limb mesenchyme, which gradually reduces from the distal limb mesenchyme as the limb reaches morphogenesis stages. *Tbx2* and *Tbx3* are also expressed from very early stages of limb development, labeling the posterior forelimb and then the anterior and posterior margins of both forelimbs and hindlimbs. In addition, *Tbx3* is also expressed in the apical ectodermal ridge (AER) throughout limb development and it is the only T-box gene to be expressed in the AER. *Tbx15* and *Tbx18* have largely overlapping expression domains mainly occupying the limb core. *T*, *Tbx1*, and *Eomes* have very restricted mesenchymal expression domains in the developing limb: *T* is expressed in the mesenchyme immediately beneath the AER, *Tbx1* in the muscle mass, and *Eomes* in the base of digit 4. This scheme summarizes the expression patterns published in Gibson-Brown et al. (1996), Farin et al. (2013), Agulnik et al. (1996), Kraus, Haenig, and Kispert (2001), Farin et al. (2008), Singh et al. (2005), Liu et al. (2003), Dastjerdi et al. (2007), Hancock, Agulnik, Silver, and Papaioannou (1999), and Russ et al. (2000).

(leg). UMS is thought to arise as a result of haploinsufficiency of *TBX3* caused by deletion, insertion, or point mutations that produce loss-of-function alleles. Genetic studies of *Tbx3* carried out in model organisms have begun to provide an understanding of the regulatory pathways of this gene during limb development and the etiology and pathogenesis of UMS.

The first reported *Tbx3* mutant mouse has phenotypes with some similarities to those seen in UMS. Homozygous mutation of *Tbx3* is embryonic lethal before E16.5, primarily due to degeneration of the yolk sac and is also accompanied by defects in limbs and mammary gland (Davenport et al., 2003). Mutant embryos have reduced forelimb and hindlimb buds, giving rise to a range of posterior skeletal defects at the level of zeugopod (ulna/

fibula) and autopod (digits 4 and 5). Consistent with the posteriorly biased limb anomalies observed in *Tbx3* mutants (Bamshad et al., 1997; Davenport et al., 2003), the AP patterning signaling molecule, Shh, positively regulates the posterior domain of *Tbx3* and this regulation has been linked to an activity in specifying AP digit identity (Tümpel et al., 2002). Accordingly, Tbx3 has been proposed to specify digit 3 in chick hindlimb by cooperatively acting with Bmp signaling and 5′ Hox genes (Suzuki et al., 2004). Similarly in the mouse limb, the posterior domains of *Tbx3* and *Tbx2* are positively regulated by Shh. In the *Shh* null background, *Tbx3* and *Tbx2* posterior expression domains are initiated but not maintained (Galli et al., 2010). Moreover, in the absence of the bHLH transcription factor Hand2 that is expressed in the posterior mesenchyme and functions upstream of *Shh*, the posterior expression domains of *Tbx3* and *Tbx2* are lost (Galli et al., 2010; Osterwalder et al., 2014) (Fig. 2). Hand2 directly interacts with a *cis*-regulatory region that is required *for Tbx3* limb expression, regulating *Tbx3* expression in the posterior and flank mesenchyme during the onset of limb bud development (Osterwalder et al., 2014). Prior to *Shh* expression in the ZPA, Hand2 prepatterns the limb AP axis by repressing the expression of the Gli zinc finger transcription factor, *Gli3* in the posterior limb and restricting it to the anterior limb. *Tbx3* plays an essential role in defining the posterior boundary of *Gli3* expression (Osterwalder et al., 2014). Negative regulation of *Tbx3* on *Gli3* was also demonstrated in chick limb by retrovirus misexpression studies. *Tbx3* can represses *Gli3* in the prospective posterior limb bud and allows expansion of *Hand2* expression, which is necessary to position *Shh* in the ZPA in the posterior limb (Rallis et al., 2005). In agreement with this model, *Hand2* is downregulated in *Tbx3* mutant mice limbs (Davenport et al., 2003; Emechebe et al., 2016). Collectively, these studies place *Tbx3* in a gene regulatory network in the posterior limb that establishes *Shh* expression in the ZPA in the posterior limb and that can help explain the posteriorly biased limb anomalies in UMS.

Tbx3 is also expressed in the anterior limb. By specifically deleting *Tbx3* in the anterior and posterior limb mesenchyme, a recent study has revealed distinct functions for *Tbx3* in these two regions (Emechebe et al., 2016). Deletion of *Tbx3* in the posterior limb caused loss of digit 5 whereas, its absence in the anterior limb mesenchyme results in preaxial polydactyly. In the anterior limb, Tbx3 can form a complex along with Gli3, Kif7 ciliary protein, and Sufu, and this complex is required for the stability of both Gli3—full length activator and Gli3—processed repressor forms (Emechebe et al., 2016).

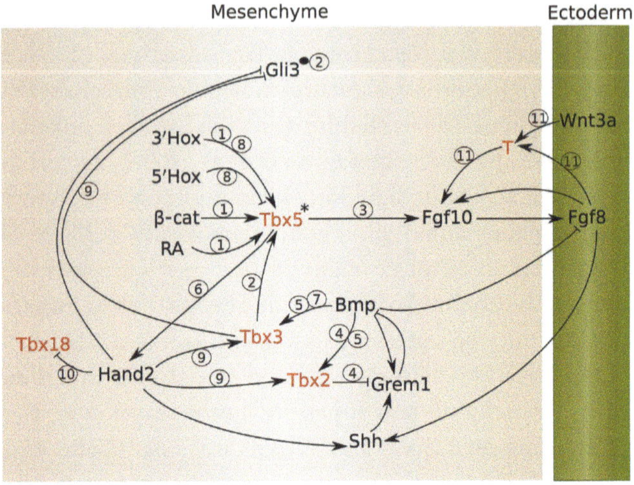

Fig. 2 Schematic summary of known *Tbx* gene interactions in the amniote limb. Retinoic acid (RA), β-catenin/TCF/LEF, and *Hox* genes cooperatively induce *Tbx5* expression in the forelimb lateral plate mesoderm, which then induces *Fgf10* in the mesenchyme (* similarly to *Tbx5* in the forelimb, in the hindlimb *Tbx4* expression is thought to be cooperatively induced by RA, β-catenin/TCF/LEF, and Hox genes, and Tbx4 induces *Fgf10* in the hindlimb mesenchyme). *Fgf10* expression is also upregulated by *Brachyury* (*T*), expressed in the subridge mesenchyme. Fgf10 induces *Fgf8* in the overlying ectoderm (apical ectodermal ridge—AER) and they establish a positive feedback loop that is important for continued limb outgrowth. *Fgf8* and *Wnt3a* expressed in the AER also facilitate *T* expression in the mesenchyme. The anterior and posterior domains of the early limb are prepatterned by mutually exclusive posterior-*Hand2* and anterior-*Gli3*-repressor (Gli3-Rep). *Hand2* activates *Tbx2* and *Tbx3* expression in the posterior limb mesenchyme and *Tbx3* is necessary to establish the posterior boundary of *Gli3*. Along with other factors, Hand2 in the mesenchyme and Fgf8 in the ectoderm, initiate *Shh* expression in posterior mesenchyme (zone of polarizing activity—ZPA). During limb initiation, Bmp signaling initiates *Grem1* expression, which is then maintained by Shh. Grem1 antagonizes Bmp signaling that is otherwise inhibitory to AER-*Fgf8*. This Shh/Grem1/Fgf8 feedback module is required to maintain and propagate ZPA-*Shh* and AER-*Fgf8* expression. BMP signaling also upregulates *Tbx2/3* expression in the limb mesenchyme. By repressing *Grem1*, Tbx2 participates in the termination of ZPA-Shh/AER-FGF signaling and thus limb outgrowth. Tbx3 acts upstream of Tbx5 in the forelimb and is required for normal Hand2 expression levels in the posterior limb. Tbx3 in the anterior limb mesenchyme regulates Gli3 stability and processing (•). This scheme summarizes the interactions reported in the following articles that correspond to the encircled number next to each interaction: ① Nishimoto, Wilde, Wood, and Logan (2015), ② Emechebe et al. (2016), ③ reviewed in Duboc and Logan (2011), ④ Farin et al. (2013), ⑤ Suzuki et al. (2004), ⑥ Agarwal et al. (2003), ⑦ Tümpel et al. (2002), ⑧ Minguillon et al. (2012), ⑨ Osterwalder et al. (2014), ⑩ Galli et al. (2010), and ⑪ Liu et al. (2003).

Another recent study has reported two new mouse *Tbx3* null alleles, a conditional "floxed" allele (*Tbx3Δflox/Δflox*) that does not produce any Tbx3 protein and another allele (*Tbx3Δ1–3/Δ1–3*) in which exons 1–3 are deleted, which produces a truncated Tbx3 protein that accumulates in the cytoplasm instead of its normal nuclear localization (Farin et al., 2013). Both deletions are embryonic lethal when homozygous. The *Tbx3Δflox/Δflox* null allele embryos survive to E13.5, while the *Tbx3Δ1–3/Δ1–3* allele rarely survive beyond E12.5, suggesting the truncated protein product is pathogenic and contributes to the earlier lethality. *Tbx3Δflox/Δflox* embryos have severe posteriorly biased limb abnormalities (absent ulna and digits 4 and 5) and heart phenotypes (Farin et al., 2013). In addition to forelimb defects, the hindlimbs were also extensively affected (Farin et al., 2013). Although the classical clinical features of UMS, as its names suggests, affects the forelimbs only, it is striking that in homozygous mouse mutants the hindlimb is also affected, indicating a role for *Tbx3* in normal hindlimb development. For reasons that remain unclear, in humans the hindlimb is not sensitive to *TBX3* mutations. Comparison of the limb defects that arise in these two mutant alleles has been used to challenge the conventional model that UMS arises through haploinsufficiency of the *TBX3* gene product and it is alternatively proposed that mutations in *TBX3* associated with UMS produce aberrant proteins with altered function. Such alteration in TBX3 protein may not only disrupt its transcriptional activity, but may also alter a proposed pre-mRNA splicing activity (Kumar et al., 2014). In an unbiased proteomic screening to identify Tbx3 interacting proteins, multiple RNA splicing factors, and RNA-binding proteins were identified, revealing a possible function of Tbx3 as splicing regulator (Kumar et al., 2014). Importantly, mutated Tbx3 proteins found in UMS were demonstrated to have different splicing function and even interfered with the endogenous splicing regulatory activity of Tbx3 (Kumar et al., 2014), adding an alternative mechanism to explain at least some of the abnormalities associated with UMS.

Most studies have focused on the roles of *Tbx3* in the limb bud mesenchyme once the limb bud has formed, but *Tbx3* is also expressed in the presumptive limb bud LPM prior to limb bud outgrowth (Emechebe et al., 2016; Rallis et al., 2005). In chick, perturbation of early *Tbx3* expression in the LPM can influence the position of the limb bud along the embryo rostrocaudal axis (Rallis et al., 2005). In mouse, deletion of *Tbx3* at prelimb bud stages results in decreased *Tbx5* expression in the LPM, defective forelimb initiation, and early limb bud morphology, indicating that Tbx3 is acting upstream of Tbx5 and can contribute to the establishment of *Tbx5* expression in the forelimb-forming LPM (Emechebe et al., 2016) (Fig. 2).

Unlike its paralog *Tbx3*, mutations in *Tbx2* have not been reported to be associated with any developmental disorder in humans. *Tbx2* homozygous mutant mouse embryos die by midgestation due to cardiovascular defects, while heterozygous mutants are apparently normal. Although no defects are reported in the forelimbs, the hindlimbs of homozygous embryos have bilateral distal duplication of digit 4 (Farin et al., 2013; Harrelson et al., 2004), consistent with *Tbx2* expression in the posterior of the limb bud, and later in the interdigital region of the autopod between digits 4 and 5 (Farin et al., 2013; Gibson-Brown et al., 1996). Duplication of digit 4 was preceded by a reduction in the number of apoptotic cells in the interdigital mesenchyme between digits 4 and 5 and prolonged expression of *Fgf4*, *Fgf9*, *Fgf17*, and *Shh*, suggesting *Tbx2* is involved in the termination of the epithelial–FGF and mesenchymal–Shh positive feedback loop (Farin et al., 2013). Accordingly, genetic reduction of FGF signaling in *Tbx2* null background reversed polydactyly in *Tbx2* mutants (even to the extent of generating oligodactyly—four digits instead of five), indicating over-activation of the FGF–Shh signaling loop as the primary cause for the polydactyly. Further molecular analysis revealed *Tbx2* as a target of BMP signaling in the limb and Tbx2 terminates the FGF–Shh loop through direct repression of *Gremlin1* in the limb mesenchyme. In contrast to *Tbx2* null mutants, *Tbx2* transgenic lines that ectopically express *Tbx2* in the entire limb mesenchyme produced embryos with oligodactyly and expanded apoptotic mesenchymal zone (Farin et al., 2013). This is also observed in *Gremlin1* mutants (Khokha, Hsu, Brunet, Dionne, & Harland, 2003; Michos et al., 2004), consistent with a model in which Tbx2 inhibition of *Grem1* terminates the *Shh–Gremlin1–FGF* module that controls limb bud outgrowth (Farin et al., 2013) (Fig. 2). Similarly, misexpression studies in the chick suggest a combination of Tbx2 and Tbx3 specify digit 4 identities in chick hindlimb through interactions with *Shh*, *Bmp2*, *Gli3*, and *Gremlin* (Suzuki et al., 2004).

Tbx2 expression in the anterior and posterior limb mesenchyme depends on the presence of the non-AER DV border ectoderm, whereas the AER–FGFs inhibit *Tbx2* expression, explaining the regulation behind its striped pattern (Nissim et al., 2007). While the significance of the anterior *Tbx2* domain is not understood, in the posterior limb, *Tbx2* positively and negatively regulates *Shh* and *Gli3*, respectively, possibly contributing to the establishment of the limb AP axis (Nissim et al., 2007; Suzuki et al.,

2004). *Tbx2* has been implicated in confining *Shh* expression to the posterior limb margin (Nissim et al., 2007).

4. *TBX4* AND *TBX5*, THE MARKERS OF THE HINDLIMB AND FORELIMB

Tbx4 and *Tbx5* represent the remaining paralogous gene pair of the *Tbx2/3/4/5* subfamily. A notable feature of these genes is their striking limb-type restricted expression patterns: *Tbx5* is only expressed in the forelimb bud while *Tbx4* is expressed in the hindlimb bud. Based on the close correlation between their limb-type restricted expression patterns and on results from gene misexpression studies in the chick (Rodriguez-Esteban et al., 1999; Takeuchi et al., 1999) and a study in mouse (Ouimette et al., 2010), *Tbx5* and *Tbx4* have been proposed to determine the morphological differences between forelimbs and hindlimbs, respectively. This model has been challenged by gene deletion–gene replacement approaches in mouse that indicate *Tbx5* and *Tbx4* have common roles in establishing the formation of forelimb and hindlimb buds and do not determine limb-type morphologies (Duboc & Logan, 2011; Minguillon, Del Buono, & Logan, 2005; Minguillon et al., 2009).

The forelimb and hindlimbs are commonly referred to as serially homologous structures as they contain elements with morphological and functional similarities, for example the humerus and femur, radius/ulna, and tibia/fibula, laid out in series from shoulder to digit tip. In addition, patterning of the forelimb and hindlimb structures is achieved in large part by recapitulation of the same gene regulatory networks and signaling pathways. *Tbx5* and *Tbx4* are the earliest markers of the forelimb and hindlimb. From prelimb initiation stages, *Tbx5* and *Tbx4* are expressed in the presumptive forelimb and hindlimb forming regions and this limb-type restricted expression continues during limb outgrowth and morphogenesis stages (Gibson-Brown et al., 1996; Isaac et al., 1998; Logan et al., 1998) (Fig. 1). Mutations in *TBX5* and *TBX4* are associated with dominant congenital disorders and in both cases they are thought to be the result of gene haploinsufficiency. Holt–Oram syndrome (HOS), characterized by heart and upper limb abnormalities, is caused by mutations in *TBX5* (Basson et al., 1997; Li et al., 1997) and Small Patella syndrome, which is associated with dysplasia of patella, pelvis, and foot, is caused by mutations in *TBX4* (Bongers et al., 2004). In both Tbx4 and Tbx5 mutant mice, heterozygous deletion does not produce limb defects failing to recapitulate the haploinsufficiency seen in humans.

Homozygous deletion of *Tbx5* in mouse (Agarwal et al., 2003; Rallis et al., 2003) and zebrafish (Ahn, Kourakis, Rohde, Silver, & Ho, 2002; Garrity, Childs, & Fishman, 2002; Ng et al., 2002), leads to a failure of forelimb formation. In the hindlimb, following deletion of *Tbx4* a smaller, abnormal hindlimb forms (Naiche & Papaioannou, 2003).

Establishing expression of *Fgf10* in the limb bud mesenchyme is essential for limb formation and continued limb outgrowth. In *Fgf10* mutants both forelimb and hindlimb buds fail to form and only residual girdle elements (scapula and pelvis) are present (Min et al., 1998; Sekine et al., 1999). A key activity of *Fgf10* is to activate *Fgf8* expression in the overlying AER and in doing so establishing a positive feedback loop of FGF signaling between the limb mesenchyme and AER that is crucial for continued limb outgrowth. *Tbx5* and *Tbx4* initiate limb bud formation by activating *Fgf10* expression in the limb mesenchyme (Duboc & Logan, 2011; Naiche & Papaioannou, 2003, 2007; Ng et al., 2002; Rallis et al., 2003). Expression of *Fgf10* is never established in the *Tbx5* mutant, which fails to form a forelimb bud (Agarwal et al., 2003; Rallis et al., 2003). Although *Tbx5* input is essential for *Fgf10* expression and subsequent forelimb bud formation, it is only required in a narrow time window during the earliest phases of limb formation. A time course of *Tbx5* conditional deletion demonstrates that limb bud formation is only affected if *Tbx5* is deleted at stages before E10.5 and beyond this stage deletion has no effect on continued *Fgf10* expression, limb outgrowth, and limb skeletal development (Hasson, Del Buono, & Logan, 2007). More recent evidence suggests *Tbx5* and *Fgf10* are involved in triggering an epithelial-to-mesenchymal transition of the somatopleure epithelium from where limb bud mesenchymal progenitors arise at the very earliest steps of limb initiation (Gros & Tabin, 2014). *Fgf10* expression is established in the hindlimbs of *Tbx4* mutant embryos but at low levels. However, it is not maintained (Naiche & Papaioannou, 2003), indicating that *Tbx4* is not exclusively required for *Fgf10* expression in mouse hindlimb mesenchyme as is the case for *Tbx5* in the forelimb. This can explain the ability of a smaller hindlimb to form in the *Tbx4* mutant (Duboc & Logan, 2011; Naiche & Papaioannou, 2003).

The correlation between expression of *Tbx5* and *Tbx4* and limb-type morphologies is demonstrated clearly by experiments in chick embryos using FGF-soaked beads to induce ectopic limbs (Isaac et al., 1998; Logan et al., 1998; Ohuchi et al., 1998). Application of FGF to cells of the interlimb flank can induce the formation of an ectopic limb. The type of limb that is formed is dependent on the location of the source of FGF. Beads placed close to the wing induce ectopic wings, beads close to the leg generate ectopic legs and most remarkably, beads placed in the center of the interlimb flank produce mosaic limbs, with wing elements present in the anterior and hindlimb elements present in the posterior of the same limb. The expression patterns of *Tbx5* and *Tbx4* in ectopic limbs matches precisely with the type of limb that ultimately emerges including limb bud induced in the center of the interlimb flank having *Tbx5* expression only in the anterior portion of the ectopic bud and *Tbx4* only in the posterior portion. Direct evidence for a role of *Tbx5* and *Tbx4* in determining limb-type morphologies has come from misexpression studies in the chick. Ectopic expression of *Tbx5* is reported to partially transform a hindlimb to look more like a forelimb and in the reciprocal experiment *Tbx4* ectopically expressed in the forelimb appears to partially transform the forelimb to look more like a hindlimb (Rodriguez-Esteban et al., 1999; Takeuchi et al., 1999). Genetic approaches in the mouse have challenged the model that *Tbx5* and *Tbx4* can determine forelimb and hindlimb morphology. By exploiting the limb-restricted conditional deletion of *Tbx5* that produces a forelimb-less embryo, various transgenic lines containing *Tbx5*, *Tbx4* and other chimeric fusion forms of these proteins were transgenically delivered to the *Tbx5* mutant forelimb region. A forelimb bud is formed in all cases irrespective of whether *Tbx5*, *Tbx4* or other fusion forms of the proteins are provided. These results are consistent with *Tbx5* and *Tbx4* having equivalent roles in establishing the limb buds but not having roles in determining the limb-type morphologies in the resulting limbs (Duboc & Logan, 2011; Minguillon et al., 2005). This gene-deletion gene-replacement strategy has also been used to test the ability of the preduplication *Tbx4/5* gene from the marine organism, amphioxus. These animals are studied as they provide an extant example of a primitive cephalochordate and have a body plan, with a primitive head and notochord with adjacent paired somites on either side but no limbs, that can provide clues to what the last common ancestor of all vertebrate looked like. In amphioxus, the *Tbx2/3/4/5* family is represented by a single gene pair, *Amphi2/3* and *Amphi4/5*. The *Amphi4/5* gene is capable of rescuing forelimb formation in the *Tbx5* conditional mutant indicating that an ancestral T-box gene from an organism with no limbs can produce a forelimb if activated in the appropriate region of the mouse (Minguillon et al., 2009). These results suggest the type of T-box gene used to produce a limb bud has no influence on the limb-type morphology of limb that ultimately forms. For limb initiation to occur the critical property of the T-box protein is that it is a transcriptional activator capable of binding regulatory elements upstream of *Fgf10* to activate expression.

4.1 Regulation of *Tbx4* and *Tbx5* Expression and the Molecular Mechanism Underlying Limb Initiation

A combination of Hox gene, Wnt/β-catenin and retinoic acid (RA) signaling inputs are necessary to regulate *Tbx5* expression in the limb-forming region (Minguillon et al., 2012; Nishimoto, Minguillon, Wood, & Logan, 2014; Nishimoto et al., 2015) (Fig. 2). *Hox* genes are homeodomain containing transcription factors necessary for proper embryonic development. In the mammalian genome, they are arranged in four clusters, named *HoxA*, *HoxB*, *HoxC*, and *HoxD*, and share redundant functions among the paralogous groups. A remarkable property of Hox genes is that their temporal and spatial expression pattern along the main body axis correlates with their order within the gene cluster, a property termed colinearity (Mallo & Alonso, 2013). A forelimb *cis*-regulatory element of *Tbx5* contains Hox-binding sites that are required for its limb-restricted expression (Minguillon et al., 2012; Nishimoto et al., 2014). The 3′ Hox paralogs (HoxPG4 and HoxPG5) expressed in rostral regions of the LPM that spans the forelimb-forming region directly, positively regulate the onset of *Tbx5* expression in the forelimb region whereas, the 5′ Hox genes (Hoxc8/9/10) expressed in more caudal regions of the LPM including the interlimb and hindlimb regions repress *Tbx5* and define the caudal boundary of *Tbx5* expression (Nishimoto et al., 2014).

RA in axial tissues, lateral to the limb-forming regions is required for limb induction and initiation (Duboc & Logan, 2011; Nishimoto & Logan, 2016). Inhibition of RA signaling in chick, mouse, and zebrafish prevents limb bud formation (Grandel et al., 2002; Mercader, Fischer, & Neumann, 2006; Mic, Sirbu, & Duester, 2004). In mice mutant for RALDH2, an enzyme essential for RA synthesis, the forelimbs do not form and this defect can be rescued by maternal dietary RA (Mic et al., 2004; Zhao et al., 2009). RA acts upstream of *Tbx5* and in the absence of RA signaling *Tbx5* is not expressed. Significantly, RA responsive elements were identified in Tbx5 forelimb regulatory element that are essential for its activity providing evidence for direct regulation of *Tbx5* expression by RA signaling (Nishimoto et al., 2015). RA acts in a feedforward loop with *Tbx5* and acts both upstream of *Tbx5* and is required in parallel with *Tbx5* for the activation of *Fgf10* in the forelimb. An equivalent relationship between RA and *Tbx4* is acting in the hindlimb (Nishimoto & Logan, 2016; Nishimoto et al., 2015).

Wnt/β-catenin signaling has also been implicated in limb initiation, patterning, and outgrowth. While candidate Wnt proteins have been identified in chick and zebrafish (Kawakami et al., 2001; Mercader et al., 2006; Ng et al., 2002), evidence for Wnt ligands expressed appropriately in limb-forming regions in the mouse remains elusive however, β-catenin conditional KOs do not form hindlimbs (Kawakami et al., 2011). In zebrafish, Wnt2b acts upstream of *Tbx5* during limb induction (Ng et al., 2002). Conflicting data in chick suggest Wnt signaling is downstream of Tbx5 during limb initiation (Takeuchi et al., 2003). Significantly, a TCF/LEF-binding site is present in the Tbx5 forelimb regulatory element and is required for its activity (Nishimoto et al., 2015).

Thus, by directly binding to the forelimb enhancer element, RA, β-catenin/TCF/LEF, and Hox genes cooperatively induce *Tbx5* expression in the LPM during forelimb induction stages (Minguillon et al., 2012; Nishimoto et al., 2014, 2015) (Fig. 2).

4.2 *Tbx5* and *Tbx4* in Soft Tissue Patterning

Deletion of *Tbx5* and *Tbx4* in the forelimb and hindlimb produces soft tissue anomalies in muscles and tendons, although continued outgrowth of the limb and formation of skeletal elements is unaffected (Hasson et al., 2010). HOS caused by mutations in human *TBX5* also display limb soft tissue abnormalities that are not associated with skeletal defects (Newbury-Ecob, Leanage, Raeburn, & Young, 1996; Spranger et al., 1997). Deletion of *Tbx5* and *Tbx4* between E9.5 and E10.5 disrupted normal muscle splitting patterns, muscle size, and the sites of individual muscle origins and insertions (Hasson et al., 2010).

Classical embryological experiments performed in chick have provided support for extrinsic cues from the surrounding muscle connective tissues (MCT) in patterning the myoblast progenitors that invaded the limb from the hypaxial compartment of nearby somites (Chevallier & Kieny, 1982; Christ, Jacob, & Jacob, 1977a, 1977b; Kardon, Harfe, & Tabin, 2003). *Tbx5* is expressed in MCT cells surrounding the muscle progenitors and in *Tbx5* mutants, the organization of MCT is affected (Hasson et al., 2010). While deletion of mesenchymal *Tbx5* in later stages of limb development resulted in defective muscle development, ablation of *Tbx5* specifically in muscle progenitors failed to cause muscle-patterning defects, revealing a noncell autonomous role of *Tbx5* in the MCT on forelimb muscle morphogenesis (Hasson et al., 2010). Equivalent late-stage deletion of *Tbx4* also produced muscle-mispatterning phenotypes in the hindlimb (Hasson et al., 2010), suggesting a similar noncell autonomous role for

Tbx4 in hindlimb muscle morphogenesis. *Tbx5* and *Tbx4* are regulating limb muscle patterning in part by controlling the expression of cell adhesion molecules, β-catenin, and N-cadherin that are required for the integrity of the MCT (Hasson et al., 2010). Interestingly, this regulation is occurring at the posttranscriptional level as levels of the proteins are seen to be reduced while levels of transcripts appear unaffected.

5. TBX1 FAMILY MEMBERS HAVE FUNCTIONS IN LIMB CHONDROGENIC PRECURSORS

Members of the Tbx1 family, *Tbx1*, *Tbx15*, and *Tbx18*, are expressed in the developing limb bud. *Tbx1* is expressed in the dorsal and ventral limb muscle masses during embryogenesis (Dastjerdi et al., 2007; Garg et al., 2001) and in adult life (de Wilde et al., 2010). Mutations in *TBX1* have been associated with DiGeorge syndrome (OMIM 188400) (Chieffo et al., 1997; Yagi et al., 2003), which is characterized by pharyngeal and heart abnormalities and cognitive impairment. There are no profound limb defects associated with DiGeorge syndrome; however, patients have been reported to suffer from leg pains and deformities such as flat feet and an inability to flex the foot (Dastjerdi et al., 2007). Homozygous deletion of *Tbx1* in mouse does not produce any reported limb defects (Dastjerdi et al., 2007). Ectopic expression and downregulation of *Tbx1* in chick affects the number of fibers formed in the dorsal and ventral muscle masses, without affecting muscle patterning (Dastjerdi et al., 2007). Although *Tbx1* has an important role in pharyngeal and facial development (Jerome & Papaioannou, 2001; Scambler, 2010), there is currently no strong evidence to support a role for Tbx1 in limb development.

Tbx15 and *Tbx18* are closely related transcription repressors that are coexpressed in the limb mesenchyme and play at least partially redundant functions during limb development (Farin et al., 2008; Singh et al., 2005; Tanaka & Tickle, 2004). To date, no human syndromes are reported for *Tbx18* and the limbs of *Tbx18* mutant mice are also apparently normal (Farin et al., 2008). A role for *Tbx18* in endochondral ossification has, however, been recently reported (Haraguchi, Kitazawa, & Kitazawa, 2015). In mouse limbs at E10.5 and E14.5, *Tbx18* is coexpressed with *Sox9* (Haraguchi et al., 2015) an early chondrogenic marker, in chondrogenic progenitors and differentiating chondrocytes. At E12.5, however, Tbx18 protein is excluded from the Sox9-positive cartilage primordia. This transient suppression of *Tbx18* is proposed to trigger the differentiation of mesenchymal cells into hypertrophic chondrocytes (Haraguchi et al., 2015).

Recently, *Tbx18* was reported as a direct transcriptional target of Hand2 in mouse limb (Osterwalder et al., 2014). Chromatin immunoprecipitation studies using a FLAG-tagged *Hand2* knock-in allele has identified binding peaks upstream of *Tbx18* and it has been proposed that Hand2 negatively regulates *Tbx18* expression (Osterwalder et al., 2014).

In humans, homozygous mutation of *TBX15* causes Cousin syndrome, which is characterized by complex craniofacial dysmorphism, hypoplasia of scapula, and pelvis and short stature (Lausch et al., 2008) (OMIM 260660). In mice, homozygous *Tbx15* mutants show an overall reduction in bone size, including the limb bones. The scapula is particularly affected with a hole forming in the center of the scapula blade (Kuijper et al., 2005; Singh et al., 2005), resembling the scapula abnormalities observed in Cousin syndrome. Mutants have reduced muscle mass, which has been attributed to a secondary effect of the skeletal phenotype although, *Tbx15* has been reported to be expressed in the glycolytic myofibres and be involved in the regulation of fiber type and muscle metabolism (Lee et al., 2015).

Analysis of compound mutants of either *Tbx15* or *Tbx18* and *Pax3* has also provided evidence of possible cooperative interaction between the forming limb muscles and skeleton. While deletion of *Tbx18* and *Pax3* individually does not produce scapula defects, *Tbx18/Pax3* compound mutants had a hole in the scapula blade similar to that seen in the single *Tbx15* mutant (Farin et al., 2008) and in *Tbx15/Pax3* double mutants the proximal scapula is almost absent (Farin et al., 2008). *Tbx15* and *Tbx18* are coexpressed in the mesenchymal precursors of the scapula, while *Pax3* is expressed in the myogenic precursor of the limb muscles. It is known that some scapular precursors do have a somatic origin unlike the rest of the appendicular skeleton which is derived from LPM (Huang, Christ, & Patel, 2006; Huang, Zhi, Patel, Wilting, & Christ, 2000) suggesting that exacerbation of *Tbx15/Tbx18* scapular phenotypes by removal of *Pax3* alleles might be a consequence of their cooperative action in the scapula precursor cells present in the somitic mesoderm (Farin et al., 2008).

6. *BRACHYURY* (*T*) HAS A ROLE IN AER MATURATION

Brachyury (*T*), the founding member of the T-box family of transcription factors is expressed in the LPM at limb induction stages and once the limb is formed, it is confined to the subridge mesoderm beneath the AER (Liu et al., 2003). Retroviral misexpression of *T* in the prospective hindlimb produces an expanded AER, both in DV and AP extents, possibly through upregulation of *Fgf10*. This results in anterior digit duplications probably due to a failure of normal reduction of interdigital mesenchyme though apoptosis, which is the consequence of a delay in AER regression. Analysis of limb defects has also been carried out on a mouse *T* mutant bred on a genetic background that extends survival to early limb bud stages, E10.5. The forelimb buds of these *T* mutants are smaller and more rounded than wild type. Molecular analysis of genes important in establishing the AER (e.g., *En1*, *Bmp4*, *Fgf8*, *Wnt7a*) indicates that the AER is induced, although less robustly than normal and AER maturation is disrupted (Liu et al., 2003). Interestingly following deletion of mesodermally expressed *T*, it is an ectodermal structure, the AER that is most profoundly affected. The results of overexpression of *T* in the chick limb and deletion in the mouse suggest that *T* may positively regulate *Fgf10* in the distal limb mesenchyme and in this way influence AER development.

7. *EOMES* (*TBR2*), A MARKER FOR DIGIT 4

Eomesodermin (*Eomes*), also known as *Tbr2*, is a member of the *Tbr1* family and in the limb it is exclusively expressed in the mesenchyme at the base of digit 4, starting from E11.5 onwards (Russ et al., 2000). As a result of this strikingly restricted expression pattern, it has been used as a molecular marker for digit 4 precursors and the loss of *Eomes* expression correlates with the loss of digit 4 (Farin et al., 2013). *Eomes* is required for trophoblast development and mesoderm formation and the mouse *Eomes* mutant embryos arrest soon after implantation (Russ et al., 2000), several days prior to limb formation and therefore it remains unclear what role if any *Eomes* plays in limb development.

8. FUTURE PERSPECTIVES

There is a relative paucity of data on the biochemical properties of T-box genes. Progress has been stymied by practical difficulties in producing soluble proteins for structural and other biochemical studies and by the lack of sufficiently good antibodies for these types of studies. The structures of the DNA-binding domains of T/BRACHYURY, TBX5 and TBX3 have been reported (Coll, Seidman, & Muller, 2002; Muller & Herrmann, 1997;

Stirnimann, Ptchelkine, Grimm, & Muller, 2010) and optimal DNA-binding sites for several T-box genes have been identified by various binding site selection methods but identifying bona fide, functionally relevant T-box-binding sites remains a challenge.

Perhaps partly as a consequence of the practical difficulties in T-box protein studies, relatively few direct targets of T-box genes are known. Some of the factors acting upstream of T-box genes expressed in the limb have been identified but much remains to be done to piece together the networks acting upstream to regulate the distinctive expression patterns of T-box genes in the limb. Given the many and varied roles of T-box genes in limb development, advances in this area will undoubtedly be important to understand how limb development is controlled and the etiology and pathogenesis of diseases affecting normal limb development.

Finally, recent studies on *Tbx3* have uncovered intriguing new roles for this protein in the cilia (Emechebe et al., 2016) suggesting we should view this family of transcription factors in a new light and look beyond the classical model of these proteins binding to regulatory sequences upstream of target genes to activate or repress transcription and consider their potential to regulate cellular events from inside and outside the nucleus.

REFERENCES

Agarwal, P., Wylie, J. N., Galceran, J., Arkhitko, O., Li, C., Deng, C., et al. (2003). Tbx5 is essential for forelimb bud initiation following patterning of the limb field in the mouse embryo. *Development, 130*, 623–633.

Agulnik, S. I., Garvey, N., Hancock, S., Ruvinsky, I., Chapman, D. L., Agulnik, I., et al. (1996). Evolution of mouse T-box genes by tandem duplication and cluster dispersion. *Genetics, 144*, 249–254.

Ahn, D. G., Kourakis, M. J., Rohde, L. A., Silver, L. M., & Ho, R. K. (2002). T-box gene tbx5 is essential for formation of the pectoral limb bud. *Nature, 417*, 754–758.

Ataliotis, P., Ivins, S., Mohun, T. J., & Scambler, P. J. (2005). XTbx1 is a transcriptional activator involved in head and pharyngeal arch development in Xenopus laevis. *Developmental Dynamics, 232*(4), 979–991.

Baala, L., Briault, S., Etchevers, H. C., Laumonnier, F., Natiq, A., Amiel, J., et al. (2007). Homozygous silencing of T-box transcription factor EOMES leads to microcephaly with polymicrogyria and corpus callosum agenesis. *Nature Genetics, 39*(4), 454–456.

Bamshad, M., Lin, R. C., Law, D. J., Watkins, W. C., Krakowiak, P. A., Moore, M. E., et al. (1997). Mutations in human TBX3 alter limb, apocrine and genital development in ulnar-mammary syndrome. *Nature Genetics, 16*, 311–315.

Basson, C. T., Bachinsky, D. R., Lin, R. C., Levi, T., Elkins, J. A., Soults, J., et al. (1997). Mutations in human TBX5 [corrected] cause limb and cardiac malformation in Holt–Oram syndrome. *Nature Genetics, 15*, 30–35.

Bjornson, C. R., Griffin, K. J., Farr, G. H., III, Terashima, A., Himeda, C., Kikuchi, Y., et al. (2005). Eomesodermin is a localized maternal determinant required for endoderm induction in zebrafish. *Developmental Cell, 9*(4), 523–533.

Bongers, E. M., Duijf, P. H., van Beersum, S. E., Schoots, J., Van Kampen, A., Burckhardt, A., et al. (2004). Mutations in the human TBX4 gene cause Small Patella syndrome. *American Journal of Human Genetics*, *74*, 1239–1248.

Burke, A. C., Nelson, C. E., Morgan, B. A., & Tabin, C. (1995). Hox genes and the evolution of vertebrate axial morphology. *Development*, *121*, 333–346.

Carlson, H., Ota, S., Campbell, C. E., & Hurlin, P. J. (2001). A dominant repression domain in Tbx3 mediates transcriptional repression and cell immortalization: Relevance to mutations in TBX3 that cause Ulnar-Mammary syndrome. *Human Molecular Genetics*, *10*(21), 2403–2413.

Carreira, S., Dexter, T. J., Yavuzer, U., Easty, D. J., & Goding, C. R. (1998). Brachyury-related transcription factor Tbx2 and repression of the melanocyte-specific TRP-1 promoter. *Molecular and Cellular Biology*, *18*, 5099–5108.

Chapman, D. L., Garvey, N., Hancock, S., Alexiou, M., Agulnik, S. I., Gibson-Brown, J. J., et al. (1996). Expression of the T-box family genes, Tbx1–Tbx5, during early mouse development. *Developmental Dynamics: An Official Publication of the American Association of Anatomists*, *206*, 379–390.

Chevallier, A., & Kieny, M. (1982). On the role of the connective tissue in the patterning of the chick limb musculature. *Wilhelm Roux's Archives of Developmental Biology*, *191*, 277–280.

Chieffo, C., Garvey, N., Gong, W., Roe, B., Zhang, G., Silver, L., et al. (1997). Isolation and characterization of a gene from the DiGeorge chromosomal region homologous to the mouse Tbx1 gene. *Genomics*, *43*, 267–277.

Christ, B., Jacob, H. J., & Jacob, M. (1977a). Experimental analysis of the origin of the wing musculature in avian embryos. *Anatomy and Embryology*, *150*, 171–186.

Christ, B., Jacob, H. J., & Jacob, M. (1977b). Experimental findings on muscle development in the limbs of the chick embryo. *Verhandlungen der Anatomischen Gesellschaft*, (71 Pt 2), 1231–1237.

Coll, M., Seidman, J. G., & Muller, C. W. (2002). Structure of the DNA-bound T-box domain of human TBX3, a transcription factor responsible for ulnar-mammary syndrome. *Structure*, *10*, 343–356.

Cousin, J., Walbaum, R., Cegarra, P., Huguet, J., Louis, J., Pauli, A., et al. (1982). Familial pelvi-scapulary dysplasia with anomalies of the epiphyses, dwarfism and dysmorphy: A new syndrome? *Archives Françaises de Pédiatrie*, *39*, 173–175.

Dastjerdi, A., Robson, L., Walker, R., Hadley, J., Zhang, Z., Rodriguez-Niedenfuhr, M., et al. (2007). Tbx1 regulation of myogenic differentiation in the limb and cranial mesoderm. *Developmental Dynamics: An Official Publication of the American Association of Anatomists*, *236*, 353–363.

Davenport, T. G., Jerome-Majewska, L. A., & Papaioannou, V. E. (2003). Mammary gland, limb and yolk sac defects in mice lacking Tbx3, the gene mutated in human ulnar mammary syndrome. *Development*, *130*, 2263–2273.

de Wilde, J., Hulshof, M. F., Boekschoten, M. V., de Groot, P., Smit, E., & Mariman, E. C. (2010). The embryonic genes Dkk3, Hoxd8, Hoxd9 and Tbx1 identify muscle types in a diet-independent and fiber-type unrelated way. *BMC Genomics*, *11*, 176.

Duboc, V., & Logan, M. P. (2011). Regulation of limb bud initiation and limb-type morphology. *Developmental Dynamics: An Official Publication of the American Association of Anatomists*, *240*, 1017–1027.

Emechebe, U., Kumar, P. P., Rozenberg, J. M., Moore, B., Firment, A., Mirshahi, T., et al. (2016). T-box3 is a ciliary protein and regulates stability of the Gli3 transcription factor to control digit number. *eLife*, *5*, e07897. http://dx.doi.org/10.7554/eLife.07897.

Farin, H. F., Bussen, M., Schmidt, M. K., Singh, M. K., Schuster-Gossler, K., & Kispert, A. (2007). Transcriptional repression by the T-box proteins Tbx18 and Tbx15 depends on Groucho corepressors. *The Journal of Biological Chemistry*, *282*(35), 25748–25759.

Farin, H. F., Ludtke, T. H., Schmidt, M. K., Placzko, S., Schuster-Gossler, K., Petry, M., et al. (2013). Tbx2 terminates shh/fgf signaling in the developing mouse limb bud by direct repression of gremlin1. *PLoS Genetics*, *9*, e1003467.

Farin, H. F., Mansouri, A., Petry, M., & Kispert, A. (2008). T-box protein Tbx18 interacts with the paired box protein Pax3 in the development of the paraxial mesoderm. *The Journal of Biological Chemistry*, *283*, 25372–25380.

Frank, D. U., Emchebe, U., Thomas, K. R., & Moon, A. M. (2013). Mouse TBX3 mutants suggest novel molecular mechanisms for Ulnar-mammary syndrome. *PLoS ONE*, *8*(7), e67841. http://dx.doi.org/10.1371/journal.pone.0067841.

Galli, A., Robay, D., Osterwalder, M., Bao, X., Benazet, J. D., Tariq, M., et al. (2010). Distinct roles of Hand2 in initiating polarity and posterior Shh expression during the onset of mouse limb bud development. *PLoS Genetics*, *6*, e1000901.

Garg, V., Yamagishi, C., Hu, T., Kathiriya, I. S., Yamagishi, H., & Srivastava, D. (2001). Tbx1, a DiGeorge syndrome candidate gene, is regulated by sonic hedgehog during pharyngeal arch development. *Developmental Biology*, *235*, 62–73.

Garrity, D. M., Childs, S., & Fishman, M. C. (2002). The heartstrings mutation in zebrafish causes heart/fin Tbx5 deficiency syndrome. *Development*, *129*, 4635–4645.

Gibson-Brown, J. J., Agulnik, S. I., Chapman, D. L., Alexiou, M., Garvey, N., Silver, L. M., et al. (1996). Evidence of a role for T-box genes in the evolution of limb morphogenesis and the specification of forelimb/hindlimb identity. *Mechanisms of Development*, *56*, 93–101.

Gibson-Brown, J. J., Agulnik, S. I., Silver, L. M., & Papaioannou, V. E. (1998). Expression of T-box genes Tbx2–Tbx5 during chick organogenesis. *Mechanisms of Development*, *74*, 165–169.

Grandel, H., Lun, K., Rauch, G. J., Rhinn, M., Piotrowski, T., Houart, C., et al. (2002). Retinoic acid signalling in the zebrafish embryo is necessary during pre-segmentation stages to pattern the anterior–posterior axis of the CNS and to induce a pectoral fin bud. *Development*, *129*, 2851–2865.

Gros, J., & Tabin, C. J. (2014). Vertebrate limb bud formation is initiated by localized epithelial-to-mesenchymal transition. *Science*, *343*, 1253–1256.

Hancock, S. N., Agulnik, S. I., Silver, L. M., & Papaioannou, V. E. (1999). Mapping and expression analysis of the mouse ortholog of Xenopus Eomesodermin. *Mechanisms of Development*, *81*, 205–208.

Haraguchi, R., Kitazawa, R., & Kitazawa, S. (2015). Epigenetic regulation of Tbx18 gene expression during endochondral bone formation. *Cell and Tissue Research*, *359*, 503–512.

Harrelson, Z., Kelly, R. G., Goldin, S. N., Gibson-Brown, J. J., Bollag, R. J., Silver, L. M., et al. (2004). Tbx2 is essential for patterning the atrioventricular canal and for morphogenesis of the outflow tract during heart development. *Development*, *131*, 5041–5052.

Hasson, P., Del Buono, J., & Logan, M. P. (2007). Tbx5 is dispensable for forelimb outgrowth. *Development*, *134*, 85–92.

Hasson, P., DeLaurier, A., Bennett, M., Grigorieva, E., Naiche, L. A., Papaioannou, V. E., et al. (2010). Tbx4 and tbx5 acting in connective tissue are required for limb muscle and tendon patterning. *Developmental Cell*, *18*, 148–156.

He, M., Wen, L., Campbell, C. E., Wu, J. Y., & Rao, Y. (1999). Transcription repression by Xenopus ET and its human ortholog TBX3, a gene involved in ulnar-mammary syndrome. *Proceedings of the National Academy of Sciences of the United States of America*, *96*, 10212–10217.

Horton, A. C., Mahadevan, N. R., Minguillon, C., Osoegawa, K., Rokhsar, D. S., Ruvinsky, I., et al. (2008). Conservation of linkage and evolution of developmental function within the Tbx2/3/4/5 subfamily of T-box genes: Implications for the origin of vertebrate limbs. *Development Genes and Evolution*, *218*, 613–628.

Hu, T., Yamagishi, H., Maeda, J., McAnally, J., Yamagishi, C., & Srivastava, D. (2004). Tbx1 regulates fibroblast growth factors in the anterior heart field through a reinforcing autoregulatory loop involving forkhead transcription factors. *Development, 131*(21), 5491–5502.

Huang, R., Christ, B., & Patel, K. (2006). Regulation of scapula development. *Anatomy and Embryology, 211*(Suppl. 1), 65–71.

Huang, R., Zhi, Q., Patel, K., Wilting, J., & Christ, B. (2000). Dual origin and segmental organisation of the avian scapula. *Development, 127*, 3789–3794.

Isaac, A., Rodriguez-Esteban, C., Ryan, A., Altabef, M., Tsukui, T., Patel, K., et al. (1998). Tbx genes and limb identity in chick embryo development. *Development, 125*, 1867–1875.

Jerome, L. A., & Papaioannou, V. E. (2001). DiGeorge syndrome phenotype in mice mutant for the T-box gene, Tbx1. *Nature Genetics, 27*, 286–291.

Kardon, G., Harfe, B. D., & Tabin, C. J. (2003). A Tcf4-positive mesodermal population provides a prepattern for vertebrate limb muscle patterning. *Developmental Cell, 5*, 937–944.

Kawakami, Y., Capdevila, J., Buscher, D., Itoh, T., Rodriguez Esteban, C., & Izpisua Belmonte, J. C. (2001). WNT signals control FGF-dependent limb initiation and AER induction in the chick embryo. *Cell, 104*, 891–900.

Kawakami, Y., Marti, M., Kawakami, H., Itou, J., Quach, T., Johnson, A., et al. (2011). Islet1-mediated activation of the beta-catenin pathway is necessary for hindlimb initiation in mice. *Development, 138*, 4465–4473.

Khokha, M. K., Hsu, D., Brunet, L. J., Dionne, M. S., & Harland, R. M. (2003). Gremlin is the BMP antagonist required for maintenance of Shh and Fgf signals during limb patterning. *Nature Genetics, 34*, 303–307.

Kraus, F., Haenig, B., & Kispert, A. (2001). Cloning and expression analysis of the mouse T-box gene Tbx18. *Mechanisms of Development, 100*, 83–86.

Kuijper, S., Beverdam, A., Kroon, C., Brouwer, A., Candille, S., Barsh, G., et al. (2005). Genetics of shoulder girdle formation: Roles of Tbx15 and aristaless-like genes. *Development, 132*, 1601–1610.

Kumar, P. P., Franklin, S., Emechebe, U., Hu, H., Moore, B., Lehman, C., et al. (2014). TBX3 regulates splicing in vivo: A novel molecular mechanism for Ulnar-mammary syndrome. *PLoS Genetics, 10*, e1004247.

Lausch, E., Hermanns, P., Farin, H. F., Alanay, Y., Unger, S., Nikkel, S., et al. (2008). TBX15 mutations cause craniofacial dysmorphism, hypoplasia of scapula and pelvis, and short stature in Cousin syndrome. *American Journal of Human Genetics, 83*, 649–655.

Lee, K. Y., Singh, M. K., Ussar, S., Wetzel, P., Hirshman, M. F., Goodyear, L. J., et al. (2015). Tbx15 controls skeletal muscle fibre-type determination and muscle metabolism. *Nature Communications, 6*, 8054.

Li, Q. Y., Newbury-Ecob, R. A., Terrett, J. A., Wilson, D. I., Curtis, A. R., Yi, C. H., et al. (1997). Holt-Oram syndrome is caused by mutations in TBX5, a member of the Brachyury (T) gene family. *Nature Genetics, 15*, 21–29.

Liu, C., Nakamura, E., Knezevic, V., Hunter, S., Thompson, K., & Mackem, S. (2003). A role for the mesenchymal T-box gene Brachyury in AER formation during limb development. *Development, 130*, 1327–1337.

Logan, M., Simon, H. G., & Tabin, C. (1998). Differential regulation of T-box and homeobox transcription factors suggests roles in controlling chick limb-type identity. *Development, 125*, 2825–2835.

Lolas, M., Valenzuela, P. D., Tjian, R., & Liu, Z. (2014). Charting Brachyury-mediated developmental pathways during early mouse embryogenesis. *Proceedings of the National Academy of Science of the United States of America, 111*(12), 4478–4483. http://dx.doi.org/10.1073/pnas.1402612111.

Mallo, M., & Alonso, C. R. (2013). The regulation of Hox gene expression during animal development. *Development, 140*, 3951–3963.

Mercader, N., Fischer, S., & Neumann, C. J. (2006). Prdm1 acts downstream of a sequential RA, Wnt and Fgf signaling cascade during zebrafish forelimb induction. *Development, 133*, 2805–2815.

Mic, F. A., Sirbu, I. O., & Duester, G. (2004). Retinoic acid synthesis controlled by Raldh2 is required early for limb bud initiation and then later as a proximodistal signal during apical ectodermal ridge formation. *The Journal of Biological Chemistry, 279*, 26698–26706.

Michos, O., Panman, L., Vintersten, K., Beier, K., Zeller, R., & Zuniga, A. (2004). Gremlin-mediated BMP antagonism induces the epithelial–mesenchymal feedback signaling controlling metanephric kidney and limb organogenesis. *Development, 131*, 3401–3410.

Min, H., Danilenko, D. M., Scully, S. A., Bolon, B., Ring, B. D., Tarpley, J. E., et al. (1998). Fgf-10 is required for both limb and lung development and exhibits striking functional similarity to Drosophila branchless. *Genes & Development, 12*, 3156–3161.

Minguillon, C., Del Buono, J., & Logan, M. P. (2005). Tbx5 and Tbx4 are not sufficient to determine limb-specific morphologies but have common roles in initiating limb outgrowth. *Developmental Cell, 8*, 75–84.

Minguillon, C., Gibson-Brown, J. J., & Logan, M. P. (2009). Tbx4/5 gene duplication and the origin of vertebrate paired appendages. *Proceedings of the National Academy of Sciences of the United States of America, 106*, 21726–21730.

Minguillon, C., & Logan, M. (2003). The comparative genomics of T-box genes. *Briefings in Functional Genomics & Proteomics, 2*, 224–233.

Minguillon, C., Nishimoto, S., Wood, S., Vendrell, E., Gibson-Brown, J. J., & Logan, M. P. (2012). Hox genes regulate the onset of Tbx5 expression in the forelimb. *Development, 139*, 3180–3188.

Muller, C. W., & Herrmann, B. G. (1997). Crystallographic structure of the T domain–DNA complex of the Brachyury transcription factor. *Nature, 389*, 884–888.

Naiche, L. A., & Papaioannou, V. E. (2003). Loss of Tbx4 blocks hindlimb development and affects vascularization and fusion of the allantois. *Development, 130*, 2681–2693.

Naiche, L. A., & Papaioannou, V. E. (2007). Tbx4 is not required for hindlimb identity or post-bud hindlimb outgrowth. *Development, 134*, 93–103.

Newbury-Ecob, R. A., Leanage, R., Raeburn, J. A., & Young, I. D. (1996). Holt–Oram syndrome: A clinical genetic study. *Journal of Medical Genetics, 33*, 300–307.

Ng, J. K., Kawakami, Y., Buscher, D., Raya, A., Itoh, T., Koth, C. M., et al. (2002). The limb identity gene Tbx5 promotes limb initiation by interacting with Wnt2b and Fgf10. *Development, 129*, 5161–5170.

Nishimoto, S., & Logan, M. P. (2016). Subdivision of the lateral plate mesoderm and specification of the forelimb and hindlimb forming domains. *Seminars in Cell & Developmental Biology, 49*, 102–108.

Nishimoto, S., Minguillon, C., Wood, S., & Logan, M. P. (2014). A combination of activation and repression by a colinear Hox code controls forelimb-restricted expression of Tbx5 and reveals Hox protein specificity. *PLoS Genetics, 10*, e1004245.

Nishimoto, S., Wilde, S. M., Wood, S., & Logan, M. P. (2015). RA acts in a coherent feed-forward mechanism with Tbx5 to control limb bud induction and initiation. *Cell Reports, 12*, 879–891.

Nissim, S., Allard, P., Bandyopadhyay, A., Harfe, B. D., & Tabin, C. J. (2007). Characterization of a novel ectodermal signaling center regulating Tbx2 and Shh in the vertebrate limb. *Developmental Biology, 304*, 9–21.

Ohuchi, H., Takeuchi, J., Yoshioka, H., Ishimaru, Y., Ogura, K., Takahashi, N., et al. (1998). Correlation of wing-leg identity in ectopic FGF-induced chimeric limbs with the differential expression of chick Tbx5 and Tbx4. *Development, 125*, 51–60.

Osterwalder, M., Speziale, D., Shoukry, M., Mohan, R., Ivanek, R., Kohler, M., et al. (2014). HAND2 targets define a network of transcriptional regulators that compartmentalize the early limb bud mesenchyme. *Developmental Cell*, *31*, 345–357.
Ouimette, J. F., Jolin, M. L., L'Honore, A., Gifuni, A., & Drouin, J. (2010). Divergent transcriptional activities determine limb identity. *Nature Communications*, *1*, 35.
Papangeli, I., & Scambler, P. (2013). The 22q11 deletion: DiGeorge and velocardialfacial syndromes and the role of TBX1, Wiley Interdisciplinary Reviews. *Developmental Biology*, *2*(3), 393–403. http://dx.doi.org/10.1002/wdev.75.
Paxton, C., Zhao, H., Chin, Y., Langner, K., & Reecy, J. (2002). Murine Tbx2 contains domains that activate and repress gene transcription. *Gene*, *283*, 117–124.
Rallis, C., Bruneau, B. G., Del Buono, J., Seidman, C. E., Seidman, J. G., Nissim, S., et al. (2003). Tbx5 is required for forelimb bud formation and continued outgrowth. *Development*, *130*, 2741–2751.
Rallis, C., Del Buono, J., & Logan, M. P. (2005). Tbx3 can alter limb position along the rostrocaudal axis of the developing embryo. *Development*, *132*, 1961–1970.
Rodriguez-Esteban, C., Tsukui, T., Yonei, S., Magallon, J., Tamura, K., & Izpisua Belmonte, J. C. (1999). The T-box genes Tbx4 and Tbx5 regulate limb outgrowth and identity. *Nature*, *398*, 814–818.
Russ, A. P., Wattler, S., Colledge, W. H., Aparicio, S. A., Carlton, M. B., Pearce, J. J., et al. (2000). Eomesodermin is required for mouse trophoblast development and mesoderm formation. *Nature*, *404*, 95–99.
Scambler, P. J. (2010). 22q11 Deletion syndrome: A role for TBX1 in pharyngeal and cardiovascular development. *Pediatric Cardiology*, *31*, 378–390.
Sekine, K., Ohuchi, H., Fujiwara, M., Yamasaki, M., Yoshizawa, T., Sato, T., et al. (1999). Fgf10 is essential for limb and lung formation. *Nature Genetics*, *21*, 138–141.
Sheeba, C. J., Andrade, R. P., & Palmeirim, I. (2016). Getting a handle on embryo limb development: Molecular interactions driving limb outgrowth and patterning. *Seminars in Cell & Developmental Biology*, *49*, 92–101.
Singh, M. K., Petry, M., Haenig, B., Lescher, B., Leitges, M., & Kispert, A. (2005). The T-box transcription factor Tbx15 is required for skeletal development. *Mechanisms of Development*, *122*, 131–144.
Sinha, S., Abraham, S., Gronostajski, R. M., & Campbell, C. E. (2000). Differential DNA binding and transcription modulation by three T-box proteins, T, TBX1 and TBX2. *Gene*, *258*, 15–29.
Spranger, S., Ulmer, H., Troger, J., Jansen, O., Graf, J., Meinck, H. M., et al. (1997). Muscular involvement in the Holt–Oram syndrome. *Journal of Medical Genetics*, *34*, 978–981.
Stirnimann, C. U., Ptchelkine, D., Grimm, C., & Muller, C. W. (2010). Structural basis of TBX5-DNA recognition: The T-box domain in its DNA-bound and -unbound form. *Journal of Molecular Biology*, *400*, 71–81.
Suzuki, T., Takeuchi, J., Koshiba-Takeuchi, K., & Ogura, T. (2004). Tbx genes specify posterior digit identity through Shh and BMP signaling. *Developmental Cell*, *6*, 43–53.
Takeuchi, J. K., Koshiba-Takeuchi, K., Matsumoto, K., Vogel-Hopker, A., Naitoh-Matsuo, M., Ogura, K., et al. (1999). Tbx5 and Tbx4 genes determine the wing/leg identity of limb buds. *Nature*, *398*, 810–814.
Takeuchi, J. K., Koshiba-Takeuchi, K., Suzuki, T., Kamimura, M., Ogura, K., & Ogura, T. (2003). Tbx5 and Tbx4 trigger limb initiation through activation of the Wnt/Fgf signaling cascade. *Development*, *130*, 2729–2739.
Tanaka, M., & Tickle, C. (2004). Tbx18 and boundary formation in chick somite and wing development. *Developmental Biology*, *268*, 470–480.
Tümpel, S., Sanz-Ezquerro, J. J., Isaac, A., Eblaghie, M. C., Dobson, J., & Tickle, C. (2002). Regulation of Tbx3 expression by anteroposterior signalling in vertebrate limb development. *Developmental Biology*, *250*, 251–262.

Yagi, H., Furutani, Y., Hamada, H., Sasaki, T., Asakawa, S., Minoshima, S., et al. (2003). Role of TBX1 in human del22q11.2 syndrome. *Lancet*, *362*, 1366–1373.
Zaragoza, M. V., Lewis, L. E., Sun, G., Wang, E., Li, L., Said-Salman, I., et al. (2004). Identification of the TBX5 transactivating domain and the nuclear localization signal. *Gene*, *330*, 9–18.
Zhao, X., Sirbu, I. O., Mic, F. A., Molotkova, N., Molotkov, A., & Kumar, S. (2009). Retinoic acid promotes limb induction through effects on body axis extension but is unnecessary for limb patterning. *Current Biology: CB*, *19*, 1050–1057.

CHAPTER THIRTEEN

T-Box Genes in Human Development and Disease

T.K. Ghosh, J.D. Brook[1], A. Wilsdon[1]
School of Life Sciences, Queens Medical Centre, University of Nottingham, Nottingham, United Kingdom
[1]Corresponding authors: e-mail address: david.brook@nottingham.ac.uk; anna.wilsdon@nottingham.ac.uk

Contents

1. Introduction 384
2. *T* (Brachyury)/Chordoma 384
3. *TBX1*/22q11.2 Deletion Syndrome (DiGeorge Syndrome/Velocardiofacial Syndrome) 390
4. *TBX3*/Ulnar–Mammary Syndrome 392
5. *TBX4*/Small Patella Syndrome 394
6. *TBX5*/Holt–Oram Syndrome/CHD 396
7. *TBX6*/Spondylocostal Dysostosis 398
8. *TBX15*/Cousin Syndrome 399
9. *TBX19*/Isolated ACTH Deficiency 400
10. *TBX20*/Congenital Heart Disease 401
11. *TBX22*/Cleft Palate With or Without Ankyloglossia, X-Linked (CPX) 403
12. Concluding Remarks 405
Acknowledgment 405
References 405

Abstract

T-box genes are important development regulators in vertebrates with specific patterns of expression and precise roles during embryogenesis. They encode transcription factors that regulate gene transcription, often in the early stages of development. The hallmark of this family of proteins is the presence of a conserved DNA binding motif, the "T-domain." Mutations in T-box genes can cause developmental disorders in humans, mostly due to functional deficiency of the relevant proteins. Recent studies have also highlighted the role of some T-box genes in cancer and in cardiomyopathy, extending their role in human disease. In this review, we focus on ten T-box genes with a special emphasis on their roles in human disease.

Current Topics in Developmental Biology, Volume 122
ISSN 0070-2153
http://dx.doi.org/10.1016/bs.ctdb.2016.08.006

1. INTRODUCTION

T-box genes encode dosage-sensitive transcription factors that regulate morphogenesis and organogenesis at an early stage of vertebrate development. Brachyury (*T*), the founder member of the T-box gene family, was cloned from the mouse no tail mutant in 1990 (Herrmann, Labeit, Poustka, King, & Lehrach, 1990), and subsequently discoveries of other orthologs and homologs across the species led to the emergence of the family known as the "T-box" family (Bollag et al., 1994). So far, 17 human T-box genes have been identified with diverse roles in development. All T-box family members carry a signature sequence known as the "T-domain," containing approximately 200 amino acids that interact with specific DNA sequences to regulate gene transcription. The T-domain also interacts with a number of other important transcription factors, which are essential for normal embryogenesis.

The majority of mutations identified in T-box genes are missense, which alter protein function (Fig. 1). However, splice-site mutations, deletions, insertions, and copy number variation are not uncommon. Mutations in regulatory regions are rare, almost certainly because less is known about T-box regulatory elements. During development T-box proteins work in combinatorial pathways involving signaling molecules, transcription factors, and cofactors. Mutations in T-box genes often disrupt these processes, leading to anatomical abnormalities. Here we discuss the roles of ten T-box genes in development and the subsequent consequences in humans when gene function is abnormal. Effects are diverse and range from isolated hormone deficiencies to recognizable syndromes affecting both the heart and limbs (Table 1 and Fig. 2).

2. *T* (BRACHYURY)/CHORDOMA

The Brachyury (*T*) gene was originally cloned in the mouse (Herrmann et al., 1990) and subsequently found to encode a novel DNA binding protein (Kispert & Hermann, 1993). It is expressed in the early-stage mesoderm and then restricted to the notochord. Homozygous *T* mutant mice show severe defects of mesoderm-derived structure and axial skeletal defects, due to apoptosis of paraxial mesoderm (Pennimpede et al., 2012; Smith, Price, Green, Weigel, & Herrmann, 1991; Wilkinson et al., 1990). Brachyury induces epithelial–mesenchymal transition. It is

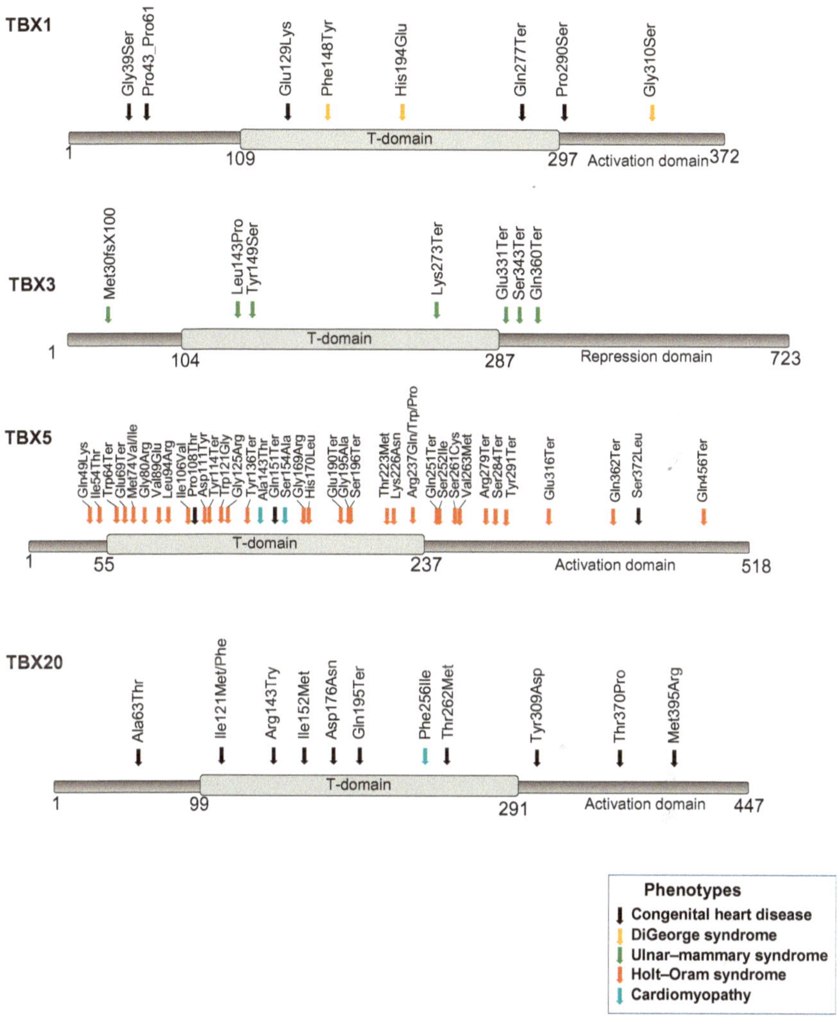

Fig. 1 Mutation spectrum in T-box genes. Missense mutations that are associated with disease phenotypes are mapped in protein domains. The changes in the protein and their associated disease phenotypes are represented in color-coded *arrows*.

overexpressed in many human tumor tissues suggesting an association between Brachyury and tumor progression, which could be potentially targeted for therapeutic use (Fernando et al., 2010; Palena et al., 2014).

A chordoma is a rare bone tumor. Previously, genetic linkage analysis in three multiplex chordoma families linked the disease locus to chromosome 7q33 (Kelley et al., 2001). Subsequently, high-resolution array-comparative

Table 1 Expression Pattern and Knockout Phenotypes of T-Box Genes in Mouse and Their Link in Human Diseases

T-Box Genes	Expression Domains	Diseases	Knockout Phenotypes in Mouse		References
			Homozygous	**Heterozygous**	
T	Early-stage mesoderm, notochord	Chordoma	Lethal, lack notochord, allantois, and primitive streak	Viable, short tail	Wilkinson, Bhatt, and Herrmann (1990), Yang et al. (2009)
TBX1	Embryonic mesoderm, pharyngeal region, and the optic vesicle	22q11.2 deletion syndrome	Embryonic lethal, aortic arch abnormalities, salivary and thymus gland reduced or absent	Viable, cardiac outflow tract anomalies, conotruncal defects	Chapman, Agulnik, Hancock, Silver, and Papaioannou (1996), Chapman, Garvey, et al. (1996), Jerome and Papaioannou (2001), Merscher et al. (2001)
TBX3	Pharyngeal arches, heart, limb, mammary gland	Ulnar–mammary syndrome/breast cancer	Embryonic lethal, yolk sac defects, mammary gland and limb abnormalities	Viable and fertile, abnormal external genitalia	Bamshad et al. (1997), Yarosh et al. (2008), Davenport, Jerome-Majewska, and Papaioannou (2003)
TBX4	Allantois, developing heart, hind limb, genital papilla	Small patella syndrome	Die at mid-gestation, stunted allantoises, apoptosis, defects in anterior/posterior patterning of the hind limb	Viable, reduced allantois growth rate	Bongers et al. (2004), Chapman, Agulnik, et al. (1996), Chapman, Garvey, et al. (1996), Naiche and Papaioannou (2003)

TBX5	Heart, forelimb, lungs, eyes	Holt–Oram syndrome/ congenital heart disease (CHD)/cardiomyopathy	Embryonic lethal, arrested cardiac development	Viable, defects in heart, forelimb, and conduction system	Chapman, Agulnik, et al. (1996), Chapman, Garvey, et al. (1996), Li et al. (1997), Bruneau et al. (2001)
TBX6	Primitive streak, paraxial mesoderm, and tail bud	Spondylocostal dysostosis	Embryonic lethal, somite abnormalities, ectopic neural tubes	Viable and fertile	Chapman and Papaioannou (1998), Sparrow et al. (2013)
TBX15	Craniofacial region, limb buds, branchial arches	Cousin syndrome	Normal, craniofacial abnormalities, skeletal abnormalities	Normal and fertile	Agulnik, Papaioannou, and Silver (1998), Lausch et al. (2008), Singh et al. (2005)
TBX19	Rostral ventral diencephalon and pituitary gland	Isolated ACTH deficiency (IAD)	Viable, low plasma ACTH levels, higher plasma glucose levels	Normal	Lamolet et al. (2001), Pulichino, Vallette-Kasic, Couture, et al. (2003), Pulichino, Vallette-Kasic, Tsai, Couture, Gauthier, Drouin (2003)
TBX20	Extraembryonic mesoderm, heart, eye anlage, motor neurons of hindbrain and spinal chord	Congenital heart disease/cardiomyopathy	Die at mid-gestation, abnormal heart morphogenesis	Viable	Kraus, Haenig, and Kispert (2001), Stennard et al. (2005), Kirk et al. (2007)
TBX22	Palatal shelves, tongue, nasal septum that fuses to the palatal shelves	Cleft palate/ ankyloglossia	Viable, submucous cleft palate and ankyloglossia	Unaffected	Braybrook et al. (2002), Bush, Lan, Maltby, and Jiang (2002), Braybrook et al. (2001), Pauws et al. (2009)

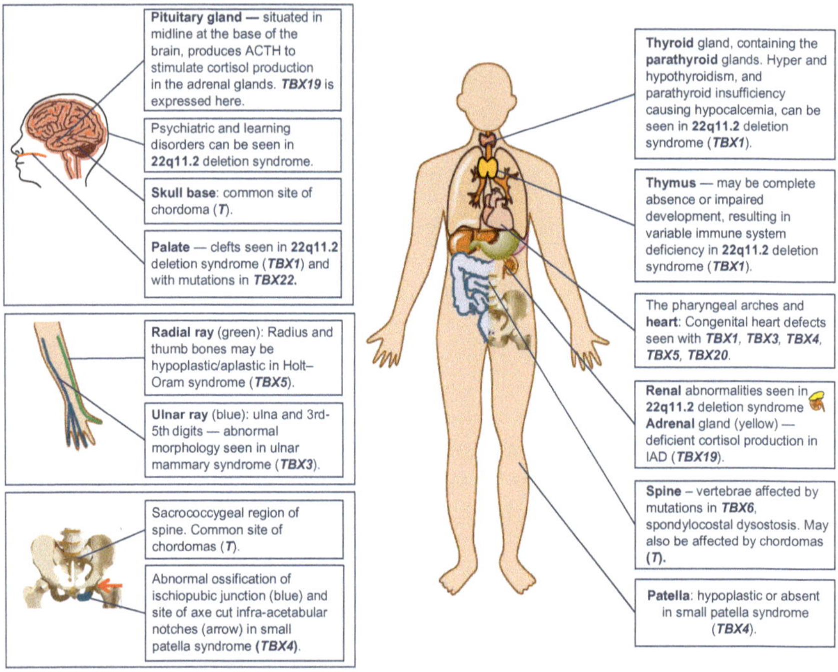

Fig. 2 Representation of some of the clinical features of TBX-related disorders in the human. This can be compared with the expression of each gene as detailed in Table 1. Variable clinical expression and reduced penetrance are seen.

genomic hybridization (array-CGH) studies identified a *T* gene duplication and thus the *T* gene as a susceptibility gene for chordoma (Yang et al., 2009). An association study including 40 individuals with chordoma and 358 ancestry-matched controls identified a common single-nucleotide variant allele (at rs2305089). This caused the nonsynonymous mutation Gly177Asp in *T* (Pillay et al., 2012). A second common variant (rs1056048) was found to be strongly associated with familial chordoma. Another common variant (rs3816300) was also strongly associated when jointly analyzed with rs2305089. The association was significantly stronger in individuals with early-age onset. Three rare variants were also observed among sporadic chordoma occurrences (Kelley et al., 2014). Chordomas have also been seen in individuals with tuberous sclerosis, with germline in combination with somatic mutations in *TSC1* and *TSC2* (Lee-Jones et al., 2004).

Chordomas are believed to originate from notochord remnants. They affect less than 0.1 per 100,000 people (Smoll, Gautschi, Radovanovic, Schaller, & Weber, 2013). They are more frequent in males than females and in Caucasians than African Americans. They tend to occur between the ages of 40 and 60 years old (Wu et al., 2010). Familial chordomas are rare but have been reported (Wang et al., 2015). The tumor is most commonly identified in the sacrum (50%), skull base, and spine (Flanagan & Yamaguchi, 2013). Presentation is related to the site of the tumor and can include neurological findings in the limbs, cauda equina syndrome (compression of the nerves at the base of the spinal cord), diplopia (double vision), and headaches.

The notochord usually disappears by 8 weeks gestation, after formation of the vertebral column has commenced (Stacchiotti, Sommer, & Consensus, 2015). Remnants of the notochord may persist in some, and it is suggested that a benign notochord cell tumor may later develop into a chordoma (Kreshak et al., 2014). Because the benign remnants do not always transform into a chordoma, it is thought that there are other factors necessary to produce a malignant tumor (Choi, Cohn, & Harfe, 2008). Histology tends to show a lobar structure with vacuolated tumor epithelial cells surrounded by fibroblast-like cells. This is illustrated in Fig. 3. Immunohistochemistry for Brachyury can be used if there is doubt over the diagnosis (Miettinen et al., 2015; Vujovic et al., 2006). Chordomas tend to invade local structures but can also metastasize to other sites including the lung, liver, and bones (Young et al., 2015). Treatment is with surgery and consideration of adjuvant radiotherapy. Unfortunately recurrence is seen in most.

Brachyury is also significantly overexpressed in breast and prostate cancer, supporting a role for this gene in cancer (Palena et al., 2014; Pinto et al., 2014).

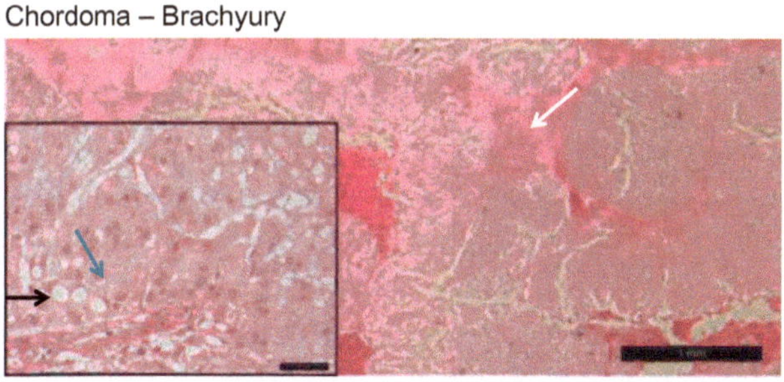

Fig. 3 Hematoxylin & eosin-stained chordoma. Physaliferous cells are identified by the presence of vacuoles within the cytoplasm (*blue arrow*). Signet ring cells (*black arrow*) resemble a signet ring, there is a large vacuole, and the nucleus is pushed to the periphery. The lobules are separated by fibrous septa (*white arrow*). *Black bars* represent 50 μm and 1 mm.

In addition, inconsistent evidence has linked specific *T* alleles and single-nucleotide polymorphisms with a susceptibility to neural tube defects (Jensen et al., 2004; Morrison et al., 1996; Speer et al., 2002).

3. *TBX1*/22Q11.2 DELETION SYNDROME (DIGEORGE SYNDROME/VELOCARDIOFACIAL SYNDROME)

Tbx1 is expressed during early embryogenesis, especially in the lung epithelium, and shows very little overlap with other T-box genes (Chapman, Garvey, et al., 1996). Human *TBX1* was identified and cloned from a skeletal muscle cDNA library. The gene maps to the center of 22q11, the region deleted in 22q11.2 deletion syndrome. Northern blot analysis suggests that human *TBX1* is expressed in adult and fetal tissues in a similar pattern to that found in the mouse (Chieffo et al., 1997). In the pharyngeal arches, *Tbx1* expression overlaps with Sonic hedgehog (*Shh*) and the latter regulates *Tbx1* expression during pharyngeal arch development (Garg et al., 2001). Forkhead proteins also directly regulate expression of *Tbx1* through a single *cis* element upstream of *Tbx1* (Yamagishi et al., 2003). The Tbx1-FGF cascade has been shown to play a pivotal role in early heart development, especially in the outflow tract (Seo & Kume, 2006). Vascular endothelial growth factor (Vegf) interacts with tbx1, and the expression of *Tbx1* is downregulated in *Vegf* mutant mice and zebrafish. Knockdown of *Vegf* significantly worsens the pharyngeal arch artery defects induced by *tbx1* knockdown (Stalmans et al., 2003). *Tbx1*-mediated outflow tract morphogenesis is also controlled by Prdm1, a DNA binding transcriptional repressor (Vincent et al., 2014). In the mouse model of 22q11.2 deletion syndrome, p53 suppression can partially rescue the mutant phenotype, suggesting a strong genetic interaction. TBX1 and p53 do not interact directly, but occupy a genetic element on the *Gbx2* promoter that is required for aortic arch and cardiac outflow tract development (Caprio & Baldini, 2014).

The 22q11.2 deletion syndrome is one of the most well-characterized deletion syndromes. The prevalence is between 1 in 4000 and 1 in 7700 people (Devriendt, Fryns, Mortier, van Thienen, & Keymolen, 1998; Goodship, Cross, LiLing, & Wren, 1998). The phenotype is variable but includes congenital heart disease (CHD), palatal abnormalities and velopharyngeal insufficiency, hypocalcemia, immune dysfunction, learning difficulties, autism, and psychiatric disorders. Penetrance is complete. It is the most common cause of CHD with developmental delay, and the most common cause of syndromic cleft palate (McDonald-McGinn et al., 2015).

A clinical diagnosis is usually made based on a combination of signs and symptoms known to be associated with 22q11.2 deletion syndrome. A facial phenotype may also be recognizable, with hooded eyelids and a pinched nose. Heart abnormalities usually affect the outflow tract and aortic arch, and tetralogy of Fallot is the most common heart defect seen. The severity of the heart defect often determines the individual's prognosis. The phenotypic spectrum is wide and may extend to include ophthalmological, genitourinary, gastrointestinal, skeletal, and cerebral abnormalities. A number of cancers have also been reported including hepatoblastoma, possibly due to hemizygous deletion of *catechol-O-methyltransferase* (McDonald-McGinn et al., 2006).

Treatment is multidisciplinary and lifelong, and is directed by the patient's phenotype. At diagnosis individuals should undergo an evaluation for abnormalities known to be associated with 22q11.2 deletion syndrome. This includes palatal assessment, blood tests including calcium, parathyroid hormone and thyroid function, an electrocardiogram, and echocardiogram to assess the heart, audiometry, ophthalmology, renal, musculoskeletal, and developmental systems reviews. Families should also be offered genetic counseling.

Almost 90% of individuals have a heterozygous microdeletion of 22q11.2 (Packham & Brook, 2003). Diagnosis is usually confirmed by fluorescent in situ hybridization (FISH), but a small number of individuals have smaller microdeletions that may be missed by FISH testing and require the more sensitive array-CGH. The deletion is de novo in 93% of affected individuals. Most commonly there is a 3-Mb deletion containing around 30 genes, with breakpoints at sites of low copy number repeats. Smaller nested deletions are also seen between other low copy repeats in the area, including a 1.5-Mb deletion with around 20 genes. These smaller atypical deletions may have a milder phenotype and are more likely to be inherited (McDonald-McGinn et al., 2015). These may also be associated with reduced incidence of CHD (Verhagen et al., 2012). The common 3-Mb and 1.5-Mb deletions both contain *TBX1*. However, there are a number of deletions reported that do not include *TBX1*, but still have a phenotype consistent with that seen in the larger deletions including *TBX1*. This may be due to a positional effect. Heterozygous mice for a *Tbx1* null mutation produce similar conotruncal defects that can be partially rescued by *TBX1* (Jerome & Papaioannou, 2001; Lindsay et al., 2001; Merscher et al., 2001).

There is also a reciprocal duplication at this locus, but the phenotypic effect of this is currently not clear as it has been identified in healthy individuals. A number of individuals with a clinical diagnosis of 22q11.2 deletion syndrome do not show any detectable defects in the common locus. A second locus at 10p13-p14 has been suggested to cause a similar phenotype (Daw et al., 1996).

Although multiple genes are included in the 22q11.2 deletion, *TBX1* is thought to be the major gene relevant to the phenotype. Point mutations in *TBX1* have been identified in three Japanese individuals with conotruncal anomalies and facial dysmorphology, who did not have a 22q11.2 deletion (Yagi et al., 2003). A frameshift mutation in *TBX1* has also been identified by targeted exome sequencing in another Japanese family with normal FISH and array-CGH analysis, in three individuals with craniofacial features of 22q11.2 deletion syndrome and hypocalcemia, two individuals with craniofacial features only, and three with a normal phenotype (Ogata et al., 2014). The *TBX1* mutation at exon 9C was predicted to produce a nonfunctional truncated protein (Ogata et al., 2014). Overall SNPs remain a rare cause of 22q11.2 deletion syndrome, and the variable expressivity of 22q11.2 deletion syndrome suggests the involvement of modifiers. *CRK-like protein* and other genes are likely to contribute an effect (Guris, Fantes, Tara, Druker, & Imamoto, 2001).

4. *TBX3*/ULNAR–MAMMARY SYNDROME

Tbx3 is expressed in the early blastocyst and subsequently in the limbs, lung mesenchyme, and mammary gland (Chapman, Garvey, et al., 1996). Apart from a canonical T-domain, it has a transcription repression domain that is highly conserved among TBX3, *Xenopus* ET, and TBX2 (He, Wen, Campbell, Wu, & Rao, 1999). TBX3 is a potent repressor of tumor suppressor genes *p19* Arf (ADP Ribosylation Factor) in the mouse and *p14* ARF in the human and thereby inhibits senescence (Brummelkamp et al., 2002). *TBX3* is overexpressed in breast cancer and represses *p14* ARF by interacting with histone deacetylases (Yarosh et al., 2008). Genetic interaction between *Tbx3* and *Bmp4* is required to define dorsoventral boundaries (Cho et al., 2006). *Tbx3* is involved in heart outflow tract and atrioventricular conduction system development (Bakker et al., 2008; Mesbah, Harrelson, Theveniau-Ruissy, Papaioannou, & Kelly, 2008). The retinoic acid signaling pathway regulates its expression during mouse limb development (Ballim, Mendelsohn, Papaioannou, & Prince, 2012). *TBX3* is also upregulated by the transforming growth factor β1 (Li, Weinberg, Zerbini, & Prince, 2013). It interacts with multiple mRNA splicing factors and RNA metabolic proteins and *TBX3* mutations hamper splicing regulatory function (Kumar et al., 2014).

Previously, linkage analysis in one kindred mapped UMS to chromosome 12q23-24.1. The region encodes two members of the T-box genes *TBX3* and *TBX5* (Bamshad et al., 1995). Subsequently a frameshift mutation resulting in a premature termination, and a splice-site mutation in the first nucleotide of intron 2 of *TBX3* were detected in two UMS families (Bamshad et al., 1997). Additional dominant mutations in *TBX3* have since been found in other UMS families (Bamshad et al., 1999; Sasaki et al., 2002). In addition, a 1.28-Mb deletion encompassing *TBX3* has been seen in a patient with features of UMS, with dysmorphic facies and mental retardation (Klopocki et al., 2006). Mice lacking *Tbx3* exhibit defects in limb and mammary gland development. In addition to UMS features, Tbx3 homozygous mice also show defects in yolk sac development (Davenport et al., 2003), which is the site of hematopoiesis, essential for survival during gestation.

Autosomal dominant UMS should be suspected in an individual with asymmetric postaxial upper limb defects; however, penetrance and expression within a family are variable (Loyal & Laub, 2014; Meneghini, Odent, Platonova, Egeo, & Merlo, 2006). Some clinical features of UMS are displayed in Fig. 4. Abnormalities can range from hypoplasia of the terminal

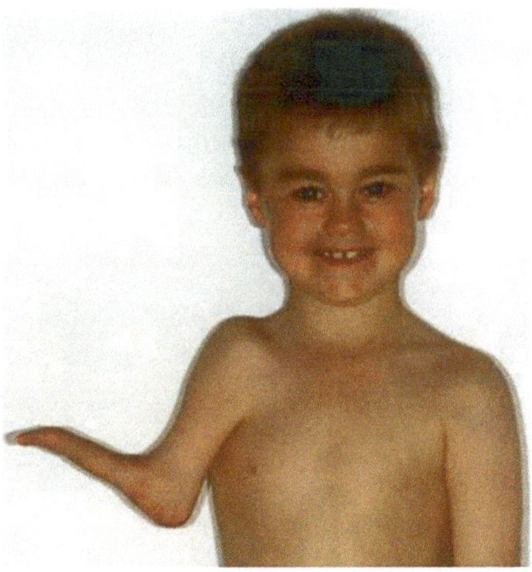

Fig. 4 Ulnar–mammary syndrome. Patient 7, aged 6, broad nasal tip, thin upper vermilion, and upslanting palpebral fissures. *Reproduced from Joss, S., Kini, U., Fisher, R., Mundlos, S., Prescott, K., Newbury-Ecob, R., et al. (2011). The face of ulnar mammary syndrome?* European Journal of Medical Genetics, 54*(3), 301–305. Copyright © 2011 Elsevier Masson SAS. All rights reserved.*

phalanx of the fifth finger only, to complete absence of the ulna and third to fifth fingers. Duplications of the digits may also occur (Bamshad, Root, & Carey, 1996) and stiffness of the elbow is a common feature (Ramirez & Kozin, 2014). The phenotype of ulnar–mammary syndrome (UMS) is variable, but also includes hypoplasia of the apocrine glands, which can result in the absence of perspiration along with reduced axillary hair. Hypoplasia of the mammary glands and areola, with reduced or absent lactation, is also seen (Bamshad et al., 1996; Gonzalez, Herrmann, & Opitz, 1976; Ramirez & Kozin, 2014). Males exhibit delayed puberty, micropenis, shawl scrotum (scrotum surrounds the penis), and cryptorchidism (undescended or maldescended testes) (Bamshad et al., 1996), and females may have an imperforate hymen (Ramirez & Kozin, 2014). Dental abnormalities include hypoplasia or absence of teeth (Bamshad et al., 1997). Cardiac malformations have also been described including ventricular septal defects (VSDs), pulmonary stenosis, and Wolff–Parkinson–White syndrome which can cause tachycardia (Linden, Williams, King, Blair, & Kini, 2009; Meneghini et al., 2006).

5. *TBX4*/SMALL PATELLA SYNDROME

Tbx4 and *Tbx5* are paralogs that diverged by gene duplication from an ancestral gene locus during early vertebrate evolution (Agulnik et al., 1996; Gibson-Brown et al., 1996). *Tbx4* is expressed in the hind limb region and determines leg identity (Rodriguez-Esteban et al., 1999; Takeuchi et al., 1999). The human counterpart was identified by in silico analysis of genomic sequences and EST database (Yi, Russ, & Brook, 2000). *Tbx4* expression is regulated by Pitx1, a paired-type homeodomain transcription factor (DeLaurier, Schweitzer, & Logan, 2006; Logan & Tabin, 1999). Deletion of *Pitx1* resulted in decreased distal expression of *Tbx4* (Takemoto et al., 2011). In *Tbx4* knockout mice, the hind limb failed to develop and the allantois and vascularization were also affected (Naiche & Papaioannou, 2003). A tight regulatory network has been established between *Tbx4/5*, *Fgf*, and the *Wnt* families (Takeuchi et al., 2003). Besides its role in hind limb development, *Tbx4* also plays a major role in the development of the respiratory system. *Tbx4* is specifically expressed in the visceral mesoderm of the lung primordium, and when ectopically expressed it induces lung bud formation by activating *Fgf10* expression (Sakiyama, Yamagishi, & Kuroiwa, 2003). During lung growth and branching Tbx4 and Tbx5 interact with Fgf10.

Mutant mice deficient in *Tbx4* and *Tbx5* show severely reduced lung branching at mid-gestation (Arora, Metzger, & Papaioannou, 2012). In these mutant mice mesenchymal markers such as *Wnt2* and *Fgf10* are also downregulated.

Mutations in *TBX4* cause small patella syndrome (SPS), an autosomal dominant skeletal dysplasia (Scott & Taor, 1979). Diagnostic features include patellar aplasia or hypoplasia, abnormal or absent ossification of the ischiopubic (pelvic bone) junctions, and axe cut infra-acetabular notches. Additional abnormalities in the feet include a sandal gap, pes planus (flat feet), and short 4th and 5th metatarsals. Facial dysmorphism has also been described, but does not appear to be a consistent feature (Bongers et al., 2004). Symptoms resulting from the abnormal or absent patellar are variable; some affected individuals are asymptomatic, whereas others may experience pain, recurrent subluxations, and subsequent arthritic changes (Bongers, van Kampen, van Bokhoven, & Knoers, 2005).

Six different heterozygous mutations in *TBX4* including one nonsense, two missense, one frameshift, one splice-site, and one exon skipping mutation have been identified in six families with SPS (Bongers et al., 2004). Other studies detected microduplications in the chromosome region 17q23.1q23.2 that segregated with autosomal dominant clubfoot in three families with variable expressivity and incomplete penetrance. This region contains *TBX4*, whose upstream regulator PITX1 had previously been implicated in clubfoot etiology (Alvarado et al., 2010; Peterson et al., 2014).

More recently it has been suggested that childhood onset pulmonary arterial hypertension (PAH) is associated with mutations in *TBX4* (Bongers et al., 2004). This finding was based on a small cohort of children with PAH, some of whom also had mental retardation and/or dysmorphic facies. Three of six children who underwent array-CGH testing had a de novo 17q23.2 deletion containing *TBX2* and *TBX4*. Targeted sequencing of *TBX4* identified three mutations in *TBX4* in three further individuals, two mutations were inherited from a parent and inheritance of the third is unknown. The individuals were subsequently examined and found to have clinical features of SPS. A further single mutation in *TBX4* was identified in an individual with adult onset pulmonary hypertension (from a group of 49 individuals). However, screening of a cohort of 23 adults with SPS did not identify anyone with PAH. Further review will be needed to determine whether PAH is a true feature of SPS.

6. *TBX5*/HOLT–ORAM SYNDROME/CHD

TBX5 is one of the most studied T-box genes. It is linked to the autosomal dominant disorder Holt–Oram syndrome (HOS), which is characterized by CHD and radial ray defects in the upper limb (Basson et al., 1997; Li et al., 1997). Somatic mutations have also been linked to isolated CHD (Reamon-Buettner & Borlak, 2004).

Tbx5 is expressed in the heart, forelimb, lungs, and eyes (Chapman, Garvey, et al., 1996). In the developing mouse and chick heart, *Tbx5* is expressed differentially throughout the cardiac crescent and linear heart tube, being stronger at posterior end and weaker at the anterior end. This asymmetric expression continues in the heart at later stages. Ventricular trabeculae, the vena cava, and the atrial aspect of the atrioventricular valves also express high levels of *Tbx5*. The pattern of *Tbx5* expression correlates well with cardiac and limb features observed in HOS (Bruneau et al., 1999). *Tbx5* is essential for forelimb bud formation and limb bud outgrowth (Rallis et al., 2003; Takeuchi et al., 1999). Mice heterozygous for *Tbx5* mutations demonstrate cardiac and forelimb abnormalities similar to those observed in HOS (Bruneau et al., 2001). TBX5 is a transcriptional activator that interacts with 8-bp DNA sequences (Ghosh et al., 2001). While limb-specific factors that interact with TBX5 are largely unknown, cardiac-specific transcription factors such as NKX2.5 (Hiroi et al., 2001), GATA4 (Gibson-Brown et al., 1996), MEF2C (Ghosh et al., 2009), and MYOCD (Wang, Cao, Wang, & Wang, 2011) are all known to interact with TBX5. TBX5 also interacts with the nucleosome remodeling and deacetylase repressor complex, and this interaction is essential for heart development (Waldron et al., 2016). TBX5 and some of its cofactors are found to be key determinants in transdifferentiation of noncardiac cells into induced-cardiomyocytes, and this could be a promising area for future heart regeneration therapy (Qian et al., 2012; Song et al., 2012).

Mutations in *TBX5* cause HOS and isolated CHD. Mutations identified in individuals with HOS/CHD range from missense, nonsense, insertion, deletion to those affecting splice sites (Akrami, Winter, Brook, & Armour, 2001; Li et al., 1997). It is likely that the disease mechanisms are related to functional deficiencies of the mutant proteins such as defective in DNA binding, protein–protein interaction, and defective transcription (Fan, Liu, & Wang, 2003; Ghosh et al., 2001). Individuals with HOS show different degrees of severity of cardiac and skeletal abnormalities. Although mutation screening is mainly focused on the coding regions, a recent study using a combination of targeted genomics, bioinformatics, and mouse genetic engineering identified a rare noncoding homozygous mutation in an enhancer ~90 kb downstream of *TBX5* in a cohort of nonsyndromic individuals with isolated atrial septal defects (ASDs) and/or VSDs (Smemo et al., 2012). The study highlights the importance of noncoding regions in disease. The single base pair mutation has been found to abrogate the expression of *Tbx5* in the mouse and zebrafish heart. Besides its role in CHD, recent studies suggest that *TBX5* loss-of-function mutations S154A and A143T may be linked to familial and sporadic dilated cardiomyopathy (Zhang et al., 2015; Zhou et al., 2015).

The main diagnostic features of HOS are radial ray abnormalities and CHD (Holt & Oram, 1960). This is a fully penetrant condition with variable expressivity. Although genotype–phenotype correlation was initially suggested by Basson et al. (1999), this was not confirmed by others (Brassington et al., 2003). The incidence is thought to be around 1 per 100,000 births (Basson et al., 1997). A few individuals who were thought to have HOS, but who have no mutation in *TBX5*, may actually have Okihiro syndrome caused by mutations in *SALL4* (Kohlhase et al., 2003). Okihiro syndrome is associated with a similar combination of heart and limb defects and the difference between the two can be subtle. Interestingly a single family has also been reported with a hemizygous deletion of both *TBX3* and *TBX5*, with features of both HOS and UMS (Borozdin et al., 2006).

The upper limb abnormalities may be bilateral and symmetrical, but are most commonly unilateral and asymmetric. Fig. 5 displays some typical radiological features. Carpal bone abnormalities are always present and limb defects can range from abnormal carpal bones with normal digits or full phocomelia (severely underdeveloped or absent long bones of the limbs, so the hands appear to be joined close to the trunk) (Basson et al., 1994; Poznanski, Gall, & Stern, 1970). The thumb is the most commonly affected structure. It may be absent, hypoplastic, triphalangeal (normally two phalanges), and finger-like (Newbury-Ecob, Leanage, Raeburn, & Young, 1996). Hypoplastic clavicles have also been demonstrated (Basson et al., 1994). Only a single patient with typical features of HOS and a *TBX5* mutation has been reported with abnormalities of the feet (hypoplasia of the phalanges, duplication of a distal phalanx) (Garavelli et al., 2008).

Seventy-five percent of affected individuals have CHD. Usually this is a septal defect, most commonly an ostium secundum ASD (60%) (Sletten & Pierpont, 1996). More complex defects including tetralogy of Fallot have been reported (Basson et al., 1997; Sletten & Pierpont, 1996; Tidake,

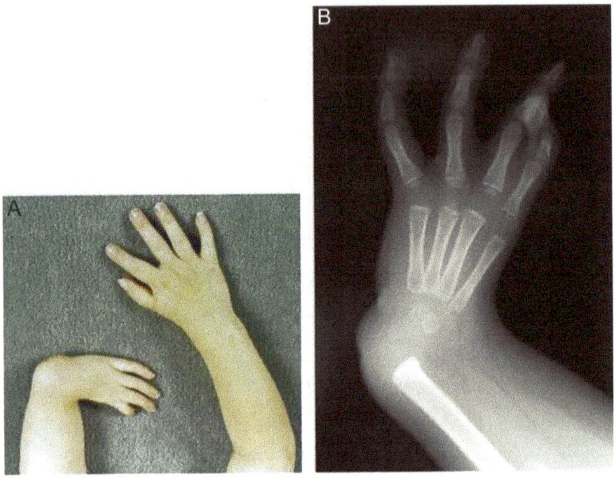

Fig. 5 Holt–Oram syndrome. (A) Clinical photograph of left radial longitudinal deficiency. Note the partial syndactyly between the index and middle fingers. (B) Radiograph of the same patient's left hand. The index ray is hypoplastic. *Reproduced from Goldfarb, C. A., & Wall, L. B. (2014). Holt-Oram syndrome.* The Journal of Hand Surgery, 39*(8), 1646–1648. Copyright © 2014 with permission from Elsevier.*

Gangurde, Shaikh, & Mahajan, 2015). Conduction disease such as bradycardia, atrial fibrillation, atrioventricular block, and sinus node dysfunction is also sometimes seen (Basson et al., 1994; Holt & Oram, 1960; Smith, Sack, & Taylor, 1979).

7. *TBX6*/SPONDYLOCOSTAL DYSOSTOSIS

Tbx6 is expressed in the primitive streak and newly recruited paraxial mesoderm during gastrulation. In the later stages of development, *Tbx6* expression is restricted to presomitic paraxial mesoderm and the tail bud (Chapman, Agulnik, et al., 1996). In zebrafish, expression of *tbx6* is restricted to ventral mesendoderm (Hug, Walter, & Grunwald, 1997). Reverse transcriptase-PCR analysis in humans suggests that it is expressed during gastrulation as well as in some adult tissues, including the testes (Papapetrou, Putt, Fox, & Edwards, 1999). Mutation of *Tbx6* in the mouse affects the differentiation of paraxial mesoderm and posterior paraxial tissue, which forms neural tube-like structures flanking the axial neural tube (Chapman & Papaioannou, 1998). The ectopic neural tube development in *Tbx6* mutant embryos is due to activation of *Sox2* expression via an enhancer element N1 in the paraxial mesoderm. An enhancer-N1-specific deletion mutation introduced into *tbx6* abolished *Sox2* expression and subsequent development of ectopic neural tube in the mesodermal compartment (Takemoto et al., 2011). Tbx6-induced somite patterning requires Notch signaling. In *Tbx6* null mutant mice Notch ligand delta-like 1 (Dll1) is absent, suggesting Dll1 is a target of *Tbx6* (White, Farkas, McFadden, & Chapman, 2003). WNT regulates transcription of the Notch ligand Dll1. In vitro, the Dll1 promoter is regulated by cooperative interaction of TCF and TBX6, and a mutation either in TBX6 or TCF sites abolishes promoter activity (Hofmann et al., 2004). A hypomorphic mutation in mouse *Tbx6* has been identified in the spontaneous mutation rib–vertebrate (that affects the morphogenesis of the axial skeleton); the phenotype is similar to that described for human birth defects (White et al., 2003).

Spondylocostal dysostosis (SCD) is a disorder of vertebral segmentation, with a number of causative genes already identified. Reported features include severe short stature due to a short trunk (height <1st centile), hemivertebrae, fused vertebrae, and scoliosis (Gucev et al., 2010). Whole exome sequencing revealed a stoploss mutation in *TBX6* that segregates with SCD in two generations of a family. The mutation significantly reduces its transcriptional activity suggesting that loss of function may be the mechanism for SCD (Sparrow et al., 2013). No other families have been reported with an associated mutation in *TBX6* as yet.

Scoliosis has also been reported in association with the 16p11.2 microdeletion syndrome, an interval which includes *TBX6*. Scoliosis is not present in every individual with a 16p11.2 deletion, and some individuals with the deletion may be entirely normal. Wu et al. carried out array-CGH and then targeted sequencing of *TBX6* in individuals with congenital scoliosis (Wu et al., 2015). They found a number of mutations and deletions of *TBX6*, but the inheritance pattern was not always known. They suggested that the reduced penetrance of scoliosis in 16p11.2 deletions might be due to other genetic factors. They showed that some people with the deletion and scoliosis also had a common T-C-A *TBX6* haplotype on the other allele. This particular allele was not present in everyone with scoliosis, and they acknowledged that there may be other haplotypes or other factors that contribute some risk.

8. *TBX15*/COUSIN SYNDROME

Autosomal recessive Cousin syndrome describes the combination of craniofacial dysmorphism, hypoplasia of the scapula and pelvis, and short stature. So far only three unrelated individuals have been reported in the literature with recessive mutations in *TBX15* (Dikoglu et al., 2013; Lausch et al., 2008). Additional clinical features include macrocephaly (Lausch et al., 2008). Frontal bossing, deep set eyes, hypertelorism (widely spaced eyes), low set dysplastic ears (abnormally shaped concha, antitragus and lobe, narrow auditory canals), a short webbed neck, and low hairline have been reported. Mixed hearing loss may also be present. Additional skeletal features include fixed flexion deformities at the elbows and bilateral dislocation of the hip. Brachydactyly, humeroradial synostosis (fusion of the humerus and radius), shortened femurs, and hypoplasia of the scapula and iliac bones are evident on radiographs (Dikoglu et al., 2013; Lausch et al., 2008). In all instances the parents were consanguineous (second and first cousins).

Targeted sequencing of *TBX15* was carried out based on similar features seen in the mouse model. This identified homozygous mutations in *TBX15* (Dikoglu et al., 2013; Lausch et al., 2008). A further two individuals described initially by Cousin et al. (1982) did not have molecular testing. They also exhibited neurodevelopmental delay which was not present in the individuals that have been reported subsequently.

9. *TBX19*/ISOLATED ACTH DEFICIENCY

TBX19 was originally identified by bioinformatic analysis of cDNA clones from ESTs (Yi et al., 1999), and *Tpit*, a mouse homolog, was cloned using a PCR-based strategy from cDNA obtained from pro-opiomelanocortin (POMC) expressing AtT-20 cells (Lamolet et al., 2001; Liu et al., 2001). Mouse *Tpit* is 94% similar in sequence to human *TBX19*. The expression of *Tpit*/*Tbx19* is highly restricted to POMC expressing cells in the pituitary (Lamolet et al., 2001).

The exclusive expression of *Tpit* in the pituitary POMC cells and the association of the *POMC* gene in early-onset obesity, adrenal insufficiency, and red hair pigmentation indicated that this gene may cause isolated adrenocorticotrophic hormone (ACTH) deficiency (IAD). Mutations ranging from nonsense, missense, point deletion, and genomic deletions cause IAD (Lamolet et al., 2001; Pulichino, Vallette-Kasic, Couture, et al., 2003). Studies of IAD families suggest the mode of transmission is recessive and the mechanism of disease is loss of function (Pulichino, Vallette-Kasic, Couture, et al., 2003). The *TPIT* mutation M86R was found in two siblings with early-onset IAD. They were compound heterozygous for M86R in exon 2 and a frame shift mutation in exon6 (782delA). However, their mother who carried the M86R mutation and father who carried the 782delA mutation were unaffected (Metherell et al., 2004). *Tbx19*-deficient mice develop IAD similar to humans (Pulichino, Vallette-Kasic, Couture, et al., 2003). Tbx19 is required for transcriptional activation of the *POMC* gene and to initiate POMC cell differentiation, but is not essential for lineage specification (Lamolet et al., 2001; Pulichino, Vallette-Kasic, Tsai, et al., 2003). Tpit cooperates with Pitx1 to activate POMC transcription (Lamolet et al., 2001).

IAD is a rare autosomal recessive disorder that can result in neonatal death (Vallette-Kasic et al., 2005). Individuals with mutations in *TBX19* have neonatal onset cortisol deficiency. Mutations in *TBX19* do not account for juvenile onset IAD, but have been found in up to 65% of individuals with early/neonatal onset (Couture et al., 2012; Metherell et al., 2004; Pulichino, Vallette-Kasic, Couture, Brue, & Drouin, 2004; Pulichino, Vallette-Kasic, Couture, et al., 2003). IAD can also result from mutations in *POMC*, the precursor to ACTH (Torchinsky, Wineman, & Moll, 2011).

Individuals with IAD demonstrate low levels of cortisol secretion from the adrenal glands, secondary to reduced levels of ACTH production by the pituitary gland. Only ACTH production is deficient and all other pituitary hormones are produced normally. Individuals are affected by severe hypoglycemia, which is complicated by seizures in around 50% of individuals. There is failure to thrive in the neonatal period; vomiting, hypotension, and affected individuals may have hepatomegaly and prolonged cholestatic jaundice (62%) (Couture et al., 2012). Pituitary imaging is usually normal (Couture et al., 2012). Postmortem findings may include bilateral adrenal hypoplasia (Malpuech et al., 1988). Other causes of congenital IAD remain to be elucidated as not all individuals have mutations in *TBX19* (Pham, Garot, Brue, & Brauner, 2011).

10. *TBX20*/CONGENITAL HEART DISEASE

The T-box gene *tbx20* was first cloned in zebrafish. This gene is an ortholog of *Drosophila melanogaster H15* and *tbx12* of *Caenorhabditis elegans*. *tbx20* is expressed in both mesoderm and ectoderm, including the heart, motor neurons, and dorsal aorta (Ahn, Ruvinsky, Oates, Silver, & Ho, 2000). The human and mouse versions were subsequently identified (Meins, Henderson, Bhattacharya, & Sowden, 2000), and both are expressed in the heart, eye, and limbs, suggesting a role in the development of these organs (Meins et al., 2000). *Neuromancer* genes (*nmr1* and *nmr2*, aka *H15* and *mid*),

Tbx20 homologs in *Drosophila*, synergistically interact with *tinman/Nkx2.5* to regulate cardiac function, cardiac structure, and rhythmicity (Qian et al., 2008). Physical interaction of Tbx20 with Nkx2.5, GATA4, and GATA5 leads to synergistic activation of genes in the developing heart (Stennard et al., 2003). *Tbx20* is coexpressed and interacts with *Tbx5* in the early stage of heart formation; however, they diverge as heart morphogenesis progresses (Brown et al., 2005; Plageman & Yutzey, 2004). Bone morphogenetic protein-2 regulates the expression of *Tbx20* (Plageman & Yutzey, 2004). *Tbx20* homozygous embryos die at mid-gestation due to abnormal heart morphogenesis, and in *Tbx20* null embryos there is an ectopic activation of *Tbx2* across the whole heart myogenic fields, suggesting a role as a repressor (Singh et al., 2005; Stennard et al., 2005). The suppressive role of *Tbx20* on *Tbx2* probably occurs via direct involvement of Bmp/Smad signaling that helps to form the atrioventricular canal (Singh et al., 2009). Bone morphogenetic protein 10 (BMP10) also regulates the expression of *Tbx20* and is important for ventricular wall development and maturation (Zhang et al., 2011). During heart development *Tbx20* is expressed in endocardial cushions and is needed for valve development, and ablation of *Tbx20* in endocardial cells leads to valve elongation defects and impaired cardiac function. Endocardial cushion maturation and valve elongation probably occur through Tbx20-Wnt pathways (Cai et al., 2013). Monocytic leukemia zinc finger protein (MOZ), a chromatin modifier, regulates *Tbx1*, and *Moz* mutant mice exhibit 22q11.2 deletion syndrome-like anomalies. A *Tbx1* transgene can rescue the heart phenotype in *Moz* mutant mice, suggesting a genetic link between *Moz* and *Tbx1* (Voss et al., 2012).

Mutations in *TBX20* lead to isolated CHD. Targeted sequencing in 352 probands with CHD revealed heterozygous germline mutations in *TBX20* in two individuals with ASDs, which appeared to segregate with other affected relatives in their families (Kirk et al., 2007). A nonsense mutation (Q195X) and a missense mutation (I152M) were identified in these families. Both mutations fall within the DNA binding domain and are predicted to disrupt its structure and function (Kirk et al., 2007). The cardiac phenotypes were variable and included dilated cardiomyopathy and mitral valve abnormalities. A third proband was identified with a mutation in *TBX20*, but this did not segregate with family members who also had CHD.

A number of other studies have performed targeted sequencing of *TBX20* and found heterozygous mutations associated with ASDs, patent foramen ovale (PFO), total anomalous pulmonary venous connection, tetralogy of Fallot, mitral valve defects, atrioventricular septal defects, VSDs, pentalogy of Fallot (an ASD is present in addition to the classical features of tetralogy of Fallot), and dilated cardiomyopathy (Liu, Shen, et al., 2008; Posch et al., 2010; Qian et al., 2008). Missense mutations in *TBX20* have also been reported in Chinese individuals with sporadic CHD (Liu, Shen, et al., 2008). This targeted sequencing study identified a number of mutations; however, there was evidence of incomplete penetrance and non-segregation in one family (Liu, Shen, et al., 2008).

Targeted sequencing has also been carried out in a population with dilated cardiomyopathy and CHD. Three missense heterozygous mutations were identified in three sporadic and two familial occurrences, although again there was evidence of reduced penetrance in two of the three families (Qian et al., 2008). Missense mutations R334Q and W349R were identified in two Caucasians families. The authors also analyzed DNA samples from 96 children with ASDs and 96 children with VSDs and identified two different mutations in exon 4. The mutation 597>G (H186P) has also been found in two unrelated children: a female child of African-American origin with an atrioventricular canal defect, secundum and primum ASDs, as well as cleft mitral valve; and a Hispanic male with pentalogy of Fallot. A second mutation 601T>C (L197P) in exon 4 was identified in an American female also with pentalogy of Fallot and with isolated ASD in another female (Qian et al., 2008).

Most identified mutations are loss of function, but a gain-of-function *TBX20* mutation I121M has been identified by screening of 170 individuals with sporadic secundum atrial septal defects (ASDII). Further investigation confirmed segregation of the mutation with CHD in a three-generation kindred. Phenotypes include a cribriform ASD, PFO, and valve defects. Functional analysis revealed that the mutant protein increased DNA binding and transcriptional activities, suggesting ASDII can be due to loss of function as well as gain of function (Posch et al., 2010). A *TBX20* promoter mutation that significantly inhibited the promoter activity has been identified in one individual with a VSD (Qiao et al., 2012).

11. *TBX22*/CLEFT PALATE WITH OR WITHOUT ANKYLOGLOSSIA, X-LINKED (CPX)

TBX22 was mapped by in silico analysis of the human X chromosome (Laugier-Anfossi & Villard, 2000). Orthologs of *TBX22* have been identified in chick (Haenig, Schmidt, Kraus, Pfordt, & Kispert, 2002), mouse (Braybrook et al., 2002), and zebrafish (Jezewski, Fang, Payne-Ferreira, &

Yelick, 2009). Mouse Tbx22 is expressed during palate and tongue development (Bush et al., 2002). TBX22 is a transcriptional repressor and typically binds key DNA consensus sequences AGGTGTGA. SUMO-1, a ubiquitin-like modifier, sumoylates TBX22 and is important for its function (Andreou et al., 2007). In NIH3T3 cells both mouse and human *TBX22* promoters are upregulated by transcription factor Mn1. In *Mn1*(−/−) mutant mice, *Tbx22* expression is downregulated, suggesting *Mn1* acts upstream of *Tbx22* (Liu, Lan, et al., 2008).

Heterozygous mutations including nonsense, splice-site, frameshift, and missense have been detected in *TBX22* in individuals with X-linked cleft palate with or without ankyloglossia (tongue tie) (Andreou et al., 2007; Liu, Lan, et al., 2008). A promoter mutation −73G>A in the *TBX22* promoter has been identified in a six-generation family with cleft palate and nasal speech. The mutation occurs in the core binding region of the transcription factor ETS-1. Both EMSA and ChIP studies suggest that the mutation markedly decreases ETS-1 binding and promoter activity, suggesting haploinsufficiency is likely to be the mechanism behind the cleft palate formation (Fu et al., 2014).

Cleft palate affects around 1 in 1500 births. Cleft lip and palate are two of the most common congenital abnormalities, which usually result in difficulty feeding, recurrent ear infections, and dental abnormalities, and individuals often require speech therapy. A cleft may present as an isolated feature (nonsyndromic) or in combination with other features (syndromic). Surgical repair is usually required. Disease expression and penetrance in X-linked cleft palate with or without ankyloglossia are variable. An individual may have only ankyloglossia, only cleft palate, or may have a more subtle phenotype of a high-arched palate, a submucous cleft, or bifid uvula. Braybrook et al. first reported mutations in *TBX22* in a number of families (Braybrook et al., 2001, 2002). This included a single individual with cleft lip and palate, suggesting that this may be part of the phenotype. In an unselected group of individuals with cleft palate, Marcano et al. performed targeted sequencing of *TBX22* and identified a number of families with pathogenic mutations (Marcano et al., 2004). Ninety-six percent of males had cleft palate or a microform of cleft palate, and 79% had ankyloglossia in addition. Four percent of males had ankyloglossia only. In female carriers 45% had ankyloglossia, 6% had a cleft palate, and 11% had both. Females may also be completely normal. There does not appear to be any significant genotype–phenotype correlation (Marcano et al., 2004).

Suphapeetiporn, Tongkobpetch, Siriwan, and Shotelersuk (2007) also performed targeted sequencing of *TBX22* in a Thai population with nonsyndromic cleft palate. They identified other novel *TBX22* mutations in four individuals, two of whom had no family history of cleft palate. There has been a single report of a family with syndromic X-linked cleft palate and a splice acceptor mutation in *TBX22* detected by targeted sequencing (Pauws et al., 2013). The family had an X-linked pattern of dysmorphic facies, coloboma (developmental defect in the eye), and sensorineural hearing impairment. They had been diagnosed with Abruzzo–Erickson syndrome previously and the authors proposed that *TBX22* mutations may also cause this condition.

12. CONCLUDING REMARKS

Identification of T-box genes and establishing their role in developmental disorders represent the initial steps in elucidating their molecular pathways. Finding their targets and regulatory networks is important to increase our understanding of the disease mechanisms and to provide a basis for future targeted therapy. A great deal of information about the regulation of T-box genes is unknown, and this is an area where mutation screening can help to understand the disease pathology in those individuals that show no polymorphism in the coding region. Another unresolved area is what determines the specificity of T-box proteins, given the fact that they bind to very similar DNA sequences. Whole exome or genome sequencing studies will help elucidate the full mutation and phenotypic spectrum of these disorders. The role of T-box genes in certain cancers may also lend themselves a direct target for cancer therapy. Thus study of T-box genes provides a valuable insight into developmental and disease processes in the human being.

ACKNOWLEDGMENT

Our research was funded by the British Heart Foundation (BHF).

REFERENCES

Agulnik, S. I., Garvey, N., Hancock, S., Ruvinsky, I., Chapman, D. L., Agulnik, I., et al. (1996). Evolution of mouse T-box genes by tandem duplication and cluster dispersion. *Genetics*, *144*, 249–254.

Agulnik, S. I., Papaioannou, V. E., & Silver, L. M. (1998). Cloning, mapping, and expression analysis of TBX15, a new member of the T-Box gene family. *Genomics*, *51*, 68–75.

Ahn, D. G., Ruvinsky, I., Oates, A. C., Silver, L. M., & Ho, R. K. (2000). Tbx20, a new vertebrate T-box gene expressed in the cranial motor neurons and developing cardiovascular structures in zebrafish. *Mechanisms of Development, 95*, 253–258.

Akrami, S. M., Winter, R. M., Brook, J. D., & Armour, J. A. (2001). Detection of a large TBX5 deletion in a family with Holt-Oram syndrome. *Journal of Medical Genetics, 38*, E44.

Alvarado, D. M., Aferol, H., McCall, K., Huang, J. B., Techy, M., Buchan, J., et al. (2010). Familial isolated clubfoot is associated with recurrent chromosome 17q23.1q23.2 microduplications containing TBX4. *American Journal of Human Genetics, 87*, 154–160.

Andreou, A. M., Pauws, E., Jones, M. C., Singh, M. K., Bussen, M., Doudney, K., et al. (2007). TBX22 missense mutations found in patients with X-linked cleft palate affect DNA binding, sumoylation, and transcriptional repression. *American Journal of Human Genetics, 81*, 700–712.

Arora, R., Metzger, R. J., & Papaioannou, V. E. (2012). Multiple roles and interactions of Tbx4 and Tbx5 in development of the respiratory system. *PLoS Genetics, 8*, e1002866.

Bakker, M. L., Boukens, B. J., Mommersteeg, M. T., Brons, J. F., Wakker, V., Moorman, A. F., et al. (2008). Transcription factor Tbx3 is required for the specification of the atrioventricular conduction system. *Circulation Research, 102*, 1340–1349.

Ballim, R. D., Mendelsohn, C., Papaioannou, V. E., & Prince, S. (2012). The ulnar-mammary syndrome gene, Tbx3, is a direct target of the retinoic acid signaling pathway, which regulates its expression during mouse limb development. *Molecular Biology of the Cell, 23*, 2362–2372.

Bamshad, M., Krakowiak, P. A., Watkins, W. S., Root, S., Carey, J. C., & Jorde, L. B. (1995). A gene for ulnar-mammary syndrome maps to 12q23-q24.1. *Human Molecular Genetics, 4*, 1973–1977.

Bamshad, M., Le, T., Watkins, W. S., Dixon, M. E., Kramer, B. E., Roeder, A. D., et al. (1999). The spectrum of mutations in TBX3: Genotype/phenotype relationship in ulnar-mammary syndrome. *American Journal of Human Genetics, 64*, 1550–1562.

Bamshad, M., Lin, R. C., Law, D. J., Watkins, W. C., Krakowiak, P. A., Moore, M. E., et al. (1997). Mutations in human TBX3 alter limb, apocrine and genital development in ulnar-mammary syndrome. *Nature Genetics, 16*, 311–315.

Bamshad, M., Root, S., & Carey, J. C. (1996). Clinical analysis of a large kindred with the Pallister ulnar-mammary syndrome. *American Journal of Medical Genetics, 65*, 325–331.

Basson, C. T., Bachinsky, D. R., Lin, R. C., Levi, T., Elkins, J. A., Soults, J., et al. (1997). Mutations in human TBX5 [corrected] cause limb and cardiac malformation in Holt-Oram syndrome. *Nature Genetics, 15*, 30–35.

Basson, C. T., Cowley, G. S., Solomon, S. D., Weissman, B., Poznanski, A. K., Traill, T. A., et al. (1994). The clinical and genetic spectrum of the Holt-Oram syndrome (heart-hand syndrome). *The New England Journal of Medicine, 330*, 885–891.

Basson, C. T., Huang, T., Lin, R. C., Bachinsky, D. R., Weremowicz, S., Vaglio, A., et al. (1999). Different TBX5 interactions in heart and limb defined by Holt-Oram syndrome mutations. *Proceedings of the National Academy of Sciences of the United States of America, 96*, 2919–2924.

Bollag, R. J., Siegfried, Z., Cebra-Thomas, J. A., Garvey, N., Davison, E. M., & Silver, L. M. (1994). An ancient family of embryonically expressed mouse genes sharing a conserved protein motif with the T locus. *Nature Genetics, 7*, 383–389.

Bongers, E. M., Duijf, P. H., van Beersum, S. E., Schoots, J., Van Kampen, A., Burckhardt, A., et al. (2004). Mutations in the human TBX4 gene cause small patella syndrome. *American Journal of Human Genetics, 74*, 1239–1248.

Bongers, E. M., van Kampen, A., van Bokhoven, H., & Knoers, N. V. (2005). Human syndromes with congenital patellar anomalies and the underlying gene defects. *Clinical Genetics, 68*, 302–319.

Borozdin, W., Bravo-Ferrer Acosta, A. M., Seemanova, E., Leipoldt, M., Bamshad, M. J., Unger, S., et al. (2006). Contiguous hemizygous deletion of TBX5, TBX3, and RBM19 resulting in a combined phenotype of Holt-Oram and ulnar-mammary syndromes. *American Journal of Medical Genetics. Part A*, *140A*, 1880–1886.

Brassington, A. M., Sung, S. S., Toydemir, R. M., Le, T., Roeder, A. D., Rutherford, A. E., et al. (2003). Expressivity of Holt-Oram syndrome is not predicted by TBX5 genotype. *American Journal of Human Genetics*, *73*, 74–85.

Braybrook, C., Doudney, K., Marcano, A. C., Arnason, A., Bjornsson, A., Patton, M. A., et al. (2001). The T-box transcription factor gene TBX22 is mutated in X-linked cleft palate and ankyloglossia. *Nature Genetics*, *29*, 179–183.

Braybrook, C., Lisgo, S., Doudney, K., Henderson, D., Marcano, A. C., Strachan, T., et al. (2002). Craniofacial expression of human and murine TBX22 correlates with the cleft palate and ankyloglossia phenotype observed in CPX patients. *Human Molecular Genetics*, *11*, 2793–2804.

Brown, D. D., Martz, S. N., Binder, O., Goetz, S. C., Price, B. M., & Smith, J. C. (2005). Tbx5 and Tbx20 act synergistically to control vertebrate heart morphogenesis. *Development*, *132*, 553–563.

Brummelkamp, T. R., Kortlever, R. M., Lingbeek, M., Trettel, F., MacDonald, M. E., & van Lohuizen, M. (2002). TBX-3, the gene mutated in ulnar-mammary syndrome, is a negative regulator of p19ARF and inhibits senescence. *The Journal of Biological Chemistry*, *277*, 6567–6572.

Bruneau, B. G., Logan, M., Davis, N., Levi, T., Tabin, C. J., Seidman, J. G., et al. (1999). Chamber-specific cardiac expression of Tbx5 and heart defects in Holt-Oram syndrome. *Developmental Biology*, *211*, 100–108.

Bruneau, B. G., Nemer, G., Schmitt, J. P., Charron, F., Robitaille, L., Caron, S., et al. (2001). A murine model of Holt-Oram syndrome defines roles of the T-box transcription factor Tbx5 in cardiogenesis and disease. *Cell*, *106*, 709–721.

Bush, J. O., Lan, Y., Maltby, K. M., & Jiang, R. (2002). Isolation and developmental expression analysis of Tbx22, the mouse homolog of the human X-linked cleft palate gene. *Developmental Dynamics: An Official Publication of the American Association of Anatomists*, *225*, 322–326.

Cai, X., Zhang, W., Hu, J., Zhang, L., Sultana, N., Wu, B., et al. (2013). Tbx20 acts upstream of Wnt signaling to regulate endocardial cushion formation and valve remodeling during mouse cardiogenesis. *Development*, *140*, 3176–3187.

Caprio, C., & Baldini, A. (2014). p53 Suppression partially rescues the mutant phenotype in mouse models of DiGeorge syndrome. *Proceedings of the National Academy of Sciences of the United States of America*, *111*, 13385–13390.

Chapman, D. L., Agulnik, I., Hancock, S., Silver, L. M., & Papaioannou, V. E. (1996). Tbx6, a mouse T-Box gene implicated in paraxial mesoderm formation at gastrulation. *Developmental Biology*, *180*, 534–542.

Chapman, D. L., Garvey, N., Hancock, S., Alexiou, M., Agulnik, S. I., Gibson-Brown, J. J., et al. (1996). Expression of the T-box family genes, Tbx1-Tbx5, during early mouse development. *Developmental Dynamics: An Official Publication of the American Association of Anatomists*, *206*, 379–390.

Chapman, D. L., & Papaioannou, V. E. (1998). Three neural tubes in mouse embryos with mutations in the T-box gene Tbx6. *Nature*, *391*, 695–697.

Chieffo, C., Garvey, N., Gong, W., Roe, B., Zhang, G., Silver, L., et al. (1997). Isolation and characterization of a gene from the DiGeorge chromosomal region homologous to the mouse Tbx1 gene. *Genomics*, *43*, 267–277.

Cho, K. W., Kim, J. Y., Song, S. J., Farrell, E., Eblaghie, M. C., Kim, H. J., et al. (2006). Molecular interactions between Tbx3 and Bmp4 and a model for dorsoventral positioning of mammary gland development. *Proceedings of the National Academy of Sciences of the United States of America*, *103*, 16788–16793.

Choi, K. S., Cohn, M. J., & Harfe, B. D. (2008). Identification of nucleus pulposus precursor cells and notochordal remnants in the mouse: Implications for disk degeneration and chordoma formation. *Developmental Dynamics: An Official Publication of the American Association of Anatomists*, *237*, 3953–3958.

Cousin, J., Walbaum, R., Cegarra, P., Huguet, J., Louis, J., Pauli, A., et al. (1982). [Familial pelvi-scapulary dysplasia with anomalies of the epiphyses, dwarfism and dysmorphy: a new syndrome? (author's transl)]. *Archives Françaises de Pédiatrie*, *39*(3), 173–175 (in French).

Couture, C., Saveanu, A., Barlier, A., Carel, J. C., Fassnacht, M., Fluck, C. E., et al. (2012). Phenotypic homogeneity and genotypic variability in a large series of congenital isolated ACTH-deficiency patients with TPIT gene mutations. *The Journal of Clinical Endocrinology and Metabolism*, *97*, E486–E495.

Davenport, T. G., Jerome-Majewska, L. A., & Papaioannou, V. E. (2003). Mammary gland, limb and yolk sac defects in mice lacking Tbx3, the gene mutated in human ulnar mammary syndrome. *Development*, *130*, 2263–2273.

Daw, S. C., Taylor, C., Kraman, M., Call, K., Mao, J., Schuffenhauer, S., et al. (1996). A common region of 10p deleted in DiGeorge and velocardiofacial syndromes. *Nature Genetics*, *13*, 458–460.

DeLaurier, A., Schweitzer, R., & Logan, M. (2006). Pitx1 determines the morphology of muscle, tendon, and bones of the hindlimb. *Developmental Biology*, *299*, 22–34.

Devriendt, K., Fryns, J. P., Mortier, G., van Thienen, M. N., & Keymolen, K. (1998). The annual incidence of DiGeorge/velocardiofacial syndrome. *Journal of Medical Genetics*, *35*, 789–790.

Dikoglu, E., Simsek-Kiper, P. O., Utine, G. E., Campos-Xavier, B., Boduroglu, K., Bonafe, L., et al. (2013). Homozygosity for a novel truncating mutation confirms TBX15 deficiency as the cause of Cousin syndrome. *American Journal of Medical Genetics. Part A*, *161A*, 3161–3165.

Fan, C., Liu, M., & Wang, Q. (2003). Functional analysis of TBX5 missense mutations associated with Holt-Oram syndrome. *The Journal of Biological Chemistry*, *278*, 8780–8785.

Fernando, R. I., Litzinger, M., Trono, P., Hamilton, D. H., Schlom, J., & Palena, C. (2010). The T-box transcription factor Brachyury promotes epithelial-mesenchymal transition in human tumor cells. *The Journal of Clinical Investigation*, *120*, 533–544.

Flanagan, A. M., & Yamaguchi, T. (2013). Chordoma. In C. D. M. Fletcher, J. A. Bridge, P. C. W. Hogendoorn, & F. Mertens (Eds.), *World Health Organization classification of tumours of soft tissue and bone*. Lyon: IARC Press.

Fu, X., Cheng, Y., Yuan, J., Huang, C., Cheng, H., & Zhou, R. (2014). Loss-of-function mutation in the X-linked TBX22 promoter disrupts an ETS-1 binding site and leads to cleft palate. *Human Genetics*, *134*, 147–158.

Garavelli, L., De Brasi, D., Verri, R., Guareschi, E., Cariola, F., Melis, D., et al. (2008). Holt-Oram syndrome associated with anomalies of the feet. *American Journal of Medical Genetics. Part A*, *146A*, 1185–1189.

Garg, V., Yamagishi, C., Hu, T., Kathiriya, I. S., Yamagishi, H., & Srivastava, D. (2001). Tbx1, a DiGeorge syndrome candidate gene, is regulated by sonic hedgehog during pharyngeal arch development. *Developmental Biology*, *235*, 62–73.

Ghosh, T. K., Packham, E. A., Bonser, A. J., Robinson, T. E., Cross, S. J., & Brook, J. D. (2001). Characterization of the TBX5 binding site and analysis of mutations that cause Holt-Oram syndrome. *Human Molecular Genetics*, *10*, 1983–1994.

Ghosh, T. K., Song, F. F., Packham, E. A., Buxton, S., Robinson, T. E., Ronksley, J., et al. (2009). Physical interaction between TBX5 and MEF2C is required for early heart development. *Molecular and Cellular Biology*, *29*, 2205–2218.

Gibson-Brown, J. J., Agulnik, S. I., Chapman, D. L., Alexiou, M., Garvey, N., Silver, L. M., et al. (1996). Evidence of a role for T-box genes in the evolution of limb morphogenesis and the specification of forelimb/hindlimb identity. *Mechanisms of Development*, *56*, 93–101.

Gonzalez, C. H., Herrmann, J., & Opitz, J. M. (1976). Studies of malformation syndromes of man XXXXIIB: Mother and son affected with the ulnar-mammary syndrome type Pallister. *European Journal of Pediatrics*, *123*, 225–235.

Goodship, J., Cross, I., LiLing, J., & Wren, C. (1998). A population study of chromosome 22q11 deletions in infancy. *Archives of Disease in Childhood*, *79*, 348–351.

Gucev, Z. S., Tasic, V., Pop-Jordanova, N., Sparrow, D. B., Dunwoodie, S. L., Ellard, S., et al. (2010). Autosomal dominant spondylocostal dysostosis in three generations of a Macedonian family: Negative mutation analysis of DLL3, MESP2, HES7, and LFNG. *American Journal of Medical Genetics. Part A*, *152A*, 1378–1382.

Guris, D. L., Fantes, J., Tara, D., Druker, B. J., & Imamoto, A. (2001). Mice lacking the homologue of the human 22q11.2 gene CRKL phenocopy neurocristopathies of DiGeorge syndrome. *Nature Genetics*, *27*, 293–298.

Haenig, B., Schmidt, C., Kraus, F., Pfordt, M., & Kispert, A. (2002). Cloning and expression analysis of the chick ortholog of TBX22, the gene mutated in X-linked cleft palate and ankyloglossia. *Mechanisms of Development*, *117*, 321–325.

He, M., Wen, L., Campbell, C. E., Wu, J. Y., & Rao, Y. (1999). Transcription repression by Xenopus ET and its human ortholog TBX3, a gene involved in ulnar-mammary syndrome. *Proceedings of the National Academy of Sciences of the United States of America*, *96*, 10212–10217.

Herrmann, B. G., Labeit, S., Poustka, A., King, T. R., & Lehrach, H. (1990). Cloning of the T gene required in mesoderm formation in the mouse. *Nature*, *343*, 617–622.

Hiroi, Y., Kudoh, S., Monzen, K., Ikeda, Y., Yazaki, Y., Nagai, R., et al. (2001). Tbx5 associates with Nkx2-5 and synergistically promotes cardiomyocyte differentiation. *Nature Genetics*, *28*, 276–280.

Hofmann, M., Schuster-Gossler, K., Watabe-Rudolph, M., Aulehla, A., Herrmann, B. G., & Gossler, A. (2004). WNT signaling, in synergy with T/TBX6, controls Notch signaling by regulating Dll1 expression in the presomitic mesoderm of mouse embryos. *Genes & Development*, *18*, 2712–2717.

Holt, M., & Oram, S. (1960). Familial heart disease with skeletal malformations. *British Heart Journal*, *22*, 236–242.

Hug, B., Walter, V., & Grunwald, D. J. (1997). Tbx6, a Brachyury-related gene expressed by ventral mesendodermal precursors in the zebrafish embryo. *Developmental Biology*, *183*, 61–73.

Jensen, L. E., Barbaux, S., Hoess, K., Fraterman, S., Whitehead, A. S., & Mitchell, L. E. (2004). The human T locus and spina bifida risk. *Human Genetics*, *115*, 475–482.

Jerome, L. A., & Papaioannou, V. E. (2001). DiGeorge syndrome phenotype in mice mutant for the T-box gene, Tbx1. *Nature Genetics*, *27*, 286–291.

Jezewski, P. A., Fang, P. K., Payne-Ferreira, T. L., & Yelick, P. C. (2009). Alternative splicing, phylogenetic analysis, and craniofacial expression of zebrafish tbx22. *Developmental Dynamics: An Official Publication of the American Association of Anatomists*, *238*, 1605–1612.

Kelley, M. J., Korczak, J. F., Sheridan, E., Yang, X., Goldstein, A. M., & Parry, D. M. (2001). Familial chordoma, a tumor of notochordal remnants, is linked to chromosome 7q33. *American Journal of Human Genetics*, *69*, 454–460.

Kelley, M. J., Shi, J., Ballew, B., Hyland, P. L., Li, W. Q., Rotunno, M., et al. (2014). Characterization of T gene sequence variants and germline duplications in familial and sporadic chordoma. *Human Genetics*, *133*, 1289–1297.

Kirk, E. P., Sunde, M., Costa, M. W., Rankin, S. A., Wolstein, O., Castro, M. L., et al. (2007). Mutations in cardiac T-box factor gene TBX20 are associated with diverse cardiac pathologies, including defects of septation and valvulogenesis and cardiomyopathy. *American Journal of Human Genetics*, *81*, 280–291.

Kispert, A., & Hermann, B. G. (1993). The Brachyury gene encodes a novel DNA binding protein. *The EMBO Journal*, *12*, 4898–4899.

Klopocki, E., Neumann, L. M., Tonnies, H., Ropers, H. H., Mundlos, S., & Ullmann, R. (2006). Ulnar-mammary syndrome with dysmorphic facies and mental retardation caused by a novel 1.28 Mb deletion encompassing the TBX3 gene. *European Journal of Human Genetics: EJHG, 14*, 1274–1279.

Kohlhase, J., Schubert, L., Liebers, M., Rauch, A., Becker, K., Mohammed, S. N., et al. (2003). Mutations at the SALL4 locus on chromosome 20 result in a range of clinically overlapping phenotypes, including Okihiro syndrome, Holt-Oram syndrome, acro-renal-ocular syndrome, and patients previously reported to represent thalidomide embryopathy. *Journal of Medical Genetics, 40*, 473–478.

Kraus, F., Haenig, B., & Kispert, A. (2001). Cloning and expression analysis of the mouse T-box gene tbx20. *Mechanisms of Development, 100*, 87–91.

Kreshak, J., Larousserie, F., Picci, P., Boriani, S., Mirra, J., Merlino, B., et al. (2014). Difficulty distinguishing benign notochordal cell tumor from chordoma further suggests a link between them. *Cancer Imaging: The Official Publication of the International Cancer Imaging Society, 14*, 4.

Kumar, P. P., Franklin, S., Emechebe, U., Hu, H., Moore, B., Lehman, C., et al. (2014). TBX3 regulates splicing in vivo: A novel molecular mechanism for ulnar-mammary syndrome. *PLoS Genetics, 10*, e1004247.

Lamolet, B., Pulichino, A. M., Lamonerie, T., Gauthier, Y., Brue, T., Enjalbert, A., et al. (2001). A pituitary cell-restricted T box factor, Tpit, activates POMC transcription in cooperation with Pitx homeoproteins. *Cell, 104*, 849–859.

Laugier-Anfossi, F., & Villard, L. (2000). Molecular characterization of a new human T-box gene (TBX22) located in xq21.1 encoding a protein containing a truncated T-domain. *Gene, 255*, 289–296.

Lausch, E., Hermanns, P., Farin, H. F., Alanay, Y., Unger, S., Nikkel, S., et al. (2008). TBX15 mutations cause craniofacial dysmorphism, hypoplasia of scapula and pelvis, and short stature in Cousin syndrome. *American Journal of Human Genetics, 83*, 649–655.

Lee-Jones, L., Aligianis, I., Davies, P. A., Puga, A., Farndon, P. A., Stemmer-Rachamimov, A., et al. (2004). Sacrococcygeal chordomas in patients with tuberous sclerosis complex show somatic loss of TSC1 or TSC2. *Genes, Chromosomes & Cancer, 41*, 80–85.

Li, Q. Y., Newbury-Ecob, R. A., Terrett, J. A., Wilson, D. I., Curtis, A. R., Yi, C. H., et al. (1997). Holt-Oram syndrome is caused by mutations in TBX5, a member of the Brachyury (T) gene family. *Nature Genetics, 15*, 21–29.

Li, J., Weinberg, M. S., Zerbini, L., & Prince, S. (2013). The oncogenic TBX3 is a downstream target and mediator of the TGF-beta1 signaling pathway. *Molecular Biology of the Cell, 24*, 3569–3576.

Linden, H., Williams, R., King, J., Blair, E., & Kini, U. (2009). Ulnar mammary syndrome and TBX3: Expanding the phenotype. *American Journal of Medical Genetics. Part A, 149A*, 2809–2812.

Lindsay, E. A., Vitelli, F., Su, H., Morishima, M., Huynh, T., Pramparo, T., et al. (2001). Tbx1 haploinsufficieny in the DiGeorge syndrome region causes aortic arch defects in mice. *Nature, 410*, 97–101.

Liu, W., Lan, Y., Pauws, E., Meester-Smoor, M. A., Stanier, P., Zwarthoff, E. C., et al. (2008). The Mn1 transcription factor acts upstream of Tbx22 and preferentially regulates posterior palate growth in mice. *Development, 135*, 3959–3968.

Liu, J., Lin, C., Gleiberman, A., Ohgi, K. A., Herman, T., Huang, H. P., et al. (2001). Tbx19, a tissue-selective regulator of POMC gene expression. *Proceedings of the National Academy of Sciences of the United States of America, 98*, 8674–8679.

Liu, C., Shen, A., Li, X., Jiao, W., Zhang, X., & Li, Z. (2008). T-box transcription factor TBX20 mutations in Chinese patients with congenital heart disease. *European Journal of Medical Genetics, 51*, 580–587.

Logan, M., & Tabin, C. J. (1999). Role of Pitx1 upstream of Tbx4 in specification of hindlimb identity. *Science, 283*, 1736–1739.
Loyal, J., & Laub, D. R. (2014). Ulnar-mammary syndrome: Clinical presentation, genetic underpinnings, diagnosis, and treatment. *Eplasty, 14*, ic35.
Malpuech, G., Vanlieferinghen, P., Dechelotte, P., Gaulme, J., Labbe, A., & Guiot, F. (1988). Isolated familial adrenocorticotropin deficiency: Prenatal diagnosis by maternal plasma estriol assay. *American Journal of Medical Genetics, 29*, 125–130.
Marcano, A. C., Doudney, K., Braybrook, C., Squires, R., Patton, M. A., Lees, M. M., et al. (2004). TBX22 mutations are a frequent cause of cleft palate. *Journal of Medical Genetics, 41*, 68–74.
McDonald-McGinn, D. M., Reilly, A., Wallgren-Pettersson, C., Hoyme, H. E., Yang, S. P., Adam, M. P., et al. (2006). Malignancy in chromosome 22q11.2 deletion syndrome (DiGeorge syndrome/velocardiofacial syndrome). *American Journal of Medical Genetics. Part A, 140*, 906–909.
McDonald-McGinn, D. M., Sullivan, K. E., Marino, B., Philip, N., Swillen, A., Vorstman, J. A. S., et al. (2015). 22q11.2 deletion syndrome. *Nature Reviews. Disease Primers, 1*, 15071.
Meins, M., Henderson, D. J., Bhattacharya, S. S., & Sowden, J. C. (2000). Characterization of the human TBX20 gene, a new member of the T-box gene family closely related to the Drosophila H15 gene. *Genomics, 67*, 317–332.
Meneghini, V., Odent, S., Platonova, N., Egeo, A., & Merlo, G. R. (2006). Novel TBX3 mutation data in families with ulnar-mammary syndrome indicate a genotype-phenotype relationship: Mutations that do not disrupt the T-domain are associated with less severe limb defects. *European Journal of Medical Genetics, 49*, 151–158.
Merscher, S., Funke, B., Epstein, J. A., Heyer, J., Puech, A., Lu, M. M., et al. (2001). TBX1 is responsible for cardiovascular defects in velo-cardio-facial/DiGeorge syndrome. *Cell, 104*, 619–629.
Mesbah, K., Harrelson, Z., Theveniau-Ruissy, M., Papaioannou, V. E., & Kelly, R. G. (2008). Tbx3 is required for outflow tract development. *Circulation Research, 103*, 743–750.
Metherell, L. A., Savage, M. O., Dattani, M., Walker, J., Clayton, P. E., Farooqi, I. S., et al. (2004). TPIT mutations are associated with early-onset, but not late-onset isolated ACTH deficiency. *European Journal of Endocrinology/European Federation of Endocrine Societies, 151*, 463–465.
Miettinen, M., Wang, Z., Lasota, J., Heery, C., Schlom, J., & Palena, C. (2015). Nuclear Brachyury expression is consistent in Chordoma, common in germ cell tumors and small cell carcinomas, and rare in other carcinomas and sarcomas: An immunohistochemical study of 5229 cases. *The American Journal of Surgical Pathology, 39*, 1305–1312.
Morrison, K., Papapetrou, C., Attwood, J., Hol, F., Lynch, S. A., Sampath, A., et al. (1996). Genetic mapping of the human homologue (T) of mouse T(Brachyury) and a search for allele association between human T and spina bifida. *Human Molecular Genetics, 5*, 669–674.
Naiche, L. A., & Papaioannou, V. E. (2003). Loss of Tbx4 blocks hindlimb development and affects vascularization and fusion of the allantois. *Development, 130*, 2681–2693.
Newbury-Ecob, R. A., Leanage, R., Raeburn, J. A., & Young, I. D. (1996). Holt-Oram syndrome: A clinical genetic study. *Journal of Medical Genetics, 33*, 300–307.
Ogata, T., Niihori, T., Tanaka, N., Kawai, M., Nagashima, T., Funayama, R., et al. (2014). TBX1 mutation identified by exome sequencing in a Japanese family with 22q11.2 deletion syndrome-like craniofacial features and hypocalcemia. *PLoS One, 9*, e91598.
Packham, E. A., & Brook, J. D. (2003). T-box genes in human disorders. *Human Molecular Genetics, 12*(Spec No 1), R37–R44.
Palena, C., Roselli, M., Litzinger, M. T., Ferroni, P., Costarelli, L., Spila, A., et al. (2014). Overexpression of the EMT driver brachyury in breast carcinomas: Association with poor prognosis. *Journal of the National Cancer Institute, 106*.

Papapetrou, C., Putt, W., Fox, M., & Edwards, Y. H. (1999). The human TBX6 gene: Cloning and assignment to chromosome 16p11.2. *Genomics, 55*, 238–241.

Pauws, E., Hoshino, A., Bentley, L., Prajapati, S., Keller, C., Hammond, P., et al. (2009). Tbx22null mice have a submucous cleft palate due to reduced palatal bone formation and also display ankyloglossia and choanal atresia phenotypes. *Human Molecular Genetics, 18*, 4171–4179.

Pauws, E., Peskett, E., Boissin, C., Hoshino, A., Mengrelis, K., Carta, E., et al. (2013). X-linked CHARGE-like Abruzzo-Erickson syndrome and classic cleft palate with ankyloglossia result from TBX22 splicing mutations. *Clinical Genetics, 83*, 352–358.

Pennimpede, T., Proske, J., Konig, A., Vidigal, J. A., Morkel, M., Bramsen, J. B., et al. (2012). In vivo knockdown of Brachyury results in skeletal defects and urorectal malformations resembling caudal regression syndrome. *Developmental Biology, 372*, 55–67.

Peterson, J. F., Ghaloul-Gonzalez, L., Madan-Khetarpal, S., Hartman, J., Surti, U., Rajkovic, A., et al. (2014). Familial microduplication of 17q23.1-q23.2 involving TBX4 is associated with congenital clubfoot and reduced penetrance in females. *American Journal of Medical Genetics. Part A, 164A*, 364–369.

Pham, L. L., Garot, C., Brue, T., & Brauner, R. (2011). Clinical, biological and genetic analysis of 8 cases of congenital isolated adrenocorticotrophic hormone (ACTH) deficiency. *PloS One, 6*, e26516.

Pillay, N., Plagnol, V., Tarpey, P. S., Lobo, S. B., Presneau, N., Szuhai, K., et al. (2012). A common single-nucleotide variant in T is strongly associated with chordoma. *Nature Genetics, 44*, 1185–1187.

Pinto, F., Pertega-Gomes, N., Pereira, M. S., Vizcaino, J. R., Monteiro, P., Henrique, R. M., et al. (2014). T-box transcription factor brachyury is associated with prostate cancer progression and aggressiveness. *Clinical Cancer Research: An Official Journal of the American Association for Cancer Research, 20*, 4949–4961.

Plageman, T. F., Jr., & Yutzey, K. E. (2004). Differential expression and function of Tbx5 and Tbx20 in cardiac development. *The Journal of Biological Chemistry, 279*, 19026–19034.

Posch, M. G., Gramlich, M., Sunde, M., Schmitt, K. R., Lee, S. H., Richter, S., et al. (2010). A gain-of-function TBX20 mutation causes congenital atrial septal defects, patent foramen ovale and cardiac valve defects. *Journal of Medical Genetics, 47*, 230–235.

Poznanski, A. K., Gall, J. C., Jr., & Stern, A. M. (1970). Skeletal manifestations of the Holt-Oram syndrome. *Radiology, 94*, 45–53.

Pulichino, A. M., Vallette-Kasic, S., Couture, C., Brue, T., & Drouin, J. (2004). Tpit mutations reveal a new model of pituitary differentiation and account for isolated ACTH deficiency. *Médecine Sciences: M/S, 20*, 1009–1013.

Pulichino, A. M., Vallette-Kasic, S., Couture, C., Gauthier, Y., Brue, T., David, M., et al. (2003). Human and mouse TPIT gene mutations cause early onset pituitary ACTH deficiency. *Genes & Development, 17*, 711–716.

Pulichino, A. M., Vallette-Kasic, S., Tsai, J. P., Couture, C., Gauthier, Y., & Drouin, J. (2003). Tpit determines alternate fates during pituitary cell differentiation. *Genes & Development, 17*, 738–747.

Qian, L., Huang, Y., Spencer, C. I., Foley, A., Vedantham, V., Liu, L., et al. (2012). In vivo reprogramming of murine cardiac fibroblasts into induced cardiomyocytes. *Nature, 485*, 593–598.

Qian, L., Mohapatra, B., Akasaka, T., Liu, J., Ocorr, K., Towbin, J. A., et al. (2008). Transcription factor neuromancer/TBX20 is required for cardiac function in Drosophila with implications for human heart disease. *Proceedings of the National Academy of Sciences of the United States of America, 105*, 19833–19838.

Qiao, Y., Wanyan, H., Xing, Q., Xie, W., Pang, S., Shan, J., et al. (2012). Genetic analysis of the TBX20 gene promoter region in patients with ventricular septal defects. *Gene, 500*, 28–31.

Rallis, C., Bruneau, B. G., Del Buono, J., Seidman, C. E., Seidman, J. G., Nissim, S., et al. (2003). Tbx5 is required for forelimb bud formation and continued outgrowth. *Development, 130*, 2741–2751.

Ramirez, R. N., & Kozin, S. H. (2014). Ulnar mammary syndrome. *The Journal of Hand Surgery, 39*, 803–805.

Reamon-Buettner, S. M., & Borlak, J. (2004). TBX5 mutations in non-Holt-Oram syndrome (HOS) malformed hearts. *Human Mutation, 24*, 104.

Rodriguez-Esteban, C., Tsukui, T., Yonei, S., Magallon, J., Tamura, K., & Izpisua Belmonte, J. C. (1999). The T-box genes Tbx4 and Tbx5 regulate limb outgrowth and identity. *Nature, 398*, 814–818.

Sakiyama, J., Yamagishi, A., & Kuroiwa, A. (2003). Tbx4-Fgf10 system controls lung bud formation during chicken embryonic development. *Development, 130*, 1225–1234.

Sasaki, G., Ogata, T., Ishii, T., Hasegawa, T., Sato, S., & Matsuo, N. (2002). Novel mutation of TBX3 in a Japanese family with ulnar-mammary syndrome: Implication for impaired sex development. *American Journal of Medical Genetics, 110*, 365–369.

Scott, J. E., & Taor, W. S. (1979). The "small patella" syndrome. *The Journal of Bone and Joint Surgery, 61-B*, 172–175.

Seo, S., & Kume, T. (2006). Forkhead transcription factors, Foxc1 and Foxc2, are required for the morphogenesis of the cardiac outflow tract. *Developmental Biology, 296*, 421–436.

Singh, M. K., Christoffels, V. M., Dias, J. M., Trowe, M. O., Petry, M., Schuster-Gossler, K., et al. (2005). Tbx20 is essential for cardiac chamber differentiation and repression of Tbx2. *Development, 132*, 2697–2707.

Singh, R., Horsthuis, T., Farin, H. F., Grieskamp, T., Norden, J., Petry, M., et al. (2009). Tbx20 interacts with smads to confine tbx2 expression to the atrioventricular canal. *Circulation Research, 105*, 442–452.

Sletten, L. J., & Pierpont, M. E. (1996). Variation in severity of cardiac disease in Holt-Oram syndrome. *American Journal of Medical Genetics, 65*, 128–132.

Smemo, S., Campos, L. C., Moskowitz, I. P., Krieger, J. E., Pereira, A. C., & Nobrega, M. A. (2012). Regulatory variation in a TBX5 enhancer leads to isolated congenital heart disease. *Human Molecular Genetics, 21*, 3255–3263.

Smith, J. C., Price, B. M., Green, J. B., Weigel, D., & Herrmann, B. G. (1991). Expression of a Xenopus homolog of Brachyury (T) is an immediate-early response to mesoderm induction. *Cell, 67*, 79–87.

Smith, A. T., Sack, G. H., Jr., & Taylor, G. J. (1979). Holt-Oram syndrome. *The Journal of Pediatrics, 95*, 538–543.

Smoll, N. R., Gautschi, O. P., Radovanovic, I., Schaller, K., & Weber, D. C. (2013). Incidence and relative survival of chordomas: The standardized mortality ratio and the impact of chordomas on a population. *Cancer, 119*, 2029–2037.

Song, K., Nam, Y. J., Luo, X., Qi, X., Tan, W., Huang, G. N., et al. (2012). Heart repair by reprogramming non-myocytes with cardiac transcription factors. *Nature, 485*, 599–604.

Sparrow, D. B., McInerney-Leo, A., Gucev, Z. S., Gardiner, B., Marshall, M., Leo, P. J., et al. (2013). Autosomal dominant spondylocostal dysostosis is caused by mutation in TBX6. *Human Molecular Genetics, 22*, 1625–1631.

Speer, M. C., Melvin, E. C., Viles, K. D., Bauer, K. A., Rampersaud, E., Drake, C., et al. (2002). T locus shows no evidence for linkage disequilibrium or mutation in American Caucasian neural tube defect families. *American Journal of Medical Genetics, 110*, 215–218.

Stacchiotti, S., Sommer, J., & Consensus, Chordoma Global. (2015). Building a global consensus approach to chordoma: A position paper from the medical and patient community. *The Lancet Oncology, 16*, e71–e83.

Stalmans, I., Lambrechts, D., De Smet, F., Jansen, S., Wang, J., Maity, S., et al. (2003). VEGF: A modifier of the del22q11 (DiGeorge) syndrome? *Nature Medicine, 9*, 173–182.

Stennard, F. A., Costa, M. W., Elliott, D. A., Rankin, S., Haast, S. J., Lai, D., et al. (2003). Cardiac T-box factor Tbx20 directly interacts with Nkx2-5, GATA4, and GATA5 in regulation of gene expression in the developing heart. *Developmental Biology, 262*, 206–224.

Stennard, F. A., Costa, M. W., Lai, D., Biben, C., Furtado, M. B., Solloway, M. J., et al. (2005). Murine T-box transcription factor Tbx20 acts as a repressor during heart development, and is essential for adult heart integrity, function and adaptation. *Development, 132*, 2451–2462.

Suphapeetiporn, K., Tongkobpetch, S., Siriwan, P., & Shotelersuk, V. (2007). TBX22 mutations are a frequent cause of non-syndromic cleft palate in the Thai population. *Clinical Genetics, 72*, 478–483.

Takemoto, T., Uchikawa, M., Yoshida, M., Bell, D. M., Lovell-Badge, R., Papaioannou, V. E., et al. (2011). Tbx6-dependent Sox2 regulation determines neural or mesodermal fate in axial stem cells. *Nature, 470*, 394–398.

Takeuchi, J. K., Koshiba-Takeuchi, K., Matsumoto, K., Vogel-Hopker, A., Naitoh-Matsuo, M., Ogura, K., et al. (1999). Tbx5 and Tbx4 genes determine the wing/leg identity of limb buds. *Nature, 398*, 810–814.

Takeuchi, J. K., Koshiba-Takeuchi, K., Suzuki, T., Kamimura, M., Ogura, K., & Ogura, T. (2003). Tbx5 and Tbx4 trigger limb initiation through activation of the Wnt/Fgf signaling cascade. *Development, 130*, 2729–2739.

Tidake, A., Gangurde, P., Shaikh, Z., & Mahajan, A. (2015). Tetralogy of Fallot with Holt-Oram syndrome: Case report and review. *Clinical Research in Cardiology: Official Journal of the German Cardiac Society, 104*, 790–793.

Torchinsky, M. Y., Wineman, R., & Moll, G. W. (2011). Severe Hypoglycemia due to isolated ACTH deficiency in children: A new case report and review of the literature. *International Journal of Pediatrics, 2011*, 784867.

Vallette-Kasic, S., Brue, T., Pulichino, A. M., Gueydan, M., Barlier, A., David, M., et al. (2005). Congenital isolated adrenocorticotropin deficiency: An underestimated cause of neonatal death, explained by TPIT gene mutations. *The Journal of Clinical Endocrinology and Metabolism, 90*, 1323–1331.

Verhagen, J. M., Diderich, K. E., Oudesluijs, G., Mancini, G. M., Eggink, A. J., Verkleij-Hagoort, A. C., et al. (2012). Phenotypic variability of atypical 22q11.2 deletions not including TBX1. *American Journal of Medical Genetics. Part A, 158A*, 2412–2420.

Vincent, S. D., Mayeuf-Louchart, A., Watanabe, Y., Brzezinski, J. A., 4th, Miyagawa-Tomita, S., Kelly, R. G., et al. (2014). Prdm1 functions in the mesoderm of the second heart field, where it interacts genetically with Tbx1, during outflow tract morphogenesis in the mouse embryo. *Human Molecular Genetics, 23*, 5087–5101.

Voss, A. K., Vanyai, H. K., Collin, C., Dixon, M. P., McLennan, T. J., Sheikh, B. N., et al. (2012). MOZ regulates the Tbx1 locus, and Moz mutation partially phenocopies DiGeorge syndrome. *Developmental Cell, 23*, 652–663.

Vujovic, S., Henderson, S., Presneau, N., Odell, E., Jacques, T. S., Tirabosco, R., et al. (2006). Brachyury, a crucial regulator of notochordal development, is a novel biomarker for chordomas. *The Journal of Pathology, 209*, 157–165.

Waldron, L., Steimle, J. D., Greco, T. M., Gomez, N. C., Dorr, K. M., Kweon, J., et al. (2016). The cardiac TBX5 interactome reveals a chromatin remodeling network essential for cardiac septation. *Developmental Cell, 36*, 262–275.

Wang, C., Cao, D., Wang, Q., & Wang, D. Z. (2011). Synergistic activation of cardiac genes by myocardin and Tbx5. *PLoS One, 6*, e24242.

Wang, K. E., Wu, Z., Tian, K., Wang, L., Hao, S., Zhang, L., et al. (2015). Familial chordoma: A case report and review of the literature. *Oncology Letters, 10*, 2937–2940.

White, P. H., Farkas, D. R., McFadden, E. E., & Chapman, D. L. (2003). Defective somite patterning in mouse embryos with reduced levels of Tbx6. *Development, 130*, 1681–1690.

Wilkinson, D. G., Bhatt, S., & Herrmann, B. G. (1990). Expression pattern of the mouse T gene and its role in mesoderm formation. *Nature*, *343*, 657–659.
Wu, N., Ming, X., Xiao, J., Wu, Z., Chen, X., Shinawi, M., et al. (2015). TBX6 null variants and a common hypomorphic allele in congenital scoliosis. *The New England Journal of Medicine*, *372*, 341–350.
Wu, Z., Zhang, J., Zhang, L., Jia, G., Tang, J., Wang, L., et al. (2010). Prognostic factors for long-term outcome of patients with surgical resection of skull base chordomas-106 cases review in one institution. *Neurosurgical Review*, *33*, 451–456.
Yagi, H., Furutani, Y., Hamada, H., Sasaki, T., Asakawa, S., Minoshima, S., et al. (2003). Role of TBX1 in human del22q11.2 syndrome. *Lancet*, *362*, 1366–1373.
Yamagishi, H., Maeda, J., Hu, T., McAnally, J., Conway, S. J., Kume, T., et al. (2003). Tbx1 is regulated by tissue-specific forkhead proteins through a common Sonic hedgehog-responsive enhancer. *Genes & Development*, *17*, 269–281.
Yang, X. R., Ng, D., Alcorta, D. A., Liebsch, N. J., Sheridan, E., Li, S., et al. (2009). T (brachyury) gene duplication confers major susceptibility to familial chordoma. *Nature Genetics*, *41*, 1176–1178.
Yarosh, W., Barrientos, T., Esmailpour, T., Lin, L., Carpenter, P. M., Osann, K., et al. (2008). TBX3 is overexpressed in breast cancer and represses p14 ARF by interacting with histone deacetylases. *Cancer Research*, *68*, 693–699.
Yi, C. H., Russ, A., & Brook, J. D. (2000). Virtual cloning and physical mapping of a human T-box gene, TBX4. *Genomics*, *67*, 92–95.
Yi, C. H., Terrett, J. A., Li, Q. Y., Ellington, K., Packham, E. A., Armstrong-Buisseret, L., et al. (1999). Identification, mapping, and phylogenomic analysis of four new human members of the T-box gene family: EOMES, TBX6, TBX18, and TBX19. *Genomics*, *55*, 10–20.
Young, V. A., Curtis, K. M., Temple, H. T., Eismont, F. J., DeLaney, T. F., & Hornicek, F. J. (2015). Characteristics and patterns of metastatic disease from chordoma. *Sarcoma*, *2015*, 517657.
Zhang, W., Chen, H., Wang, Y., Yong, W., Zhu, W., Liu, Y., et al. (2011). Tbx20 transcription factor is a downstream mediator for bone morphogenetic protein-10 in regulating cardiac ventricular wall development and function. *The Journal of Biological Chemistry*, *286*, 36820–36829.
Zhang, X. L., Qiu, X. B., Yuan, F., Wang, J., Zhao, C. M., Li, R. G., et al. (2015). TBX5 loss-of-function mutation contributes to familial dilated cardiomyopathy. *Biochemical and Biophysical Research Communications*, *459*, 166–171.
Zhou, W., Zhao, L., Jiang, J. Q., Jiang, W. F., Yang, Y. Q., & Qiu, X. B. (2015). A novel TBX5 loss-of-function mutation associated with sporadic dilated cardiomyopathy. *International Journal of Molecular Medicine*, *36*, 282–288.

INDEX

Note: Page numbers followed by "*f*" indicate figures, and "*t*" indicate tables.

A

Adult DG, Tbr1 and Tbr2
  expression, 286–287
  functions, 293–294
Adult SVZ, Tbr1 and Tbr2
  expression, 287–288
  functions, 294–295
Alary muscle-derived cells (AMDCs), 183, 185
Alary muscles (AMs), 185
  embryonic development, 183–185
  Org-1 expression in, 183
  TARMs and, 180–181
Amphi4/5 gene, 368–369
Amygdala, Tbr1 and Tbr2
  expression, 288
  functions, 295
Animal–vegetal axis, 126
Ankyloglossia, 403–405
Anterior–posterior (AP) axis, 94–95
Anterior visceral endoderm (AVE), 98–99
Apical ectodermal ridge (AER), 356–357
  maturation, Brachyury in, 373–374
  Tbx3, 356–357, 362*f*
Apical IPs (IP-a), 284–285
Ascidians, T-box genes
  brachyury and temporal control, 59–66
  maternal T and tail development, 76–77
  Tbx1/10, 74–75
  Tbx2/3, 66–68, 81–82
  Tbx6, 68–70
  Tbx6b/Tbx6c, 70–74
  Tbx15/18/22, 75–76
  Tbx20, 77–78
  and metamorphosis, 58*f*
  transcriptional regulation, 78–82
Atrial septum, 206–207
Autism spectrum disorders (ASDs), 305
Autosomal signaling elements (ASEs), 45

B

Baf60a, 226
Basal forebrain, Tbr1 and Tbr2
  expression, 288
  functions, 295
Basal IPs (IP-b), 284–285
Basal/outer RGPs (bRGPs), 284
Bifid, 330–331
Bladder development, *Tbx2/Tbx3*, 269–270
Bmp-signaling, 265–266, 269
Bone morphogenetic protein (BMP), 356–357
Bone morphogenetic protein 10 (BMP10), 401–402
Brachyenteron, 163–166
Brachyury, 384
  activators and repressor, 79–81
  in AER maturation, 373–374
  cloning in mouse, 384–390, 386–387*t*
  controls posterior fate restrictions, 132–134
  heterologous expression assays, 13*f*
  metazoan T-box genes, 6
  in morphogenetic movements, 15–17
  notochord specification, 59–66
  premetazoan and early-metazoan, 10–14
  short-tail phenotype, 249–250
  Sox2 and, 142*f*
  unicellular regulatory network, 18*f*
Brain
  Tbr1
    expression, 280–283
    functions, 290–296
    regulators, 298–299*t*
  Tbr2 regulators, 298–299*t*
Branchiostoma floridae, 7
Brinker (Brk), 320–321, 325–326

C

Caenorhabditis elegans, T-box genes
  blastomere identity, 38–39, 38*f*
  caenorhabditis genes, 36*t*
  developmental functions, 46–47
  DNA-binding specificity, 34–37
  epithelial cells, 42*f*
  gene regulatory network, 44*f*

Caenorhabditis elegans, T-box genes (*Continued*)
genomic locations, 32*f*
in hermaphrodites and males, 45–46
hindgut, 42–45, 42*f*
hypodermis, 42–45
multiple sequence alignment, 35*f*
and nematodes, 28–34
nuclear localization, 48
pharyngeal development, 37–40, 38*f*
phylogenetic tree, 33–34*f*
posttranslational mechanisms, 47–48
SEA-1, 45–46
stereotyped development, 28
Tbx-1, 41–43
Tbx-2, 39–41
Tbx-8, 43–45
Tbx-9, 43–45
Cajal–Retzius cells, 293
Callosal fibers, 291–293
Capsaspora owczarzaki, 4–5, 11–14, 12*f*
Cardiac conduction system development, TBX5, 207
Cardiac development, 228–234
Cardiac morphologic development, TBX5
atrial septum, 206–207
ventricular septum, 206
Cardioblasts (CBs)
alignment, 169–170
formation, 172–174
progenitors, 175–176
subtypes, 177
Cardiogenesis, *Drosophila*, 163–166
Cardiogenic mesoderm (CM), 169–170
Cardiopharyngeal lineage, 233–234
β-Catenin signaling, 370–371
Caudal-related homeobox2 (Cdx2), 95–96
Caudal visceral mesoderm (CVM), 163–166
CBs. *See* Cardioblasts (CBs)
CEH-39, 45–46
Cerebellum, Tbr1 and Tbr2
expression, 289
functions, 295
Cerebral cortex
downstream target genes regulation, 303–305
Tbr1, 290–293, 300*t*
Tbr2, 300*t*
Tbx1, 236–237
CHARGE syndrome, 226–227
ChIP-seq, 224–225
Chordoma, 384–390, 386–387*t*
Chordoneural hinge, 119–120, 129, 138
Chromatin interactions, 226–227
Ciona, 57–58, 58*f*
C. Brachyury, 59–64
C. intestinalis, 61*t*, 62*f*
Circular visceral muscle founder cells (ciVM-FC), 167–169
Cis-regulatory modules (CRMs), 64, 65*f*
ciVM-FC. *See* Circular visceral muscle founder cells (ciVM-FC)
Cloaca
endodermal infolding of, 248
Tbx20, 265–266
Conditional knockout (cKO), *Tbr2*, 290–291
Congenital anomalies of the kidney and urinary tract (CAKUT), 260–261
Congenital heart disease (CHD), 396–398, 401–403
Cortical glutamatergic neurogenesis, Tbr2 and Tbr1, 284–286
Cortical plate (CP), 282, 292*f*
Cousin syndrome, 399–400
Creolimax fragrantissima, 17
Ctenophora, 6–7
Ctip1, 302
Ctip2, 302
CVM. *See* Caudal visceral mesoderm (CVM)

D

Deep cerebellar nuclei (DCN), 289
22q11.2 Deletion syndrome, 223–224, 390–392
Dentate migratory stream (DMS), 286
Dentate neuroepithelium (DNe), 286
DiGeorge syndrome, 223–224, 262–263, 390–392
Distal visceral endoderm (DVE), 98–99
DNA-binding properties, 224–226, 225*t*
DNA-binding specificity, 34–37
Dorsocross (Doc), 325–328, 339

cardiac specification, 172–174
CVM development, 163–166
Tbx20-related genes, 175–177
Drosophila
limb development
embryonic origin, 316–319, 317–318*f*
in haltere development, 340–341, 341*f*
leg development, 320–323, 321*f*
omb, 315–316, 320–323
orthology, 314–316
pattern formation, 319
wing development, 325–340
wing disc, 323–325, 325*f*
mesoderm development, 162–163, 162*t*, 171*f*
adult musculature, 182–185
cardioblasts, 169–170, 172–177
cardiogenesis, 169–177
chromatin immunoprecipitation, 172–174
CVM development, 163–166
Dorsocross genes, 163–166, 172–177
embryonic somatic muscle, 177–182, 179–180*f*
larval somatic musculature, 178
midline gene, 181–182
org-1 expression, 178–181
TVM development, 166–169

E

Embryonic AP axis, *Eomes*, 98–99
Embryonic midbrain, Tbr2, 290
Embryonic somatic muscle, *Drosophila*, 177–182, 179–180*f*
Embryonic stem (ES) cells, 95, 141–142
Eminentia thalami, Tbr1/Tbr2, 288
Endoderm (DE) cells, 102
Endoderm progenitors, 94
Endothelial cells (ECs), Tbx1, 234–236
Enhancer
in *Ciona*, 63–64
Ci-snail muscle, 75–76
notochord, 80*f*
Tbx6-binding sites, 73–74
Eomesodermin (Eomes)
during AVE/DVE induction and gastrulation, 103–104*f*
controls anterior fate restrictions, 131–132
embryonic AP axis, 98–99
gastrulation onset, 109–110
germ layers formation
functions, 99–106
transcriptional activities, 106–109
metazoan, 6–7
mRNA, 101
in trophectoderm, 95–98
Xenopus, 99–106, 100*f*
Eosin-stained chordoma, 389*f*
Epithelial-to-mesenchymal transition (EMT), 120–121
Excretory system
in adult fish and amphibia, 246
in mouse and man, 246–247
Tbx18 in, 255*f*
vertebrates, 252–253*t*
Extracellular matrix (ECM), 316

F

Facial dysmorphism, 395
Fate restriction
Brachyury controls posterior, 132–134
Eomes controls anterior, 131–132
mesoderm, 133, 135–141
Tbx6 subfunctionalization, 134–135
Fgf10, 356–357, 364*f*, 368, 373–374, 394–395
Fibroblast growth factor (FGF) signals, 356–357
Filasterea, 2–5, 7
Fluorescent in situ hybridization (FISH), 391
Forelimb, 367–372
Fusion-competent myoblasts (FCMs), 178

G

Gastrulation
around mid-gastrulation, 129
cell movements of, 118
continuation of, 119
Eomes, 127–129
germ layer segregation, 119–120
onset during *Eomes*, 109–110
Germ layers, T-box factors, 118
determination of regional identity, 140
distinct subtypes, 131–141
dynamic expression patterns, 127*f*

Germ layers, T-box factors (*Continued*)
Eomesodermin, 121–122
evolution of, 121–122
gastrulation, 118–122, 130
genome-wide chromatin profiling, 122–123
Holmdahl's blastema, 118–120
implications for reverse engineering ES cells, 141–142
induction and maintenance of, 135
marginal zone, 126–127
mesodermal subtypes, 135–141
specification, 139
spinal cord, 118, 129, 133, 140–142
subtype restriction, 138
Vogt's predetermination model, 118–120
zones of, 126–131
Glutamatergic neurogenesis, Tbr1/Tbr2, 284
Glycogen synthase kinase 3 (GSK-3), 302
Gonadal mesoderm development, 163–169
Gremlin1, 366
Groucho-family corepressor, 47

H

Halocynthia, 57–58, 82
Haploinsufficiency, 227
Heart development
Tbx5
atrial septum, 206–207
cardiac conduction system development, 207
embryonic and adult, 198–200
extracardiac expression, 200
gene regulatory network, 209–213
Holt–Oram syndrome, 203–205
mammalian, 208
transcriptional regulation, 200–202
ventricular septum, 206
zebrafish, 208–209
Tbx6b and Tbx6c, 73–74
Hematoxylin, 389*f*
Hermaphrodite-specific neurons (HSNs), 40–41
Heterotopic grafts, 119–120
H15 expression, 320–322
High mobility group (HMG)-box transcription factor, 302
Hindlimb, 367–372
H3K4, 227
H3K27, 227
H3K79, 302–303
Holmdahl's blastema, 118–120
Holozoa, 4–6
Holt–Oram syndrome (HOS), TBX5, 367–368, 396–398, 398*f*
animal model, 205
haploinsufficiency, 203–205
Homeostasis, *Tbx2/Tbx3*, 269–270
Hox genes, 370
Hydra magnipapillata, 2–3
Hydronephrosis, 259–261

I

Ichthyosporea, 4–5
Induced TSCs (iTSCs), 97–98
Inner cell mass (ICM), 95–96
In situ hybridization (ISH), 282
Intermediate progenitors (IPs), 284
Iroquois gene complex *(Iro-C)*, 331
Isolated ACTH deficiency, 400–401

J

Jun N-terminal kinase (JNK), 328–329

K

Kidney development, 246–249
developmental regulators, 249–252
metanephric kidney, 246–247
pronephros, 246
Tbx1, 262–264
Tbx2, 266–270
Tbx3, 266–270
Tbx18, 253–262
Tbx20, 264–266

L

Lateral adult muscle precursors (lAMPs), 180–181
Lateral body wall muscles, 181–182
Lateral plate mesoderm (LPM), 356–357
Limb development, 356–357
bud initiation, 356–357
chondrogenic precursors, 372–373
hindlimb and forelimb, 367–372
paralogous genes, 361–367
T-box genes expressed in, 358–360*t*
Tbx2/3/4/5 subfamily, 357–361

Lim homeobox protein 1 (Lhx1), 99
Longitudinal visceral musculature (loVM), 163–166

M

MAB-9, 41–43
Mammalian Tbx5, 208
Maternal T, and tail development, 76–77
Mesenchymal cells, Tbx18, 247–248, 257–259, 261–262
Mesp1/2 gene induction, 106
Metanephric kidney, 246–247
Metazoan T-box genes, 6
  Brachyury, 6, 10–14
  Eomes/Tbrain class, 6–7
  Tbx1/15/20 class, 9
  Tbx2/3, 8
  Tbx 4/5 class, 8–9
  Tbx6 class, 9–10
  Tbx7 class, 7
  Tbx8 class, 7–8
Mga transcript, 282–283
Mid, in *Drosophila* wing development, 325–328, 339–340
Ministeria vibrans, 4–5
MLS-1, 41–43
Mnemiopsis leyidi, 8–10
Morphogenesis
  body wall muscle cells, 43–45
  in *Drosophila* wing imaginal disc, 317–318*f*
  transcription regulation to, 228
Motor control, Tbr1, 296
Mouse behavior, Tbr1, 296
Muscle development
  Tbx1/10, 74–75
  Tbx6b/Tbx6c, 70–73
  Tbx15/18/22, 75–76, 80*f*
Mutations
  T-box genes, 384, 385*f*
  TBX1, 392
  TBX3, 392–393
  TBX4, 395
  TBX5, 396–397
  TBX6, 398–399

N

Nematodes, 28–29
Nephric duct, 247–248
Neural retina
  T-box gene expression in, 280–283
  Tbr1 in, 295–296
Neuronal development, 40–41
Northern blot analysis, 269, 390
Notochord
  brachyury and temporal control, 59–66
  Tbx2/3 brachyury function in, 66–68
Nuclear export signal (NES), 124–125
Nuclear localization sequences (NLS), 197
Nucleosome remodeling and deacetylase (NuRD) complex, 212

O

Oikopleura dioica, 59–63
Olfactory bulbs (OBs), Tbr1 and Tbr2
  expression, 287–288
  functions, 294–295
Optomotor-blind (omb), 325–328
  ectopic expression, 322–323
  ectopic wing pair development, 338
  functions in wing development, 328–338
  high dosage sensitivity, 329–330
  loss and gain of function, 329*f*
  posterior wing vein L5 specification, 337–338
  in wing disc anterior compartment, 332, 333*f*
  wing disc A/P boundary, 336–337
  in wing disc posterior compartment, 332–335, 335*f*
  in wing hinge development, 330–331
Organogenesis, 266–268
Outflow tract (OFT), 229–231

P

Pharyngeal apparatus (PA), 223–224, 228–234
Pharyngeal arch arteries, 228
Pharyngeal gill slits (PGS), 57–58
Pharyngeal muscles, Tbx1/10, 74–75
Phasmid B neurons (PHBs), 40–41
Phylogeny, 2–3
Porifera-sister hypothesis, 6–7
Premetazoan
  Brachyury homologs, 10–14
  T-box genes, 3–6
Prepulse inhibition (PPI), 236–237
Primitive streak (PS), 102
Pronephric duct, 246

Pronephric tubules, 247–248
Pronephros, 246, 268–269
Proopiomelanocortin (POMC), 282–283, 400
Protein interactions, 226–227
Ptychodera flava, 7
Pulmonary arterial hypertension (PAH), 395

R

Radial glial progenitors (RGPs), 284
Retina
  Tbr1 subfamily members in, 295–296
  Tbr2 in, 289
Retinal ganglion cells (RGCs), 289
Retinoic acid (RA), 370
RNAi screens, 46–47
Rostral migratory stream (RMS), 287
RT-PCR analysis, Tbx1, 263

S

Scoliosis, 399
SEA-1, 36–37, 45–46
Second heart field (SHF), 228–233
Selex, 224–225
Sensory adaptation, 40–41
Setd7, 226
Sex combs reduced (Scr), 321–322
Small patella syndrome (SPS), 394–395
Smoothened (Smo), 332
Soft tissue patterning, Tbx5 and Tbx4, 371–372
Somatic gonadal precursors (SGPs), 169
Somatosensory (S1) cortex, 237
Somitogenesis, 134–135
Sonic Hedgehog (Shh), 356–357
Sox transcription factor casanova (cas/sox32), 101–102
Spinal cord, 118, 129, 133, 140–142
Spondylocostal dysostosis (SCD), 398–399
Steric hindrance, 123–124
Stromal cell, *Tbx18*, 261–262
Subfunctionalization process, 14, 134–135
Subgranular zone (SGZ), 286–287
SUMOylation, 47–48
Sycon ciliatum, 7, 11–15
Systematic evolution of ligands by exponential enrichment (SELEX), 122–124

T

Tailbud
  amphibian, 119
  chordoneural hinge tissue, 119
  outgrowth, 135
  stages, 120, 123*f*, 127*f*, 137*f*
  tissue, 118
  Vogt's predetermination model, 119–120
TBE. *See* T-box-binding element (TBE)
T-bet, 282
  mutations, 305
  in OB and adult SVZ, 287–288
T-box-binding element (TBE), 34–37, 249–251, 254–256
T-box genes
  ascidians
    before and after metamorphosis, 58*f*
    brachyury and temporal control, 59–66
    Tbx1/10, 74–75
    Tbx2/3, 66–68, 81–82
    Tbx6, 68–70
    Tbx6b/Tbx6c, 70–74
    Tbx15/18/22, 75–76
    Tbx20, 77–78
    maternal T and tail development, 76–77
    transcriptional regulation, 78–82
  Brachyury, 249–251
  Caenorhabditis elegans
    blastomere identity, 38–39, 38*f*
    caenorhabditis genes, 36*t*
    developmental functions, 46–47
    DNA-binding specificity, 34–37
    epithelial cells, 42*f*
    gene regulatory network, 44*f*
    genomic locations, 32*f*
    hindgut, 42–45, 42*f*
    hypodermis, 42–45
    multiple sequence alignment, 35*f*
    and nematodes, 28–34
    nuclear localization, 48
    pharyngeal development, 37–40, 38*f*
    phylogenetic tree, 33–34*f*
    posttranslational mechanisms, 47–48
    SEA-1, 45–46
    sex determination, 45–46
    stereotyped development, 28
    Tbx-1, 41–43

Tbx-2, 39–41
Tbx-8, 43–45
Tbx-9, 43–45
in developing brain and neural retina, 280–283
downstream target genes regulation, 303–305
Drosophila limb development
embryonic origin, 316–319, 317–318*f*
in haltere development, 340–341, 341*f*
leg development, 320–323, 321*f*
omb, 315–316, 320–323
orthology, 314–316
pattern formation, 319
wing development, 325–340
wing disc, 323–325, 325*f*
Drosophila mesoderm development, 162–163, 162*t*, 171*f*
adult musculature, 182–185
cardioblasts, 169–170, 172–177
cardiogenesis, 169–177
chromatin immunoprecipitation, 172–174
CVM development, 163–166
embryonic somatic muscle, 177–182, 179–180*f*
larval somatic musculature, 178
midline gene, 181–182
org-1 expression, 178–181
TVM development, 166–169
excretory system of vertebrates, 252–253*t*
function
embryonic roles, 126–141
genetic regulatory network, 137*f*
germ layers, 121–122
mesodermal subtypes, 135–141
kidney and urinary tract, 246–249
developmental regulators, 249–252
Tbx1, 262–264
Tbx2, 266–270
Tbx3, 266–270
Tbx18, 253–262
Tbx20, 264–266
metazoan, 6
mutation
in brain, 305
spectrum in, 385*f*
phylogenetic analysis, 251*f*, 266
phylogenetic classification, 2–3
premetazoan, 3–6
regulation of, 17–20, 296–303
Tbr1 subfamily, 283–290
transcription factors, 122–125
Tbr1, 282
binding partners of, 303
brain distributions, 283*f*
in cerebral cortex, 300*t*
cortical phenotypes in, 292*f*
domain organization, 281*f*
expression
adult DG, 286–287
adult SVZ, 287–288
amygdala, 288
basal forebrain, 288
in cerebellum, 289
cortical glutamatergic neurogenesis, 284–286
eminentia thalami, 288
glutamatergic neurogenesis, 284
OB, 287–288
functions
adult DG, 293–294
adult SVZ neurogenesis, 294–295
amygdala development, 295
in basal forebrain, 295
in cerebellum development, 295
cerebral cortex development, 290–293
in mouse behavior and motor control, 296
in neural retina development, 295–296
OB development, 294–295
multiple protein sequence alignment, 281*f*
regulation of, 302–303
regulators in brain, 298–299*t*
subfamily members, 281*f*
in T-box domain, 280–282
transcription factor cascade including, 285*f*
Tbr2, 282
adult DG, 293–294
conditional ablation of, 294
cortical phenotypes in, 292*f*
in developing cerebral cortex, 300*t*
expression
adult DG, 286–287
adult SVZ, 287–288
amygdala, 288
basal forebrain, 288

Tbr2 (*Continued*)
cerebellum, 289
cortical glutamatergic neurogenesis, 284–286
embryonic midbrain, 290
eminentia thalami, 288
glutamatergic neurogenesis, 284
OB, 287–288
retina, 289
regulation, 297–302
regulators in brain, 298–299*t*
T-box domain, 280–282
transcription factor cascade including, 285*f*
Tbx, 28
in amniote limb, 364*f*
EOMES, 106–109
expression domains, 362*f*
Tbx1, 41–43, 223–224, 314–315, 386–387*t*, 390–392
cardiac and great artery abnormalities, 230*f*
cardiopharyngeal region, 262–263
and cerebral cortex development, 236–237
consensus binding sequence, 224–225
DNA-binding properties, 224–226
embryonic somatic muscle, 177–182
histone modifiers, 226, 237–238
kidney development, 262–264
in limb chondrogenic precursors, 372–373
in mouse embryos, 232*f*
PA and cardiac development, 228–234
protein and chromatin interactions, 226–227
roles in ECs, 234–236
RT-PCR analysis, 263
transcription regulation
to morphogenesis, 228
strong repressor, 227
and vascular development, 234–236
and VEGF receptors, 236*f*
Tbx1/10, pharyngeal muscles, 74–75
Tbx1/15/20, 9
Tbx2, 39–41
bladder development and homeostasis, 269–270
Bmp-signaling, 265–266, 269
limb development, 357–361
Northern blot analysis, 269
paralogous genes, 361–367
Xenopus pronephros development, 268–269
Tbx2/3
brachyury function in notochord, 66–68
classes and function, 8
expression pattern, 81–82
Tbx3, 386–387*t*, 392–394
bladder development and homeostasis, 269–270
limb development, 357–361
paralogous genes, 361–367
Tbx4, 386–387*t*, 394–395
hindlimb and forelimb, 367–372
limb development, 357–361
regulation, 370–371
in soft tissue patterning, 371–372
Tbx 4/5, 8–9
Tbx5, 196–198, 386–387*t*, 396–398
cardiac conduction system development, 207
cardiac morphologic development
atrial septum, 206–207
ventricular septum, 206
expression
embryonic and adult heart, 198–200
extracardiac expression, 200
transcriptional regulation, 200–202
hindlimb and forelimb, 367–372
Holt–Oram syndrome
animal model, 205
haploinsufficiency, 203–205
limb development, 357–361
mammalian, 208
regulation, 370–371
regulatory network
inhibition of noncardiomyocyte, 212
positive transcriptional activation, 209–212, 210*f*
targets of, 213
in soft tissue patterning, 371–372
zebrafish, 208–209
Tbx6, 386–387*t*, 398–399
classes and function, 9–10

lineage-specific duplication in ascidians, 68–69
subfunctionalization to control both fate restrictions, 134–135
Tbx6a, 69–70
Tbx6b
autoregulation and crossregulation, 82
in heart development, 73–74
muscle development, 70–73
Tbx6c
autoregulation and crossregulation, 82
in heart development, 73–74
muscle development, 70–73
Tbx7, 7, 29
Tbx8, 7–8, 43–45
Tbx-8/tbx-9, 29
Tbx-9, 43–45
Tbx15, 386–387*t*, 399–400
Tbx15/18/22, 75–76, 80*f*
Tbx18, 253–257, 282–283
eh1-motif deletion, 254–256
in excretory system, 255*f*
ink injection experiments, 259
kidney development, 261–262
in limb mesenchyme, 372–373
mesenchymal cells, 257–259, 261–262
ureter development, 257–261
whole exome sequencing, 260–261
Xenopus cDNA, 253–254
Tbx19, 282–283, 386–387*t*, 400–401
Tbx20, 41–42, 77–78, 314–315, 386–387*t*, 401–403
bladder/cloacal development, 264–266
Drosophila melanogaster, 264–265
embryonic somatic muscle, 177–182
zebrafish cloaca development, 265–266
Tbx22, 41–42, 386–387*t*, 403–405
Tbx30, 29
Tbx-33, 46–47
Tbx-34, 46
Tbx-35, 40
Tbx-36, 46–47
Tbx-37, 39–40
Tbx-38, 39–40
Tbx-41, 46
Tbx-42, 29
T-domain, 384
TEA domain family member4 (Tead4), 95–96
Temporal control, notochord specification, 59–66
Thalamo-cortical axons, 291–293
Transcription factors
dosage-sensitive, 384
ETS-1, 404
paired-type homeodomain, 394–395
Tribolium, 314–315
Trophectoderm (TE) cells, 94–98
development, 94–95
intrauterine, 95–96
Trophoblast stem cells (TSCs), 95–96
Trunk visceral mesoderm (TVM), 163–169, 172–174

U

Ulnar–Mammary syndrome (UMS), 268–269, 361–363, 392–394, 393*f*
Ultrabithorax (Ubx), 338
UNC-37, 47
Unipolar brush cells (UBCs), 289
Ureter development, *Tbx18*, 257–261
Ureteric bud, 247–248
Urinary tract, T-box genes, 246–249
developmental regulators, 249–252
Tbx1, 262–264
Tbx2, 266–270
Tbx3, 266–270
Tbx18, 253–262
Tbx20, 264–266

V

Vascular development, Tbx1, 234–236
Vegfr2, 235–236
Vegfr3, 235–236
Velocardiofacial syndrome, 262–263, 390–392
Ventral longitudinal muscles (VLMs), 170–171, 182–185
Ventricular septum, 206
Ventricular zone (VZ), 284
VE promotes formation, *Eomes* in, 98–99
Vesicular glutamate transporter (VGluT), 294
Visceral endoderm (VE), 94–95
Visceral mesoderm development, 163–169

VLMs. *See* Ventral longitudinal muscles (VLMs)
Vogt's predetermination model, 118–120

W

Whole exome sequencing, Tbx18, 260–261
Wnt5a, 233
Wnt-signaling, 258, 370–371

X

X-chromosome signaling elements (XSEs), 45
Xenopus, 99–106, 100*f*
  embryos, 11, 136*f*
  endoderm development, 99–101
  pronephros, 268–269
  Tbx18 cDNA, 253–254
X-linked cleft palate (CPX), 403–405
X-ray crystallography, 122–123

Z

Zebrafish *tbx5a*/*tbx5b*, 208–209
Zone of polarizing activity (ZPA), 356–357

FSC
www.fsc.org
MIX
Papier | Fördert
gute Waldnutzung
FSC® C083411

Kolibri 360 GmbH
Ferdinand-Jühlke-Straße 7
99095 Erfurt, Deutschland
produktsicherheit@kolibri360.de